Progress in Mathematics

Volume 248

Infinite Groups: Geometric, Combinatorial and Dynamical Aspects

Laurent Bartholdi
Tullio Ceccherini-Silberstein
Tatiana Smirnova-Nagnibeda
Andrzej Zuk
Editors

Birkhäuser Verlag
Basel · Boston · Berlin

Editors:

Laurent Bartholdi
Institut de Mathématiques
EPFL
1015 Lausanne
Switzerland
e-mail:
laurent.bartholdi@epfl.ch

Tatiana Smirnova-Nagnibeda
Section de mathématiques
Université de Genève
1211 Geneva 4
Switzerland
e-mail:
tatiana.smirnova-nagnibeda@math.unige.ch

Tullio Ceccherini-Silberstein
Dipartimento di Ingegneria
Università del Sannio
C.so Garibaldi 108
82100 Benevento
Italy
e-mail: tceccher@mat.uniroma1.it

Andrzej Zuk
Institut de Mathématiques
Université Paris 7
175 rue du Chevaleret
75013 Paris
France
e-mail: zuk@math.jussieu.fr

2000 Mathematics Subject Classification 20Fxx, 37Axx, 46Lxx, 20E18, 20P05, 43A07, 55R35

A CIP catalogue record for this book is available from the Library of Congress, Washington D.C., USA

Bibliographic information published by Die Deutsche Bibliothek
Die Deutsche Bibliothek lists this publication in the Deutsche Nationalbibliografie; detailed bibliographic data is available in the Internet at <http://dnb.ddb.de>.

ISBN 3-7643-7446-2 Birkhäuser Verlag, Basel – Boston – Berlin

© 2005 Birkhäuser Verlag, P.O. Box 133, CH-4010 Basel, Switzerland
Part of Springer Science+Business Media
Printed on acid-free paper produced of chlorine-free pulp. TCF ∞
Printed in Germany
ISBN-10: 3-7643-7446-2   e-ISBN: 3-7643-7447-0
ISBN-13: 978-3-7643-7446-4

9 8 7 6 5 4 3 2 1   www.birkhauser.ch

Contents

Introduction

This book is a collection of selected papers on recent trends in the study of infinite discrete groups. The authors who contributed to this volume were invited plenary speakers at the International Conference on Group Theory: "Geometric, Combinatorial and Dynamical Aspects of Infinite Groups" which was held in Gaeta, Italy, in June 1-6, 2003. The exceptional quality of their lectures and the great interest they arose among the almost two hundred conference participants from over twenty countries gave us, the organisers of the conference, the idea of this book. We are most grateful to the authors who have all responded with wonderful enthusiasm to our request for a contribution.

This volume is far from being a book of conference proceedings. It is rather a one-of-a-sort collection of papers which aims at reflecting the present state of the art in a most active area of research at the intersection of several branches of mathematics.

Topics discussed in the book include geometric and combinatorial group theory; ergodic theory and operator algebras; low-dimensional topology; groups acting on trees and their boundaries; random walks; amenability; growth; groups and fractals; L^2-cohomology.

The articles collected in this volume should be of interest to specialists not only in group theory but also in such areas of mathematics as dynamical systems, geometry, topology, operator algebras, probability theory and combinatorics. The broad spectrum of topics covered should also present an exciting opportunity for graduate students and young researchers working in any of these areas who are willing to put their research in a wider mathematical perspective.

On the other hand, there is a strong unity among the diverse topics in this book. This reflects, we believe, a characteristic feature of the theory of infinite discrete groups, which progresses much through its interactions with other areas of mathematics. This is one of the great attractions of this subject, for us, as well as for many of our colleagues.

The idea of the conference stemmed from the spectacular dynamism of this subject over the past twenty years, and the occasion was offered by the fiftieth anniversary of Slava Grigorchuk. Slava's research has deeply influenced group theory in the recent years. His impact on our own mathematical development has been tremendous. We would like to dedicate this book to him.

Last but not the least, we are very grateful to Birkhäuser for publishing this volume in the "Progress in Mathematics" series, and in particular to Dr. Thomas Hempfling whose expert advice and encouragement during the preparation of the volume have been greatly appreciated.

Laurent Bartholdi, Tullio Ceccherini-Silberstein and Tatiana Smirnova-Nagnibeda

Odessa, July 2005

The editors of this volume express their gratitude to the following institutions for the generous financial support of the International Conference on Group Theory: "Geometric, Combinatorial and Dynamical Aspects of Infinite Groups", held in Gaeta, Italy, in June 1-6, 2003:

Cofinanziamento MURST "Analisi Armonica" (Prof. Elena Prestini);
Cofinanziamento MURST "Gruppi, grafi e geometrie" (Prof. Dina Ghinelli);
Comune di Gaeta;
Dipartimento di Ingegneria, Università degli Studi del Sannio (Benevento);
Dipartimento di Matematica "G. Castelnuovo", Università di Roma "La Sapienza";
Dipartimento di Matematica, Università di Roma Tre;
Dipartimento di Metodi e Modelli Matematici, Università di Roma "La Sapienza";
GNAMPA, Istituto Nazionale di Alta Matematica "Francesco Severi";
GNSAGA, Istituto Nazionale di Alta Matematica "Francesco Severi";
National Science Foundation (grant number DMS 0307231);
Università degli Studi di Roma "La Sapienza";
Università degli Studi del Sannio (Benevento).

Progress in Mathematics, Vol. 248, 1–14

Parafree Groups

Gilbert Baumslag

Abstract. A group is termed parafree if it is residually nilpotent and has the same nilpotent quotients as a given free group. The object of this paper is, in the main, to survey some of the known results about parafree groups, to record several new ones and to discuss some old and new open problems.

Mathematics Subject Classification (2000). 20F14, 20F05, 20E26, 20E05, 20F18, 20E06, 20E10.

Keywords. Parafree group, free (non-abelian) group, residually nilpotent group, lower central series, lower central sequence, one-relator group, HNN extension.

1. Introductory remarks

This is a slightly expanded version of a talk that I gave at the Gaeta International Group Theory Meeting in June, 2003.

More than 35 years ago, Hanna Neumann asked whether free groups can be characterised in terms of their lower central series. Parafree groups grew out of an attempt to answer her question.

My current interest in these groups arose from two sources.

The first involves the beginning of a systematic study of groups defined by a single relation. I will very briefly discuss this in **5** and list some related problems in **12**, #11; I hope to develop this further in a subsequent paper.

The second source of my recent interest in parafree groups arose out of some ongoing joint work with Sean Cleary [9] and even more recently, joint work with Cleary and George Havas [10]. Despite the fact that parafree groups were introduced many years ago, many questions about them have remained unanswered.

Following on my remark about one–relator groups above, I have taken this opportunity to formulate several of these questions for one–relator parafree groups in **12** leaving further exploration for another time.

Any investigation of parafree groups can itself be viewed as the beginnings of a study of finitely generated residually torsion-free nilpotent groups. Indeed one might look for similarities between finitely generated residually nilpotent groups which share the same lower central sequences (see section 2 for the relevant definitions). The paper [8] examines questions involving the existence of large numbers of finitely generated groups with the same lower central sequences and introduces some associated invariants. I will discuss this a tiny bit further in the sequel. My major concern here will be with the properties that parafree groups have in common with the prototypical parafree groups, the free groups.

2. Notation and definitions

As usual we denote the conjugate of the element x by the element y, where x and y are elements in a group G, by x^y and the commutator $x^{-1}y^{-1}xy$ of x and y by $[x, y]$. The lower central series

$$G = \gamma_1(G) \geq \gamma_2(G) \geq \ldots \gamma_n(G) \ldots$$

is defined inductively by

$$\gamma_{n+1}(G) = gp([x, y] \mid x \in \gamma_n(G), y \in G).$$

G is termed residually nilpotent if

$$\bigcap_{n=1}^{\infty} \gamma_n(G) = 1;$$

equivalently, given any non-trivial element $g \in G$, there exists a normal subgroup N of G such that $g \notin N$ with G/N nilpotent. More generally, if $\mathcal{P}$ is a property or class of groups, then G is termed residually $\mathcal{P}$ if, given any non-trivial element $g \in G$, there exists a normal subgroup N of G such that $g \notin N$ with $G/N \in \mathcal{P}$. One can associate with the lower central series of a group G its lower central sequence:

$$G = G/\gamma_2(G), G/\gamma_3(G), \ldots, G/\gamma_n(G), \ldots .$$

We say that two groups G and H have the same lower central sequence if $G/\gamma_n(G) \cong H/\gamma_n(H)$ for every $n \geq 1$.

We denote the minimal number of generators of the group G by $d(G)$.

3. Some general remarks

The study of the lower central series has received a great deal of attention in work on the fundamental groups of three dimensional manifolds, in knot theory, in the study of links and their invariants and in homology cobordism. The papers referenced below by Bellis [13], Cochran [15], Cochran and Harvey [16], Cochran and Orr [17], Levine [25] and Orr [33] and the open problems gathered together by Kirby [24], provide a guide to the literature dealing with some of these topics. The question as to whether the second homology group of a finitely generated parafree group vanishes has turned out to have a number of interesting geometric consequences and those interested in this aspect of parafree groups should consult these papers, in particular Cochran and Orr [17]. Not only do parafree groups arise in the study of three manifolds, they arise also in the work of Bousfield and Kan [14] in homotopy theory, in homological group theory in the work of Stallings [36] and that of Stammbach [38].

4. Magnus' theorem and parafree groups

Wilhelm Magnus [28] proved, in 1936, that free groups are residually torsion-free nilpotent. Three years later Magnus [29] obtained the following characterisation of free groups.

Theorem 1. *Suppose that G is an n–generator group ($n < \infty$) and suppose that G has the same lower central sequence as a free group of rank n. Then G is free.*

The possibility that free groups can be characterised by their lower central series was originally raised by Hanna Neumann.

Hanna Neumann's question. *Suppose that G is a finitely generated residually nilpotent group with the same lower central sequence as a free group. Is G free?*

The answer to this question is no, i.e., there exist non–free parafree groups, i.e., as noted in the abstract, residually nilpotent groups with the same lower central sequences as a free group, which are not free. The rank of a parafree group is defined to be the minimal number of generators of its factor derived group.

The following definition underscores one of the primary differences between free groups and non–free parafree groups.

Definition. *Let P be a parafree group. Then the deviation $\delta(P)$ of P is defined by*

$$\delta(P) = d(P) - (rank\ of\ P).$$

The finitely generated parafree groups of deviation 0 are the finitely generated free groups. A key notion in working with parafree groups is that of a parabasis.

Definition. *A subset X of a parafree group P is termed a parabasis if X freely generates P modulo $\gamma_2(P)$. The elements in a parabasis are termed para–primitives.*

So an element in a parafree group G is a para–primitive if it is not a proper power modulo the commutator subgroup of G.

5. One-relator groups

Many problems about one–relator groups have turned out to be very difficult. In order to make a start on solving some of them we have chosen to introduce the notion of *height* for a one–relator group. To begin with we have the following

Definition *Let* $G =< a, \ldots ; r >$ *be a one-relator group. G is said to be of height 0 if r is a power of a primitive element (in the underlying free group on the given generators $a, \ldots$ of G).*

Suppose that $G =< a, \ldots ; r >$ is a one–relator group. If r is a power of a primitive, then we define $\hat{G} = G$. If r is not a power of a primitive, then G can be embedded in a one-relator group $\hat{G}$ which is obtained from G by adjoining a root to one of its generators in such a way to ensure that it satisfies the following conditions. First, $\hat{G}$ is an HNN–extension of a one-relator group $B = \beta(\hat{G})$. Second, the length of the given defining relator of B is shorter than the length of the defining relator of G (see [27] and [32]). This allows us to define a sequence

$$G_1 = \hat{G}, B_1 = \beta(G_1), G_2 = \hat{B}_1, \ldots, B_j = \beta(G_j), G_{j+1} = \hat{B}_j, \ldots$$

of one–relator groups whose defining relators decrease in length and which terminates with the first G_ℓ of height 0. The length ℓ of such a series depends on a number of choices that are available at every step in the production of the series. We term the mimimum length of such a series for G its *height*.

So the one–relator groups of height 0 are either free or the free product of a finite cyclic group and a free group. Thus they can be considered as well known. Here we suggest that a systematic study of one–relator groups be initiated by focussing attention on the one-relator groups of height 1. These groups are much more complicated than the one–relator groups of height 0. They are very special subgroups of one–relator groups which are HNN extensions of the free product of a free group and a finite cyclic group. A family of examples of this kind are HNN–extensions of the form:

$$G =< a_1, \ldots, a_h, b_1, \ldots, b_i, \ldots, c_1, \ldots, c_j, t; r(a_1, \ldots, a_h, b_1, \ldots, b_i, c_1, \ldots, c_j)^e = 1,$$

$$t^{-1}a_1 t = a_2, \ldots, t^{-1}a_{h-1} t = a_h, t^{-1}b_1 t = b_2, \ldots, t^{-1}b_{i-1} t = b_i, \ldots,$$

$$t^{-1}c_1 t = c_2, \ldots, t^{-1}c_{j-1} t = b_j >;$$

here r is a cyclically reduced, primitive element in the free group on

$$a_1, \ldots, a_h, b_1, \ldots, b_i, \ldots, c_1, \ldots, c_j$$

which involves all of the generators

$$a_1, \ldots, a_h, b_1, \ldots, b_i, \ldots, c_1, \ldots, c_j.$$

Many one–relator parafree groups have this structure, for instance

$$G =< a, s, t; a = [s, a][s, t] > .$$

It is unclear to me whether there exist one-relator groups of height 2 or more, but this is surely the case.

6. Similarities

Many of the theorems that will be discussed in this and the following two sections are contained in [3] and [7].

Parafree groups have many properties in common with free groups. The first of these corresponds to the fact that the free group of rank two contains free subgroups of every countable rank.

Theorem 2. *Every countable parafree group can be embedded in a parafree group of rank two.*

This is Theorem 3.3 of [7].

G. Higman and B.H. Neumann [22] proved that the Frattini subgroup of a free group is trivial (cf. e.g., [23]). Here we have the

Theorem 3. *The Frattini subgroup of a parafree group is trivial.*

This is Corollary 1 of Theorem 7.1 of [7].

Of course the two-generator subgroups of free groups are free. The same is true of parafree groups (Theorem 4.2 of [7]).

Theorem 4. *The two–generator subgroups of parafree groups are free.*

Since the parafree group $G =< a, b, c; a^2b^2c^3 >$ is not free, Theorem 4 cannot be extended to three–generator subgroups.

It may be worth–while to indicate how one can prove that a given parafree group is not free. In fact there are many ways of doing so. For instance by considering the types of equations that cannot hold in free groups. This touches on the celebrated problem of Tarski, which I will not go into here. For our purposes it suffices to appeal to the work of Schützenberger [35], who proved that if a, b, c generate a free group and if $a^\ell b^m c^n = 1$, where ℓ, m, n are all at least 2, then a, b, c all commute. Another way of proving that G is not free, starts out with the observation that if we add the relations $a^2 = 1$, $b^2 = 1$, $c^3 = 1$ to the defining relations of G, the resultant factor group of G is the free product of a group of order 2, a second group of order 2 and a group of order 3. Now, by Grushko's theorem [19], the minimal number of generators of a free product of freely indecomposable groups is the sum of the minimal number of generators of each of the factors. So G is a 3-generator group that cannot be generated by fewer than 3 elements. Thus if G were free, it would have to be a free group of rank 3. Hence its abelianization would be free abelian of rank 3, which is not the case. Yet another way is to count the number of subgroups of a given finite index or the number of homomorphisms onto small finite groups. We shall see how this works in **9** where the procedure distinguishes one parafree group from another and can also be used to show that these groups are not free.

Non-cyclic free groups are residually free on 2-generators (Benjamin Baumslag [1]). There is a corresponding theorem for parafree groups (Theorem 5.2 of [7]).

Theorem 5. *Non–cyclic parafree groups are residually parafree of rank two.*

Magnus [28] proved that free groups can be embedded in the group of units of a power series ring. The same is true of parafree groups.

Theorem 6. *Every parafree group P can be embedded in the group of units of a power series ring over $\mathbb{Z}$.*

This is Theorem 3.1 of [7].

Finally, O. Schreier [34] proved that if F is a non-abelian free group and if N is a non-trivial normal subgroup of infinite index in F, then $d(N) = \infty$. Recently I was able to prove a very much weaker theorem for parafree groups, which depends on a theorem of Brian Hartley [21].

Theorem 7. *Let P be a non–abelian parafree group and let N be a non–trivial normal subgroup of P. If P/N is an infinite, residually torsion–free nilpotent group, then $d(N) = \infty$.*

7. Lie rings

There is a simple characterization of parafree groups. In order to explain, let me recall the definition of the Lazard Lie ring of a group.

Let G be a group. Then the Lazard Lie ring $L(G)$ of the group G is defined as follows. First, additively

$$L(G) = \bigoplus_{n=1}^{\infty} \gamma_n(G)/\gamma_{n+1}(G).$$

Then $L(G)$ becomes a Lie ring on setting

$$[a\gamma_{i+1}(G), b\gamma_{j+1}(G)] = [a,b]\gamma_{i+j+1}(G) \quad (a \in \gamma_i(G),\ b \in \gamma_j(G)).$$

Now Magnus [28] has proved that the Lazard Lie ring of a free group is free. It follows immediately that a residually nilpotent group is parafree if and only if its Lazard Lie ring is free. Witt [40] proved that the subrings of free Lie rings are again free. It is possible to deduce from this theorem of Witt that the 2–generator subgroups of parafree groups are free. In fact one can deduce that a number of special subgroups of parafree groups are free on appealing to Witt's theorem.

8. Differences

Free groups of finite rank have zero deviation, emphasising one of the differences between free groups and non–free parafree groups cf., e.g., Theorem 2.2 [7]):

Theorem 8. *There exist parafree groups of finite rank $n > 1$ and every countable deviation.*

Free groups are characterized by their ranks. However this is not the case for parafree groups as we have already seen from Theorem 8. Indeed

Theorem 9. *There exist continuously many countable parafree groups.*

Here is a sketch of a proof of Theorem 9. For each odd prime p, let

$$G_p = < a, b, c; a^2 b^p c^p = 1 > .$$

Then the G_p turn out to be parafree of rank 2. It follows that they are freely indecomposable. Moreover it follows on examining the subgroups of index 2, that $G_p \cong G_q$ only if $p = q$. Now for each increasing sequence σ of odd primes, let G_σ be the free product of the G_p, where p ranges over the primes in the sequence σ. It is easy enough to prove that the free product of parafree groups embeds in an associated power series ring (see the discussion in **11**), which allows one to prove that the free product of parafree groups is parafree. Thus the groups G_σ are parafree. Moreover, the G_σ can then be distinguished from each other by noting, using Grushko's theorem [19], that their finitely generated free factors differ from one another since this is true of the G_p. Since the number of sequences σ can be put into a one–one–one correspondence with the reals, this proves Theorem 9.

M. Takahasi [39] proved that any properly ascending union of subgroups of a free group of boundedly finite rank is finite. Again, this is not the case for parafree groups (Theorem 2.2 of [7]).

Theorem 10. *There exist parafree groups of rank two which are properly ascending unions of free groups of rank two (and hence are locally free).*

9. Distinguishing parafree groups by computing the number of homomorphisms into finite groups

There are three existing families of parafree groups that are not easy to distinguish one from another. These families can be described as follows:

$$G_{i,j} = < a, b, c | a = [c^i, a][c^j, b] > \quad (i, j \text{ positive integers}),$$

$$H_{i,j} = < a, s, t | a = [a^i, t^j][s, t] > \quad (i, j \text{ positive integers})$$

$$K_{i,j} = < a, s, t | a^i [s, a] = t^j > \quad (i, j \text{ relatively prime integers}).$$

In order to differentiate these groups from one another, we have followed the lead of Lewis and Liriano [26] and also here the work of Matei and Suciu [31] and computed the numbers of homomorphisms onto some small finite groups (see Baumslag, Cleary and Havas [10]). The result of these computations is the following

Theorem 11. *None of the 254 distinct parafree groups in the three families $G_{i,j}$, $H_{i,j}$ and $K_{i,j}$ with $1 \leq i, j \leq 10$ are isomorphic. They can be distinguished from each other by counting epimorphisms onto small simple groups.*

Fine, Rosenberger and Stille [18] have solved the isomorphism problem for various other parafree groups using so–called Nielsen methods.

10. Some new parafree groups

Recently Baumslag and Cleary [9] have constructed some new parafree groups. These are described below.

Theorem 12. *Let F be free on $a, b, \dots$, let $w \in \gamma_2(F)$ and suppose that $f \in F$ is not a proper power modulo $\gamma_2(F)$. Then for every $n > 1$*

$$P = <a, b, \dots, t; t^n = fw>$$

is parafree. Moreover if $w \in V(F)$, V a variety of groups where the free groups are residually torsion-free nilpotent, then $P/V(P)$ is free in V.

Theorem 13. *Let F be free on $a, b, \dots, s, t$ and let w be a word in these generators that involves a and s, but not t. If w lies in the derived group of F, then the one–relator group*

$$P = <a, b, \dots, s, t; a = w[s, t, \dots, t]>$$

is parafree. Moreover if $w \in V(F)$, where V is any given variety of groups, then $P/V(P)$ is free in V.

The following is an example of the type described in Theorem 13:

$$P = <a, s, t; a = [s, a][s, t]> .$$

As noted before in the discussion of one-relator groups P is a one–relator group of height 1. It is parafree of rank two and has turned out to be of interest elsewhere [13].

Theorem 14. *Let F be a free group, freely generated by $a, b, \dots$. Suppose that $w \in V(F)$, where V is any given variety of groups. If $\overline{F}$ is an isomorphic copy of F and if w belongs to the derived group of F, then the amalgamated product*

$$P = \{F * \overline{F}; aw = \overline{aw}\}$$

is parafree. Moreover if $w \in V(F)$, then $P/V(P)$ is free in V.

Here is an example of one of the groups in the family described in Theorem 14.

$$P =< a, b, c, d; a^2b^3 = c^2d^3 >;$$

P is parafree of rank three.

The main difficulty involved in proving that a given group is parafree is the verification that it is residually nilpotent. The following theorem of P.C. Wong [41] can often be used to overcome this problem.

Theorem 15. *Let G be an infinite cyclic extension of a free group N. If $N/\gamma_2(N)$, viewed as a module over G/N, is free, then G is residually torsion–free nilpotent.*

11. Using power series to obtain some even newer parafree groups

Let $\Xi =<< \xi_1, \ldots, \xi_q >>$ be the ring of integral power series in the non–commuting variables $\xi_1, \ldots, \xi_q$. We recall that the elements $\xi \in \Xi$ are formal power series in the given variables, which take the form

$$\xi = \xi(0) + \xi(1) + \cdots + \xi(n) + \ldots,$$

where $\xi(0) \in \mathbb{Z}$ and ξ_n is a finite sum of linearly independent integral multiples of monomials of the form $\xi_{i_1} \ldots \xi_{i_n}$ of degree n. Two such monomials are equal if and only if they are identical. We refer to the ξ_j as free generators of Ξ.

Theorem 16. *Let G and H be parafree and let $g \in G$ and $h \in H$. Suppose that g is not a proper power modulo $\gamma_2(G)$ and also that h is not a proper power modulo $\gamma_2(H)$. Then the amalgamated product*

$$P = \{G * H; g = h\}$$

is parafree.

Theorem 16 is proved by using the fact that each of G and H can be respectively embedded in the power series rings

$$\Xi =<< \xi_1, \ldots, \xi_q >> \quad and \quad \Omega =<< \omega_1, \ldots, \omega_r >>$$

in such a way that g maps to $1 + \xi_1$ and h maps to $1 + \omega_1$. If we now form the power series ring

$$\Lambda =<< \xi_1, \ldots, \xi_q, \omega_2, \ldots, \omega_r >>$$

obtained from Ξ and Ω by identifying ω_1 with ξ_1, this gives rise to an embedding of P into Λ, which essentially proves Theorem 16.

Here is an example of a parafree group of the type given in Theorem 16:

$$P =< a, b, c, d, e, f; a^2b^2c^3, d^2e^2f^3, abc = def > .$$

The next new theorem that I want to record here is

Theorem 17. *Let G be parafree and let $g \in G$ be an element of G which is not a proper power modulo $\gamma_2(G)$. Then the HNN–extension*

$$E =< G, t; t^{-1}gt = g >$$

is discriminated by G. Hence E is residually torsion–free nilpotent and its subgroup

$$gp(t^{-i}Gt^i \mid i = 0, \pm 1, \pm 2, \dots),$$

the result of forming infinitely many doubles of G, is parafree.

Here we say that a group H is discriminated by a group K if given any finite set of non–trivial elements of H, there exists a homomorphism of H into K which maps each of the given elements into non-trivial elements of K. The proof of Theorem 17 then follows along the following lines. First we embed G in a power series ring

$$\Xi =<< \xi_1, \dots, \xi_q >>$$

in such a way that g maps onto $1 + \xi_1$. Next we form the power series ring

$$\Sigma =<< \xi_1, \dots, \xi_q, \tau; \xi_1 \tau = \tau \xi_1 >>$$

and then extend the embedding of G into Ξ to what we prove is an embedding of E into Σ. Finally the proof that E is discriminated by G makes use of a special instance of what has become known now as the "big powers" property. A group P is a "big powers group", if given any element u of P which is not a proper power and any finite set $\{a_1, \dots, a_k\}$ of elements of P which do not commute with u, then for all sufficiently large values of the $\mid n_i \mid$,

$$a_1 u^{n_1} a_2 u^{n_2} \dots a_k u^{n_k} \neq 1.$$

Whether parafree groups are big powers groups is not yet known. This property was introduced by Baumslag [2] where it was proved that free groups are big powers groups and was subsequently extended by Baumslag and Short [12] to hyperbolic groups.

12. Some open problems

One might suspect that two finitely generated residually torsion-free nilpotent groups with the same lower central sequences have many properties in common. This has been partly borne out by the fact that a number of properties of free groups are shared also by parafree groups. The problems that are being put forward here, not all of which are new, should be viewed in this light. We will assume that the groups discussed in Problems 1, 2, 3 and 4 are finitely generated and residually torsion-free nilpotent. If two such groups have the same lower central sequences, then we will say that they are of the same genus. The following problems follow on naturally from the discussion of parafree groups detailed above.

1. Suppose that G and H are of the same genus. If the second integral homology group of G is trivial, is the same true of H? The special case where G is free, a problem that I raised many years ago and is now referred to as the parafree conjecture [17], is still open as are many of the other problems that follow.

2. Suppose G and H are of the same genus. If G is finitely presented, is H finitely presented?

3. Suppose that G and H are of the same genus. If the Frattini subgroup of G is trivial, is the same true of H? Here this suggests also the question as to whether the Frattini subgroup of a non-abelian one-relator group is trivial and in particular whether non–abelian one–relator groups of height 1 have trivial Frattini subgroups.

4. Suppose that G and H are finitely presented and of the same genus. If G has a solvable conjugacy problem, is this also true of H?

    It is perhaps appropriate to mention that there exist finitely presented residually torsion-free nilpotent groups with unsolvable conjugacy problem [8]. In addition, there exists a recursive family of finitely presented, residually torsion-free nilpotent groups with unsolvable isomorphism problem [11].

5. Suppose that a finitely generated, residually finite group G has the same finite images as a free group. Is G free? Is G parafree?

    It is easy to find parafree groups which do not have the same finite images as a free group. For example,

$$G = < a, b, c; a^2 b^3 c^3 >$$

is parafree. G has the same lower central series as a free group F of rank two. However let N be the normal closure in G of the elements b and c. N can be presented in the form

$$N = < w, x, y, z; w^3 x^3 y^3 z^3 > .$$

Hence $N/[N,N]N^3$ is a direct product of 4 groups of order 3 and hence cannot be generated by 3 elements. Notice that $G/[N,N]N^3$ is a finite group of order 162. Now any subgroup of index 2 in a 2-generator group can be generated by 3 elements. It follows that $G/[N,N]N^3$ cannot be generated by 2 elements since $N/[N,N]N^3$ cannot be generated by 3 elements. So G and F have different finite images.

6. Let N be a non–trivial normal subgroup of a parafree group. If N is of infinite index, is $d(N) = \infty$?

7. If N is a normal subgroup of a parafree group P and if $N/\gamma_2(N)$ is torsion–free, is $P/\gamma_2(N)$ torsion–free? If, in addition, P/N is torsion–free, is $P/\gamma_2(N)$ a $\mathcal{U}$-group, i.e., a group in which extraction of nth root is unique whenever it exists?

8. Are parafree groups linear?

9. Are parafree groups automatic? Hyperbolic? It should be noted that the isomorphism problem for torsion–free hyperbolic groups has been solved by Sela [37].

10. Is the intersection of two finitely generated subgroups of a parafree group finitely generated?

11. In closing this discussion of open problems, I want here to raise some very special questions about one-relator groups of height 1. These are special cases of some known open problems about one-relator groups. I have not thought deeply, indeed at times not at all, about these problems, which I am here putting forth in the hope that their investigation will shed some light on the nature of one–relator groups as a whole.

Is the conjugacy problem solvable for one-relator groups of height 1?

Is every one-relator group of height 1 with non–trivial elements of finite order residually finite?

Is every one-relator group of height 1 residually finite?

Are two residually finite one–relator groups of height 1 with the same finite images isomorphic?

Is a one–relator group of height 1 with the same finite images as a finitely generated free group free? Parafree?

Is the isomorphism problem for one-relator groups of height 1 solvable?

Is there an algorithm to decide if a one-relator group of height 1 is residually nilpotent and hence parafree?

And so on.

References

[1] Benjamin Baumslag, Residually free groups, *Proc. London Math. Soc.* (3) **17**, 402–418.

[2] Gilbert Baumslag, On generalized free products, *Math. Z.* **78** (1962), 423–438.

[3] Gilbert Baumslag, Groups with the same lower central sequence as a relatively free group. I. The groups, *Trans. Amer. Math. Soc.* **129** (1967), 308–321.

[4] Gilbert Baumslag, Some groups that are just about free, *Bull. Amer. Math. Soc.* **73** (1967), 621–622.

[5] Gilbert Baumslag, More groups that are just about free. *Bull. Amer. Math. Soc.* **74** (1968), 752–754.

[6] Gilbert Baumslag, On the residual nilpotence of certain one-relator groups, *Comm. Pure Appl. Math.* **21** (1968), 491–506.

[7] Gilbert Baumslag, Groups with the same lower central series as a relatively-free group, II, *Trans. Amer. Math. Soc.* **142** (1969), 507–538.

[8] Gilbert Baumslag, Finitely generated residually torsion–free nilpotent groups. I. *Journal Austral. Math. Soc.* Ser. A. **67** (1999), 289–317.

[9] Gilbert Baumslag, and Sean Cleary, Parafree one-relator groups, *Journal of Group Theory*, to appear (2005).

[10] Gilbert Baumslag, Sean Cleary, and George Havas, Experimenting with infinite groups, *Experimental Math.* **13** (2004), 495–502.

[11] Gilbert Baumslag, and C.F. Miller, The isomorphism problem for finitely generated residually torsion–free nilpotent groups. In preparation.

[12] Gilbert Baumslag and Hamish Short, Unpublished.

[13] P. Bellis, Realizing homology boundary links with arbitrary patterns *Trans. American Math. Soc.* **320** (1998), 87–100.

[14] A.K. Bousfield, and D.M. Kan, Homotopy limits, completions and localizations, *Lecture Notes in Mathematics*, 134, Springer–Verlag, Berlin–New York (1972).

[15] T.D. Cochran, Derivatives of links: Milnor's concordance invariants and Massey's product, volume 84 Memoir #427, American Mathematical Society, Providence, R.I., 1990

[16] T.D. Cochran and Shelly Harvey, Homology and derived series of groups, Submitted *J. American Math. Soc.*.

[17] T.D. Cochran and Kent E. Orr, Stability of Lower Central Series of Compact 3-Manifold Groups. *Topology* **37** no. 3 (1998), 497–526.

[18] Benjamin Fine, Gerhard Rosenberger and Michael Stille, The isomorphism problem for a class of para-free groups. *Proc. Edinburgh Math. Soc. (2)*, **40** no. 3 (1997), 541–549.

[19] I.A. Grushko, Ueber die Basen einem freien Produkes von Gruppen, *Mat. Sb.* **8** (1940), 169–182.

[20] M.A. Gutierrez, Homology and completions of groups, *J. Algebra* **51** (1978), 354–366.

[21] B. Hartley, The residual nilpotence of wreath products, *Proc. London Math. Soc. (3)* **20** (1970), 365–372.

[22] G. Higman and B.H. Neumann, On two questions of Ito, *J. London Math. Soc.* **29** (1954), 84–88.

[23] Ilya Kapovich, The Frattini subgroup for subgroups of hyperbolic groups, *J. Group Theory* **6** no. 1 (2003), 115–126.

[24] R. Kirby, Problem list, new version, 1995, preprint, Univ. California, Berkeley, California.

[25] J. Levine, Finitely presented groups with long lower central series, *Israel J. Math.* **73** no. 1 (1991), 57–64.

[26] Robert H. Lewis and Sal Liriano, Isomorphism classes and derived series of certain almost-free groups. *Experiment. Math.*, **3** no. 3 (1994), 255–258.

[27] Wilhelm Magnus, Ueber diskontinuierliche gruppen mit einer definerden relation (der freiheitssatz). *J. Reine Agnew. Math,* **163** (1930), 141–165.

[28] Wilhelm Magnus, Beziehungen zwischen Gruppen und Idealen in einem speziellen Ring, *Math. Ann.* **111** (1935), 259–280.

[29] Wilhelm Magnus, Ueber freie Faktorgruppen und frei Untergruppen gegebener Gruppen, *Monatshafte fuer Math. und Phys.* **47** (1939), 307–313.

[30] Wilhelm Magnus, A. Karrass and D. Solitar, Combinatorial Group Theory: Presentations of Groups in Terms of Generators and Relations, Interscience Publishers, John Wiley and Sons.

[31] Daniel Matei and Alexander Suciu, Counting homomorphisms onto finite solvable groups, *Math. ArXiv math. GR/0405122*.

[32] D.I. Moldavanskii, Certain subgroups of groups with one defining relation, *Sibirsk. Math. Zh.* **8** (1967), 1370–1384.

[33] Kent Orr, New link invariants, *Comment. Math. Helvetici* **62** (1987), 542–560.

[34] O. Schreier, Die Untergruppen der freien Gruppen, *Abh. Math. Sem. Univ. Hamburg* **5** (1927), 161–183.

[35] M.P. Schützenberger, Sur l'équation $a^{2+n} = b^{2+n}c^{2+p}$ dans un groupe libre, *C.R. Acad. Sci. Paris* **248** (1959), 2435–2436.

[36] J. Stallings, Homology and central series of groups, *Journal of Algebra*, **12** (1965), 170–181.

[37] Z. Sela, The isomorphism problem for hyperbolic groups. I. *Ann. of Math. (2)* **141** (1995), 217–283.

[38] Urs Stammbach, Homology in Group Theory, *Lecture Notes in Mathematics*, 359, Springer–Verlag, Berlin, Heidelberg, New York, 1973.

[39] M. Takahasi, Note on chain conditions in free groups, *Osaka Math. J.* **1** (1951), 221–255.

[40] Ernst Witt, Treue Darstellung Liescher Ringe, *J. Reine Angew. Math.* **117** (1937), 152–160.

[41] P.C. Wong, On cyclic extensions of parafree groups, New York University Graduate School, 1978.

Acknowledgments

This work was supported in part by NSF grants #02-02382 and #02-15942.

Gilbert Baumslag
Department of Mathematics
City College of New York
City University of New York, New York, NY 10031
USA

e-mail: gilbert@groups.sci.ccny.cuny.edu

Progress in Mathematics, Vol. 248, 15–30

The Finitary Andrews–Curtis Conjecture

Alexandre V. Borovik, Alexander Lubotzky and Alexei G. Myasnikov

To Slava Grigorchuk as a token of our friendship.

Abstract. The well known Andrews–Curtis Conjecture [2] is still open. In this paper, we establish its finite version by describing precisely the connected components of the Andrews–Curtis graphs of finite groups. This finite version has independent importance for computational group theory. It also resolves a question asked in [5] and shows that a computation in finite groups cannot lead to a counterexample to the classical conjecture, as suggested in [5].

Mathematics Subject Classification (2000). Primary 20E05; Secondary 20D99.

Keywords. Andrews–Curtis conjecture, generators, finite group.

1. Andrews–Curtis graphs

Let G be a group and G^k be the set of all k-tuples of elements of G.

The following transformations of the set G^k are called *elementary Nielsen transformations (or moves)*:

(1) $(x_1, \ldots, x_i, \ldots, x_k) \longrightarrow (x_1, \ldots, x_i x_j^{\pm 1}, \ldots, x_k)$, $i \neq j$;

(2) $(x_1, \ldots, x_i, \ldots, x_k) \longrightarrow (x_1, \ldots, x_j^{\pm 1} x_i, \ldots, x_k)$, $i \neq j$;

(3) $(x_1, \ldots, x_i, \ldots, x_k) \longrightarrow (x_1, \ldots, x_i^{-1}, \ldots, x_k)$.

Elementary Nielsen moves transform generating tuples of G into generating tuples. These moves together with the transformations

(4) $(x_1, \ldots, x_i, \ldots, x_k) \longrightarrow (x_1, \ldots, x_i^w, \ldots, x_k)$, $w \in S \cup S^{-1} \subset G$,

where S is a fixed subset of G, form a set of *elementary Andrews–Curtis transformations relative to S* (or, shortly, AC_S-moves). If $S = G$ then AC-moves transform n-generating tuples (i.e., tuples which generate G as a normal subgroup) into n-generating tuples. We say that two k-tuples U and V are AC_S-equivalent, and write $U \sim_S V$, if there is a finite sequence of AC_S-moves which transforms U into V. Clearly, $\sim_S$ is an equivalence relation on the set G^k of k-tuples of elements from G. In the case when $S = G$ we omit S in the notations and refer to AC_S-moves simply as to AC-moves.

We slightly change notation from that of [5]. For a subset $Y \subset G$ we denote by $gp_G(Y)$ the normal closure of Y in G, by $d(G)$ the minimal number of generators of G, and by $d_G(G)$ the minimal number of normal generators of G. Now, $d_G(G)$ coincides with $nd(G)$ of [5].

Let $N_k(G)$, $k \geqslant d_G(G)$, be the set of all k-tuples of elements in G which generate G as a normal subgroup:

$$N_k(G) = \{ (g_1, \ldots, g_k) \mid gp_G(g_1, \ldots, g_k) = G \}.$$

Then the *Andrews–Curtis graph* $\Delta_k^S(G)$ of the group G with respect to a given subset $S \subset G$ is the graph whose vertices are k-tuples from $N_k(G)$ and such that two vertices are connected by an edge if one of them is obtained from another by an elementary AC_S-transformation. Again, if $S = G$ then we refer to $\Delta_k^G(G)$ as to the Andrews–Curtis graph of G and denote it by $\Delta_k(G)$. Clearly, if S is a generating set of G then the graph $\Delta_k^S(G)$ is connected if and only if the graph $\Delta_k(G)$ is connected. Observe, that if S is finite then $\Delta_k^S(G)$ is a regular graph of finite degree.

The famous Andrews–Curtis conjecture [2] can be stated in the following way.

AC-Conjecture: *For a free group F_k of rank $k \geqslant 2$, the Andrews–Curtis graph $\Delta_k(F_k)$ is connected.*

There are some doubts whether this well known old conjecture is true. Indeed, Akbulut and Kirby [1] suggested a series of potential counterexamples for $k = 2$:

$$(u, v_n) = (xyxy^{-1}x^{-1}y^{-1}, x^n y^{-(n+1)}), \quad n \geqslant 2. \tag{1}$$

In [5], it has been suggested that one may be able to confirm one of these potential counterexamples by showing that for some homomorphism $\phi : F_2 \to G$ into a finite group G the pairs (u^ϕ, v_n^ϕ) and (x^ϕ, y^ϕ) lie in different connected components of $\Delta_2(G)$. Notice that in view of [16] the group G in the counterexample cannot be soluble.

Our main result describes the connected components of the Andrews–Curtis graph of a finite group. As a corollary we show that (u^ϕ, v_n^ϕ) and (x^ϕ, y^ϕ) lie in the same connected components of $\Delta_2(G)$ for every finite group G and any homomorphism $\phi : F_2 \to G$, thus resolving the question from [5].

Theorem 1.1. *Let G be a finite group and $k \geqslant \max\{d_G(G), 2\}$. Then two tuples U, V from $N_k(G)$ are AC-equivalent if and only if they are AC-equivalent in the abelianisation $\mathrm{Ab}(G) = G/[G, G]$, i.e., the connected components of the AC-graph $\Delta_k(G)$ are precisely the preimages of the connected components of the AC-graph $\Delta_k(\mathrm{Ab}(G))$.*

Notice that, for the abelian group $A = \mathrm{Ab}(G)$, a normal generating set is just a generating set and the non-trivial Andrews–Curtis transformations are Nielsen moves (1)–(3). Therefore the vertices of $\Delta_k(A)$ are the same as these of the *product replacement graph* $\Gamma_k(A)$ [7, 18]: they are all generating k-tuples of A. The only difference between $\Gamma_k(A)$ and $\Delta_k(A)$ is that the former has edges defined only by 'transvections' (1)–(2), while in the latter the inversion of components (3) is also allowed. The connected components of product replacements graphs $\Gamma_k(A)$ for finite abelian groups A have been described by Diaconis and Graham [7]; a slight modification of their proof leads to the following observation

Fact 1.2 (Diaconis and Graham [7]). *Let A be a finite abelian group and*

$$A = Z_1 \times \cdots \times Z_d$$

its canonical decomposition into a direct product of cyclic groups such that $|Z_i|$ divides $|Z_j|$ for $i < j$. Then

(a) *If $k > d$ then $\Delta_k(A)$ is connected.*
(b) *If $k = d \geqslant 2$, fix generators $z_1, \ldots, z_d$ of the subgroups $Z_1, \ldots, Z_d$, correspondingly. Let $m = |Z_1|$. Then $\Delta_d(A)$ has $\phi(m)/2$ connected components (here $\phi(n)$ is the Euler function). Each of these components has a representative of the form*

$$(z_1^\lambda, z_2, \ldots, z_d), \quad \lambda \in (\mathbb{Z}/m\mathbb{Z})^*.$$

Two tuples

$$(z_1^\lambda, z_2, \ldots, z_d) \quad \text{and} \quad (z_1^\mu, z_2, \ldots, z_d), \quad \lambda, \mu \in (\mathbb{Z}/m\mathbb{Z})^*,$$

belong to the same connected component if and only if $\lambda = \pm\mu$.

Taken together, Theorem 1.1 and Fact 1.2 give a complete description of components of the Andrews–Curtis graph $\Delta_k(G)$ of a finite group G.

Notice that in an abelian group A

$$(xyxy^{-1}x^{-1}y^{-1}, x^n y^{-(n+1)}) \;\sim\; (xy^{-1}, x^n y^{-(n+1)})$$
$$\sim\; (xy^{-1}, x^{n-1}y^{-n})$$
$$\vdots$$
$$\sim\; (yx^{-1}, y^{-1})$$
$$\sim\; (x, y)$$

so for every homomorphism $\phi : F_2 \to G$ as above the images (u^ϕ, v_n^ϕ) and (x^ϕ, y^ϕ) are AC equivalent in the abelianisation of G, hence they lie in the same connected component of $\Delta_2(G)$.

The following corollary of Theorem 1.1 leaves no hope of finding an counterexample to the Andrews–Curtis conjecture by looking at the connected components of the Andrews–Curtis graphs of finite groups.

Corollary 1.3. *For any $k \geqslant 2$, and any epimorphism $\phi : F_k \to G$ onto a finite group G, the image of $\Delta_k(F_k)$ in $\Delta_k(G)$ is connected.*

One may try to reject the AC-conjecture by testing AC-equivalence of the tuples (u, v_n) and (x, y) in the *infinite* quotients of the group F_2. To this end we introduce the following definition.

Definition: *We say that a group G satisfies the* generalised Andrews–Curtis conjecture *if for any $k \geqslant \max\{d_G(G), 2\}$ tuples $U, V \in N_k(G)$ are AC-equivalent in G if and only if their images are AC-equivalent in the abelianisation* $\mathrm{Ab}(G)$.

Problem: *Find a group G which does not satisfy the generalised Andrews–Curtis conjecture.*

It will be interesting to look, for example, at the Grigorchuk group [8, 9]. It is a finitely generated residually finite 2-group G which is just-infinite, that is, every normal subgroup has finite index. Therefore the generalised Andrews–Curtis conjecture holds in every proper factor group of G by Theorem 1.1. What might be also relevant, the conjugacy problem in the Grigorchuk group is solvable [13, 19, 3]. This makes the Grigorchuk group a very interesting testing ground for the generalised Andrews–Curtis conjecture.

2. Relativised Andrews–Curtis graphs and black-box groups

Following [5], we also introduce a relativised version of the Andrews–Curtis transformations of the set G^k for the situation when G admits some fixed group of operators Ω (that is, a group Ω which acts on G by automorphisms); we shall say in this situation that G is an Ω-*group*[1]. In that case, we view the group G as a subgroup of the natural semidirect product $G \cdot \Omega$ of G and Ω. In particular, the set of $AC_{G\Omega}$-moves is defined and the set G^k is invariant under these moves. In particular, if N is a normal subgroup of G, we view N as a G-subgroup in the sense of this definition. As we shall soon see, $AC_{G\Omega}$-moves appear in the product replacement algorithm for generating pseudo-random elements of a normal subgroup in a black box finite group.

For a subset $Y \subset G$ of an Ω-group G we denote by $gp_{G\Omega}(Y)$ the normal closure of Y in $G \cdot \Omega$, and by $d_{G\Omega}(G)$ the minimal number of normal generators of G as a normal subgroup of $G \cdot \Omega$.

Let $N_k(G, \Omega)$, $k \geqslant d_{G\Omega}(G)$, be the set of all k-tuples of elements in G which generate G as a normal Ω-subgroup:

$$N_k(G, \Omega) = \{ (g_1, \ldots, g_k) \mid gp_{G\cdot\Omega}(g_1, \ldots, g_k) = G \}.$$

Then the *relativised Andrews–Curtis graph* $\Delta_k^\Omega(G)$ of the group G is the graph whose vertices are k-tuples from $N_k(G, \Omega)$ and such that two vertices are connected by an edge if one of them is obtained from another by an elementary $AC_{G\Omega}$-transformation.

A *black box group* G is a finite group with a device ('oracle') which produces its (pseudo)random (almost) uniformly distributed elements; this concept is of crucial importance for computational group theory, see [10]. If the group G is given by generators, the so-called *product replacement algorithm* [6, 18] provides a

---

[1]We shall use the terms Ω-subgroup, normal Ω-subgroup, Ω-simple Ω-subgroup, etc. in their obvious meaning.

very efficient and practical way of producing random elements from G; see [14] for a likely theoretical explanation of this (still largely empirical) phenomenon in terms of the (conjectural) Kazhdan's property (T) [11] for the group of automorphisms of the free group F_k for $k > 4$. In the important case of generation of random elements in a normal subgroup G of a black box group Ω, the following simple procedure is a modification of the product replacement algorithm: start with the given tuple $U \in N_k(G, \Omega)$, walk randomly over the graph $\Delta_k^\Omega(G)$ (using the 'oracle' for Ω for generating random $AC_{G\Omega}$-moves and return randomly chosen components v_i of vertices V on your way. See [4, 5, 12] for a more detailed discussion of this algorithm, as well as its further enhancements.

Therefore the understanding of the structure—and ergodic properties—of the Andrews–Curtis graphs $\Delta_k^\Omega(G)$ is of some importance for the theory of black box groups.

The following results are concerned with the connectivity of the relativised Andrews–Curtis graphs of finite groups.

Theorem 2.1. *Let G be a finite Ω-group which is perfect as an abstract group, $G = [G, G]$. Then the graph $\Delta_k^\Omega(G)$ is connected for every $k \geqslant 2$.*

Of course, this result can be immediately reformulated for normal subgroups of finite groups:

Corollary 2.2. *Let G be a finite group and $N \triangleleft G$ a perfect normal subgroup. Then the graph $\Delta_k^G(N)$ is connected for every $k \geqslant 2$.*

We would like to record another immediate corollary of Theorem 2.1.

Corollary 2.3. *Let G be a perfect finite group, $g_1, \ldots, g_k$, $k \geqslant 2$ generate G as a normal subgroup and $\phi : F_k \longrightarrow G$ an epimorphism. Then there exist $f_1, \ldots, f_k \in F_k$ such that $\phi(f_i) = g_i$, $i = 1, \ldots, k$, and $f_1, \ldots, f_k$ generate F_k as a normal subgroup.*

Note that if we take $g_1, \ldots, g_k$ as a set of generators for G, then in general we cannot pull them back to a set $f_1, \ldots, f_k$ of generators for F_k, an example can be found in $G = \mathrm{Alt}_5$, the alternating group on 5 letters [17].

In case of non-perfect finite groups we prove the following theorem.

Theorem 2.4. *Let G be a finite Ω-group. Then the graph $\Delta_k^\Omega(G)$ is connected for every $k \geqslant d_{G\Omega}(G) + 1$.*

Note this is not true for $k = d_{G\Omega}(G)$, e.g. for when G is abelian.

Corollary 2.5. *Let G be a finite group and $N \triangleleft G$ a normal subgroup. Then the graph $\Delta_k^G(N)$ is connected for every $k \geqslant d_G(N) + 1$.*

These results lead us to state the following conjecture.

Relativised Finitary AC-Conjecture: *Let G be a finite Ω-group and $k = d_{G\Omega}(G) \geqslant 2$. Then two tuples U, V from $N_k(G, \Omega)$ are $AC_{G\Omega}$-equivalent if and only if they are $AC_{\Omega Ab(G)}$-equivalent in the abelianisation $Ab(G) = G/[G, G]$, i.e., the connected components of the graph $\Delta_k^{\Omega}(G)$ are precisely the preimages of the connected components of the graph $\Delta_k^{\Omega}(Ab(G))$.*

Theorem 1.1 confirms the conjecture when $G = \Omega$.

3. Elementary properties of AC-transformations

Let G be an Ω-group. From now on for tuples $U, V \in G^k$ we write $U \sim_G V$, or simply $U \sim V$, if the tuples U, V are $AC_{G\Omega}$-equivalent in G.

Lemma 3.1. *Let G be an Ω-group, N a normal Ω-subgroup of G, and $\phi : G \to G/N$ the canonical epimorphism. Suppose $(u_1, \ldots, u_k)$ and $(v_1, \ldots, v_k)$ are two k-tuples of elements from G. If*

$$(u_1^{\phi}, \ldots, u_k^{\phi}) \sim_{G/N} (v_1^{\phi}, \ldots, v_k^{\phi})$$

then there are elements $m_1, \ldots, m_k \in N$ such that

$$(u_1, \ldots, u_k) \sim_G (v_1 m_1, \ldots, v_k m_k).$$

Moreover, one can use the same system of elementary transformations (after replacing conjugations by elements $gN \in G/N$ by conjugations by elements $g \in G$).

Proof. Straightforward. □

Lemma 3.2. *Let G be an Ω-group. If $(w_1, \ldots, w_k) \in G^k$ then for every i and every element $g \in gp_{G\Omega}(w_1, \ldots, w_{i-1}, w_{i+1}, \ldots, w_k)$*

$$(w_1, \ldots, w_k) \sim_G (w_1, \ldots, w_i g, \ldots, w_k).$$

Proof. Obvious. □

4. The N-Frattini subgroup and semisimple decompositions

Definition 1. *Let G be an Ω-group. The N-Frattini subgroup of G is the intersection of all proper maximal normal Ω-subgroups of G, if such exist, and the group G, otherwise. We denote it by $W(G)$.*

Observe, that if G has a non-trivial finite Ω-quotient then $W(G) \neq G$.

An element g in an Ω-group G is called *non-N-generating* if for every subset $Y \subset G$ if $gp_G(Y \cup \{g\}) = G$ then $gp_G(Y) = G$.

Lemma 4.1.

(1) *The set of all non-N-generating elements of an Ω-group G coincides with $W(G)$.*

(2) *A tuple $U = (u_1, \ldots, u_k)$ generates G as a normal Ω-subgroup if and only if the images $(\bar{u}_1, \ldots, \bar{u}_k)$ of elements $u_1, \ldots, u_k$ in $\bar{G} = G/W(G)$ generate $\bar{G}$ as normal Ω-subgroup.*

(3) *$G/W(G)$ is an Ω-subgroup of an (unrestricted) Cartesian product of Ω-simple Ω-groups (that is, Ω-groups which do not have proper non-trivial normal Ω-subgroups).*

(4) *As an abstract group, $G/W(G)$ is a subgroup of an (unrestricted) Cartesian product of characteristically simple groups. In particular, if G is finite then $G/W(G)$ is a product of simple groups.*

Proof. (1) and (2) are similar to the standard proof for the analogous property of the Frattini subgroup.

To prove (3) let N_i, $i \in I$, be the set of all maximal proper normal Ω-subgroups of G. The canonical epimorphisms $G \to G/N_i = G_i$ give rise to a homomorphism $\phi : G \to \overline{\prod}_{i \in I} G_i$ of G into the unrestricted Cartesian product of Ω-groups G_i. Clearly, $\ker \phi = W(G)$. So $G/W(G)$ is an Ω-subgroup of the Cartesian product of Ω-simple Ω-groups G_i.

To prove (4) it suffices to notice that $G_i = G/N_i$ has no Ω-invariant normal subgroups, hence is characteristically simple. $\qquad\Box$

To study the quotient $G/W(G)$ we need to recall a few definitions. Let

$$G = \prod_{i \in I} G_i$$

be a direct product of Ω-groups. Elements $g \in G$ are functions $g : I \to \bigcup G_i$ such that $g(i) \in G_i$ and with finite support $supp(g) = \{i \in I \mid g_i \neq 1\}$. By $\pi_i : G \to G_i$ we denote the canonical projection $\pi_i(g) = g(i)$, we also denote $\pi_i(g) = g_i$. Sometimes we identify the group G_i with its image in G under the canonical embedding $\lambda_i : G_i \to G$ such that $\pi_i(\lambda_i(g)) = g$ and $\pi_j(\lambda_i(g)) = 1$ for $j \neq i$.

An embedding (and we can always assume it is an inclusion) of an Ω-group H into the Ω-group G

$$\phi : H \hookrightarrow \prod_{i \in I} G_i \tag{2}$$

is called a *subdirect decomposition* of H if $\pi_i(H) = G_i$ for each i (here H is viewed as a subgroup of G). The subdirect decomposition (2) is termed *minimal* if $H \cap G_i \neq \{1\}$ for any $i = 1, \ldots, n$, where both G_i and H are viewed as subgroups of G. It is easy to see that given a subdirect decomposition of H one can obtain a minimal one by deleting non-essential factors (using Zorn's lemma).

Definition 2. *An Ω-group G admits a finite semisimple decomposition if $W(G) \neq G$ and $G/W(G)$ is a finite direct product of Ω-simple Ω-groups.*

The following lemma shows that any minimal subdirect decomposition into simple groups is, in fact, a direct decomposition.

Lemma 4.2. *Let* $\phi : G \to \prod_{i \in I} G_i$ *be a minimal subdirect decomposition of an* Ω-*group* G *into* Ω-*simple* Ω-*groups* G_i, $i \in I$. *Then* $G = \prod_{i \in I} G_i$.

Proof. Let $K_i = G \cap G_i$, $i \in I$. It suffices to show that $K_i = G_i$. Indeed, in this event $G \geqslant \prod_{i \in I} G_i$ and hence $G = \prod_{i \in I} G_i$..

Fix an arbitrary $i \in I$. Since ϕ is minimal there exists a non-trivial $g_i \in K_i$. For an arbitrary $x_i \in G_i$ there exists an element $x \in G$ such that $\pi_i(x) = x_i$. It follows that $g_i^x = g_i^{x_i} \in K_i$. Hence $K_i \geqslant gp_{G_i\Omega}(g_i) = G_i$, as required. $\qquad \square$

Lemma 4.3. *If an* Ω-*group* G *has a finite semisimple decomposition then it is unique (up to a permutation of factors).*

Proof. Obvious. $\qquad \square$

Obviously, an Ω-group G admits a finite semisimple decomposition if and only if $W(G)$ is intersection of finitely many maximal normal Ω-subgroups of G. This implies the following lemma.

Lemma 4.4. *A finite* Ω-*group admits a finite semisimple decomposition.*

5. Connectivity of Andrews–Curtis graphs of perfect finite groups

Recall that a group G is called perfect if $[G, G] = G$.

Lemma 5.1. *Let an* Ω-*group* G *admits a finite semisimple decomposition:*

$$G/W(G) = G_1 \times \cdots \times G_k.$$

Then G *is perfect if and only if all* Ω-*simple* Ω-*groups* G_i *are non-abelian.*

Proof. Obvious. $\qquad \square$

We need the following notations to study normal generating tuples in an Ω-group G admitting finite semisimple decomposition. If $g \in \prod_{i \in I} G_i$ then by $supp(g)$ we denote the set of all indices i such that $\pi_i(g) \neq 1$.

Lemma 5.2. *Let* $G = \prod_{i \in I} G_i$ *be a finite product of* Ω-*simple non-abelian* Ω-*groups. If* $g \in G$ *then* $gp_{G\Omega}(g) \geqslant G_i$ *for any* $i \in supp(g)$.

Proof. If $g \in G$ and $g_i = \pi_i(g) \neq 1$, then there exists $x_i \in G_i\Omega$ with $[g_i, x_i] \neq 1$. Hence $1 \neq [g, x_i] = [g_i, x_i] \in gp_{G\Omega}(g) \cap G_i$. Since G_i is Ω-simple it coincides with the nontrivial normal Ω-subgroup $gp_{G\Omega}(g) \cap G_i$, as required. $\qquad \square$

Let $G/W(G) = \prod_{i \in I} G_i$ be the canonical semisimple decomposition of an Ω-group G. For an element $g \in G$ by $\bar{g}$ we denote the canonical image $gW(G)$ of g in $G/W(G)$ and by $supp(g)$ we denote the support $supp(\bar{g})$ of $\bar{g}$. $\qquad \square$

Lemma 5.3. *Let* G *be a finite perfect* Ω-*group and* $G/W(G) = \prod_{i \in I} G_i$ *be its canonical semisimple decomposition. Then a finite set of elements* $g_1, \ldots, g_m \in G$ *generates* G *as a normal* Ω-*subgroup if and only if*

$$supp(g_1) \cup \cdots \cup supp(g_m) = I.$$

Proof. It follows from Lemma 5.2 and Lemma 4.1.   □

Proof of Theorem 2.1. We can now prove Theorem 2.1 which settles the Relativised Finitary AC-Conjecture in affirmative for finite perfect Ω-groups.

Let G be a finite perfect Ω-group, $\overline{G} = G/W(G)$, and $\overline{G} = \prod_{i \in I} G_i$ be its canonical semisimple decomposition. Fix an arbitrary $k \geqslant 2$.

CLAIM 1. Let $U = (u_1, \ldots, u_k) \in N_k(G, \Omega)$. Then there exists an element $g \in G$ with $supp(g) = I$ such that

$$(u_1, \ldots, u_k) \sim_G (g, u_2, \ldots, u_k).$$

Indeed, by Lemma 4.1 the tuple U generates G as a normal subgroup if and only if its image $\overline{U}$ generates $\overline{G}$ as a normal subgroup. Lemma 3.1 shows that it suffices to prove the claim for the Ω-group $\overline{G}$ (recall that $supp(g) = supp(\bar{g})$). So we can assume that $G = \prod_{i \in I} G_i$. Since $U \in N_k(G, \Omega)$, Lemma 5.3 implies that

$$supp(u_1) \cup \cdots \cup supp(u_k) = I.$$

Let $i \in I$ and $i \notin supp(u_1)$. Then there exists an index j such that $i \in supp(u_j)$. By Lemma 5.2, $gp_{G\Omega}(u_j) \geqslant G_i$. So there exists a non-trivial $h \in gp_{G\Omega}(u_j)$ with $supp(h) = \{i\}$. By Lemma 3.2, $U \sim (u_1h, u_2, \ldots, u_k) = U^*$ and $supp(u_1h) = supp(u_1) \cup \{i\}$. Now the claim follows by induction on the cardinality of $I \setminus supp(u_1)$. In fact, one can bound the number of elementary AC-moves needed in Claim 1. Indeed, since G_i is non-abelian Ω-simple there exists an element $x \in G\Omega$ such that $u_j^x \neq u_j$. Then the element h above can be taken in the form $h = u_j^x u_j^{-1}$, and only four moves are needed to transform U into U^*. This proves the claim.

CLAIM 2. Every k-tuple $U_1 = (g, u_2, \ldots, u_k)$ with $supp(g) = I$ is AC-equivalent to a tuple $U_2 = (g, 1, \ldots, 1)$.

By Lemma 5.3 g generates G as a normal Ω-subgroup. Now the claim follows from Lemma 3.2.

CLAIM 3. Every two k-tuples $U_2 = (g, 1, \ldots, 1)$ and $U_3 = (h, 1, \ldots, 1)$ from $N_k(G, \Omega)$ are AC-equivalent.

Indeed, U_2 is AC-equivalent to $(g, 1, \ldots, 1, g)$. By Lemma 3.2 the former one is AC-equivalent to $(h, \ldots, 1, g)$, which is AC-equivalent to $(h, 1, \ldots, 1)$, as required.

The theorem follows from Claims 1, 2, and 3.   □

6. Arbitrary finite groups

Lemma 6.1. *Let*

$$G = G_1 \times \cdots \times G_s \times A \tag{3}$$

be a direct decomposition of an Ω-group G into a product of non-abelian Ω-simple Ω-groups G_i, $i = 1, \ldots s$, and an abelian Ω-group A. Then, assuming $G \neq 1$,

$$d_{G\Omega}(G) = \max\{d_{A\Omega}(A), 1\}.$$

Proof. Put $S(G) = G_1 \times \cdots \times G_s$. Since A is a quotient of G then $d_{G\Omega}(G) \geqslant d_{A\Omega}(A)$. Therefore, $d_{G\Omega}(G) \geqslant \max\{d_{A\Omega}(A), 1\}$. On the other hand, if g generates $S(G)$ as a normal Ω-subgroup (such g exists by Lemma 5.3) and $a_1, \ldots, a_{d_\Omega(A)}$ generate A then we claim that the tuple of elements from G:

$$(ga_1, a_2, \ldots, a_{d_\Omega(A)})$$

generates G as a normal Ω-subgroup. Indeed, let $g = g_1 \cdots g_s$ with $1 \neq g_i \in G_i$. Since G_i is non-abelian then g_i is not central in G_i and hence there exists $h_i \in G_i$ such that $[g_i, h_i] \neq 1$. It follows that if $h = h_1 \ldots h_s$ then $[g, h] \neq 1$ and $supp([g, h]) = \{1, \ldots, n\}$. In particular, $[g, h]$ belongs to $N = gp_{G\Omega}(ga_1, a_2, \ldots, a_{d_\Omega(A)})$ and generates $S(G)$ as a normal Ω-subgroup. Therefore, $S(G) \subset N$ and hence $a_1, \ldots, a_{d_\Omega(A)} \in N$, which implies that $G = N$. This shows that $d_{G\Omega}(G) = \max\{d_\Omega(A), 1\}$, as required. □

Proof of Theorem 2.4. Let G be a minimal counterexample to the statement of the theorem. Then G is not perfect. G is also non-abelian by Fact 1.2. Put $t = d_{G\Omega}(G)$ and $k \geqslant t+1$. Let M be a minimal non-trivial normal Ω-subgroup of G. It follows that $M \neq G$, and the theorem holds for the Ω-group $\overline{G} = G/M$. Obviously, $k > d_{G\Omega}(G) \geqslant d_{\overline{G}\Omega}(\overline{G})$, hence the AC-graph $\Delta_k^\Omega(\overline{G})$ is connected. Fix any tuple $(z_1, \ldots, z_t) \in N_t(G, \Omega)$. If $(y_1, \ldots, y_k)$ is an arbitrary tuple from $N_k(G, \Omega)$ then the k-tuples $(\bar{y}_1, \ldots, \bar{y}_k)$ and $(\bar{z}_1, \ldots, \bar{z}_t, 1, \ldots, 1)$ are AC-equivalent in $\overline{G}$. Hence by Lemma 3.1 there are elements $m_1, \ldots, m_k \in M$ such that

$$(y_1, \ldots, y_k) \sim (z_1 m_1, \ldots, z_t m_t, m_{t+1}, \ldots, m_k).$$

We may assume that one of the elements $m_{t+1}, \ldots, m_k$ in distinct from 1, say $m_k \neq 1$. Indeed, if $m_{t+1} = \ldots = m_k = 1$ then the elements $z_1 m_1, \ldots, z_t m_t$ generate G as a normal Ω-subgroup, hence applying AC-transformations we can get any non-trivial element from M in the place of m_k. Since M is a minimal normal Ω-subgroup of G it follows that M is the $G\Omega$-normal closure of m_k in G, in particular, every m_i is a product of conjugates of $m_k^{\pm 1}$. Applying AC-transformations we can get rid of all elements m_i, $i = 1, \ldots, m_t$, in the tuple above. Hence,

$$(z_1 m_1, \ldots, z_t m_t, m_{t+1}, \ldots, m_k) \sim (z_1, \ldots, z_t, 1, \ldots, 1, m_k).$$

But $(z_1, \ldots, z_t) \in N_t(G, \Omega)$, hence

$$(z_1, \ldots, z_t, 1, \ldots, m_k) \sim (z_1, \ldots, z_t, 1, \ldots, 1).$$

We showed that any k-tuple $(y_1, \ldots, y_k) \in N_k(G, \Omega)$ is AC-equivalent to the fixed tuple $(z_1, \ldots, z_t, 1, \ldots, 1)$. So the AC-graph $\Delta_k^\Omega(G)$ is connected and G is not a counterexample. This proves the theorem. □

7. Proof of Theorem 1.1

We denote by $\tilde{g}$ the image of $g \in G$ in the abelinisation $Ab(G) = G/[G, G]$.

We systematically, and without specific references, use elementary properties of Andrews–Curtis transformations, Lemmas 3.1 and 3.2.

Suppose Theorem 1.1 is false. Consider a counterexample G of minimal order for a given $k \geqslant d_G(G)$. For a given k-tuple $(g_1, \ldots, g_k) \in N_k(G)$ we denote by $\mathcal{C}(g_1, \ldots, g_k)$ the set

$$\{(h_1, \ldots, h_k) \in N_k(G) \mid (\tilde{g}_1, \ldots, \tilde{g}_k) \sim (\tilde{h}_1, \ldots, \tilde{h}_k) \ \& \ (g_1, \ldots, g_k) \not\sim (h_1, \ldots, h_k)\}$$

Put

$$\mathcal{D} = \{(g_1, \ldots, g_k) \in N_k(G) \mid \mathcal{C}(g_1, \ldots, g_k) \neq \emptyset\}.$$

Then the set $\mathcal{D}$ is not empty. Consider the following subset of $\mathcal{D}$:

$$\mathcal{E} = \{(g_1, \ldots, g_k) \in \mathcal{D} \mid |gp_G(g_2, \ldots, g_k)| \ \text{is minimal possible}\}.$$

Finally, consider the subset $\mathcal{F}$ of $\mathcal{E}$:

$$\mathcal{F} = \{(g_1, \ldots, g_k) \in \mathcal{E} \mid |gp_G(g_1)| \ \text{is minimal possible}\}$$

In order to prove the theorem it suffices to show that G is abelian.

Fix an arbitrary tuple $(g_1, \ldots, g_k) \in \mathcal{F}$ and an arbitrary tuple $(h_1, \ldots, h_k) \in \mathcal{C}(g_1, \ldots, g_k)$. Denote $G_1 = gp_G(g_1)$ and $G_2 = gp_G(g_2, \ldots, g_k)$.

The following series of claims provides various inductive arguments which will be in use later.

Notice that the minimal choice of g_1 and $g_2, \ldots, g_k$ can be reformulated as

CLAIM 1.1 *Let $f_1 \in G_1$, $f_2, \ldots, f_k \in G_2$ such that $(f_1, f_2, \ldots, f_k) \in \mathcal{C}(g_1, \ldots, g_k)$. Then*

$$gp_G(f_1) = G_1 \ and \ gp_G(f_2, \ldots, f_k) = G_2.$$

CLAIM 1.2 *Let $f_1 \in G$, $f_2, \ldots, f_k \in G_2$ such that $(f_1, f_2, \ldots, f_k) \in \mathcal{C}(g_1, \ldots, g_k)$. Then*

$$gp_G(f_2, \ldots, f_k) = G_2.$$

CLAIM 1.3 *Let M be a non-trivial normal subgroup of G. Then*

$$(h_1, \ldots, h_k) \sim (g_1 m_1, \ldots, g_{k-1} m_{k-1}, g_k m_k)$$

for some $m_1, \ldots, m_k \in M$.

Indeed, obviously

$$(h_1 M, \ldots, h_k M), (g_1 M, \ldots, g_k M) \in N_k(G/M).$$

Moreover, since

$$(\tilde{g}_1, \ldots, \tilde{g}_k) \sim (\tilde{h}_1, \ldots, \tilde{h}_k)$$

there exists a sequence of AC-moves $t_1, \ldots, t_n$ (where each t_i is one of the transformations (1)–(4), with the specified values of w in the case of transformations (4)) and elements $c_1, \ldots, c_k \in [G, G]$ such that

$$(h_1, \ldots, h_k) t_1 \cdots t_k = (g_1 c_1, \ldots, g_k c_k)$$

Therefore
$$(h_1M, \ldots, h_kM)t_1 \cdots t_k = (g_1c_1M, \ldots, g_kc_kM)$$
Since $c_iM \in [G/M, G/M]$ for every $i = 1, \ldots, k$ this shows that the images of the tuples $(h_1M, \ldots, h_kM)$ and $(g_1M, \ldots, g_kM)$ are AC-equivalent in the abelianisation $Ab(G/M)$. Now the claim follows from the fact that $|G/M| < |G|$ and the assumption that G is the minimal possible counterexample.

The following claim says that the set $\mathcal{C}(g_1, \ldots, g_k)$ is closed under $\sim$.

CLAIM 1.4 *If* $(e_1, \ldots, e_k) \in \mathcal{C}(g_1, \ldots, g_k)$ *and* $(f_1, \ldots, f_k) \sim (e_1, \ldots, e_k)$ *then* $(f_1, \ldots, f_k) \in \mathcal{C}(g_1, \ldots, g_k)$

Now we study the group G in a series of claims.

CLAIM 2. $G = G_1 \times G_2$.

Indeed, it suffices to show that $G_1 \cap G_2 = 1$. Assume the contrary, then $M = G_1 \cap G_2 \neq 1$ and by Claim 1.3
$$(h_1, \ldots, h_k) \sim (g_1m_1, \ldots, g_{k-1}m_{k-1}, g_km_k)$$
for some $m_1, \ldots, m_k \in M$. By Claim 1.4
$$(g_1m_1, \ldots, g_{k-1}m_{k-1}, g_km_k) \in \mathcal{C}(g_1, \ldots, g_k)$$
By Claim 1.1,
$$gp_G(g_1m_1) = G_1, \quad gp_G(g_2m_2, \ldots, g_km_k) = G_2$$
and we can represent the elements $m_2, \ldots, m_k \in G_1 \cap G_2$ as products of conjugates of g_1m_1, therefore deducing that
$$(g_1m_1, g_2m_2 \ldots, g_km_k) \sim (g_1m_1, g_2, \ldots, g_k).$$
Since $m_1 \in gp_G(g_2, \ldots, g_k)$, we conclude that
$$(g_1m_1, g_2, \ldots, g_k) \sim (g_1, g_2, \ldots, g_k),$$
and therefore
$$(h_1, \ldots, h_k) \sim (g_1, \ldots, g_k),$$
a contradiction. This proves the claim. □

CLAIM 3. $[G_2, G_2] = 1$. *In particular,* $G_2 \leqslant Z(G)$.

Indeed, assume the contrary. Then $M = [G_2, G_2] \neq 1$ and by Claim 1.3
$$(h_1, \ldots, h_k) \sim (g_1m_1, \ldots, g_km_k), \quad m_1, \ldots, m_k \in M \leqslant G_2.$$
By virtue of Claims 1.4 and 1.2, $gp_G(g_2m_2, \ldots, g_km_k) = G_2$ and hence $m_1 \in gp_G(g_2m_2, \ldots, g_km_k)$. It follows that
$$(g_1m_1, g_2m_2, \ldots, g_km_k) \sim (g_1, g_2m_2 \ldots, g_km_k).$$
Therefore it will be enough to prove
$$(g_1, g_2m_2, \ldots, g_km_k) \sim (g_1, g_2, \ldots, g_k).$$

We proceed as follows, systematically using the fact that $g_2, \ldots, g_k$ and all their conjugates commute with all the conjugates of g_1.

We start with a series of Nielsen moves which lead to

$$(g_1, g_2 m_2 \ldots, g_k m_k) \quad \sim \quad (g_1, g_1 \cdot g_2 m_2, g_3 m_3, \ldots, g_k m_k)$$
$$\sim \quad (g_1 \cdot m_2, g_1 g_2 m_2, g_3 m_3, \ldots, g_k m_k).$$

The last transformation is the key for the whole proof and requires some explanation. Since m_2 belongs to

$$[G_2, G_2] = [gp_G(g_2 m_2, \ldots, g_k m_k), gp_G(g_2 m_2, \ldots, g_k m_k)],$$

m_2 can be expressed as a word

$$w(x_2, \ldots, x_k) = (x_{i_1}^{f_1})^{\varepsilon_1} \cdots (x_{i_l}^{f_l})^{\varepsilon_l}$$

where $x_i = g_i m_i$, $i = 2, \ldots, k$, $f_j \in G$ and the word w is balanced for each variable x_i, that is, for each $h = 2, \ldots, k$, the sum of exponents for each x_h is zero:

$$\sum_{i_j = h} \varepsilon_j = 0.$$

Moreover, since $G = G_1 \times G_2$, we can choose $f_j \in G_2$, whence commuting with $g_1 \in G_1$. Therefore

$$w(g_1 x_2, x_3, \ldots, x_k) = w(x_2, x_3, \ldots, x_k)$$

and

$$w(g_1 g_2 m_2, g_3 m_3, \ldots, g_k m_k) = m_2.$$

Hence, by several consecutive multiplications by appropriate conjugates of $g_1 g_2 m_2$ and $g_i m_i$, $i = 3, \ldots, k$, we can produce the factor m_2 in the leftmost position in the tuple. We now continue:

$$(g_1 m_2, g_1 g_2 m_2, g_3 m_3, \ldots, g_k m_k) \quad \sim \quad (g_1 m_2, g_1 g_2 m_2 \cdot (g_1 m_2)^{-1}, g_3 m_3, \ldots, g_k m_k)$$
$$= \quad (g_1 m_2, g_2, g_3 m_3, \ldots, g_k m_k).$$

Again by Claims 1.4 and 1.2 $G_2 = gp_G(g_2, g_3 m_3, \ldots, g_k m_k)$. Since $m_2 \in G_2$,

$$(g_1 m_2, g_2, g_3 m_3, \ldots, g_k m_k) \quad \sim \quad (g_1, g_2, g_3 m_3, \ldots, g_k m_k).$$

Next we want to kill m_3. Present m_3 as a balanced word in $g_2, g_3 m_3, \ldots, g_k m_k$ conjugated by elements $f_i \in G_2$. Note that they all commute with g_1. As before,

$$m_3 = w(g_2, g_1 g_3 m_3, g_4 m_4, \ldots, g_k m_k)$$

(and, actually, $m_3 = w(g_2, y_3, \ldots, y_k)$ where y_i are arbitrarily chosen from $g_i m_i$ or $g_1 g_i m_i$, $i = 3, \ldots, k$.).

Thus we have:

$$(g_1, g_2, g_3 m_3, \ldots, g_k m_k) \quad \sim \quad (g_1, g_2, g_1 g_3 m_3, g_4 m_4, \ldots, g_1 g_k m_k)$$
$$\sim \quad (g_1 m_3, g_2, g_1 g_3 m_3, g_4 m_4, \ldots, g_k m_k)$$
$$\sim \quad (g_1 m_3, g_2, g_3, g_4 m_4, \ldots, g_k m_k)$$
$$\sim \quad (g_1, g_2, g_3, g_4 m_4, \ldots, g_k m_k)$$

(the last transformation uses the fact that $gp_G(g_2, g_3, g_4m_4, \ldots, g_km_k) = G_2$ by Claims 1.4 and 1.2).

One can easily observe that we can continue this argument in a similar way until we come to $(g_1, g_2, \ldots, g_k)$ - contradiction, which completes the proof of the claim. $\quad\square$

CLAIM 4.
$$[G_1, G_1] = 1.$$

Let $[G_1, G_1] \neq 1$. For a proof, take a minimal non-trivial normal subgroup M of G which lies in $[G_1, G_1]$. Again, by Claim 1.3, we conclude that
$$(h_1, \ldots, h_k) \sim (g_1m_1, g_2m_2, \ldots, g_km_k)$$
for some $m_1, \ldots, m_k \in M$. We assume first that $M \leqslant gp_G(g_1m_1)$. Then
$$(g_1m_1, g_2m_2 \ldots, g_km_k) \sim (g_1m_1, g_2, \ldots, g_k)$$
and $gp_G(g_1m_1) = gp_G(g_1)$ by Claims 1.4 and 1.1. In particular,
$$M \leqslant [gp_G(g_1m_1), gp_G(g_1m_1)] = [gp_G(g_2g_1m_1), gp_G(g_2g_1m_1)],$$
where the last equality follows from the observation that $g_2 \in Z(G)$. We shall use this in further transformations:
$$\begin{aligned}(g_1m_1, g_2, \ldots, g_k) &\sim (g_1m_1, g_2g_1m_1, g_3, \ldots, g_k) \\ &\sim (g_1, g_2g_1m_1, g_3, \ldots, g_k) \\ &\sim (g_1, g_2, g_3, \ldots, g_k).\end{aligned}$$
This shows that $(h_1, \ldots, h_k) \sim (g_1, \ldots, g_k)$ - contradiction. Therefore we can assume that $M \not\subseteq gp_G(g_1m_1)$ and hence $M \cap gp_G(g_1m_1) = 1$. We claim that not all of the elements $m_2, \ldots, m_k$, are trivial. Otherwise
$$(h_1, \ldots, h_k) \sim (g_1m_1, g_2, \ldots, g_k),$$
and we can repeat the previous argument and come to a contradiction. So we assume, with out loss of generality, that $m_2 \neq 1$.

If M is non-abelian then
$$M = [M, M] = [gp_G(m_2), gp_G(m_2)] = [gp_G(g_2m_2), gp_G(g_2m_2)]$$
and
$$\begin{aligned}(g_1m_1, g_2m_2, g_3m_3, \ldots, g_km_k) &\sim (g_1, g_2m_2, g_3m_3, \ldots, g_km_k) \\ &\sim (g_1, g_2, g_3, \ldots, g_k);\end{aligned}$$
we use in the last transformation that $gp_G(g_1) = G_1 \geqslant M$.

Therefore we can assume that M is abelian. Since $M \cap gp_G(g_1m_1) = 1$ we conclude that $[M, gp_G(g_1m_1)] = 1$. But then $[M, gp_G(g_1)] = 1$. In particular, $M \leqslant Z(G)$ and the subgroup $[gp_G(g_1m_1), gp_G(g_1m_1)] = [gp_G(g_1), gp_G(g_1)]$ contains M. But this is a contradiction with $M \cap gp_G(g_1m_1) = 1$. This proves the claim. $\quad\square$

FINAL CONTRADICTION. Claims 3 and 4 now yield that G is abelian, as required.

□

Final comments

The referee has kindly called to our attention that the result of Myasnikov [16] (mentioned in the Introduction) was also proved independently in 1978 by Wes Browning (unpublished).

References

[1] S. Akbut and R. Kirby, 'A potential smooth counterexample in dimension 4 to the Poincare conjecture, the Schoenflies conjecture, and the Andrews–Curtis conjecture', *Topology* **24** (1985), 375–390.

[2] J. J. Andrews and M. L. Curtis, 'Free groups and handlebodies', *Proc. Amer. Math. Soc.* **16** (1965), 192–195.

[3] L. Bartholdi, R. I. Grigorchuk and Z. Sunik, 'Branch groups', in *Handbook of Algebra*, vol. 3 (M. Hazelwinkel, ed.), 2003.

[4] A. V. Borovik, 'Centralisers of involutions in black box groups', *Computational and Statistical Group Theory* (R. Gilman et al., eds.), Contemporary Mathematics **298** (2002), 7–20; math.GR/0110233.

[5] A. V. Borovik, E. I. Khukhro, A. G. Myasnikov, 'The Andrews–Curtis Conjecture and black box groups', *Int. J. Algebra and Computation* **13** no. 4 (2003), 415–436; math.GR/0110246.

[6] F. Celler, C. Leedham-Green, S. Murray, A. Niemeyer and E. O'Brien, 'Generating random elements of a finite group', *Comm. Algebra* **23** (1995), 4931–4948.

[7] P. Diaconis and R. Graham, 'The graph of generating sets of an abelian group', *Colloq. Math.* **80** (1999), 31–38.

[8] R. I. Grigorchuk,'Degrees of growth of finitely generated groups and the theory of invariant means', *Math. USSR – Izv.* **25** no. 2 (1985), 259–300.

[9] R. I. Grigorchuk, 'Just infinite branch groups', in *New Horizons in pro-p-groups* (M. P. F. du Sautoy, D. Segal and A. Shalev, eds.), Birkhäuser, Boston, 2000, 121–179.

[10] W. Kantor and A. Seress, 'Black box classical groups', *Memoirs Amer. Math. Soc.* **149** no. 708, Amer. Math. Soc., Providence, RI, 2000.

[11] D. A. Kazhdan, 'On the connection of the dual space of a group with the structure of its closed subgroups', *Funkcional. Anal. i Prilozh.* **1** (1967), 71–74.

[12] C. R. Leedham-Green and S. H. Murray, 'Variants of product replacement', *Computational and statistical group theory* (R. Gilman et al., eds.), Contemp. Math. **298** (2002), 97–104.

[13] Y. G. Leonov, 'The conjugacy problem in a class of 2-groups', *Mat. Zametki* **64** no. 4 (1998), 573—583.

[14] A. Lubotzky and I. Pak, 'The product replacement algorithm and Kazhdan's property (T)', *J. Amer. Math. Soc.* **14** (2001), 347–363.

[15] K. Mueller, 'Probleme des einfachen Homotopietyps in niederen Dimensionen und ihre Behandlung mit Mitteln der topologischen Quantenfeldtheorie', Ph. D. Thesis, Frankfurt.

[16] A. G. Myasnikov, 'Extended Nielsen transformations and the trivial group', *Math. Notes* **35** no. 3–4 (1984), 258–261.

[17] B. H. Neumann and H. Neumann, 'Zwei Klassen characteristischer Untergruppen und ihre Faktorgruppen', *Math. Nachr.* **4** (1951), 106–125.

[18] I. Pak, 'What do we know about the product replacement algorithm', in *Groups and Computation III* (W. Kantor and A. Seress, eds.), DeGruyter, Berlin, 2001, pp. 301–348.

[19] A. V. Rozhkov, 'The conjugacy problem in an automorphism group of an infinite tree', *Mat. Zametki* **64** no. 4 (1998), 592–597.

Alexandre V. Borovik
School of Mathematics
The University of Manchester
PO Box 88
Manchester M60 1QD
United Kingdom
e-mail: `borovik@manchester.ac.uk`
URL: `http://www.ma.umist.ac.uk/avb/`

Alexander Lubotzky
Department of Mathematics
Hebrew University
Givat Ram, Jerusalem 91904
Israel
e-mail: `alexlub@math.huji.ac.il`

Alexei G. Myasnikov
Department of Mathematics
The City College of New York
New York, NY 10031
USA
e-mail: `alexeim@att.net`
URL: `http://home.att.net/~alexeim/index.htm`

Progress in Mathematics, Vol. 248, 31–55

Cuts in Kähler Groups

Thomas Delzant and Misha Gromov

Abstract. We study fundamental groups of Kähler manifolds via their cuts or relative ends.

Mathematics Subject Classification (2000). 32Q15, 20F65, 57M07.

Keywords. Fundamental groups of Kähler manifolds, small cancelation groups.

1. A group G is called Kähler if it serves as the fundamental group $\pi_1(V)$ of a compact Kähler manifold V. Equivalently, such a group G appears as a discrete free co-compact isometry group of a complete simply connected Kähler manifold X – the Galois group acting on the universal covering of V denoted X. To keep the perspective (compare [De-Gr]) we indicate possible generalization of this setting.

(a) Dropping "free", i.e. allowing discrete actions with fixed points (having finite stabilizers).

(b) Replacing "co-compact" by a weaker smallness condition on the quotient X/G, e.g. by requiring X/G to have finite volume (or slow volume growth) combined with a sufficiently simple geometry at infinity in the spirit of the following two examples.

(b') Complete (and natural non-complete) Kähler metrics on quasi-projective varieties V and the corresponding metrics on coverings of such V.

(b") Complete Kähler metrics on X with bounded geometry i.e. with curvature bounded from above and the injectivity radius from below.

(c) Admitting (closed) non-discrete isometry groups G of X.

(d) Allowing singular spaces V and X.

(e) Replacing Kähler groups by Kähler groupoïds that are leaf-wise Kähler foliations with transversal measures ($[Gr]_{FPP}$).

1.1. Central Problem. Identify the constraints imposed by the Kähler nature of the space X on its asymptotic metric invariants and then express these constraints in terms of some algebraic properties of G.

If V is projective algebraic then the structure of the profinite completion of its fundamental group G is accounted for by finite, and hence algebraic, coverings of V. Here we are primely concerned with infinite coverings and issuing "transcendental" constraints on G that are not expressible (at least not directly) in terms of subgroups of finite index in G and/or finite dimensional representation of G.

1.2. The basic examples of Kähler groups are surface groups that are the fundamental groups of Riemann surfaces S, i.e. complex algebraic curves, and Cartesian products of surface groups, e.g. free Abelian groups of even rank. Less obvious examples are provided by discrete co-compact groups acting on Hermitian symmetric spaces, such as the unit ball in $\mathbb{C}^n$ with the Bergman metric, for instance.
Let $V \subset \mathbb{P}^N$ be a projective algebraic manifold of dimension n. Due to the Lefschetz theorem, if $V' = V \cap P$ is the intersection of V by a transverse projective subspace of dimension $N - n + 2$, then the inclusion $V' \subset V$ induces an isomorphism on the fundamental groups, $\pi_1(V') \simeq \pi_1(V)$. However, such a surface -called Lefschetz surface- usually is more complicated than the original V; to see this try to visualize such an (hyper)surface in the Cartesian product of three Riemann surfaces of positive genera, $V' \subset V = S_1 \times S_2 \times S_3$.

1.3. The above central problem is accompanied by its relative version: determine characteristic features of homomorphisms between Kähler groups induced by holomorphic maps f.
Observe in this regard, that an arbitrary proper holomorphic map $f : V \to W$ factors trough a surjective holomorphic map $f' : V \to W'$ with non empty connected fibers $f^{-1}(v), w \in W$, followed by a finite-to-one map $W' \to W$. The connectedness of the fibers makes f' surjective on the fundamental groups ; furthermore, if W is non-singular and f is onto, then the map $W' \to W$, being a ramified covering, sends the fundamental group of W' (and, hence of V) onto a subgroup of finite index in $\pi_1(W)$ (here W is assumed compact). In particular, if V fibers over a non-singular curve (Riemann surface) of genus g, i.e. admits a surjective holomorphic map to such curve S, then the fundamental group of V surjects onto a surface group $\pi_1(S')$ of genus $g' \geq g$.
A particular seemingly innocuous instance of the relative problem concerns subgroups in the products of surface groups, $G \subset \pi_1(W = S_1 \times S_2, \ldots, \times S_N)$. When is such G Kähler ? When does there exist, for some choice of conformal structures in S_i, an algebraic sub-variety $V \subset W$ (singular or non-singular) of a given dimension n such that the image of the fundamental group of V in $\pi_1(W)$ equals G? (Notice that for a general non-Lefschetz V the inclusion homomorphism is not injective on the fundamental groups.) This question seems non-trivial already for $n = 1$ where it has a purely topological counterpart: find a real surface V in W such that the projections of V to all S_i's are ramified coverings and such that the image of $\pi_1(V)$ in $\pi_1(W)$ equals G.
We shall see later on that some algebraic condition (existence of cuts) on an general Kähler group makes it a subgroup in a product of surface groups.

2. The Kähler nature of X becomes metrically discernible when X is harmonically mapped into a (globally) H-non positive space. Here are the necessary definitions.

2.1. The energy density E of a smooth map f between Euclidean balls at a point x is defined as one half of the squared norm of its differential D at x,

$$E(x) = \frac{1}{2} trace D^* D(x).$$

This generalizes to maps where the differential exists on a dense set of points x', e.g. to Lipschitz maps f, as $\lim\sup_{x' \to x} E(x')$.

2.2. For a Lipschitz map between arbitrary metric spaces, $f : X \to Y$, the (Euclidean-like) energy density is defined as the infimum of those e such that, for every two Euclidean balls B_1 and B_2 and arbitrary 1-Lipschitz maps $B_1 \to X$ and $Y \to B_2$, where the center 0 of B_1 goes to x, the composed map $B_1 \to B_2$ has $E(0) \le e$.

2.3. If the space X is endowed with a measure then the energy of an f, denoted $E(f)$ is defined as the integral of the energy density with this measure, where for Riemannian, e.g. Kähler manifolds one uses the ordinary Riemannian measure for this purpose (see $[\text{Gr}]_{FPP}$).

2.4. A map f is called harmonic if it is locally energy minimizing, i.e minimizing under variations of f which are non trivial on small balls in X.

2.5. If X is a Riemann surface, i.e. a 1-dimensional complex manifold, then the energy of an arbitrary map as well as the harmonicity obviously are conformal invariants, i.e. are independent of the Kähler metric compatible with the complex structure. This allows one to define *pluriharmonic* maps from an arbitrary complex space to a metric space as those f whose restrictions to all holomorphic curves in X are harmonic.

One knows (this is easy, at least for Riemannian targets) that every pluriharmonic map of a Kähler manifold is harmonic but for $dim X = n > 1$ most harmonic maps are not pluriharmonic. For example a real valued function f on a Kähler manifold is pluriharmonic if and only if its gradient (vector field) is Hamiltonian, i.e. preserving the symplectic part ω of the Kähler metric while harmonicity amounts to preservation of the corresponding volume form ω^n under the gradient flow of f.

2.6. If a space X is properly (e.g. discretely) acted upon by a group G and the energy density $E(x)$ of some map f is G-invariant, then the G-energy of f is defined as the integral of E descended to the quotient space X/G. This applies, in particular, to G-equivariant (harmonic and non-harmonic) maps between G-spaces.

2.7. A metric G-space Y, i.e. a space Y isometrically acted upon by G, is called (globally) H-non-positive if every G-equivariant harmonic map f of finite G-energy from an arbitrary Kähler G-manifold X to Y is pluriharmonic.

This definition (albeit provisional) is justified by the following fundamental (and amazing)

2.8. Hodge Lemma (see [ABCKT]). *Flat Hilbertian manifolds are H-non-positive.*

Recall that "flat Hilbertian" signifies that Y is locally isometric to a finite or infinite dimensional Hilbert space.

2.9. Basic corollary. *Let X be a complete Kähler G-manifold and suppose the group G is represented by isometries of a Hilbert space Y. Then every harmonic G-equivariant map $f : X \to Y$ of finite G-energy is pluriharmonic.*

Remarks.

2.10. The relevant actions on the Hilbert space Y above are *affine* rather than linear. For example, they may be free and discrete (Haagerup property, see [CCJV]).

2.11. If the actions of G on X and Y are discrete, then the G-energy equals the ordinary energy of the corresponding map between the quotient spaces. In particular, this energy is necessarily finite if the action of G on X is co-compact.

2.12. In some cases (e.g. if X is simply connected and Y is a Hilbert space) every pluriharmonic map from X to Y analytically extends to a holomorphic map from X to a suitable complexification of Y.

2.13. The existence of a G-equivariant harmonic map often (but not always) comes cheap: for example, it follows in many cases from the existence of a continuous G-equivariant map with finite G-energy. But the issuing pluriharmonic (and even more so holomorphic) map carries a much higher price tag and the presence of such a map imposes strong geometric restrictions on the manifolds and groups in question.

The idea of Hermitian sectional curvature, as well as the following non-linear Hodge Lemma has been discovered by Y.T. Siu ([Siu]), and further developed in [He],[C-T],[Sam] and [Gr-Sc] ; see [ABCKT] for additional informations and references.

2.14. Non-linear Hodge lemma (see [ABCKT]). *The H-non-positivity property, remains valid for the following non-flat target spaces Y.*

(A) *Metric graphs, e.g. trees (including $\mathbb{R}$-trees with no local finiteness condition).*
(B) *Euclidean buildings (these generalize trees).*
(C) *Riemannian and Hilbertian symmetric spaces of non-positive sectional curvature.*
(D) *Riemannian and Hilbertian manifolds with point-wise 1/4-pinched negative sectional curvature, e.g. Riemann surfaces with negative sectional curvature.*
(E) *Riemannian and Hilbertian manifolds with non-positive Hermitian sectional curvature. (Riemann surfaces with non-positive sectional curvature, symmetric spaces and 1/4-pinched manifolds fall into this category.)*
(F) *Metric spaces locally isometric to finite and infinite Cartesian products of the above (A)–(E).*

The maximal class of known H-non positive spaces Y admits a local characterization saying in effect that these are locally $CAT(0)$, their non-singular loci have non-positive Hermitian sectional curvature and the singularities of Y are quasi-regular in the sense of [Gr-Sc], i.e. they have no more negativity (of singular sectional) curvature than Euclidean buildings do. Observe that this class is closed under scaling and Cartesian products but it is unclear how stable this class (and H-negativity in general) is under conventional limits of metric spaces.

2.15. Examples. Let V be a compact connected Kähler manifold and Y a compact Riemannian manifold of non-positive sectional curvature. Then every continuous map $f_0 : V \to Y$ is homotopic to a (essentially unique) harmonic map f and this f is pluriharmonic if Y is H-non-positive.

(a) For instance, if Y is a flat torus (where the harmonicity and pluriharmonicity of f does not depend on a choice of a flat metric compatible with the affine structure) then one obtains a harmonic, hence pluriharmonic, map f homotopic to a given f_0, where this f is unique up to a toral translation. This applies, in particular, to the the Jacobian (torus) of V, that is $J(V) = H_1(V;\mathbb{R})/H_1(V;\mathbb{Z})$, where one concludes to the existence of, a unique up to a translation, pluriharmonic Abel-Jacobi map $f_0 : V \to J(V)$ that induces the identity isomorphism on the 1-dimensional real homology (for the canonical identification of the 1-dimensional homology $H_1(V;\mathbb{R})$ with the homology of the Jacobian $J(V)$). Furthermore, according to the Albanese-Abel-Jacobi theorem, there exists a unique invariant complex structure on $J(V)$ for which f_0, called Albanese map, is holomorphic and such that every holomorphic map from V to a flat Kähler manifold A with Abelian fundamental group (compact complex torus) factors via the Albanese map followed by an affine holomorphic map $J(V) \to A$. The existence of the complex structure on the Jacobian makes the first Betti number of V even : this is the first basic constraint on the fundamental group of V.

(b) If Y has constant negative curvature then harmonic maps $f : V \to Y$, besides being pluriharmonic, necessarily have rank (of their differentials at all point in V) at most two ([Sam]). Moreover, every harmonic map $V \to Y$ of rank 2 factors via a holomorphic map of V to a hyperbolic (i.e. of genus¿1) Riemann surface, $V \to S \to Y$, by a theorem of Sampson ([Sam]) see also [ABCKT],[C-T] for generalizations.

(c) If Y is itself Kähler and moreover, has constant Hermitian curvature (i.e. covered by the unit ball in $\mathbb{C}^n$ with the Bergman metric) and if a harmonic map f has $rank > 2$ at some point in V then, by Siu's theorem, f is either holomorphic or anti-holomorphic. (A map f is anti-holomorphic if it maps each holomorphic curve $C \subset V$ to a holomorphic curve $C' \subset Y$ and the maps $C \to C'$ are conformal and orientation reversing for the canonical orientation on holomorphic curves).

2.16. The above (a), (b), (c) remain valid for infinite dimensional (Hilbertian) manifolds Y (with accordingly constant curvatures), where one may needs a certain stability conditions (depending on Y, V and the homotopy class) that ensure the existence of a harmonic map in a given homotopy class of maps $f : V \to Y$.
Let x_0 a base-point in V, and let $G = \pi_1(X, x_0)$ acts on the universal cover $\tilde{Y}$ by $f_* : G \to \pi_1(Y, f(x_0))$. Let $g_1, \ldots g_r$ be a generating system of G. One says ([Gr]$_{RWRG}$ 3.7.A') that the action is *stable* if for every $K > 0$, any sequence

$y_n \in \tilde{Y}$ s.t for all $1 \leqslant i \leqslant r, d(g_i y_n, y_n) < K$ admits a convergent subsequence. This property only depends on the action of G on $\tilde{Y}$ induced by f_*.

We shall return on this stability condition latter on, but even in the absence of stability one still can obtain harmonic maps starting from maps of finite energy and by applying an energy minimizing process. Such a process, which may diverge in ordinary sense, often (essentially always) converges in a generalized sense where the target space Y need to be eventually replaced by a suitable limit of pointed spaces (Y, y_i) ([Mo],[Ko-Sc], $[\text{Gr}]_{RWRG}$). In particular, N. Mok [Mo] proved that, if the fundamental group of a Kähler manifold X does not satisfy Kazhdan T property, then the universal cover of X carries a non constant holomorphic function to some Hilbert space, equivariant for some affine isometric representation of the fundamental group ; consequently the universal cover of X support a non constant holomorphic function with bounded differential (and thus, of at most linear growth).

Basic examples of infinite dimensional symmetric spaces have been introduced by P. de la Harpe [Ha].

2.17. Strict H-negativity. This means, by definition, that Y satisfies the conclusion of the above (b) : every harmonic map $V \to Y$ (or, in general, every harmonic H-equivariant with finite H-energy map $X \to Y$) of rank $\geqslant 2$ is of rank 2 and factors via a holomorphic map to a hyperbolic Riemann surface S followed by a harmonic map $S \to Y$.

The basic examples of such Y's are Riemannian manifold with strictly negative Hermitian curvature. These include strictly $\frac{1}{4}$-pinched manifold by a Siu-Sampson-Hernandez theorem (see [ABCKT], Chap.6 and references therein).

Furthermore, the piecewise Riemannian spaces built of simplexes of negative Hermitian curvature with geodesic faces and with the links of all faces of diameter $> \pi$ (compare the regularity assumption in [Gr-Sc]) are H-negative. Moreover, the strictness of negativity is needed not everywhere but only on a "sufficiently large" part of Y. For example, every 2-dimensional $CAT(0)$-polyhedron with the above assumption on the links is H-strictly negative, provided its fundamental group is hyperbolic.

It follows, for instance, that if Y is obtained by ramified covering of a Euclidean 2-dimensional building Y_0 (e.g. the product of two graphs), where the ramification locus lies away from the 1-skeleton of Y_0, and meets all 2-simplexes in Y_0, then every harmonic map of a Kähler manifold to Y factors via a holomorphic map to a Riemann surface. Notice that the singularities of Y at the ramification points have links of diameters$> \pi$ (in fact $\geq \pi$ but these can be smoothed and therefore, the H-negativity does not suffer. Similarly, the ramified covers of manifolds of H-non-positive (H-negative) curvature along totally geodesic submanifolds of codimension two are H-non-positive (H-negative). In fact, the presence of ramification enhances H-negativity. For example, if a complex surface Y_0 of constant Hermitian curvature< 0 is ramified over a totally real geodesic surface, then every harmonic map of X to the resulting $Y \to Y_0$ that transversally meets the ramification locus factors via a holomorphic map to some $S \to Y$. In particular, the fundamental group of Y itself is non-Kähler. Similarly one sees that the majority of ramified coverings of Abelian varieties over unions of (mutually intersecting) flat real codimension two sub-tori have non-Kähler fundamental groups.

3. Let V be a compact Riemannian manifold with fundamental group G and X be a covering of V with the fundamental group $H \subseteq G$. The existence/non-existence of a non-constant harmonic function f on X with finite energy, i.e. with a square integrable differential, depends only on G and $H \subseteq G$ but not on X per se.

Examples.

3.1. If $H = \{id\}$, i.e. X equals the universal covering of V, then the existence of such f is equivalent by De Rham-Hodge theory to non-vanishing of the reduced 1-dimensional cohomology of V and/or of G with coefficients in the regular representation, $H^1(V; l_2(G)) = H^1(G; l_2(G)) \neq 0$.

3.2. On reduced cohomology. When defining cohomology with infinite dimensional coefficients one may factorize the kernel of d by the closure of the image of d, where the resulting cohomology is referred to as reduced. For example, a square integrable closed 1-form a on X represents a non-zero reduced cohomology class (in $H^1(V; l_2(G))$) if and only if there exists a square integrable 1-cycle (closed 1-current) b on X such that $a(b) \neq 0$. In what follows the cohomology is understood as reduced unless otherwise stated.

One can express this property in terms of the *stability* of a certain unitary action.

Stability and reduced cohomology. *Let $\mathcal{H}$ be a Hilbert space. Let $\rho : G \to Isom(\mathcal{H})$ be some affine isometric action of G, and let $\pi : G \to U(\mathcal{H})$ its unitary part. Then the following properties are equivalent :*

-i *The affine action ρ is stable in the sense of* 2.16.

-ii *The unitary representation π has no almost fixed vector, in other words there is no sequence ξ_n of unit vectors such that $\|\rho(g_i)\xi_n - \xi_n\| \to 0$ for all $1 \leqslant i \leqslant r$*

-iii *The unitary representation π has no fixed vectors, and $H^1(G, \pi) = H^1(G, \pi)$.*

Proof. Suppose ρ is not stable. Then there exists a constant K and a sequence y_n s.t for all $1 \leqslant i \leqslant r$, $\|\rho(g_i)y_n - y_n\| \leqslant K$, but y_n has no convergent subsequence. Therefore $\|\pi(g_i)y_n - y_n\| \leqslant K'$ for some constant K'. As the sequence y_n has no convergent subsequence for the weak topology, it is unbounded. Thus the sequence $\xi_n = \frac{y_n}{\|y_n\|}$ is an almost fixed vector. Conversely, if ξ_n is an almost fixed vector, and π has no fixed vector $\max_i \|\rho(g_i)\xi_n - \xi_n\| = \varepsilon_i \to 0$. Let $y_n = \varepsilon_n^{-1}\xi_n$, then $\|\rho(g_i)y_n - y_n\| \leqslant 1$, but y_n is not bounded. This proves i$\Leftrightarrow$ii.

Let us prove i$\Rightarrow$ iii : as G is finitely generated $Z^1(G, \pi)$ has a structure of Hilbert space. If π has no fixed vector, the boundary map $\beta : \mathcal{H} \to Z^1(G, \pi)$ is injective ; if $Im\beta$ is not closed, there exists a sequence y_n such that $\rho(g_i)y_n - y_n \to b(g_i)$ where b is some 1-boundary not homologous to zero. Thus the sequence y_n has no convergent subsequence and ρ is not stable. The implication iii$\Rightarrow$ii is due to

Guichardet (Lemma page 48 in [HV]) : if $H^1(G,\pi) = H^1(G,\pi)$, $B^1(G,\pi)$ is closed in $Z^1(G,\pi)$ and is a Hilbert space. The map β is thus an isomorphism of Hilbert spaces, and if $\|\beta(\xi_n)\| \to 0$ then $\|\xi_n\| \to 0$. $\qquad\square$

3.3. Stability at infinity. Let X be a complete connected Riemannian manifold. An *end* of X is a non compact connected component of the complementary of some non empty relatively compact open set B with smooth boundary (for instance a ball). If E is an end, one defines (see [Gri]) its capacity : $cap(E) = \inf_{\varphi \in \Phi} \int_M |\nabla\varphi|^2$, where Φ is the set of smooth maps such that $0 \leqslant \varphi \leqslant 1$, $\varphi|_{E^c} = 0$, and $\varphi = 1$ outside a compact subset of E. If $x \in M$ let $c(x,R)$ be the capacity of the complementary of $B(x,R)$.

Definition. The manifold X is stable at infinity if, as $R \to \infty$, $c(x,R) \to \infty$ uniformly in x.

3.4. Example (See [Gri] Thm. 8.1). Suppose that X satisfies an isoperimetric inequality : for every compact domain $A \subset X$, one has $(vol_{n-1}(\partial A)) > f(vol_n A)$, and suppose that the integral $\int^{+\infty} \frac{dt}{f^2(t)}$ is convergent. Then for every ball $c(x,R) > (\int_{vol(B(x,R))}^{+\infty} \frac{dt}{f^2(t)})^{-1}$, and therefore X is stable at infinity if $volB(x,R)) \to \infty$ uniformly in X.

Recall also the fundamental result of Eells Sampson, also valid for harmonic maps with values in trees ([Gr-Sc] 2.4).

3.5. *Suppose M has bounded geometry and a lower bound ρ on the injectivity radius, then, there exists a constant c s.t. if u is an harmonic map :*
$$\sup_{x \in B(x,\rho/2)} |\nabla u| < c(\int_{B(x,\rho)} |\nabla u|^2 dx)^{1/2}$$

The stability condition insures the existence of *proper* harmonic maps.

3.6. *Let E be some end. If $cap(E) > 0$, there exists a non constant harmonic map $u : E \to [0,1[$. If furthermore M has bounded geometry, a lower bound ρ on the injectivity radius, and is stable at infinity, there exists a proper harmonic map $u : E \to [0,1[$.*

Proof. Suppose $cap(E) > 0$, and let $F : E \to [0,+\infty[$ be a proper C^∞ map with $F^{-1}(0) = \partial E$, and such that $F^{-1}\{n\}$ is smooth for all n. For all n let u_n be the solution of the Dirichlet problem $u|_{F^{-1}(0)} = 0$, $u|_{F^{-1}(n)} = 1$, with minimal energy $e(u) = \int_E |\nabla u_n|^2$. The sequence of harmonic maps u_n is uniformly Lipschitz (3.5), therefore converges to some harmonic function $u : X \to [0,1]$, this convergence is uniform on each compact subset, and the sequence $|\nabla u_n|$ also converges uniformly to $|\nabla u|$ on each compact set.

Let D be the Dirichlet space of function $f : E \to \mathbb{R}$ with $f|_{\partial E} = 0$ and $\|f\| = \int_E |\nabla f|^2 < +\infty$. If $m \geqslant n$, u_m is the projection of u_n to the affine subspace $f|_{F(x)\geqslant m} = 1$, and u_m is also the projection of 0 on this space, thus $\|u_n - u_m\|^2 = 2\|u_n\|^2 + 2\|u_m\|^2 - 4\|\frac{u_n+u_m}{2}\|^2 \leqslant 2\|u_n\|^2 - 2\|u_m\|^2$.

The sequence $\|u_n\|^2$ is decreasing, hence converges to some e. Thus $\|u_n - u_m\|^2 \to 0$ as $n \to \infty$, $n \leqslant m$. Choose n so that $\|u_n\|^2 - c \leqslant \varepsilon$. Then if $m \geqslant n$, $\int_{n \leqslant F(x)} |\nabla u_m|^2 \leqslant \varepsilon$.

By definition if $e(u_n) \to 0$, the capacity of E is zero. Therefore if $c(E) \neq 0$, $c(f_n) \to e \neq 0$, and $e(u) \geqslant \int_{n \geqslant F(x)} |\nabla u|^2 \geqslant e - \varepsilon$, thus $e(u) = e$ and u is a non constant harmonic map.

Let us suppose now that M has bounded geometry, a lower bound on the injectivity radius and is stable at infinity, and let us prove that in fact u is a *proper* map $E \to [0,1[$. If not there exists a sequence $x_k \to \infty$, and $u(x_k) \to \alpha < 1$. Let $\beta = \frac{1+\alpha}{2}$. Choose R such that for all x, $cap(B(x,R)) > (\frac{1}{1 - \frac{\alpha+\beta}{2}})^2 e(u)$. Let $\varepsilon = \frac{\alpha - \beta}{1000 Rc}$, and choose a compact set K so that for all n, $\int_{K^c} |\nabla u_n|^2 \leqslant \varepsilon$.

Choose $k_1 \geqslant k_0$ large enough so that $B(x_{k_1}, R) \subset K^c$. Choose n_1 large enough so that all the function u_n are harmonic on $B(x_{k_1}, R)$ for $n \geqslant n_1$, and such that $u_n(x_{k_1}) \leqslant \frac{3\alpha+1}{4}$ for all $n \geqslant n_0$.

On the set $\Omega = \{x/d(x,K) \geqslant \rho\}$, the function u_n are c.$\varepsilon = \frac{\alpha - \beta}{1000 R}$ Lipschitz by 3.5. Thus, for all $n \geqslant n_0$, one has $|x - x_k| \leqslant R \Rightarrow |u_n(x) - u_n(x_{k_1})| \leqslant \frac{\alpha - \beta}{1000}$, and $u_n(x) \leqslant \frac{\alpha + \beta}{2}$.

On the ball $B(x_{k_1}, R)$ the function $v_n = \frac{u_n - \frac{\alpha+\beta}{2}}{1 - \frac{\alpha+\beta}{2}}$ is negative, and this function is 1 outside a compact set of M. Therefore using $\max(v_n, 0)$ to evaluate the capacity, one gets $cap(B(x_n, R)) \leqslant (\frac{1}{1 - \frac{\alpha+\beta}{2}})^2 e(u_n)$, contradiction. $\qquad\square$

3.7. Recall that X has more than one end if the complementary of some open relatively compact set B has several connected non compact component. This is equivalent to the existence of a proper surjective C^∞ function f_0 from X onto the open interval $]-1,1[$. Denote by $E[f_0]$ the energy of the proper homotopy class $[f_0]$ of f_0 that is the infimum of the energies of the maps properly homotopic to f_0. The following version of the Dirichlet Riemann Kelvin theorem will be useful.

DRK Theorem. *If X has bounded geometry, a lower bound on its injectivity radius and is stable at infinity, there exists a unique proper harmonic map $f : X \to]-1,1[$ in the same proper homotopy class than f_0.*

The existence follows from the argument of 3.6 ; the uniqueness follows from the convexity of the function $t \to e(t f_0 + (1-t) f_1)$ (which is also valid for maps f_i with values in trees, [Gr-Sc], prop.4.1).

3.8. Let $G = \pi_1(M)$ be the fundamental group of a compact manifold M, and H a subgroup such $X = \tilde{M}/H$ has several ends. The stability of the representation $l^2(G/H)$ is equivalent to the non amenability of the Shreier graph $Cay(G)/H$, where $Cay(G)$ is the Cayley graph of G relative to a fixed system of generators Σ. This is equivalent also to the fact that X (or $Cay(G)/H$) satisfies a strong isoperimetric inequality $(vol_{n-1}(\partial A)) > k(vol_n A)$. This linear isoperimetric inequality implies that X is stable at infinity. In this case we say that (G/H) is *stable*.

More concretely, if $R_n \to \infty$, and $B(x_n, R_n)$ is a family of balls in this graph with bounded capacity, there exists a family of functions f_n such that $f_n|_{B(x_n,R_n)} = 1$, $f_n = 0$ outside a compact set and $e(f_n)$ is bounded by some constant C. Then $u_n = \frac{f_n}{\|f_n\|} \in l^2(G/H)$ is an almost invariant vector, as $e(u_n) = \frac{1}{\|f_n\|^2} \Sigma_{g\in\Sigma} \|f_n - f_n \circ g\|^2 \leqslant \frac{C}{vol(B(x_n,R_n))} \to 0$.

Therefore the stability of the representation $l^2(G/H)$ is *stronger* that the stability at infinity of X/H. For example, if $G = \mathbb{Z}^n$, $H = \{e\}$, $l^2(G/H)$ is not stable but, if $n \geqslant 3$, $\mathbb{R}^n$ is stable at infinity as it satisfies the isoperimetric inequality $vol_{n-1}(\partial A) > k(vol_n A)^{\frac{n-1}{n}}$, and $2\frac{n-1}{n} > 1$.

3.9. Let X be a Riemannian manifold with bounded geometry discretely acted upon by a group G with $H^1(G; l_2(G)) \neq 0$. If X/G has finite volume, then X supports a non-zero exact square integrable harmonic 1-form, i.e. the differential of a non-constant harmonic function f with finite energy (see [Ch-Gr1]).

The above concept of capacity defined via the L_2-norm of the gradient (quadratic energy) extends to all L_p, where the most studied case after $p = 2$ is that of the conformally invariant energy for $p = dim_{\mathbb{R}}(V)$. (This conformal p equals 2 for Riemann surfaces.) What is badly missing for $p \neq 2$ is a Hodge lemma.

3.10. Questions. Let X be an infinite (not necessarily Galois) covering of a compact manifold V with the fundamental group $H \subset \pi_1(V)$. When does a homotopy class of maps $X \to Y$ have a representative f_0 of finite L_p-energy for a given $p \in [1, \infty]$? Here, Y may be some standard metric space, e.g. a flat torus; another possibility is where $Y = X$ and f_0 is homotopic to identity. Among examples one singles out aspherical spaces V, e.g. those with non-positive curvature, and finitely generated groups H such as $H = \mathbb{Z}^n$, for instance.

A closely related question concerns the structure of the L_p-subspaces in the cohomology of X, denoted $L_pH^*(X) \subset H^*(X; \mathbb{R})$, of the cohomology classes of X realizable by closed L_p-forms on X. For example, can such subspace be irrational in the case where $H^*(Y; \mathbb{R})$ comes with a natural $\mathbb{Q}$-structure, e.g. for aspherical X with finitely generated Abelian fundamental group H? (This subspace in the aspherical case is determined solely by $G = \pi_1(V)$, the subgroup $H \subset G$ and the number p).

3.11. If the group G "branches", i.e. has at least three (and hence, infinitely many) ends, then its Cayley graph is stable (at infinity). This can be seen, for example, by exhibiting square integrable 1-cycles b that flow (as currents) from a given non-empty open subset of ends to another, say from ∂_- to ∂_+, where the underlying X is some manifold or polyhedron with a co-compact discrete action of G such as the Cayley graph of G, see for instance [ABCKT] page 50. This stability implies the existence of a harmonic function on X of finite energy separating ∂_- from ∂_+ and thus, the non-vanishing of $H^1(G; l_2(G))$. Notice, one does not use here the Stallings theorem on groups with an infinity of ends. In fact Stallings theorem follows from the existence of a harmonic function separating two complementary open ends, see p.228 in $[Gr]_{HG}$.

3.12. The non-vanishing of the first l_2-cohomology group $H^1(G; l_2(G))$ in other known examples is established with a use of Atiyah-Euler-Poincaré formula applied to the first two l^2-Betti numbers of infinite Galois G-coverings X' of connected 2-polyhedra V',

$$l_2b^2(V') - l_2b^1(G) = \chi(V')$$

where $\chi(V')$ is the ordinary Euler characteristics,

$$\chi(V') = b^2(V') - b^1(V') + 1.$$

For instance, the inequality $\chi(V') < 0$ for some V' with the fundamental group G yields non-vanishing of $l_2b^1(G)$ and hence, of $H^1(G; l_2(G))$, which amounts to the existence of a non-zero harmonic square integrable 1-cocycle in X'.

A more convincing example is given by G obtained by adding l relations R_i to the free product of k infinite groups G_p, such that the natural homomorphisms $G_p \to G$ are injective. Then, by applying the l^2-Mayer-Vietoris sequence, like in [Ch-Gr 2], one checks that the first l_2-Betti number satisfies $l_2b^1(G) \geq k - l - 1$. If the maps $G_p \to G$ have *infinite* image H_p, the same result applies : indeed, G is the quotient of the free products of H_p by the images S_i of the relations R_i in this free product. This infinite image condition can be, probably, removed with a use of Romanovskii Freiheitssatz ; on the other hand, the injectivity holds for generic relations added to free products, by small cancellation theory over free products [L-S].

4. Clean Functions and Maps

4.1. Let f be a pluri-harmonic *function* on a complex manifold X or, more generally, a pluri-harmonic map of X to a *metric graph* Y e.g. to a tree. Then there exists a unique holomorphic 1-codimensional foliation $\mathcal{F}$ on X such that f is constant on the leaves of $\mathcal{F}$. We call f clean if the leaves L of $\mathcal{F}$ are closed and say that f is properly clean if the leaves are compact. Notice that "clean" $\Longrightarrow$ "properly clean" if f is proper and thus has compact level sets.

Let use note that, if *some* leaf of $\mathcal{F}$ is compact, and if X is *Kähler*, complete and with bounded geometry, then f is properly clean. Furthermore, the leaves have uniformly bounded volume and diameter.

Proof. (compare [Gr-Sc] p.240). Let us check that the set Y of points s.t the leaf through this point is compact is an open set. If L is a compact leaf, as the restriction of idf to L is 0, L has a neighborhood where $idf = dg$ is exact. In this neighborhood the foliation $\mathcal{F}$ is defined by an holomorphic function F : thus Y is open. Let $Y' \subset Y$ be the set of compact non-singular leaves. On each component of Y' the homology class $[L_x]$ of the leaf through x is constant. If $x \in Y \setminus Y'$, L_x is a singular fiber, and its homology class is $\frac{1}{m}[L_y]$, where L_y is a non singular leaf close to x and $m \in \mathbb{N}$ is the multiplicity. Let Y_1 be a connected component of the open set Y. Let us check that Y_1 is also closed. As Y_1 is connected, and as Y' has codimension 2 in Y, $[L_x] = \frac{1}{m_x}[L]$ for L a generic leaf in Y_1. In particular the volumes of the leaves in Y_1 are uniformly bounded. Let $x_n \to x^*$ be a converging sequence of points of Y_1.

As X is Kähler with bounded geometry, the leaves L_{x_n} have uniformly bounded diameter R. Thus, for n large enough L_{x_n} is in the compact set $K = B(x^*, R+1)$. Let $(B(z_i, \eta)_{1 \leqslant i \leqslant k}$ be a finite cover of this compact set by closed balls of radius η, such that on each $B(z_i, \eta)$ the foliation is defined by a function F_i, and such that the balls $B(z_i, \eta/100)$ cover K. If L is a leaf of our foliation, and L pass through some $y \in B(z_i, \eta/100)$, the volume of the connected component of $L \cap B(z_i, \eta/2)$ through y is $\geqslant \alpha$ for some universal constant α. From the bound on the volume of $[L_{x_n}]$, we deduce that there exists a uniform bound on the number of connected component of L_{x_n} through $B(z_i, \eta/100)$. Thus, one can extract a subsequence such that all of these components converge to the leaf through x which is therefore compact $\square$

In the special case where $Y = \mathbb{R}$, the function f locally serves as the real part of a holomorphic function on X whose level sets define the above foliation. Globally, one has a holomorphic 1-form, say a on X with the real part df. This form becomes exact on some Abelian covering $\tilde{X}$ of X; therefore, the lift of f to $\tilde{X}$ becomes clean. As for the function f itself, it can be clean without a being necessarily exact; in fact, f is clean iff a represents a multiple of a rational (possibly non-zero) cohomology class on X.

4.2. If a harmonic function $f : X \to \mathbb{R}$ is clean then (by an easy argument) there exists, a Riemann surface S, a surjective holomorphic map with connected fibers $h : X \to S$ and a harmonic function $S \to \mathbb{R}$ such that f equals the composed map $X \to S \to \mathbb{R}$. Moreover, this remains true with an arbitrary one-dimensional target space Y in place of $\mathbb{R}$.

If f is properly clean then, clearly, the factorization map $h : X \to S$ is proper. Furthermore, the group H of holomorphic automorphisms of X sends (compact!) leaves to leaves. Indeed, if $g \in H$ and L is a compact leaf, then $h|_{g.L}$ is holomorphic from a compact manifold to a Stein manifold hence it is constant, and $g.L$ is another leaf. Thus the group H acts on S and the map h is equivariant. For example, if X serves as a Galois covering of a compact manifold V, then the Galois group G acts on S and V fibers (i.e. admits a surjective holomorphic map with connected fibers) over a Riemann surface. (See [ABCKT] for details and references).

4.3. If $rank H^1(X; \mathbb{R}) < 2$, e.g. if X is simply connected, then, by the above, every pluriharmonic function f on X is clean. Furthermore, if X is Kähler and f has finite energy then some (generic) leaf L_0 has finite volume (due to the co-area formula, see $[Gr]_{GFK}$). If X is complete and has bounded geometry then "finite volume" $\implies$ "compact" for complete holomorphic submanifolds in X. It follows that L_0 is compact; consequently all leaves L are compact (4.1). (All one needs here of the bounded geometry is a slow decay of the convexity radius of X).

4.4. Corollary *Let a countable group G discretely act on a Kähler manifold X such that the quotient X/G has finite volume. If $H^1(G, l_2(G)) \neq 0$ while $H^1(X, \mathbb{R}) = 0$ (e.g. X is simply connected) and and if X has bounded geometry (e.g. the action is co-compact) then X/G fibers over a Riemann surface: there is a holomorphic* *action of G on a Riemann surface S and an equivariant holomorphic map h from X onto S with compact connected fibers.*

This corollary applies, for instance, to non-singular projective varieties V and yields the following (see $[\text{Gr}]_{GFK}$):

4.5. Theorem. *If the fundamental group G of V has $l_2 b^1 \neq 0$ or, equivalently, $H^1(G; l_2(G)) \neq 0$, then G is commensurable to a surface group.*

4.6. If V is singular then the above considerations can be applied to a nonsingular G-equivariant resolution of the universal covering X of V (induced from a resolution of V), say to X, where the corresponding map $X \to S$ necessarily factors via a holomorphic map $X \to S$.

4.7. The simplest way to handle a smooth quasi-projective variety $V_0 = V \setminus W$ would be by constructing a complete Kähler metric on V_0 such that the induced metric on the universal covering X_0 of V_0 had bounded geometry. By the Hironaka theorem, one may assume that V_0 is smooth and W is a divisor with normal crossings where handy candidates for the desired metric come readily; yet, one has to check that the curvatures of such metrics are bounded.

5. Cleanness and branching

Let f be a proper harmonic function $X \to]-1, +1[$ with finite energy separating two open ends on a Kähler manifold X as in 3.4. Cleanness and proper cleanness, as was mentioned earlier, are equivalent for such f; more significantly one has the following :

5.1. *First cleanness criterion.* If X has at least three ends then f is properly clean. This follows from L^2-version of the Castelnuovo de Franchis theorem discussed in [ABCKT] pp. 60-62.

5.2. There is a more general geometric version of this result. Let T be the tree obtained by joining several (finite or infinite number of) copies of the segment $[0, 1[$ at 0 and let a group H isometrically act on T, while fixing the origin 0, so that T/H is a finite tree, perhaps reduced to $[0, 1[$. (Such an action amounts to permuting the copies of $[0, 1[$ with finitely many orbits. This action is not necessarily discrete, not even proper for infinite H since the action fixes 0).

Second cleanness criterion. Let H discretely and isometrically act on a Kähler manifold X and $f : X \to T$ be a surjective H-equivariant pluriharmonic map of H-finite energy. If the tree T has at least three branches (i.e. there are at least three copies of $[0, 1[$ in the above construction) and f is H-proper, i.e. the corresponding map $X/H \to T/H$ is proper, then f is clean; moreover, f is H-properly clean, i.e. the holomorphic leaves L from X go to compact (complex analytic) subsets in X/H under the quotient map $X \to X/H$.
This follows from the argument on pp 239-40 in [Gr-Sch], (see also [Si]) : note that as the number of branches of the tree is at least three, the foliation defined by f

must have a singular leaf (i.e a leaf whose image is a branch point of the tree). This leaf is compact modulo H and therefore (4.1)f is properly clean.

5.3. In order to apply the second criterion one needs a useful version of the DRK-theorem. As X/H is not a manifold but an orbifold, we must modify our assumption on the geometry of X/H. We suppose that X as bounded geometry, a lower bound on the injectivity radius ρ. We also assume that there exist a $r < \rho$ such that for every x in X the ball $B(x,r)$ is the quotient of $B(x,r)$ by a finite group of bounded cardinality. This r plays the role of the injectivity radius in the orbifold case. The definition of capacity and uniform stability at infinity of 3.3 remains valid in this case, as well as the Ells-Sampson Theorem.

Theorem. *If X/H is uniformly stable at infinity, then there exists a proper H-equivariant harmonic map $u : X \to T$ with finite non-zero H-energy.*

Proof. We explain the modifications needed in the proof of 3.6. Let $T - \{0\} =]0,1[\times\Sigma$ where Σ is some set, and let T the metric completion of T obtained by adjoining a point 1 to each interval $[0,1[$. Let $F : T/H \to [0,+1[$ be the folding map, so that $u : F \circ f : X/H \to [0,1[$ is proper. For every n,let $U_n : X \to T$ be the unique solution of the Dirichlet problem $U_n|_{f^{-1}([1/n,1[\times\sigma} = 1 \times \sigma$, U_n is H invariant, and of minimal energy. By the argument of 3.6, U_n converges to some harmonic map $U : X \to T$. It remains to check that $U(X) \subset T$ and U is proper. If $U(x) = 1 \times \sigma$ for some point x, by the maximum principle, U is constant of zero energy, but the G-energy of f is the energy of U, hence the capacity of X/H is zero and this manifold would not be stable at infinity.

Hence the harmonic map U sends X in T. In order to prove that U is proper, let us check that if $x_n \in X$ is such that $f(x_n) \to 1 \times \sigma$, then $U(x_n) \to 1 \times \sigma$. If not, there exists an α such that $U(x_n)$ remains on the complement of $H \times [\alpha,1[\times\sigma$. Let $e(U)$ be the energy of this harmonic map U, and let $\beta = \frac{1+\alpha}{2}$.

Let $u_n(resp.u) : X/H \to [0,1[$ be the map induced from $F \circ U_n(resp. F \circ U)$. Choose R such that for all $x \in X/H$, $cap(B(x,R)) \geqslant (\frac{1}{1 - \frac{\alpha+\beta}{2}})^2 e(u)$. Let $\varepsilon = \frac{\alpha-\beta}{1000Rc}$, and choose a set $K \subset X/H$ so that for all n, $\int_{K^c} |\nabla u_n|^2 \leqslant \varepsilon$. As the image of K by u is compact, one can choose $k_1 \geqslant k_0$ large enough so that $B(x_{k_1}, R) \subset K^c$. Choose n_1 large enough so that all the functions U_n are harmonic on $B(x_{k_1}, R)$ for $n \geqslant n_1$, and such that $u_n(x_{k_1}) \notin [\frac{3\alpha+1}{4}, 1[\times\Sigma$ for all $n \geqslant n_0$. On the set $\Omega = \{x/d(x,K) \geqslant R\}$,the function U_n are $c.\varepsilon = \frac{\alpha-\beta}{1000R}$ Lipschitz. Thus, for all $n \geqslant n_0$, one has $|x-x_k| \leqslant R \Rightarrow |U_n(x)-U_n(x_{k_1})| \leqslant \frac{\alpha-\beta}{1000}$, and $U_n(x) \notin [\frac{\alpha+\beta}{2}, 1[\times\Sigma \leqslant \frac{\alpha+\beta}{2}$. Let $v_n = \frac{u_n - \frac{\alpha+\beta}{2}}{1 - \frac{\alpha+\beta}{2}}$ if $u_n(x) \in [\frac{\alpha+\beta}{2}, 1] \times \sigma$, and 0 otherwise : this function is 1 outside a compact set of X/H, and 0 on $B(x_n, R)$. Therefore using v_n to evaluate the capacity, one gets $cap(B(x_n, R)) \leqslant (\frac{1}{1 - \frac{\alpha+\beta}{2}})^2 e(u_n)$, contradiction.   □

5.4. Hyperbolic example. Let H be a quasi-convex subgroup in a hyperbolic group G such that ∂G is connected and the limit set $\partial(H) \subset \partial(G)$ divides the ideal boundary $\partial(G)$ into at least three components. Then every Riemannian manifold

X with a discrete isometric co-compact action of G admits the above H-equivariant harmonic map f to Y. We postpone the proof to 6.3 and 6.6. Furthermore, if X is Kähler, 5.2 proves that f is H-properly clean.

5.5. The desired branching, i.e. the strict inequality $card(\pi_0(\partial(G)\backslash\partial(H))) > 2$, can be always achieved in the hyperbolic case by enlarging the subgroup H according to the following simple version of the "ping-pong" lemma.

Lemma ($[Gr]_{HG}5.3.C_1$, see also [Ar].) *Let H be a quasi-convex subgroup in a non-elementary hyperbolic group G where $card(\pi_0(\partial(G) \backslash \partial(H))) = 2$. Let f be an hyperbolic element such that no power of f is in H. Then for some power f^k of f the subgroup H' of G generated by f^k and H is quasi-convex; if F is the finite subgroup of H of elements commuting with f^k, H' is the free product $H' = H *_F < f^k \times F >$ amalgamated along F. Furthermore, $card(\pi_0(\partial(G)\backslash\partial(H'))) = \infty$.*

Note that the group H' contains H as well as all its conjugates $f^{nk}Hf^{-nk}$. In the boundary of G all the boundaries of these groups are disjoint (see 6.2 below) and each of them cuts ∂G in (at least) two components, therefore $\partial G\backslash\partial H'$ has infinity many of connected components.

The above lemma can be generalized to many other (e.g. "nearly hyperbolic") groups where H is contained in a larger subgroup H' that usually cuts G into more than three pieces ; yet, the overall picture remains unclear.

5.6. The above cleanness criteria deliver holomorphic fibrations of open Kähler manifolds over Riemann surfaces with compact connected fibers, denoted $f : X \to S$. In the cases of interest such an X comes as a (possibly non-Galois) covering of a compact (e.g. projective algebraic) manifold V, say $p : X \to V$, where one can induce the fibration f from a surface fibration of a finite covering of V according to the following simple lemma. (For a similar statement, see [Ca] 1.2.3 p. 490, or [Ko] Prop. 1.2.11).

Lemma. *Given the above $f : X \to S$ and $p : X \to V$. Then, the normalizer G' of the image of the fundamental group of a generic fiber $f^{-1}(x_0)$ is of finite index in $\pi_1(V, p(x_0)) = G$. The finite covering $V' \to V$ of group G' fibers over a compact Riemann surface S' where the images of the fibers under the covering map $V' \to V$ equal the p-images $p(L) \subset V$ of the f-fibers $L \subset X$.*

Proof. Let us identify X with the quotient $f : X = \tilde{V}/H \to S$, for $H \subset G = \pi_1(V)$. Let y_0 be some pre-image of x_0 and N be the image of $\pi_1(f^{-1}(x_0), y_0)$ in $H \subset G$. In order to prove that the normalizer G_1 of N is of finite index in G, it is enough to show that N has only a finite number of conjugate in G. Indeed, the conjugate gNg^{-1} is represented by the image in X of the fiber of F through some point in $p^{-1}(p(y_0)) = G/H$. But all the fibers of f are analytic submanifolds with the *same volume*, in a Kähler manifold of bounded geometry. So their fundamental groups (at the point $g.x_0$) are generated by loops of uniformly bounded length (independent of g). Therefore the images of these groups in $\pi_1(V, p(x_0))$ is generated by elements of bounded length, and can only take a finite number of values. Thus $\hat{V}/N$ is a Galois cover of the finite cover V' of V of group G', and 4.2 applies.   □

6. Cutting Groups by Subgroups

Given a subspace in a proper geodesic metric space, $X_0 \subset X$ let $X_{-r} \subset X$ be the set of points $x \in X$ with distance $\geq r$ from X_0. We take the projective limit for $r \to \infty$ of the projective system of the sets of connected components of X_{-r} and call this limit the space of (relative) ends of $X|X_0$, denoted $Ends(X|X_0)$. Observe that if X is proper (bounded sets in X are relatively compact) and X_0 is bounded then the space $Ends(X|X_0)$ equals the ordinary space of ends $Ends(X)$.

If X is acted upon by a group H we take an orbit X_0 of H and set $Ends(X|H) = Ends(X|X_0)$. If H serves as a subgroup of a finitely generated group G we apply the above to $X = Cayl(G)$ that is the Cayley graph of G and abbreviate by putting $Ends(G|H) = Ends(Cayl(G)|H)$, and $Ends(G/H) = Ends(Cayl(G)/H)$.

6.1. Definitions. Say that $X_0 \subset X$ cuts X (at infinity) if $X|X_0$ has at least two ends, i.e. $card(Ends(X|X_0)) > 1$. In future, the noun "cut" may refers to the fact that X is being cut by X_0 or to an actual division of $Ends(X|X_0)$ into two (or more) non-empty open subsets.

The disjoint union $X \cup Ends(X|X_0)$ carries a natural topology. Thus for every subset $X' \subset X$ one can take its closure in $X \cup Ends(X|X_0)$ and then intersect this closure with $Ends(X|X_0)$. We denote the resulting subset by $\partial_{end}(X') \subset Ends(X|X_0)$ and say that X_0 cuts X' in X if $card\partial_{end}(X') > 1$.

We say that a subgroup $H \subset G$ cuts a group G if $card(Ends(G|H)) > 1$ and an H-cut is called branched if $card(Ends(G|H)) > 2$. A cut of H by G is called stable if the Schreier graph $(Cayl(G)/H)$ is uniformly stable (at infinity) in the sense of 3.3. For this it is enough that $l^2(G/H)$ is stable by 3.8.

If H cuts G, H acts on the set of relative ends $Ends(G|H)$. One can distinguish Schreier cuts where the disconnectedness persists under the action of H, i.e. where the action of H on $Ends(G|H)$ is not transitive. Note that if H cuts G, the cut is a Shreier cut if and only if the Schreier graph $(Cayl(G)/H)$ is disconnected at infinity (see also [CCJV] where such cuts are called "walls"). In particular if $Ends(G|H)$ is finite there exists a subgroup H' of finite index in H such that $Ends(G/H') = Ends(G|H) = Ends(G|H')$, and H' is a Schreier cut of G.

Example. Let S be a compact Riemann surface of genus $\geqslant 2$, C be a simple closed curve separating S in two connected components $S^{\pm}$. If $G = \pi_1(S, x_0)$ operates on the hyperbolic plane D, $H = \pi_1(S^+, x_0)$, then D/H is connected at infinity, whereas $Ends(D|Hx_0)$ is infinite, thus H cuts G at infinity, but in fact is transitive on the set of relative ends (see also 6.3).

Historical remarks. The set of ends of a topological space has been introduced by Freudenthal ; for homogeneous space of Lie groups, it has been firstly studied by Borel to prove that there are no action of a Lie group on a simply connected manifold which is 4-transitive [Bo]. After Stalling's famous paper on the structure of groups with an infinity of ends, C. Houghton [Ho] and P. Scott [Sc] began to study ends of pairs of groups.

6.2. Induced cuts. Cuts and their properties (obviously) lift under *surjective* homomorphisms $G \to G$: if H cuts G the so does the pullback $H \subset G$ of H to G; furthermore, the invariance (by the action of H) and stability pass from H-cuts to H-cuts, in other words if H is a Shreier cut of G then H is a Schreier cut of G. Cuts also pass to subgroups $G' \subset G$. In fact, if a finitely generated $G' \subset G$ is cut in G by a subgroup $H \subset G$, then G' is also cut by the intersection (subgroup) $H \cap G' \subset G'$ as follows from the following simple :

Lemma. *Given subgroups G' and H in a finitely generated group with G endowed with the word metric, there exists a function $\epsilon(\delta)$ with $\epsilon(\delta) \to_{\delta \to +\infty} +\infty$, such that the intersection of the δ-neighborhoods of G' and H in the Cayley graph of G is contained in the ϵ-neighborhood of the intersection $G' \cap H$.*

Proof. As the ball in G of radius 2δ is finite, there exists an $r < \infty$ (depending on G' and H) such that if some $g \in G'$ and $h \in H$ are s.t. $|g - h| < 2\delta$,then there exists a pair $(g_0, h_0) in G' \times H$ in the ball of radius r of G s.t. $g^{-1}h = g_0^{-1}h_0$. Then $k = gg_0^{-1} = hh_0^{-1} \in G' \cap H$. This h is at distance $< r$ of g, hence the result with $\epsilon = r + \delta$. $\square$

Thus arbitrary (non necessary surjective) homomorphisms $G_1 \to G$ induce cuts in G_1 from those in G.

6.3. Convex hyperbolic cuts. Let X be a proper geodesic δ-hyperbolic space. Recall (see for instance [CDP]) that subset $Y \subset X$ is *quasi-convex* if there exists a constant A s.t. for every pair $y, y' \in Y$ and any point z in a geodesic segment $[y, y']$, the distance of z to Y is $\leqslant A$. It is known that if Y is A-quasi convex, and $B > A + 100\delta$ the set $Y^{+B} = \{x/d(x, Y) \leqslant B\}$ is 100δ-quasi-convex.

Let H be a group of isometry acting on X. Recall that H is *quasi-convex cocompact* if there exists a geodesic subspace $Y \subset X$ which is quasi-convex and such that the action of H on Y is discrete co-compact. This is equivalent to the fact that the orbit $H.x_0$ of any point is quasi-convex. In the case where X is the hyperbolic space of constant curvature, a *quasi-convex co-compact* group is a geometrically finite group without parabolics.

If H is quasi convex co-compact, it is an hyperbolic group, and its boundary ∂H embeds as a closed subset in ∂X ; it is also the limit set of the action of H on X. It is known (see [Coo]) that the action of H on $\partial X/\partial Y$ is discrete co-compact. One says that a X is *thin* if there exists a constant B s.t. every point in X is at a distance $\leqslant B$ of a bi-infinite geodesic.

Lemma. *Suppose that H is quasi-convex co-compact in some thin proper geodesic hyperbolic space X. Let $Y = H.x_0$, so that Y is a quasi-convex subset of X. The set of relative ends $Ends(X|Y)$ is the set of "connected components" of $\partial X \backslash \partial Y$.*

Remark. It is possible that $\partial X \backslash \partial Y$is not locally connected. Thus, the expression "connected component" needs an explanation. If $\mathcal{O}$ is an open cover of $\partial X \backslash \partial Y$ by open subsets, let $|\mathcal{O}|^0$ be set of connected components of the nerve of this cover. If $\mathcal{O}'$ is finer than $\mathcal{O}$, $|\mathcal{O}'|^0$ projects onto $|\mathcal{O}|^0$. A limit point of this projective system is a *connected component*. If $\mathcal{O}$ is such a cover, if one replace each $O \in \mathcal{O}$ by $O = \cup_{O' \in n(0)} O'$, where $n(O)$ is the set of O' such that O' belongs to the same connected component of $|\mathcal{O}|$, we get a new cover by *disjoints* open sets having the same set of connected components. So we can restrict our attention to covers by *disjoint* open subsets.

Proof of the lemma. Changing the value of the hyperbolicity constant δ, one may suppose that X is δ-hyperbolic, δ-thin and that Y is δ-quasi-convex. One chooses a H-equivariant projection of $p : X \to Y$ such that $d(x, p(x)) = \min_{y \in Y} d(x, y)$. A ray $[x, y]$ is called a vertical ray if y is a projection of x on Y. If $w \in \partial X / \partial Y$, a ray $\rho = [y, w[$ is called vertical if for every $x \in \rho$,the point y is a projection of x on Y. By properness and δ hyperbolicity every $w \in \partial X / \partial Y$ is the end of some ray, and two such rays are 10δ close one to each other. One can extend p to $\partial X / \partial Y$ by choosinf once for all and for every ω a vertical ray $\rho_w = [y, \omega[$ which ends at ω, and setting $p(w) = y$. This choice can be make equivariant. If $w, w' \in \partial X / \partial Y$, let $\rho_w(t), \rho'_w(t) : [0, \infty[\to X$ be two geodesic parameterizations of the rays $[p(\omega), \omega[, [p(w'), w'[$. One sets $< w, w' >_Y = \max\{t / d(\rho(t), \rho'(t)) \leqslant 10\delta\}$.
Let us first prove :
(1) as X is thin every point x s.t. $d(x, Y) > 200\delta$ is at the distance $\leqslant 100\delta$ to some vertical ray $[y, w[$ with $y \in Y$.
Choose some bi-infinite geodesic $]w, w'[$ s.t. $d(x,]w, w'[) \leqslant \delta$. By δ-quasi convexity, w or w' do not belong to ∂Y. If $w' \in \partial Y$, and p is a projection of w on Y, so that $]p, w[$ is a vertical ray, the hyperbolicity proves that $]w, w'[$ is 10δ close to $]w, p[\cup]p, w'[$ therefore x is 11δ close to the vertical ray $]w, p[$. If neither w nor w' are in ∂Y, let p and p' be projections of w, w' on Y, so that $]p, w[$ and $]p', w'[$ are vertical rays. The hyperbolicity proves that $]w, w'[$ is 10δ close to $]w, p[\cup]p, p'[\cup]p', w'[$. By quasi-convexity, x cannot be close to $]p, p'[$, and is therefore close to one of the two vertical rays $]w, p[,]p', w'[$.
Le C be a component of Y_{-r} s.t. there exists a point in C with $d(x_0, C) > r + 10000\delta)$. Let $O(C) = \{w: \rho_w \cap C$ is not compact$\}$. Let us check that $O(C) \neq \emptyset$. By (1) x is a the distance $\leqslant 100\delta$ of some geodesic ray $[y, w[$ with $y = p(w) \in Y$. Let $x' \in [y, w[$ with $d(x, x') \leqslant 100\delta$. Then $d(x', y) \geqslant r$ and therefore $[x', w[\subset C$. It is easy to see that $O(C)$ is an open set, and that the collection $\mathcal{O}_r$ of all these sets is a cover of $\partial X \backslash \partial Y$. These sets are *disjoint* and $\mathcal{O}_r$ is an open cover of $\partial X \backslash \partial Y$ by disjoint sets. Thus $|O_r|$ is the set of connected components of Y_{-r} which are the image of a connected component of $Y_{-r'}$, $r' > r + 100\delta$ under the natural projection $Y^0_{-r'} \to Y^0_{-r}$.
Let $\mathcal{O} = (O_i)_{i \in I}$ be a H-invariant, H-finite cover of $\partial X / \partial Y$ by non empty *disjoint* open subsets. In order to conclude, its is enough to prove the following : there exists an r s.t. $\mathcal{O}_r$ is finer than $\mathcal{O}$.
As H is *co-compact* in $\partial X \backslash \partial Y$, and as our cover H-invariant, there exists an $s > 0$ s.t. for every w there exist an i s.t if $< w, w' >_Y > s \Rightarrow w, w' \in O_i$.

Let us choose such an s. Let $r = s + 1000\delta$. Let C be some component of $\mathcal{O}_r$. For each x in C, let $O(x)$ be the (non-empty, due to (1)) set of endpoints of vertical rays $\rho = [y, \omega[$s.t. $d(x, \rho) \leqslant 100\delta$. If $d(x, x') \leqslant \delta$, then the product $< \omega_x, \omega_{x'} >$is bigger than s. By connexity there exist an i s.t. for all $x \in C$, $\omega_x \in O_i$. If $w \in O(C)$ there exist a ray $[y, \omega[$ which contains a point x of C. Then $w \in O_i$, and $O(C) \subset O_i$. □

A particular instance of this is the Cayley graph X of a non-elementary word hyperbolic group $G \supset H$.

6.4. Full systems of convex cuts. Say that a (usually infinite but with finitely many mutually non-conjugate members) collection of convex subgroups H_i in a word hyperbolic group G fully cuts G if for every pair of distinct points in the ideal boundary $\partial(G)$ there is some H_{i_0} among H_i whose limit set $\partial(H_{i_0}) \subset \partial(G)$ separates these points, i.e. they lie in different connected components of the complement $\partial(G) \setminus \partial(H_{i_0})$. (This definition can be extended to general groups and spaces but we are mostly concerned with convex cuts in hyperbolic groups and in $CAT(0)$-spaces).

Examples. (a) If some immersed compact totally geodesic hypersurfaces W_j cut a compact manifold V of negative curvature into simply connected pieces then the conjugates of the fundamental (sub)groups of W_j's fully and convexly cuts $\pi_1(V)$. (This, with an appropriate definition, remains valid for arbitrary $CAT(0)$-spaces).

(b) If G is a reflection group then the isotropy subgroups of the walls provide a full system of convex Schreier (as in 6.1) cuts of G (where G does not even has to be hyperbolic for this matter, see [BJS]).

(c) The above generalizes to cubical $CAT(0)$-polyhedra and their isometry groups (see [Sa] [CCJJV]). In particular hyperbolic groups co-compactly acting on such polyhedra admit full systems of convex Shreier cuts.

(d) There are compact (arithmetic) n-manifolds for all $n > 1$ of constant negative curvature with a full system of Shreier hyperplane cuts of its universal covering and hence, of their fundamental groups. (It is unlikely that the fundamental group of each n-manifold of constant negative curvature admits a convex cut for $n > 2$ but no counter example seems to be known even for $n = 3$.)

(e) Dani Wise (see [Wi]) has shown that many small cancellation groups G, including geometric $C'(1/6)$-groups, admit full systems of convex Schreier cuts H_i with at most finitely many mutually non-conjugate among them. In conjunction with Sageev's theorem his result provides, for all such G, a cubical $CAT(0)$-polyhedra with fundamental groups G assuming G has no torsion.

6.5. Stability of hyperbolically induced cuts. If a f.g subgroup $A \subset G$ in a word hyperbolic group G is cut by a quasi-convex subgroup $H \subset G$ then the induced cut of A by $H \cap A$ is stable unless A is virtually cyclic. It follows that the cut of an arbitrary finitely generated group G_1 induced from a convex cut of a hyperbolic group by a homomorphisms $G_1 \to G$ is *stable* (hence G/H is uniformly stable at infinity) except for the virtually cyclic image case.

Proof. (Compare [K] for a discussion of the case where A is quasi-convex.) In order to prove this proposition it is enough to show:

Proposition. *Let A be a subgroup of a hyperbolic group G, and H be a quasi-convex subgroup in G. If $A/A \cap H$ and $A \cap H$ are infinite, or if A is non elementary and $A \cap H$ is finite, then$A$ contains a free group F such that F meets no A-conjugate of $A \cap H$, i.e. F freely operates in $A/A \cap H$.*

If A is non elementary and $A \cap H$ is finite, as$A$ contains free subgroups, the result is obvious. So we may assume that that $A/A \cap H$ and $A \cap H$ are infinite.

In order to prove this proposition, we think of G as a uniform convergence group on its boundary ∂G. Our proof is therefore also valid if G is a *geometrically finite* convergence group on a compact set M provided that H is *fully quasi-convex* in the sense of Dahmani [Da], i.e. H is quasi-convex and meets each parabolic subgroup of G either in a finite group of a subgroup of finite index.

Recall that $\partial^2(G)$ denotes the set of *distinct* pairs of elements in ∂G. As H is quasi-convex it is hyperbolic, and its limit set is equivariantly homeomorphic to ∂H.Let $\Lambda^2(A)$ be the closure in $\partial^2(G)$of the set of pairs (a^+, a^-) of fixed points of hyperbolic elements in A. If A is quasi-convex, $\Lambda^2(A) = \partial^2(A)$.

Lemma. $\Lambda^2(A) \cap \partial^2 H$ *is of empty interior in* $\Lambda^2(A)$.

Before proving this lemma let us recall basic facts about quasi-convex subgroups (for a proof also valid in the case of geometrically finite convergence groups and fully quasi-convex subgroups, see [Da]).

Proposition. *Let H be a quasi-convex subgroup of G, and let $g_n \in G/H$ be an infinite sequence of distinct elements.*

   i) *The intersection $\cap g_n \partial H$ is empty.*
   ii) *Furthermore, if g_n is a representative of g_n of minimal length mod H, i.e. $d(g_n, H) = d(g_n, e)$, then $d(g_n, e) \to \infty$. Suppose that $g_n \to \alpha \in \partial G$. Then $g_n \partial H \to \alpha$ as well. In particular the set $\mathcal{L} = \cup_n g_n \partial H^2$is closed in $\partial^2 G$.*
   iii) *Let H_1, H_2 be two quasi-convex subgroups of G, then $H_1 \cap H_2$ is quasi-convex and $\partial H_1 \cap \partial H_2 = \partial(H_1 \cap H_2)$. Furthermore $\partial H_1 \cap \partial H_2$ is of empty interior in ∂H_1 unless $H_1 \cap H_2$ is of finite index in H_1.* □

Proof of the lemma. Suppose first that A is quasi-convex . Assume that the lemma is false ; as A is quasi-convex, the set of pairs $(a^+, a^-), a \in A$, a hyperbolic is dense in $\partial^2 A$, $([Gr]_{HG}8.2.G)$. Thus, in this case, $\Lambda^2(A) = \partial^2(A)$. But $\partial A \cap \partial H = \partial(A \cap H)$, and this set is of empty interior in ∂A, i.e nowhere dense, in ∂A, unless $A/A \cap H$ is finite.

Let A be not necessary quasi-convex and assume again the lemma is false. Then there exists an hyperbolic element u in A s.t (u^+, u^-)belongs to the interior of $\Lambda^2(A) \cap \partial^2 H$. Let $u = u_1, u_2, \ldots \ldots u_n \ldots$.be the list of hyperbolic elements of A, and let n be a fixed integer. For n_i large enough, $(u_i^{n_i})_{1 \leqslant i \leqslant n}$ generate a free q.c group A_n : if the lemma is false, $\Lambda^2(A_n) \cap \partial^2 H$ is not of empty interior in $\Lambda^2(A_n)$. Thus $A_n \cap H$ is of finite index in A_n, and every element of A has a power in H.

Therefore $\Lambda^2(A) \subset \partial^2(H)$. Suppose $A \cap H$ is not of finite index in A, and let a_n be an infinite sequence in $A/A \cap H$. We may assume that $a_n \partial(H)$ converges to some point $\alpha \in \partial G$, but this is impossible as $a_n(\partial(H)^2) \supset a_n \Lambda^2(A) = \Lambda^2(A)$.   □

Proof of the proposition. Let $\mathcal{L} = \cup_{a \in A} a \partial H^2 = \cup_{a \in A}(\partial a H a^{-1})^2$. This set is closed in $\partial^2(G)$, therefore its intersection with $\Lambda^2(A)$ is closed, and applying the lemma to the family $(aHa^{-1})_{a \in A}$, nowhere dense in $\Lambda^2(A)$. Therefore, we can choose an hyperbolic element $a \in A$ such that $(a^+, a^-) \in \mathcal{L}^c$. Choose $m \in A$ be some element s.t. $ma^+ \neq a^\pm$ (for instance any element of infinite order in the infinite group $A \cap H$). For N large enough, the group $< a^N, ma^N m^{-1} >$ has its limit set in the neighbourhood of the four points a^+, a^-, ma^+, ma^-, and no element of this group is conjugate in H.   □

6.6. Implementation of cuts by maps into trees. A Schreier cut, i.e. a partition of the space of ends of X/H for a Riemannian H-manifold X (or a general geodesic space for this matter, e.g., the Cayley graph of a group G) into two open subsets, say ∂_- and ∂_+, can be implemented by a proper function $f_0 : X/H \to]0,1[$ with finite energy (compare 3.4) which lifts to an H-invariant function on X with finite H-energy.

The latter function can be defined for general non-Schreier cuts with $]0,1[$ replaced by a tree Y as in 5.2. Namely, for each $r > 0$ we denote by $Comp_r$ the set of connected components of the subset $X_{-r} \subset X$ of points within distance$> r$ from an H-orbit $X_0 \subset X$ of a base point x_0 in X (compare 6.1) and denote by $c : X_{-r} \to Comp_r$ the tautological map. We assign a copy of $[0,1[$ to each point c of $Comp_r$, denoted $[0,1[_c$, and choose a proper monotone function $d : [0, \infty[\to [0,1[$ that vanishes on $[0.r]$. Then we construct the map f_r from X to the tree Y obtained by identifying the copies $[0,1[_c, c \in Comp_r$ of $[0,1[$ at 0 as the composition of the maps c, $dist(., X_0)$ and d, that is each $x \in X$ goes to $d(dist(x, X_0)) \in [0,1[_c \subset Y$ for $c = c(x)$.

This map f_r is H-proper as well as H-invariant and it can be easily adjusted to have finite H-energy. Among the branches $[0,1[_c$ of Y not all are essential, i.e. totally covered by the image of f_r. The non-essential branches can be removed by retracting them to the root 0 of Y; as $r \to \infty$ the number of essential branches converges to $cardEnds(X|H)$.

7. Cuts in Kähler groups

7.1. A Riemann surface S and its fundamental group can be cut in many ways and these cuts pass to complex manifolds fibered over S. Conversely, by combining the above and 5.6 (compare [De-Gr]), one conclude to the following:

7.2. Cut Kähler Theorem. *Let the fundamental group G of a Kähler manifold V be cut by a subgroup $H \subset G$, where this cut, call it C, satisfies the following two conditions.*

(1) *The cut C is stable.*
(2) *The cut C is branched, i.e. $card(Ends(G|H)) > 2$.*

Then C is virtually induced from a Riemann surface: a finite cover V of V admits a surjective holomorphic map to a Riemann surface with connected fibers, $V \to S$, such that the pullback to $\pi_1(V) \subset \pi_1(V)$ of some subgroup in $\pi_1(S)$ equals $H \cap \pi_1(V)$. In particular, the kernel of the induced homomorphism $\pi_1(V) \to \pi_1(S)$ is contained in $H \cap \pi_1(V)$.

Remarks. (a) The stability condition is violated, for instance, for cuts of Abelian groups G; yet, the conclusion of the theorem, when properly (and obviously) modified, holds in this case. But, it remains unclear how the general non-stable picture looks like.

(b) It seems that the desired cleanness does not truly need the branching condition (introduced solely for cleanness sake) but it is unclear how to remove or significantly relax it in the general case. However, branchings come cheap in the hyperbolic case (see 5.5) that brings along the following corollary where there is no explicit reference to any branching.

(c) In the case of a Shreier cut, this result has been proved by Napier and Ramachandran [N-R]

Corollary. *If a Kähler group is hyperbolic and admits a convex cut then it is commensurable to a surface group. Moreover, let $h : G \to G_0$ be a homomorphism where G is a Kähler (not necessarily hyperbolic) group, G_0 is a hyperbolic one admitting a full system of convex cuts, e.g. a Wise small cancellation group and $h(G)$ is not virtually cyclic. Then the restriction of h to a subgroup $G' \subset G$ of finite index factors through an epimorphism $G' \to \pi_1(S')$ induced by a holomorphic map $V' \to S'$ followed by a homomorphism $\pi_1(S) \to G_0$, where V is the finite covering of V with the fundamental group G' and where S' is a Riemann surface.*

Proof. From a convex cut of $h(G)$ by a subgroup H, we construct a convex branch cut of $h(G)(5.5)$ by a group H', s.t. $H' \supset H$. Thus $h^{-1}(H')$ is a branched stable (6.5) cut of G, and we get a proper holomorphic map with connected fibers $f : X = \tilde{V}/h^{-1}(H') \to S$ for some non compact Riemann surface S. By 5.6 one gets a finite cover V' of V and an holomorphic map to a compact Riemann surface $f' : V' \to S'$, s.t the fibers of f' are the images of the fibers of f. But the image of the fundamental group of the generic fiber of f' is normal in $G' = \pi_1(G)$. Its image is a normal subgroup of $h(G)$ contained in the convex subgroup H'. But no convex group in a hyperbolic group contains an infinite normal subgroup of infinite index, therefore $h(S')$ is of finite index in $h(G)$. $\square$

7.3. Remarks. (a) Probably, most hyperbolic groups, including the majority of small cancellation ones admitting convex cuts, have $l_2 b^1 = 0$ and thus, at this point conjecturally, the above applies to a much wider class of groups than the $l_2 b^1$-theorem (see 4.5).

(b) There are Kähler hyperbolic groups that admit non-convex cuts but no convex ones. In fact, by a construction of D. Kazhdan, there are compact Kähler manifold of constant Hermitian curvature of any dimension n with infinite 1-dimensional homology groups and hence, with cuts induced from Z. These have no convex cuts for $n > 1$ by the above Corollary (or by a direct application of Grauert's solution to the Levi problem).

7.4. sbc-Groups. Consider all stable branched cuts of a group G and denote by $K = K_{sbc} \subset G$ the intersection of the subgroups $H \subset G$ implementing these cuts. We call this K the sbc-kernel of G and say that it has finite type if there finitely many subgroups among H's such the intersection of all conjugates of these equals K. We say that G is of sbc-type if K equals the identity element id in G, where "finite sbc-type" means the finiteness of the type of $K = id$ of stability, branching and finiteness conditions).

Example. A finitely generated subgroup G in the product of surface groups is of finite sbc-type, unless it admits a splitting $G = G_0 \times \mathbb{Z}$.

The above Theorem yields that the following converse to this example.

7.5. sbc-Theorem. *If the sbc-kernel K of a Kähler group G has finite type, then a finite covering $V \to V$ admits a holomorphic map to a finite product of Riemann surfaces $f : V \to W = (S_1 \times S_2, \ldots, \times S_N)$ where the kernel of the induced homomorphism of the fundamental groups equals $K \cap \pi_1(V) \subset \pi_1(V)$. Moreover, one can choose a Galois covering $V \to V$ and a G-equivariant map f for some holomorphic action of the Galois group G of $V \to V$ on W.*

sbc-Corollary. *Let G be a torsion free Kähler group with no non-trivial Abelian normal subgroups. Then G admits a subgroup G of finite index isomorphic to a subgroup in the product of N surface groups if and only if G is of finite sbc-type.*

Remark. The minimal N in this Corollary (obviously) equals the maximum of the ranks of the free Abelian subgroups in G. There are only finitely many S_i for a given V but the relations between these S_i for different finite coverings V seems rather obscure.

7.6. Cut-Kähler conjecture. Probably, the stability, branching and finiteness conditions are not truly needed and the above theorem could be generalized as follows. Let K_c denote the intersection of all cutting subgroups in G. Then a finite covering of V admits a holomorphic map $f : V \to W$, where W is a flat Kähler torus bundle over the product of a several Riemann surfaces S_i and where the the kernel of the induced homomorphism of the fundamental groups equal $K_c \cap \pi_1(V)$. (If V is algebraic then W is a product of S_i's with an Abelian variety). We shall return to this problem in [De-Gr].

Acknowledgments. The authors are particularly grateful to the referee for a multitude of remarks, suggestions and questions, and to Prof. N. Ramachandran for useful comments.

References

[ABCKT]   J. Amoros, M. Burger, K. Corlette, D. Kotschik, D. Toledo, *Fundamental groups of compact Kähler manifolds*, Mathematical surveys and monographs, vol 44, AMS 1996.

[Ar]   Arzhantseva, G. N., On quasiconvex subgroups of word hyperbolic groups. *Geom. Dedicata* **87** no. 1–3 (2001), 191–208.

[Bo]   A. Borel, Les bouts des espaces homogènes de groupes de Lie. *Ann. of Math. (2)* **58** (1953), 443–457.

[BJS]   M. Bozejko, T. Januszkiewicz, R. Spazier, Infinite Coxeter groups do not have Kazdhan's T property. *J. Operator theory* **19** no. 1 (1988), 63–67.

[Ca]   F. Campana, Connexité abélienne des variétés kählériennes compactes. *Bull. Soc. Math. France* **126** no. 4 (1998), 483–506.

[C-T]   J.A. Carlson, D. Toledo, Harmonic mappings of Kähler manifolds to locally symmetric spaces. *Inst. Hautes Études Sci. Publ. Math.* **69** (1989), 173–201.

[Ch-Gr1]   J. Cheeger, M. Gromov, On the characteristic numbers of complete manifolds of bounded curvature and finite volume. *Differential geometry and complex analysis*, 115–154, Springer, Berlin, 1985.

[Ch-Gr2]   J. Cheeger, M. Gromov, L^2-cohomology and group cohomology. *Topology* **25** no. 2 (1986), 189–215.

[CCJJV]   P.-A. Cherix, M. Cowling, P. Jolissaint, P. Julg, A. Valette, *Groups with the Haagerup property*, Progress in Math. **197**, Birkhäuser, 2001.

[Co]   M. Coornaert, Sur le domaine de discontinuité pour les groupes d'isométries d'un espace métrique hyperbolique. *Rend. Sem. Fac. Sci. Univ. Cagliari* **59** no. 2 (1989), 185–195.

[Da]   F. Dahmani, Combination of convergence groups, Geometry and Topology, Vol. 7 (2003) Paper no. 27, pages 933–963.

[De-Gr]   T. Delzant, M. Gromov, in preparation.

[GR]   G. Grauert, R. Remmert, *Coherent analytic sheaves*. Grundlehren der Mathematischen Wissenschaften **265**, Springer-Verlag, Berlin, 1984.

[Gri]   A. Grigor'yan, Analytic and geometric background of recurrence and non-explosion of the Brownian motion on Riemannian manifolds. *Bull. Amer. Math. Soc. (N.S.)* **36** no. 2 (1999), 135–249.

[Gr$_{FPP}$]   M. Gromov, Foliated Plateau problem II. Harmonic maps of foliations. *Geom. Funct. Anal.* **1** no. 3 (1991), 253–320.

[Gr$_{GFK}$]   M. Gromov, Sur le groupe fondamental d'une variété Kählerienne. *C.R. Acad. Sci I* **308** (1989), 67–70.

[Gr$_{HG}$]   M. Gromov, Hyperbolic groups.

[Gr$_{RWRW}$]   M. Gromov, Random walk in random groups. *Geom. Funct. Anal.* **13** no. 1 (2003), 73–146.

[Gr-Sc]   M. Gromov, R. Schoen, Harmonic maps into singular spaces and p-adic superrigidity for lattices in groups of rank one. *Inst. Hautes Études Sci. Publ. Math.* **76** (1992), 165–246.

[Ha]   P. de la Harpe, *Classical Banach-Lie algebras and Banach-Lie groups of operators in Hilbert space,* Lecture Notes in Mathematics **285**, Springer, 1972.

[He]   L. Hernandez, Kähler manifolds and $\frac{1}{4}$-pinching. *Duke Math. J.* **62** no. 3 (1991), 601–611.

[Ho]   C. H. Houghton, Ends of locally compact groups and their coset spaces. *J. Austral. Math. Soc.* **17** (1974), 274–284.

[K]   I. Kapovich, The non amenability of Schreier graph for infinite index quasi-convex subgroup of hyperbolic group. *L'enseignement des mathématiques,* to appear.

[Ko]   J. Kollar. *Shafarevich maps and automorphic forms,* Princeton University Press, 1995.

[K-S]   N. J. Korevaar, R. M. Schoen, Global existence theorems for harmonic maps to non-locally compact spaces. *Comm. Anal. Geom.* **5** no. 2 (1997), 333–387.

[L-S]   R.C. Lyndon, P.E. Schupp. *Combinatorial group theory.* Springer, 1977.

[Mo]   Mok, Ngaiming, Harmonic forms with values in locally constant Hilbert bundles. Proceedings of the Conference in Honor of Jean-Pierre Kahane (Orsay, 1993). *J. Fourier Anal. Appl.,* Special Issue (1995), 433–453.

[N-R]   T. Napier, N. Ramachandran, Hyperbolic Kähler manifolds and holomorphic mappings to Riemann surfaces. *GAFA* **11** (2001), 382–406.

[Sa]   M. Sageev, Ends of group pairs and non positively curved complexes. *Proc. London. Math. Soc.* **71** no. 3 (1995), 585–617.

[Sam]   J. H. Sampson, Applications of harmonic maps to Kähler geometry. Complex differential geometry and nonlinear differential equations (Brunswick, Maine, 1984), 125–134, Contemp. Math. **49**, Amer. Math. Soc., Providence, RI, 1986.

[Sc]   P. Scott, Ends of pairs of groups. *J. Pure Appl. Algebra* **11** no. 1–3 (1977/78), 179–198.

[Si]   C. Simpson, Lefschetz theorems for the integral leaves of a holomorphic one-form. *Compositio Math.* **87** no. 1 (1993), 99–113.

[Siu]   Y.T. Siu, The complex-analyticity of harmonic map and the strong rigidity of compact Kähler manifolds. *Annals of Math.* **112** (1980), 73–111.

[W]   D. Wise, Cubulating small cancellation groups, *GAFA*, to appear.

[Wo]   W. Woess, *Random walks on infinite graphs and groups.* Cambridge Tracts in Mathematics **138**, Cambridge University Press, 2000.

Thomas Delzant
Institut de Recherche Mathématique Avancée, Université Louis Pasteur, 7 rue René Descartes, 67084 Strasbourg, France
e-mail: delzant@math.u-strasbg.fr

Misha Gromov
Institut des Hautes Études Scientifiques, 35 route de Chartres, 91140 Bures-sur-Yvette, France
e-mail: gromov@ihes.fr

Progress in Mathematics, Vol. 248, 57–116

Algebraic Mapping-Class Groups of Orientable Surfaces with Boundaries

Warren Dicks and Edward Formanek

To Slava Grigorchuk on the occasion of his 50th birthday.

Abstract. Let $S_{g,b,p}$ denote a surface which is connected, orientable, has genus g, has b boundary components, and has p punctures. Let $\Sigma_{g,b,p}$ denote the fundamental group of $S_{g,b,p}$.

We define the algebraic mapping-class group of $S_{g,b,p}$, denoted by $\mathrm{Out}_{g,b,p}$, and observe that topologists have shown that $\mathrm{Out}_{g,b,p}$ is naturally isomorphic to the topological mapping-class group of $S_{g,b,p}$.

We study the algebraic version

$$1 \to \check{\Sigma}_{g,b,p} \to \mathrm{Out}_{g,b\perp 1,p} \to \mathrm{Out}_{g,b,p} \to 1$$

of Mess's exact sequence that arises from filling in the interior of the $(b+1)$st boundary component of $S_{g,b+1,p}$.

Here $\mathrm{Out}_{g,b\perp 1,p}$ is the subgroup of index $b+1$ in $\mathrm{Out}_{g,b+1,p}$ that fixes the $(b+1)$st boundary component.

If (g,b,p) is $(0,0,0)$ or $(0,0,1)$, then $\check{\Sigma}_{g,b,p}$ is trivial. If (g,b,p) is $(0,0,2)$ or $(1,0,0)$, then $\check{\Sigma}_{g,b,p}$ is infinite cyclic. In all other cases, $\check{\Sigma}_{g,b,p}$ is the fundamental group of the unit-tangent bundle of a suitably metrized $S_{g,b,p}$, and, hence, $\check{\Sigma}_{g,b,p}$ is an extension of an infinite cyclic, central subgroup by $\Sigma_{g,b,p}$.

We give a description of the conjugation action of $\mathrm{Out}_{g,b\perp 1,p}$ on $\check{\Sigma}_{g,b,p}$ in terms of the following three ingredients: an easily-defined action of $\mathrm{Out}_{g,b\perp 1,p}$ on $\Sigma_{g,b+1,p}$; the natural homomorphism $\Sigma_{g,b+1,p} \to \check{\Sigma}_{g,b,p}$; and, a twisting-number map $\Sigma_{g,b+1,p} \to \mathbb{Z}$ that we define.

The work of many authors has produced aesthetic presentations of the orientation-preserving mapping-class groups $\mathrm{Out}^+_{g,b,p}$ with $b+p \leqslant 1$, using the DLH generators. Within the program of giving algebraic proofs to algebraic results, we apply our machinery to give an algebraic proof of a relatively small part of this work, namely that the kernel $\check{\Sigma}_{g,0,0}$ of the map

$$\mathrm{Out}^+_{g,1,0} \to \mathrm{Out}_{g,0,0}$$

is the normal closure in $\mathrm{Out}^+_{g,1,0}$ of Matsumoto's A-D word (in the DLH generators).

From the algebraic viewpoint, $\mathrm{Out}_{g,1,0}$ is the group of those automorphisms of a rank-$2g$ free group which fix or invert a given genus g surface relator, $\mathrm{Out}_{g,0,0}$ is the group of outer automorphisms of the genus g surface group, and $\check{\Sigma}_{g,0,0}$ is the kernel of the natural map between these groups. What we study are presentations for $\check{\Sigma}_{g,0,0}$, both as a group and as an $\mathrm{Out}_{g,1,0}$-group, and related topics.

Mathematics Subject Classification (2000). Primary 20F34; Secondary 20E05, 20E36, 20F36, 57M60, 58A30.

Keywords. Surface, algebraic mapping-class group, unit-tangent bundle.

Contents

1. Topological background

Let $\mathbb{N}$ denote the set of finite cardinals. Let $(g, b, p) \in \mathbb{N}^3$ be fixed throughout the article. Let $S_{g,b,p}$ denote a connected, orientable surface which has genus g, and b boundary components, and p punctures. Let α be a self-homeomorphism of $S_{g,b,p}$, and let $\partial\alpha$ denote the induced self-homeomorphism of the boundary $\partial S_{g,b,p}$.

We shall now describe restrictions we wish to impose on α. If α is orientation-preserving, we want $\partial\alpha$ to act as the identity on each component of $\partial S_{g,b,p}$ that is carried to itself. But we want to allow certain orientation-reversing maps and maps that permute boundary components; this will give a better interplay between algebra and topology. We consider a referential S^1, say the complex numbers of modulus 1, and endow S^1 with an orientation-reversing map of order two, say complex conjugation. Together with the identity map, this gives a group of two distinguished self-homeomorphisms of S^1. We further specify a homeomorphism between each component of $\partial S_{g,b,p}$ and S^1, and thus obtain a b-to-1 map $\partial S_{g,b,p} \to S^1$. Then α is *admissible* if $\partial\alpha$ forms part of a commuting square in which two of the sides are the specified b-to-1 map $\partial S_{g,b,p} \to S^1$, and the remaining side is one of our two distinguished self-homeomorphisms of S^1. Thus, if $b \geqslant 1$, we allow exactly $2b!$ possibilities for $\partial\alpha$. To avoid awkward exceptions, we think of the empty set as having two self-maps, one of which is orientation-preserving and the other is orientation-reversing. This peculiar sort of convention will be useful throughout, in algebraic, as well as topological, contexts.

The surface $S_{g,b,p}$ is formed by deleting b open discs and p points from an orientable surface of genus g. A point is a degenerate closed disc. For our purposes, it is often convenient to view the $b + p$ discs as being distinguished, rather than deleted. From this viewpoint, we are looking at self-homeomorphisms of the orientable surface of genus g which act on the $b + p$ discs in one of $2b!p!$ possible ways. Thus, for example, we shall speak of the permutation of the set of punctures induced by a self-homeomorphism of $S_{g,b,p}$.

The group of all admissible self-homeomorphisms of $S_{g,b,p}$ contains a normal subgroup consisting of those self-homeomorphisms which can be homotoped to the identity map through self-homeomorphisms which act as the identity on the boundary and on the set of punctures. The resulting quotient group is called the *mapping-class group* of $S_{g,b,p}$, denoted $\mathcal{MC}_{g,b,p}$. Thus the elements of $\mathcal{MC}_{g,b,p}$ are isotopy classes of admissible self-homeomorphisms of $S_{g,b,p}$.

Let $\mathcal{MC}^+_{g,b,p}$ denote the index-two subgroup of $\mathcal{MC}_{g,b,p}$ consisting of isotopy classes of orientation-preserving admissible self-homeomorphisms. We can choose an orientation-reversing element of order two in $\mathcal{MC}_{g,b,p}$, and express $\mathcal{MC}_{g,b,p}$ as a semidirect product $\mathcal{MC}_{g,b,p} \simeq \mathcal{MC}^+_{g,b,p} \rtimes C_2$; we write C_2 to denote a cyclic, multiplicative group of order two.

Terminology is not standardized in this area and this can be confusing; for example, some authors call $\mathcal{MC}^+_{g,0,b}$ the mapping class group of $S_{g,b,0}$, and call $\mathcal{MC}_{g,0,b}$ the extended mapping class group of $S_{g,b,0}$.

Let us return to $S_{g,b,p}$. Recall that the $b + p$ discs can be viewed as either distinguished or deleted. We pass to a different surface if one of the discs is, in one viewpoint, made undistinguished, or, in the other viewpoint, filled in. Depending as the disc is open or closed, we call this process *elimination of a boundary component* or *elimination of a puncture*. Another well-behaved process is that of undistinguishing, or filling in, all but one point of an open disc. We call this process *converting a boundary component to a puncture*. Converting a boundary component to a puncture can also be viewed as contracting the closure of a distinguished open disc to a distinguished point. Notice that converting a boundary component to a puncture and then eliminating the puncture is equivalent to eliminating the boundary component; this simple observation will be used frequently.

Each of the three processes mentioned above, converting a boundary component to a puncture, eliminating a puncture, and eliminating a boundary component, determines a surjective partial homomorphism of mapping-class groups,

$$\mathcal{MC}_{g,b,p} \rightsquigarrow \mathcal{MC}_{g,b-1,p+1},$$
$$\mathcal{MC}_{g,b,p} \rightsquigarrow \mathcal{MC}_{g,b,p-1},$$
$$\mathcal{MC}_{g,b,p} \rightsquigarrow \mathcal{MC}_{g,b-1,p},$$

respectively. In each case the domain has finite index in $\mathcal{MC}_{g,b,p}$, namely the stabilizer of the disc involved. Moreover, the kernel has been calculated explicitly and is given in [13, Sections 2.8 and 6.3]. Thus each process gives an exact sequence of groups. (We shall be giving algebraic analogues of these exact sequences.)

We are particularly interested in the third case. Suppose then that we eliminate a boundary component of $S_{g,b+1,p}$. The resulting exact sequence of groups has the form

$$1 \to \check{\Sigma}_{g,b,p} \to \mathcal{MC}_{g,b\perp1,p} \to \mathcal{MC}_{g,b,p} \to 1.$$

Here $\mathcal{MC}_{g,b\perp1,p}$ denotes the set of isotopy classes of admissible self-homeomorphisms of $S_{g,b+1,p}$ which carry the $(b + 1)$st boundary component to itself. The description of $\check{\Sigma}_{g,b,p}$ is as follows. If (g, b, p) is $(0, 0, 0)$ or $(0, 0, 1)$, then $\check{\Sigma}_{g,b,p}$ is trivial. If (g, b, p) is $(0, 0, 2)$ or $(1, 0, 0)$, then $\check{\Sigma}_{g,b,p}$ is infinite cyclic. In all other cases, $\check{\Sigma}_{g,b,p}$ is the fundamental group of the unit-tangent bundle of a suitably metrized $S_{g,b,p}$, and hence an extension of an infinite cyclic, central subgroup by $\Sigma_{g,b,p}$. The only explicit references we have found for this exact sequence are the preprint by Mess [22], and the portion of the survey article [13, Section 6.3] which is based on [22], and these deal with the case $(b, p) = (0, 0)$. This case is important, and we now discuss it. Thus, we have a surjective map

$$\phi_g \colon \mathcal{MC}_{g,1,0} \twoheadrightarrow \mathcal{MC}_{g,0,0}$$

which has the same kernel, $\check{\Sigma}_{g,0,0}$, as the induced map

$$\phi_g^+ \colon \mathcal{MC}_{g,1,0}^+ \twoheadrightarrow \mathcal{MC}_{g,0,0}^+.$$

For $g \in \{0, 1\}$, these maps are well understood: ϕ_0 is an isomorphism of groups of order two, and ϕ_1^+ is equivalent to the quotient map from the braid group on three strings to the braid group on three strings modulo the square of its infinite cyclic center.

Now suppose that $g \geqslant 2$. Here, $\check{\Sigma}_{g,0,0}$ is an extension of an infinite cyclic, central, subgroup by $\Sigma_{g,0,0}$. Wajnryb [27] showed that, as a normal subgroup of $\mathcal{MC}^+_{g,1,0}$, $\check{\Sigma}_{g,0,0}$ is generated by a single element. The main part of Wajnryb's article gives a presentation of $\mathcal{MC}^+_{g,1,0}$ on the DLH generators, and the foregoing step is then used to show that adding one more relator yields a presentation of $\mathcal{MC}^+_{g,0,0}$. Wajnryb [28] later gave an elementary, self-contained proof, in which the DLH-length of the added relator dropped from $32g^2 + 8g - 26$ to $8g^2 + 6$. Subsequently, Matsumoto [20] introduced what we call *the A-D relator*, of DLH-length $8g^2 - 10g + 2$; it has a simple description in terms of centers of Artin groups. He used the theory of miniversal deformations of singularities to prove that adding the A-D relator to a DLH-presentation of $\mathcal{MC}^+_{g,1,0}$ yields a DLH-presentation of $\mathcal{MC}^+_{g,0,0}$. Labruère-Paris [14, Proposition 2.12(i)] have given a topological proof.

The complementary part of Matsumoto's article gives an aesthetic DLH-presentation of $\mathcal{MC}^+_{g,1,0}$ in terms of Artin groups; the proof uses Wajnryb's DLH-presentation and computer verifications, and the only values of g involved are 2 and 3.

2. Outline

In this article, we present some of the foregoing topological results in an algebraic setting.

In Sections 3 and 4, and Theorem 9.6, we describe $\mathcal{MC}_{g,b,p}$ as a certain group, $\mathrm{Out}_{g,b,p}$, of outer automorphisms. This description is well-known in the case where $b = 0$, and easily deduced in general; our definition is closely related to that recently given by Levitt [16, Section 4].

In Section 5, we briefly review Dehn-twist automorphisms, and in Section 6 we briefly review Artin groups, and recall the aesthetic presentations of $\mathrm{Out}^+_{g,b,p}$, $b + p \leqslant 1$.

In Sections 7–13, we give algebraic proofs of some isomorphisms, exact sequences, and semidirect product decompositions, which arise from eliminating boundaries and punctures.

In Section 14, we give an interesting description of the action of $\mathrm{Out}_{g,b\perp 1,p}$ on the normal subgroup $\check{\Sigma}_{g,b,p}$. This action makes transparent Wajnryb's result that $\check{\Sigma}_{g,0,0}$ is the normal closure of a single element in $\mathrm{Out}^+_{g,1,0}$.

Section 15 reviews some classic examples.

Section 16 recalls how to derive a DLH-presentation of $\mathrm{Out}^+_{g,0,1}$ from a DLH-presentation of $\mathrm{Out}^+_{g,1,0}$.

Now suppose that $g \geqslant 2$. In Section 17, we use our algebraic machinery to give an algebraic proof that adding the A-D relator to a DLH-presentation of $\mathrm{Out}^+_{g,1,0}$ gives a DLH-presentation of $\mathrm{Out}^+_{g,0,0}$.

In Sections 19 and 20, we use many of our results to construct a partial embedding $\mathrm{Out}_{g-1,1,1} \rightsquigarrow \mathrm{Out}_{g,1,0}$ that sheds some light on the A-D relator.

3. Surface groups and related groups

3.1. Notation. Following Bourbaki, we write $\mathbb{N} := \{0, 1, 2, 3, \ldots\}$.

For group elements x, y, we write $\bar{x} := x^{-1}$, $x^y := \bar{y}xy$, and $[x, y] := \bar{x}\,\bar{y}xy$. We let $[x]$ denote the conjugacy class of x, usually thought of as a cyclic word in a specified generating set.

All actions will be on the right, generally written as exponents, and compositions are to be read from left to right. We write $x^{-\alpha} = \bar{x}^\alpha$ and $x^{n\alpha} = (x^n)^\alpha$.

We interpret $\prod_{i=m}^{n} x_i$ to mean $x_m x_{m+1} x_{m+2} \cdots x_{n-1} x_n$, where this is understood to be 1 if $m > n$; we write $\coprod_{i=m}^{n} x_i$ to mean $x_m x_{m-1} x_{m-2} \cdots x_{n+1} x_n$, where this is understood to be 1 if $m < n$. $\qquad\square$

Recall that $(g, b, p) \in \mathbb{N}^3$.

3.2. Definitions. We define the (g, b, p)-*Euler characteristic* as $\chi_{g,b,p} = 2 - 2g - b - p$ and we fix

$$X_{g,b,p} = \{x_i, y_i \mid 1 \leqslant i \leqslant g\} \cup \{z_j \mid 1 \leqslant j \leqslant b\} \cup \{t_k \mid 1 \leqslant k \leqslant p\}.$$

This set will be involved in the presentations of the surface groups. The symbols z_j and t_k correspond to simple closed curves which start and finish at the base point of the surface and isolate the jth boundary component and the kth puncture, respectively.

We will be interested in collapsing a boundary component to a puncture and then filling in the puncture, and it will simplify the algebra involved if we have the boundary components arranged in ascending order followed by the punctures arranged in descending order. Where we wish to blur the distinction between boundary components and punctures, it will be useful to set

$$z_{b+k} = t_{p+1-k} \ (1 \leqslant k \leqslant p), \quad t_{p+j} = z_{b+1-j}, \ (1 \leqslant j \leqslant b), \tag{3.2.1}$$

and then we write

$$X_{g,b+p} = \{x_i, y_i \mid 1 \leqslant i \leqslant g\} \cup \{z_j \mid 1 \leqslant j \leqslant b+p\} = X_{g,b,p}.$$

In the free group on $X_{g,b,p}$, we set

$$w_{g,b,p} = \prod_{i=1}^{g} [x_i, y_i] \cdot \prod_{j=1}^{b} z_j \cdot \coprod_{k=p}^{1} t_k,$$

and also write

$$w_{g,b+p} = \prod_{i=1}^{g} [x_i, y_i] \cdot \prod_{j=1}^{b+p} z_j = w_{g,b,p}.$$

We define the (g, b, p)-*surface group* as

$$\Sigma_{g,b,p} = \langle X_{g,b,p} \mid w_{g,b,p} \rangle,$$

and also write $\Sigma_{g,b+p} = \langle X_{g,b+p} \mid w_{g,b+p} \rangle = \Sigma_{g,b,p}$. $\qquad\square$

3.3. Remarks. (i) Notice that $\Sigma_{g,b,p}$ is trivial if and only if $2g + b + p \leqslant 1$, that is $(g, b, p) = (0, 0, 0)$, $(0, 1, 0)$ or $(0, 0, 1)$. These three values correspond to $S_{g,b,p}$ being a sphere, a closed disc, and an open disc, respectively. We call these three values the *trivial cases*. The remaining values of (g, b, p) will be called the *nontrivial cases*.

(ii) Similarly, $\Sigma_{g,b,p}$ is abelian if and only if $2g + b + p \leqslant 2$. As well as the trivial cases, we have $2g + b + p = 2$, that is $(g, b, p) = (1, 0, 0)$, $(0, 2, 0)$, $(0, 1, 1)$ or $(0, 0, 2)$. These four values correspond to $S_{g,b,p}$ being a torus, a closed annulus, a punctured closed disc, and a punctured open disc (or open annulus), respectively.

(iii) If $2g + b + p \geqslant 3$, then $\Sigma_{g,b,p}$ has trivial center.
Denoting the center of $\Sigma_{g,b,p}$ by Ctr, we have

$$\Sigma_{g,b,p}/\operatorname{Ctr} = \begin{cases} \Sigma_{g,b,p} & \text{if} \quad 2g + b + p \neq 2, \\ 1 & \text{if} \quad 2g + b + p \leqslant 2. \end{cases}$$

(iv) If $b + p \geqslant 1$, then $w_{g,b,p}$ can be extended to a basis of the free group on $X_{g,b,p}$; here, $\Sigma_{g,b,p}$ is a free group of rank $2g + b + p - 1 = 1 - \chi_{g,b,p}$. We will make frequent use of this fact. □

3.4. Definitions. In the tensor-ring of the abelianization of $\Sigma_{g,b+p}$, let $\Omega_{g,b+p}$ denote the (infinite cyclic) additive subgroup generated by

$$\begin{cases} 1 & \text{if} \quad (g, b + p) \in \{(0, 0), (0, 1)\}, \\ \displaystyle\sum_{i=1}^{g} (x_i \otimes y_i - y_i \otimes x_i) & \text{if} \quad b + p = 0 \text{ and } g \geqslant 1, \\ \displaystyle\sum_{j=1}^{b+p} z_j & \text{if} \quad b + p \geqslant 1 \text{ and } (g, b + p) \neq (0, 1). \end{cases}$$

There is then an isomorphism

$$\operatorname{sign}\colon \operatorname{Aut}(\Omega_{g,b+p}) \xrightarrow{\sim} \{1, -1\},$$

and any $\alpha \in \operatorname{Aut}(\Omega_{g,b+p})$ acts on $\Omega_{g,b+p}$ by multiplication by $\operatorname{sign}(\alpha)$. □

We now define some groups closely related to $\Sigma_{g,b,p}$, and later they will be seen to arise naturally.

3.5. Definition. We write

$$\tilde{\Sigma}_{g,b,p} = \begin{cases} 1 & \text{if} \quad (g, b, p) = (1, 0, 0) \text{ or } (0, 0, 2), \\ \Sigma_{g,b,p} & \text{otherwise.} \end{cases}$$

We view $\tilde{\Sigma}_{g,b,p}$ as a quotient group of $\Sigma_{g,b,p}$.

Observe that $(g, b + p) = (0, 2)$ is the only case where "$\tilde{\Sigma}_{g,b+p}$" is not defined. □

3.6. Definition. Let σ be a new symbol, and let $\check{\Sigma}_{g,b,p}$ denote the group presented with generating set $X_{g,b,p} \cup \{\sigma\}$ and relations saying the following:

(i) σ is central;

(ii) $w_{g,b,p} = \sigma^{\chi_{g,b,p}}$ (or $[w_{g,b,p}] = \sigma^{2-2g-b-p}$ where $w_{g,b,p}$ can be written as a cyclic word because σ is central);

(iii) $\sigma = 1$ (or $\check{\Sigma}_{g,b,p} = \{1\}$) if (g,b,p) is $(0,0,0)$ or $(0,0,1)$;

(iv) $X_{g,b,p} = \{1\}$ if (g,b,p) is $(1,0,0)$ or $(0,0,2)$.

Here "$\check{\Sigma}_{g,b+p}$" is not defined for $(g,b+p) = (0,1)$ or $(0,2)$.

We identify the quotient group $\check{\Sigma}_{g,b,p}/\langle\sigma\rangle$ with $\widetilde{\Sigma}_{g,b,p}$.   □

3.7. Remarks. (i) We summarize the seven cases where $\Sigma_{g,b,p}$ is abelian:

$\Sigma_{0,0,0} = 1$	$\check{\Sigma}_{0,0,0} = 1$	$\widetilde{\Sigma}_{0,0,0} = 1$
$\Sigma_{0,0,1} = 1$	$\check{\Sigma}_{0,0,1} = 1$	$\widetilde{\Sigma}_{0,0,1} = 1$
$\Sigma_{0,1,0} = 1$	$\check{\Sigma}_{0,1,0} = 1$	$\widetilde{\Sigma}_{0,1,0} = \langle \sigma \mid \ \rangle$
$\Sigma_{0,0,2} = \langle t_1 \mid \ \rangle$	$\check{\Sigma}_{0,0,2} = 1$	$\widetilde{\Sigma}_{0,0,2} = \langle \sigma \mid \ \rangle$
$\Sigma_{0,1,1} = \langle z_1 \mid \ \rangle$	$\check{\Sigma}_{0,1,1} = \langle z_1 \mid \ \rangle$	$\widetilde{\Sigma}_{0,1,1} = \langle z_1, \sigma \mid [z_1,\sigma] \ \rangle$
$\Sigma_{0,2,0} = \langle z_1 \mid \ \rangle$	$\check{\Sigma}_{0,2,0} = \langle z_1 \mid \ \rangle$	$\widetilde{\Sigma}_{0,2,0} = \langle z_1, \sigma \mid [z_1,\sigma] \ \rangle$
$\Sigma_{1,0,0} = \langle x_1, y_1 \mid [x_1,y_1] \ \rangle$	$\check{\Sigma}_{1,0,0} = 1$	$\widetilde{\Sigma}_{1,0,0} = \langle \sigma \mid \ \rangle.$

(ii) If $g \geqslant 2$, then $\check{\Sigma}_{g,0,0}$ contains the universal central extension of $\Sigma_{g,0,0}$ as a subgroup of index $2g - 2 = -\chi_{g,0,0}$.

(iii) Suppose $b + p \geqslant 1$, so $\widetilde{\Sigma}_{g,b,p}$ is a free group. We have $\check{\Sigma}_{0,0,1} = \widetilde{\Sigma}_{0,0,1} = 1$. If $(g,b,p) \neq (0,0,1)$, then $\check{\Sigma}_{g,b,p} \simeq \widetilde{\Sigma}_{g,b,p} \times \langle \sigma \mid \ \rangle$.   □

3.8. Definitions. We now introduce a set of new symbols, $E_b = \{e_j \mid 1 \leqslant j \leqslant b\}$. Intuitively, e_j corresponds to a simple curve in $S_{g,b,p}$ joining the base point of the surface to the basepoint of the jth boundary component. Formally, the e_j are elements of a groupoid having $\Sigma_{g,b,p}$ as basepoint group. However, we shall treat the e_j group theoretically to avoid dealing with groupoids.

Let

$$\Sigma_{g,b,p} * E_b = \langle X_{g,b,p} \cup E_b \mid w_{g,b,p}\rangle = \Sigma_{g,b,p} * \langle E_b \mid \ \rangle.$$

We define

$$\text{Aut}(\Sigma_{g,b,p} * E_b, \ \Sigma_{g,b,p}, \ \{z_j^{\pm e_j}\}_{j=1}^b, \ \{[t_k]^{\pm 1}\}_{k=1}^p, \ \Omega_{g,b+p}), \qquad (3.8.1)$$

or, in abbreviated form, $\text{Aut}_{g,b,p}$, as follows.

In the nontrivial cases, $\text{Aut}_{g,b,p}$ is defined to be the group consisting of those automorphisms of $\Sigma_{g,b,p} * E_b$ which map each of the following three sets onto itself:

$$\Sigma_{g,b,p}, \quad \{z_j^{e_j}, \bar{z}_j^{e_j} \mid 1 \leqslant j \leqslant b\}, \quad \{[t_k], [\bar{t}_k] \mid 1 \leqslant k \leqslant p\}.$$

It can be shown that $\mathrm{Aut}_{g,b,p}$ then acts on the (infinite cyclic) group $\Omega_{g,b+p}$ introduced in Definitions 3.4.

In the trivial cases, $\mathrm{Aut}_{g,b,p}$ is defined as $\mathrm{Aut}(\Omega_{g,b+p})$ and is understood to act trivially on $\Sigma_{g,b,p} * E_b$.

In all cases then, $\mathrm{Aut}_{g,b,p}$ acts on $\Sigma_{g,b,p} * E_b$ and on $\Omega_{g,b+p}$. Moreover, an element of $\mathrm{Aut}_{g,b,p}$ is completely specified by its actions on $\Sigma_{g,b,p} * E_b$ and $\Omega_{g,p+b}$; the former is determinative in the nontrivial cases and the latter in the trivial cases.

We define sign: $\mathrm{Aut}_{g,b,p} \to \{+1, -1\}$ to be the (surjective) composite

$$\mathrm{Aut}_{g,b,p} \to \mathrm{Aut}(\Omega_{g,b+p}) \xrightarrow{\sim} \{+1, -1\}.$$

The preimage of $+1$ is denoted $\mathrm{Aut}^+_{g,b,p}$; its elements are said to be *positive*. The preimage of -1 is denoted $\mathrm{Aut}^-_{g,b,p}$; its elements are said to be *negative*. □

We shall be working with $\mathrm{Aut}_{g,b,p}$ throughout the article. Notation similar to (3.8.1) will be used later, and the correct interpretation should always be obvious.

3.9. Remark. Suppose α is an element of $\mathrm{Aut}^+_{g,b,p}$ which sends $z_j^{e_j}$ to $z_{j'}^{e_{j'}}$, for some $1 \leqslant j, j' \leqslant b$.

Clearly, α carries $[z_j]$ to $[z_{j'}]$.

Writing $A_j = e_j^\alpha$ and $Z_j = z_j^\alpha$, we have $Z_j^{A_j} = z_{j'}^{e_{j'}}$ and $z_j^\alpha = Z_j = z_{j'}^{e_{j'} \bar{A}_j}$.

Now suppose that α fixes z_j. Then

$$z_j = z_j' = Z_j^{A_j \bar{e}_j} = z_j^{A_j \bar{e}_j}.$$

In the nontrivial cases, the centralizer of z_j in the free group $\Sigma_{g,b,p} * E_b$ is $\langle z_j \rangle$, hence $A_j \bar{e}_j \in \langle z_j \rangle$, hence $A_j \in \langle z_j \rangle e_j$. That is, $e_j^\alpha \in \langle z_j \rangle e_j$. This holds in the trivial cases also. □

3.10. Definitions. We now describe several elements of $\mathrm{Aut}_{g,b,p}$ that will be used separately at various points throughout the article: Dehn-twist automorphisms α_i, β_i, and γ_i; boundary Dehn-twist automorphism σ_j; braid automorphisms τ_j, μ_k; and, a negative involution ζ. In Remarks 5.1, we will indicate the connection with Dehn twists.

(i) α_i, β_i, Dehn-twist automorphisms.

For $1 \leqslant i \leqslant g$, we define α_i, $\beta_i \in \mathrm{Aut}^+_{g,b,p}$ by

$$\alpha_i : \begin{cases} x_i & \mapsto & \overline{y}_i x_i, \\ w & \mapsto & w \quad \text{for all } w \in E_b \cup X_{g,b,p} - \{x_i\}, \end{cases}$$

$$\beta_i : \begin{cases} y_i & \mapsto & x_i y_i, \\ w & \mapsto & w \quad \text{for all } w \in E_b \cup X_{g,b,p} - \{y_i\}. \end{cases}$$

(ii) γ_i, Dehn-twist automorphisms.

For $1 \leqslant i \leqslant g-1$, we define $\gamma_i \in \mathrm{Aut}^+_{g,b,p}$ by

$$\gamma_i : \begin{cases} x_i & \mapsto & \overline{x}_{i+1} y_{i+1} x_{i+1} \overline{y}_i x_i, \\ y_i & \mapsto & \overline{x}_{i+1} y_{i+1} x_{i+1} y_i \overline{x}_{i+1} \overline{y}_{i+1} x_{i+1} = y_i^{\overline{y}_{i+1}^{x_{i+1}}}, \\ x_{i+1} & \mapsto & x_{i+1} y_i \overline{x}_{i+1} \overline{y}_{i+1} x_{i+1}, \\ w & \mapsto & w \quad \text{for all } w \in E_b \cup X_{g,b,p} - \{x_i, y_i, x_{i+1}\}. \end{cases}$$

If $g \geqslant 1$ and $b + p \geqslant 1$, we define γ_0, $\gamma_g \in \mathrm{Aut}^+_{g,b,p}$ by

$$\gamma_0 : \begin{cases} x_1 & \mapsto & x_1 z_{b+p} \overline{x}_1 \overline{y}_1 x_1, \\ z_{b+p} & \mapsto & \overline{x}_1 y_1 x_1 z_{b+p} \overline{x}_1 \overline{y}_1 x_1 = z_{b+p}^{\overline{y}_1^{x_1}}, \\ e_{b+p} & \mapsto & \overline{x}_1 y_1 x_1 \overline{z}_{b+p} e_{b+p} \quad \text{if } p = 0, \\ w & \mapsto & w \quad \text{for all } w \in (X_{g,b,p} \cup E_b) - \{x_1, z_{b+p}, e_{b+p}\}. \end{cases}$$

$$\gamma_g : \begin{cases} x_g & \mapsto & \overline{z}_1 \overline{y}_g x_g, \\ y_g & \mapsto & \overline{z}_1 y_g z_1 = y_g^{z_1}, \\ z_1 & \mapsto & \overline{z}_1 \overline{y}_g z_1 y_g z_1 = z_1^{y_g z_1}, \\ e_1 & \mapsto & \overline{z}_1 \overline{y}_g e_1 \quad \text{if } b \geqslant 1, \\ w & \mapsto & w \quad \text{for all } w \in (X_{g,b,p} \cup E_b) - \{x_g, y_g, z_1, e_1\}. \end{cases}$$

(iii) σ_j, boundary Dehn-twist automorphisms.

For $1 \leqslant j \leqslant b$, we define $\sigma_j \in \mathrm{Aut}^+_{g,b,p}$ by

$$\sigma_j : \begin{cases} e_j & \mapsto & \overline{z}_j e_j, \\ w & \mapsto & w \quad \text{for all } w \in X_{g,b,p} \cup E_b - \{e_j\}. \end{cases}$$

(iv) τ_j, (boundary) braid automorphisms.

For $1 \leqslant j < b$, we define $\tau_j \in \mathrm{Aut}^+_{g,b,p}$ by

$$\tau_j : \begin{cases} z_j & \mapsto & z_{j+1}, \\ e_j & \mapsto & e_{j+1}, \\ z_{j+1} & \mapsto & z_j^{z_{j+1}}, \\ e_{j+1} & \mapsto & \overline{z}_{j+1}\overline{z}_j e_j, \\ w & \mapsto & w \quad \text{for all } w \in (X_{g,b,p} \cup E_b) - \{z_j, e_j, z_{j+1}, e_{j+1}\}. \end{cases}$$

(v) μ_k, (puncture) braid automorphisms.

For $1 \leqslant k < p$, we define $\mu_k \in \mathrm{Aut}^+_{g,b,p}$ by

$$\mu_k : \begin{cases} z_{b+k} & \mapsto & z_{b+k+1}, \\ z_{b+k+1} & \mapsto & z_{b+k}^{z_{b+k+1}}, \\ w & \mapsto & w \quad \text{for all } w \in E_b \cup X_{g,b,p} - \{z_{b+k}, z_{b+k+1}\}. \end{cases}$$

(vi) ζ, a negative involution.

Recall (3.2.1), and set

$$w_{g,j} = \prod_{i=1}^{g}[x_i, y_i] \cdot \prod_{j'=1}^{j} z_{j'}, \quad 0 \leqslant j \leqslant b+p.$$

Even in the trivial cases, we define an order-two element ζ of $\mathrm{Aut}^-_{g,b,p}$ which acts on $\Sigma_{g,b+p} * E_b$ by

$$\zeta : \begin{cases} x_i & \mapsto & y_{g+1-i} & \text{for } 1 \leqslant i \leqslant g, \\ y_i & \mapsto & x_{g+1-i} & \text{for } 1 \leqslant i \leqslant g, \\ z_j & \mapsto & \overline{z}_j^{\,\overline{w}_{g,j-1}} & \text{for } 1 \leqslant j \leqslant b+p, \\ e_j & \mapsto & w_{g,j-1}e_j & \text{for } 1 \leqslant j \leqslant b. \end{cases}$$

It is not difficult to show that $\mathrm{Aut}_{g,b,p} = \mathrm{Aut}^+_{g,b,p} \rtimes \langle \zeta \rangle$.

We call the α_i, β_i and γ_i (distinguished) *Dehn-twist automorphisms*. We call the σ_j *boundary Dehn-twist automorphisms*. We call the τ_j and μ_k *braid automorphisms*. □

4. The algebraic mapping-class group

4.1. Definition. Consider an element w of $\Sigma_{g,b,p}$. There is an associated element $\tilde{w}$ of $\mathrm{Aut}^+_{g,b,p}$ which (right) conjugates each element of $X_{g,b,p}$ by w, and left multiplies each element of E_b by w^{-1}. It can be shown that we have a homomorphism

$$\Sigma_{g,b,p} \to \mathrm{Aut}_{g,b,p}, \quad w \mapsto \tilde{w}. \tag{4.1.1}$$

Moreover the image of (4.1.1), viewed as a quotient of $\Sigma_{g,b,p}$, is isomorphic to the group $\widetilde{\Sigma}_{g,b,p}$ introduced in Definition 3.5. The image of (4.1.1) will again be denoted $\widetilde{\Sigma}_{g,b,p}$.

It is not difficult to see that $\widetilde{\Sigma}_{g,b,p}$ is normal in $\mathrm{Aut}_{g,b,p}$, and that the action of $\mathrm{Aut}_{g,b,p}$ on $\widetilde{\Sigma}_{g,b,p}$ by conjugation agrees with the action of $\mathrm{Aut}_{g,b,p}$ on $\widetilde{\Sigma}_{g,b,p}$ induced from the action on $\Sigma_{g,b,p}$.

We define $\mathrm{Out}_{g,b,p} = \mathrm{Aut}_{g,b,p}/\widetilde{\Sigma}_{g,b,p}$, and denote the quotient map by

$$\mathrm{Aut}_{g,b,p} \to \mathrm{Out}_{g,b,p}, \quad \phi \mapsto \check{\phi}.$$

We call $\mathrm{Out}_{g,b,p}$ the *algebraic (g,b,p)-mapping-class group*.

We define $\mathrm{Out}^+_{g,b,p} = \mathrm{Aut}^+_{g,b,p}/\widetilde{\Sigma}_{g,b,p}$, a subgroup of index two in $\mathrm{Out}_{g,b,p}$. ☐

4.2. Remark. There is a natural map $\mathcal{MC}_{g,b,p} \to \mathrm{Out}_{g,b,p}$ which carries $\mathcal{MC}^+_{g,b,p}$ to $\mathrm{Out}^+_{g,b,p}$. In Theorem 9.6, we shall see that these maps are isomorphisms. For $b = 0$ this was already known, and the case $b \geqslant 1$ will quickly be reduced to the case $b = 0$. See also Remarks 11.3(ii).

Notice that $\mathrm{Out}_{g,b,p} = \mathrm{Out}^+_{g,b,p} \rtimes \langle \check{\zeta} \rangle$. Recall that a similar result holds for the topological mapping-class groups. ☐

4.3. Notation. Let Sym_b denote the group of permutations of $\{j \in \mathbb{N} \mid 1 \leqslant j \leqslant b\}$.

We have a natural homomorphism $\mathrm{Aut}_{g,b,p} \to \mathrm{Sym}_b \times \mathrm{Sym}_p$ arising from the action of $\mathrm{Aut}_{g,b,p}$ on $\{z_j^{\pm e_j}\}_{j=1}^b \cup \{[t_k]^{\pm 1}\}_{k=1}^p$. It is not difficult to show that this homomorphism is surjective.

Suppose that $(b_1, \ldots, b_m)$ and $(p_1, \ldots, p_n)$ are sequences of positive integers which sum to b and p, respectively. These determine an embedding

$$\prod_{j=1}^m \mathrm{Sym}_{b_j} \times \prod_{k=1}^n \mathrm{Sym}_{p_k} \to \mathrm{Sym}_b \times \mathrm{Sym}_p,$$

and we denote the image by $\mathrm{Sym}_{b_1 \perp \ldots \perp b_m} \times \mathrm{Sym}_{p_1 \perp \ldots \perp p_n}$. The preimage of the latter in $\mathrm{Aut}_{g,b,p}$ will be denoted

$$\mathrm{Aut}_{g,b_1 \perp \ldots \perp b_m, p_1 \perp \ldots \perp p_n}.$$

In the case where the b_j and p_k are all 1, we write $\mathrm{Aut}_{g,1^{\perp b},1^{\perp p}}$.

We leave it to the reader to define notation similar to the foregoing with Aut replaced by Out, Aut^+, and Out^+.

We remark that $\mathrm{Aut}_{g,b,p}$ acts on $\Omega_{g,b+p} \times \{[z_j]^{\pm 1}\}_{j=1}^{b+p}$. Here the kernel of the action is $\mathrm{Aut}^+_{g,1^{\perp b},1^{\perp p}}$, and the image of the action is isomorphic to $C_2 \times \mathrm{Sym}_{b \perp p}$. A similar statement holds for $\mathrm{Out}_{g,b,p}$. We call $\mathrm{Out}^+_{g,1^{\perp b},1^{\perp p}}$ the *pure algebraic (g,b,p)-mapping-class group*.

For $b \geqslant 1$, we define $\mathrm{Aut}_{g,b-1\perp \hat{1},p}$ to be the subgroup of $\mathrm{Aut}_{g,b,p}$ consisting of the elements of $\mathrm{Aut}_{g,b,p}$ which fix e_b, and, hence fix $z_b^{\pm 1}$.

For $p \geqslant 1$, we define $\mathrm{Aut}_{g,b,p-1\perp \hat{1}}$ to be the subgroup of $\mathrm{Aut}_{g,b,p}$ consisting of the elements of $\mathrm{Aut}_{g,b,p}$ which fix $t_p^{\pm 1}$. ☐

We shall give a simpler description of these subgroups in Remarks 8.2, and we shall see the following isomorphisms in Sections 7 and 10.

$$\mathrm{Aut}_{g,b,p\perp\hat{1}} \quad\simeq\quad \mathrm{Aut}_{g,b\perp\hat{1},p} \quad\twoheadrightarrow\quad \mathrm{Out}_{g,b\perp1,p}.$$

$$\mathrm{Aut}_{g,b,p\perp\hat{1}}/\langle \tilde{t}_{p+1}\rangle \quad\simeq\quad \mathrm{Aut}_{g,b\perp\hat{1},p}/\langle \sigma_{b+1}\tilde{z}_{b+1}^{-1}\rangle \quad\twoheadrightarrow\quad \mathrm{Out}_{g,b\perp1,p}/\langle \breve{\sigma}_{b+1}\rangle.$$

$$\|$$

$$\mathrm{Aut}_{g,b,p\perp\hat{1}}/\langle \tilde{t}_{p+1}\rangle \quad\twoheadrightarrow\quad \mathrm{Out}_{g,b,p\perp1}.$$

$$\|$$

$$\mathrm{Aut}_{g,b,p\perp\hat{1}}/\langle \tilde{t}_{p+1}\rangle \quad\twoheadrightarrow\quad \mathrm{Aut}_{g,b,p}.$$

5. Dehn twists and the DLH generators

In Definitions 3.10, we described elements α_i, β_i and γ_i, and called them Dehn-twist automorphisms. We now explain their connection with Dehn twists. We will use inverses of Dehn twists, since our actions are on the right.

5.1. Remarks. Let us return to Definitions 3.2. Let

$$X'_{g,b+p} = \{u_i \mid 1 \leqslant i \leqslant 2g\} \cup \{z_j \mid 1 \leqslant j \leqslant b+p\}.$$

In the free group on $X'_{g,b+p}$, let

$$w'_{g,b+p} = \prod_{i=1}^{g} u_{2i-1} \cdot \prod_{i=2g}^{1} \overline{u}_i \cdot \prod_{i=1}^{g} u_{2i} \cdot \prod_{j=1}^{b+p} z_j.$$

It is straightforward to show that $\Sigma_{g,b,p}$ is presented as $\langle X'_{g,b+p} \mid w'_{g,b+p}\rangle$; in detail, if we use the convention that $y_0 = 1$, then, for $1 \leqslant i \leqslant g$,

u_{2i-1}	corresponds to	$y_{i-1}\overline{x}_i\overline{y}_i x_i,$
u_{2i}	corresponds to	$\overline{x}_i,$
$\overline{u}_{2i}$	corresponds to	$x_i,$
$\displaystyle\prod_{i'=2i}^{1} \overline{u}_{i'} \cdot \prod_{i'=1}^{i} u_{2i'}$	corresponds to	$y_i.$

The u_i will be mentioned again in Example 15.6.

It is not difficult to express the maps of Definitions 3.10 in terms of the new generators. For $1 \leqslant i \leqslant g$,

$$\alpha_i : \begin{cases} u_{2i} \;\longmapsto\; \displaystyle\prod_{i'=2i-1}^{1} \overline{u}_{i'} \cdot \prod_{i'=1}^{i} u_{2i'}, \\[2ex] w \;\longmapsto\; w \quad \text{for all } w \in E_b \cup X'_{g,b,p} - \{u_{2i}\}. \end{cases}$$

For $1 \leqslant i \leqslant g-1$,

$$\beta_i : \begin{cases} u_{2i-1} & \mapsto & u_{2i-1}u_{2i}, \\ u_{2i+1} & \mapsto & \overline{u}_{2i}u_{2i+1}, \\ w & \mapsto & w \quad \text{for all } w \in E_b \cup X'_{g,b,p} - \{u_{2i-1}, u_{2i+1}\}. \end{cases}$$

$$\gamma_i : \begin{cases} u_{2i} & \mapsto & u_{2i}u_{2i+1}, \\ u_{2i+2} & \mapsto & \overline{u}_{2i+1}u_{2i+2}, \\ w & \mapsto & w \quad \text{for all } w \in E_b \cup X'_{g,b,p} - \{u_{2i}, u_{2i+2}\}. \end{cases}$$

Also

$$\beta_g : \begin{cases} u_{2g-1} & \mapsto & u_{2g-1}u_{2g}, \\ w & \mapsto & w \quad \text{for all } w \in E_b \cup X'_{g,b,p} - \{u_{2g-1}\}. \end{cases}$$

If $g \geqslant 1$ and $b+p \geqslant 1$, then

$$\gamma_0 : \begin{cases} u_1 & \mapsto & \overline{u}_1 \overline{z}_{b+p} u_1 z_{b+p} u_1 = u_1^{z_{b+p}u_1}, \\ u_2 & \mapsto & \overline{u}_1 \overline{z}_{b+p} u_2, \\ z_{b+p} & \mapsto & \overline{u}_1 \overline{z}_{b+p} u_1 = z_{b+p}^{u_1}, \\ e_{b+p} & \mapsto & \overline{u}_1 \overline{z}_{b+p} e_{b+p} \quad \text{if } p = 0, \\ w & \mapsto & w \quad \text{for all } w \in (X_{g,b,p} \cup E_b) - \{u_1, u_2, z_{b+p}, e_{b+p}\}. \end{cases}$$

$$\gamma_g : \begin{cases} u_{2g} & \mapsto & \displaystyle\coprod_{i=2g-1}^{1} \overline{u}_i \cdot \prod_{i=1}^{g} u_{2i} \cdot z_1, \\[2ex] z_1 & \mapsto & \displaystyle\overline{z}_1 \cdot \coprod_{i=g}^{1} \overline{u}_{2i} \cdot \prod_{i=1}^{2g} u_i \cdot z_1 \cdot \coprod_{i=2g}^{1} \overline{u}_i \cdot \prod_{i=1}^{g} u_{2i} \cdot z_1, \\[2ex] e_1 & \mapsto & \displaystyle\overline{z}_1 \cdot \coprod_{i=g}^{1} \overline{u}_{2i} \cdot \prod_{i=1}^{2g} u_i \cdot e_1 \quad \text{if } b \geqslant 1, \\[2ex] w & \mapsto & w \quad \text{for all } w \in (X_{g,b,p} \cup E_b) - \{u_{2g}, z_1, e_1\}. \end{cases}$$

Let us now discuss how these maps can be induced by Dehn twists.

We take a polygon with $4g + 3b + 3p$ edges, and label the edges clockwise with the letters from the word $w_{q,b+p}$, where, for $1 \leqslant j \leqslant b+p$, we express z_j as a product of three "letters", $e_j \cdot z_j^{e_j} \cdot \overline{e}_j$. Let $S_{g,b+p,0}$ be the surface obtained from the polygon by identifying each pair of edges with the same label, respecting the orientation. Thus we have a graph, with $1 + b + p$ vertices and $2g + 2b + 2p$ edges, embedded in $S_{g,b+p,0}$ as a one-skeleton. The boundary components are labelled with the $z_j^{e_j}$.

Consider any self-homeomorphism α of $S_{g,b+p,0}$ which permutes the vertices and the boundary components. Then α induces an automorphism of the "groupoid of the graph", and, hence, an automorphism of $\Sigma_{g,b+p} * E_{b+p}$. Moreover, the automorphism of $\Sigma_{g,b+p} * E_{b+p}$ lies in $\mathrm{Aut}_{g,b+p,0}$. If it lies in $\mathrm{Aut}_{g,b\perp p,0}$, then, on restricting to $\Sigma_{g,b+p} * E_b$, we get an element of $\mathrm{Aut}_{g,b,p}$; this amounts to converting the last p boundary components to punctures, which we will return to in Section 9.

It can be shown that, for $1 \leqslant i \leqslant g-1$, β_i is induced by the self-homeomorphism described as follows. We cut $S_{g,b+p,0}$ just to the left of the curve labelled u_{2i} and then clockwise around the base point until arriving where we started. A small tube crossed by the edges labelled u_{2i-1} and u_{2i+1} is thus cut into two pieces. We next twist the left-hand piece $360°$ in the direction determined by u_{2i}, and then we reattach the two pieces of the tube. We say that β_i is *the Dehn-twist automorphism for the word* u_{2i} *with respect to the presentation* $\langle X'_{g,b+p} \mid w'_{g,b+p} \rangle$. More generally, α_i, β_i, γ_i, σ_j and τ_j^2 are Dehn-twist automorphisms for the words given as follows.

	corresponding word	restraint
α_i	$\coprod\limits_{i'=i-1}^{1} \overline{u}_{2i'} \cdot \prod\limits_{i'=1}^{2i-1} u_{i'}$	$i = 1,..,g$
β_i	u_{2i}	$i = 1,..,g$
γ_i	u_{2i+1}	$i = 1,..,g-1$
γ_0	$\coprod\limits_{j=b+p-1}^{1} \overline{z}_j \cdot \coprod\limits_{i=g}^{1} \overline{u}_{2i} \cdot \prod\limits_{i=1}^{2g} u_i \cdot \coprod\limits_{i=g}^{2} \overline{u}_{2i-1}(= z_{b+p}u_1)$	$g \geqslant 1, b+p \geqslant 1$
γ_g	$\coprod\limits_{i=g}^{1} \overline{u}_{2i-1} \cdot \coprod\limits_{j=b+p}^{2} \overline{z}_j \; (= \coprod\limits_{i=2g}^{1} \overline{u}_i \cdot \prod\limits_{i=1}^{g} u_{2i} \cdot z_1)$	$g \geqslant 1, b+p \geqslant 1$
σ_j	$\coprod\limits_{j'=j-1}^{1} \overline{z}_{j'} \cdot \coprod\limits_{i=g}^{1} \overline{u}_{2i} \cdot \prod\limits_{i=1}^{2g} u_i \cdot \coprod\limits_{i=g}^{1} \overline{u}_{2i-1} \cdot \coprod\limits_{j'=b+p}^{j+1} \overline{z}_{j'}(= z_j)$	$j = 1,..,b$
τ_j^2	$\coprod\limits_{j'=j-1}^{1} \overline{z}_{j'} \cdot \coprod\limits_{i=g}^{1} \overline{u}_{2i} \cdot \prod\limits_{i=1}^{2g} u_i \cdot \coprod\limits_{i=g}^{1} \overline{u}_{2i-1} \cdot \coprod\limits_{j'=b+p}^{j+2} \overline{z}_{j'}(= z_j z_{j+1})$	$j = 1,..,b-1$

The given corresponding word determines a portion of the counter-clockwise boundary of the polygon, and hence an oriented curve in the surface. We find that, on travelling slightly to the left of the curve and then clockwise around the basepoint, we cross only edges labelled with letters which do not occur in the word. $\square$

5.2. Definition. In Theorem 9.6, we will see that the natural map

$$\mathcal{MC}^+_{g,b,p} \to \mathrm{Out}^+_{g,b,p}$$

is an isomorphism. By [13, Section 4.2], together with Remarks 5.1, it follows that, if $b + p \leqslant 1$, then $\mathrm{Out}^+_{g,b,p}$ is generated by

$$\{\breve{\alpha}_i\}_{i=1}^{\min\{2,g\}} \cup \{\breve{\beta}_i\}_{i=1}^{g} \cup \{\breve{\gamma}_i\}_{i=1}^{g-1}; \qquad (5.2.1)$$

we call (5.2.1) the set of *Dehn-Lickorish-Humphries generators*, or DLH *generators*, of $\mathrm{Out}^+_{g,b,p}$. □

5.3. Remark. For $g \geqslant 3$ and $1 \leqslant i \leqslant g-2$, Humphries' Identity [11] states that the following holds in $\mathrm{Out}_{g,b,p}$: if

$$h_i := \breve{\beta}_i \breve{\gamma}_i \breve{\beta}_{i+1} \breve{\gamma}_{i+1} \breve{\beta}_{i+2} \breve{\alpha}_{i+1} \breve{\beta}_{i+1} \breve{\gamma}_i \breve{\gamma}_{i+1} \breve{\beta}_{i+1} \breve{\alpha}_{i+1} \breve{\beta}_i \breve{\gamma}_i \breve{\beta}_{i+1} \breve{\gamma}_{i+1} \breve{\beta}_{i+2},$$

then $\breve{\alpha}_i h_i = h_i \breve{\alpha}_{i+2}$; this allows one to eliminate generators. The interested reader can verify Humphries' Identity algebraically by checking that, in $\mathrm{Aut}^+_{3,0,\hat{1}}$,

$$\delta_1 := \beta_1\gamma_1\beta_2\gamma_2\beta_3 : \begin{cases} x_1 &\mapsto y_3x_3x_2\overline{y}_1x_1, \\ y_1 &\mapsto y_3x_3x_2\overline{y}_1x_1y_3x_3x_2y_1\overline{x}_2\overline{x}_3\overline{y}_3, \\ x_2 &\mapsto y_3x_3\overline{y}_2x_2y_1\overline{x}_2\overline{x}_3\overline{y}_3, \\ y_2 &\mapsto y_3x_3\overline{y}_2x_2y_3x_3y_2\overline{x}_3\overline{y}_3, \\ x_3 &\mapsto x_3y_2\overline{x}_3\overline{y}_3, \\ y_3 &\mapsto x_3y_3, \\ t_1 &\mapsto t_1, \end{cases}$$

$$\delta_2 := \beta_2\gamma_1\gamma_2\beta_2 : \begin{cases} x_1 &\mapsto y_2x_2\overline{y}_1x_1, \\ y_1 &\mapsto y_2x_2y_1\overline{x}_2\overline{y}_2, \\ x_2 &\mapsto \overline{x}_3y_3x_3\overline{y}_2y_1\overline{x}_2\overline{y}_2, \\ y_2 &\mapsto \overline{x}_3y_3x_3\overline{y}_2y_1\overline{x}_2\overline{y}_2\overline{x}_3y_3x_3x_2y_2\overline{x}_3\overline{y}_3x_3, \\ x_3 &\mapsto x_3x_2y_2\overline{x}_3\overline{y}_3x_3, \\ y_3 &\mapsto y_3, \\ t_1 &\mapsto t_1, \end{cases}$$

$$\delta_3 := \delta_1\alpha_2\delta_2\alpha_2\delta_1 : \begin{cases} x_1 &\mapsto \overline{y}_3\overline{x}_1\overline{x}_2\overline{x}_3\overline{y}_3, \\ y_1 &\mapsto \overline{y}_3\overline{x}_1\overline{x}_2\overline{x}_3\overline{y}_3x_3x_2x_1y_3, \\ x_2 &\mapsto \overline{y}_3\overline{x}_1x_2x_1y_3, \\ y_2 &\mapsto \overline{y}_3\overline{x}_1y_2\overline{x}_3\overline{y}_3x_3x_1y_3, \\ x_3 &\mapsto \overline{y}_3x_1y_3, \\ y_3 &\mapsto \overline{y}_3y_1\overline{x}_2\overline{y}_2x_2y_2\overline{x}_3\overline{y}_3x_3y_3, \\ t_1 &\mapsto t_1, \end{cases}$$

$$\alpha_1\delta_3 = \delta_3\alpha_3 : \begin{cases} x_1 &\mapsto \overline{y}_3\overline{x}_1\overline{x}_2\overline{x}_3, \\ y_1 &\mapsto \overline{y}_3\overline{x}_1\overline{x}_2\overline{x}_3\overline{y}_3x_3x_2x_1y_3, \\ x_2 &\mapsto \overline{y}_3\overline{x}_1x_2x_1y_3, \\ y_2 &\mapsto \overline{y}_3\overline{x}_1y_2\overline{x}_3\overline{y}_3x_3x_1y_3, \\ x_3 &\mapsto \overline{y}_3x_1y_3, \\ y_3 &\mapsto \overline{y}_3y_1\overline{x}_2\overline{y}_2x_2y_2\overline{x}_3\overline{y}_3x_3y_3, \\ t_1 &\mapsto t_1. \end{cases}$$

□

6. Artin diagrams and Matsumoto's relators

Let us recall some background on Artin groups and diagrams.

6.1. Definitions. A *diagram* (V, r) consists of a set V together with a function r which assigns, to each two-element subset $\{x, y\}$ of V, a value

$$r_{\{x,y\}} \in \mathbb{N} \cup \{\infty\}.$$

Such a diagram can be depicted as the complete graph with vertex set V in which the edge joining two distinct vertices x and y has label $r_{\{x,y\}}$. One then applies the convention that if $r_{\{x,y\}}$ is "small", then the edge and label are replaced with $r_{\{x,y\}}$ unlabelled edges joining x and y. Thus, if $r_{\{x,y\}} = 0$, there is no edge joining x and y.

Let (V, r) be a diagram.

The *Artin group of* (V, r), denoted $\mathrm{Artin}[(V, r)]$, is the group presented with generating set V, and relations saying that, for each two-element subset $\{x, y\}$ of V,

$$(xy)^m = (yx)^m \quad \text{where } m = \frac{r_{\{x,y\}} + 2}{2}, \quad \text{if } r_{\{x,y\}} < \infty \text{ and } r_{\{x,y\}} \text{ is even,}$$

and

$$(xy)^m x = (yx)^m y \quad \text{where } m = \frac{r_{\{x,y\}} + 1}{2}, \quad \text{if } r_{\{x,y\}} < \infty \text{ and } r_{\{x,y\}} \text{ is odd.}$$

Thus, if $r_{\{x,y\}} = \infty$, then no relation is imposed; and, if $r_{\{x,y\}} < \infty$, then exactly one relation, that does not depend on the order in $\{x, y\}$, is imposed, and each side of the relation has length $r_{\{x,y\}} + 2$.

Brieskorn-Saito [6] showed that the center of $\mathrm{Artin}[(V, r)]$ is a free abelian group, and described a distinguished basis, which we denote Z-$\mathrm{Artin}[(V, r)]$, and which we now recall. Where Z-$\mathrm{Artin}[(V, r)]$ consists of a single element, we do not distinguish between the element and the set.

If n is a positive integer, and X_n is one of the diagrams in Table 6.1.1, with vertex set $\{a_1, a_2, \ldots, a_n\}$, then Z-$\mathrm{Artin}[X_n] = (\prod_{i=1}^{n} a_i)^{c(X_n)}$ where $c(X_n)$ is as given by Table 6.1.1 (see next page).

For any $\phi \in \mathrm{Sym}_n$, replacing $\prod_{i=1}^{n} a_i$ with $\prod_{i=1}^{n} a_{i\phi}$ gives another expression for the distinguished central element.

We shall understand that Z-$\mathrm{Artin}^m[X_n]$ means $(\mathrm{Z\text{-}Artin}[X_n])^m$.

If (V, r) is any diagram, then we view edges labelled zero as having been deleted, and we consider the connected components. Then Z-$\mathrm{Artin}[(V, r)]$ is the set of those Z-$\mathrm{Artin}[X]$ such that X is a connected component of (V, r) that has one of the forms in Table 6.1.1. $\qquad \square$

$A_n \ (n \geqslant 1)$ $\qquad$ $a_1 - a_2 - a_3 - a_4 - \cdots - a_{n-1} - a_n$

$B_n \ (n \geqslant 2)$ $\qquad$ $a_1 \overset{2}{-} a_2 - a_3 - a_4 - \cdots - a_{n-1} - a_n$

$D_n \ (n \geqslant 3)$ $\qquad$ $\begin{array}{c} a_1 \\ | \\ a_2 - a_3 - a_4 - a_5 - \cdots - a_{n-1} - a_n \end{array}$

E_6 $\qquad$ $\begin{array}{c} a_1 \\ | \\ a_2 - a_3 - a_4 - a_5 - a_6 \end{array}$

E_7 $\qquad$ $\begin{array}{c} a_1 \\ | \\ a_2 - a_3 - a_4 - a_5 - a_6 - a_7 \end{array}$

E_8 $\qquad$ $\begin{array}{c} a_1 \\ | \\ a_2 - a_3 - a_4 - a_5 - a_6 - a_7 - a_8 \end{array}$

F_4 $\qquad$ $a_1 - a_2 \overset{2}{-} a_3 - a_4$

H_3 $\qquad$ $a_1 - a_2 \overset{3}{-} a_3$

H_4 $\qquad$ $a_1 - a_2 \overset{3}{-} a_3 - a_4$

$I_2(r) \ (r \geqslant 1)$ $\qquad$ $a_1 \overset{r}{-} a_2$

X_n	A_1	$A_n, n \geqslant 2$	$B_n, n \geqslant 2$	$D_{2g-1}, g \geqslant 2$	$D_{2g}, g \geqslant 2$
$c(X_n)$	1	$n+1$	n	$4g-4$	$2g-1$

X_n	E_6	E_7	E_8	F_4	H_3	H_4	$I_2(2s-1), s \geqslant 1$	$I_2(2s), s \geqslant 1$
$c(X_n)$	12	9	15	6	5	15	$2s+1$	$s+1$

TABLE 6.1.1. Diagrams X_n and values $c(X_n)$.

6.2. Definitions. Let G be a group.

An *Artin diagram in* G is a diagram (V, r) in which V is a family of elements of G, and the relations of $\mathrm{Artin}[(V, r)]$ hold in G.

Let (V, r) be an Artin diagram in G.

Thus there is an induced map $\mathrm{Artin}[(V, r)] \to G$.

If this map is an inclusion, then we view $\mathrm{Artin}[(V, r)]$ as a subgroup of G; if the map is an isomorphism, we write $G = \mathrm{Artin}[(V, r)]$, and say that G is an *Artin group*.

By Z-$\mathrm{Artin}[(V, r)]$ *in* G we mean the image of Z-$\mathrm{Artin}[(V, r)]$ in G, viewed as a family of elements of G. If there is only one element in the family, we treat it as an element of G rather than as a one-element family. $\qquad\square$

6.3. Example. It is not difficult to show that

$$
\begin{array}{ccccccccc}
\alpha_1 & & \alpha_2 & & \alpha_3 & & & \alpha_{g-1} & & \alpha_g \\
| & & | & & | & & & | & & | \\
\gamma_0\!-\!\beta_1\!-\!\gamma_1\!-\!\beta_2\!-\!\gamma_2\!-\!\beta_3\!-\!\gamma_3\!-\cdots-\!\gamma_{g-2}\!-\!\beta_{g-1}\!-\!\gamma_{g-1}\!-\!\beta_g\!-\!\gamma_g \\
|\infty & & & & & & & & & |\infty \\
\mu_{p-1}\!-\!\mu_{p-2}\!-\cdots-\!\mu_2\!-\!\mu_1 & & & & \tau_{b-1}\!-\!\tau_{b-2}\!-\cdots-\!\tau_2\!-\!\tau_1
\end{array}
$$

$$
\begin{array}{ccccccc}
& \not{\infty} \; |\infty & \not{\infty}\;|\infty & \not{\infty} & \not{\infty} & |\infty\;\not{\infty}\;|\infty \\
\sigma_b & \sigma_{b-1} & \sigma_{b-2} & \cdots & \sigma_2 & \sigma_1
\end{array}
$$

is an Artin diagram in $\mathrm{Aut}_{g,b,p}$. Recall that an edge labelled ∞ carries no information. $\qquad\qquad\Box$

6.4. Examples. We can now describe presentations of $\mathrm{Out}^+_{g,1,0}$ obtained through the work of many authors; see [20].

In Proposition 7.1(iii), we shall see that $\mathrm{Out}^+_{g,1,0}$ is the group of automorphisms of the free group $\Sigma_{g,1}$ which fix the surface relator z_1.

(i) $g = 0$. $\mathrm{Out}^+_{0,1,0} = 1$. See Example 15.2.

(ii) $g = 1$. $\mathrm{Out}^+_{1,1,0} = \mathrm{Artin}\begin{bmatrix} \breve{\alpha}_1 \\ | \\ \breve{\beta}_1 \end{bmatrix}$. See Remarks 15.8(vi).

(iii) $g = 2$. A presentation for $\mathrm{Out}^+_{2,1,0}$ is obtained from

$$
\mathrm{Artin}\begin{bmatrix} \breve{\alpha}_1 & & \breve{\alpha}_2 \\ | & & | \\ \breve{\beta}_1\!-\!\breve{\gamma}_1\!-\!\breve{\beta}_2 \end{bmatrix}
$$

by imposing what we will call *the A-A-relation*, or, with more precision, the $ZA_5\text{-}Z^2A_4$-*relation*:

$$
\mathrm{Z\text{-}Artin}\begin{bmatrix} \breve{\alpha}_1 & & \breve{\alpha}_2 \\ | & & | \\ \breve{\beta}_1\!-\!\breve{\gamma}_1\!-\!\breve{\beta}_2 \end{bmatrix} \; (= (\textstyle\prod \text{generators})^6)
$$

$$(6.4.1)$$

$$
= \;\; \mathrm{Z\text{-}Artin}^2\begin{bmatrix} \breve{\beta}_2 \\ | \\ \breve{\beta}_1\!-\!\breve{\gamma}_1\!-\!\breve{\alpha}_2 \end{bmatrix} \; (= (\textstyle\prod \text{generators})^{10}).
$$

See [20] and Remarks 18.1.

(iv) For $g \geqslant 3$, a presentation for $\mathrm{Out}^+_{g,1,0}$ is obtained from

$$
\mathrm{Artin}\begin{bmatrix} \breve{\alpha}_1 & & \breve{\alpha}_2 \\ | & & | \\ \breve{\beta}_1\!-\!\breve{\gamma}_1\!-\!\breve{\beta}_2\!-\!\breve{\gamma}_2\!-\!\breve{\beta}_3\!-\cdots-\!\breve{\gamma}_{g-1}\!-\!\breve{\beta}_g \end{bmatrix}
$$

by imposing both the A-A-relation (6.4.1) and *the E-E-relation* (or ZE_7-ZE_6-re-lation):

$$\text{Z-Artin} \begin{bmatrix} \begin{matrix} \breve{\alpha}_1 & & \breve{\alpha}_2 \\ | & & | \\ \breve{\beta}_1 - \breve{\gamma}_1 - & \breve{\beta}_2 - \breve{\gamma}_2 - \breve{\beta}_3 \end{matrix} \end{bmatrix} \left(= \left(\textstyle\prod \text{generators} \right)^9 \right)$$

$$(6.4.2)$$

$$= \quad \text{Z-Artin} \begin{bmatrix} \begin{matrix} & \breve{\alpha}_2 & \\ & | & \\ \breve{\beta}_1 - \breve{\gamma}_1 - & \breve{\beta}_2 - \breve{\gamma}_2 - \breve{\beta}_3 \end{matrix} \end{bmatrix} \left(= \left(\textstyle\prod \text{generators} \right)^{12} \right).$$

See [20] and Remarks 18.2. □

6.5. Examples. In Corollary 10.2, we shall see that $\text{Out}^+_{g,0,1}$ is the group of positive automorphisms of the surface group $\Sigma_{g,0}$. A presentation of $\text{Out}^+_{g,0,1}$ is obtained from the presentation of $\text{Out}^+_{g,1,0}$ given in Examples 6.4 by imposing *the A-relation* (or $Z^2 A_{2g}$-relation):

$$\text{Z-Artin}^2 \begin{bmatrix} \begin{matrix} \breve{\alpha}_1 \\ | \\ \breve{\beta}_1 - \breve{\gamma}_1 - \breve{\beta}_2 - \cdots - \breve{\gamma}_{g-1} - \breve{\beta}_g \end{matrix} \end{bmatrix} = 1 \qquad (6.5.1)$$
$$\left(= \left(\textstyle\prod \text{generators} \right)^{4g+2} \right).$$

See [27] and Section 16.

For $g = 0$, the $Z^2 A_{2g}$-relation is vacuous, and $\text{Out}^+_{0,0,1} = 1 = \text{Out}^+_{0,1,0}$. See Example 15.1. □

6.6. Examples. We now describe a presentation of $\text{Out}^+_{g,0,0}$, the group of positive outer automorphisms of the surface group $\Sigma_{g,0}$.

(i) For $g \leqslant 1$, $\text{Out}^+_{g,0,0} = \text{Out}^+_{g,0,1}$. See Examples 15.1 and 15.2.

(ii) For $g \geqslant 2$, a presentation of $\text{Out}^+_{g,0,0}$ is obtained from the presentation of $\text{Out}_{g,1,0}$ given in Examples 6.4 by imposing *the A-D-relation* (or $Z^{2g-2} A_1$-ZD_{2g-1}-relation):

$$\text{Z-Artin}^{2g-2} \begin{bmatrix} \breve{\alpha}_1 \end{bmatrix} = \text{Z-Artin} \begin{bmatrix} \begin{matrix} & \breve{\alpha}_2 & \\ & | & \\ \breve{\gamma}_1 - \breve{\beta}_2 - \breve{\gamma}_2 - \breve{\beta}_3 - \cdots - \breve{\gamma}_{g-1} - \breve{\beta}_g \end{matrix} \end{bmatrix} \qquad (6.6.1)$$
$$\left(= \left(\textstyle\prod \text{generators} \right)^{4g-4} \right).$$

See [20] and Section 16. □

7. Isomorphisms

Recall that, for $b \geqslant 1$, Definitions 3.10 give the boundary Dehn-twist automorphism $\sigma_b \in \text{Aut}_{g,b,p}$.

7.1. Proposition. *Suppose that $b \geqslant 1$.*

(i) *$\Sigma_{g,b,p}$ is free of rank $2g + b + p - 1$.*

(ii) *$\text{Aut}_{g,b-1\perp 1,p} = \text{Aut}_{g,b-1\perp\hat{1},p} \ltimes \Sigma_{g,b,p}$.*

(iii) *The natural map $\text{Aut}_{g,b-1\perp\hat{1},p} \to \text{Out}_{g,b-1\perp 1,p}$ is an isomorphism, and it carries $\sigma_b \tilde{z}_b^{-1}$ to $\breve{\sigma}_b$.*

(iv) *The natural map $\text{Aut}_{g,b-1\perp\hat{1},p} \to \text{Aut}_{g,b-1,p\perp\hat{1}}$ is an isomorphism, and it carries $\sigma_b \tilde{z}_b^{-1}$ to $\tilde{t}_{p+1}^{-1}$.*

(v) *The natural map $\text{Aut}_{g,b-1,p\perp\hat{1}} \to \text{Out}_{g,b-1\perp 1,p}$ is an isomorphism, and it carries $\tilde{t}_{p+1}^{-1}$ to $\breve{\sigma}_b$.*

Proof. (i) is clear.

(ii) Consider any $\alpha \in \text{Aut}_{g,b-1\perp 1,p}$. Then α fixes $z_b^{e_b}$. By Remark 3.9, there is a unique $s_\alpha \in \Sigma_{g,b,p}$ such that $e_b^\alpha = s_\alpha e_b$. Now $\alpha \tilde{s}_\alpha$ fixes e_b. Moreover, $\tilde{\Sigma}_{g,b,p} = \Sigma_{g,b,p}$, since $b \geqslant 1$. Now (ii) is clear.

(iii) It follows that $\text{Aut}_{g,b-1\perp\hat{1},p} \to \text{Out}_{g,b-1\perp 1,p}$, $\alpha \mapsto \breve{\alpha}$, is an isomorphism. It is clear that $\sigma_b \tilde{z}_b^{-1}$ is mapped to $\breve{\sigma}_b$.

(iv) We note that

$$\text{Aut}_{g,b-1\perp\hat{1},p} = \text{Aut}(\Sigma_{g,b,p} * E_b, \Sigma_{g,b,p}, \{z_j^{\pm e_j}\}_{j=1}^b, e_b, \{[t_k]^{\pm 1}\}_{k=1}^p, \Omega_{g,b+p})$$
$$= \text{Aut}(\Sigma_{g,b,p} * E_b, \Sigma_{g,b,p}, \{z_j^{\pm e_j}\}_{j=1}^{b-1}, z_b^{\pm 1}, e_b, \{[t_k]^{\pm 1}\}_{k=1}^p, \Omega_{g,b+p}).$$

The latter can be identified with

$$\text{Aut}(\Sigma_{g,b,p} * E_{b-1}, \Sigma_{g,b,p}, \{z_j^{\pm e_j}\}_{j=1}^{b-1}, z_b^{\pm 1}, \{[t_k]^{\pm 1}\}_{k=1}^p, \Omega_{g,b+p}),$$

since there is a bijection between the group of all automorphisms of $\Sigma_{g,b,p} * E_{b-1}$ and the group of automorphisms of

$$\Sigma_{g,b,p} * E_b = (\Sigma_{g,b,p} * E_{b-1}) * \langle e_b \rangle$$

which fix e_b and map $\Sigma_{g,b,p} * E_{b-1}$ to itself. Under this bijection $\sigma_b \tilde{z}_b^{-1}$ corresponds to $\tilde{z}_b^{-1}$. On applying the relabelling $z_b = t_{p+1}$, we get $\text{Aut}_{g,b-1,p\perp\hat{1}}$, and this proves (iv).

(v) is a consequence of (iii) and (iv). $\qquad \square$

By applying the braid automorphisms, we can obtain similar results with any other boundary component fixed, although more notation would be required to state them.

Proposition 7.1(iii) shows that, if $b \geqslant 1$, then the index b subgroup $\text{Out}_{g,b-1\perp 1,p}$ of $\text{Out}_{g,b,p}$ is isomorphic to an automorphism group. This corresponds to contracting the simple curve e_b and using the base point of the bth boundary component as the base point of the surface.

If $p \geqslant 1$, then we can use the pth puncture as the base point of the surface and get the following, using only elementary group theory.

7.2. Proposition. *If $p \geqslant 1$, then* $\mathrm{Out}_{g,b,p-1\perp 1} \simeq \mathrm{Aut}_{g,b,p-1\perp\hat{\imath}} / \langle \tilde{t}_p \rangle$.

Proof. $\mathrm{Aut}_{g,b,p-1\perp\hat{\imath}} / \langle \tilde{t}_p \rangle = \mathrm{Aut}_{g,b,p-1\perp\hat{\imath}} / (\mathrm{Aut}_{g,b,p-1\perp\hat{\imath}} \cap \tilde{\Sigma}_{g,b,p})$

$$\simeq (\mathrm{Aut}_{g,b,p-1\perp\hat{\imath}} \cdot \tilde{\Sigma}_{g,b,p}) / \tilde{\Sigma}_{g,b,p}$$

$$= \mathrm{Aut}_{g,b,p-1\perp 1} / \tilde{\Sigma}_{g,b,p}$$

$$= \mathrm{Out}_{g,b,p-1\perp 1} . \qquad \square$$

8. Endomorphisms forced to be automorphisms

The following will be useful.

8.1. Theorem. *Suppose that $b + p \geqslant 1$, and let G be an arbitrary group.*

Let α be an endomorphism of the free product $\Sigma_{g,b+p} * G$ such that the induced map, from the set of sets of $\Sigma_{g,b+p} * G$-conjugacy classes to itself, fixes the set $\{[z_j], [\bar{z}_j]\}_{j=1}^{b+p}$ of $\Sigma_{g,b+p} * G$-conjugacy classes.

Then α acts on $\Sigma_{g,b+p}$ as the composition of an automorphism of $\Sigma_{g,b+p}$ followed by conjugation by an element c of $\Sigma_{g,b+p} * G$.

Moreover, either $\Sigma_{g,b+p}$ is trivial, or, for each $j \in \{1, \ldots, b+p\}$, the coset $\Sigma_{g,b+p}c$ is uniquely determined by $z_j^\alpha \in \Sigma_{g,b+p}^c$.

Proof. It suffices to prove the first claim, since the second claim then follows as in Remark 3.9.

Since the relation $w_{g,b,p} = 1$ is respected, and α acts on the abelianization of $\Sigma_{g,b+p} * G$, and α fixes $\{[z_j], [\bar{z}_j]\}_{j=1}^{b+p}$, we see that α fixes or interchanges $\{[z_j]\}_{j=1}^{b+p}$ and $\{[\bar{z}_j]\}_{j=1}^{b+p}$.

We recall ζ and the puncture braid automorphisms of Definitions 3.10. It is then clear that, by precomposing α with a suitable sequence of automorphisms of $\Sigma_{g,b+p} * G$ which carry $\Sigma_{g,b+p}$ to itself, we may assume that α fixes $\{[z_j]\}_{j=1}^{b+p}$ and that α fixes each element of $\{[z_j]\}_{j=1}^{b+p}$.

By postcomposing α with conjugation by a suitable element of $\Sigma_{g,b+p} * G$, we may further assume that α fixes z_{b+p}.

Thus we can write

$$x_i^\alpha = X_i, \qquad y_i^\alpha = Y_i, \qquad z_j^\alpha = z_j^{C_j}, \qquad C_{b+p} = 1,$$

$$\bar{z}_{b+p} = \prod_{i=1}^{g} [x_i, y_i] \cdot \prod_{j=1}^{b+p-1} z_j, \qquad \bar{z}_{b+p}^\alpha = \prod_{i=1}^{g} [X_i, Y_i] \cdot \prod_{j=1}^{b+p-1} z_j^{C_j}.$$

Let H denote $\Sigma_{g,b+p}$ viewed as a free group with distinguished basis

$$\{x_i, y_i \mid 1 \leqslant i \leqslant g\} \cup \{z_j \mid 1 \leqslant j \leqslant b+p-1\}, \qquad (8.1.1)$$

and let $F = H * G$. Now
$$H^\alpha = \langle X_i, Y_i, z_j^{C_j} \mid 1 \leqslant i \leqslant g,\ 1 \leqslant j \leqslant b+p-1 \rangle,$$

and we shall show that H^α contains (8.1.1), and hence contains H.

For any (right) derivation $\partial \colon F \to \mathbb{Z}[F]$, we have

$$\overline{z}_{b+p}^{\partial} = \sum_{i=1}^{g} (x_i^{\partial} \cdot y_i \cdot (1 - \overline{y}_i^{x_i y_i}) \cdot \prod_{i'=i+1}^{g} [x_{i'}, y_{i'}] \cdot \prod_{j=1}^{b+p-1} z_j)$$
$$+ \sum_{i=1}^{g} (y_i^{\partial} \cdot (1 - x_i^{y_i}) \cdot \prod_{i'=i+1}^{g} [x_{i'}, y_{i'}] \cdot \prod_{j=1}^{b+p-1} z_j) \qquad (8.1.2)$$
$$+ \sum_{j=1}^{b+p-1} (z_j^{\partial} \cdot \prod_{j'=j+1}^{b+p-1} z_{j'}),$$

and, since α fixes z_{b+p},

$$\overline{z}_{b+p}^{\partial} = \overline{z}_{b+p}^{\alpha\partial} = \sum_{i=1}^{g} (X_i^{\partial} \cdot Y_i \cdot (1 - \overline{Y}_i^{X_i Y_i}) \cdot \prod_{i'=i+1}^{g} [X_{i'}, Y_{i'}] \cdot \prod_{j=1}^{b+p-1} z_j^{C_j})$$
$$+ \sum_{i=1}^{g} (Y_i^{\partial} \cdot (1 - X_i^{Y_i}) \cdot \prod_{i'=i+1}^{g} [X_{i'}, Y_{i'}] \cdot \prod_{j=1}^{b+p-1} z_j^{C_j})$$
$$+ \sum_{j=1}^{b+p-1} (C_j^{\partial} \cdot (1 - z_j^{C_j}) \cdot \prod_{j'=j+1}^{b+p-1} z_{j'}^{C_{j'}}) \qquad (8.1.3)$$
$$+ \sum_{j=1}^{b+p-1} (z_j^{\partial} \cdot C_j \cdot \prod_{j'=j+1}^{b+p-1} z_{j'}^{C_{j'}}).$$

Applying the map $\mathbb{Z}[F] \to \mathbb{Z}[F/H^\alpha]$, $r \mapsto r \cdot H^\alpha$, to (8.1.3) gives

$$\overline{z}_{b+p}^{\partial} \cdot H^\alpha = 0 + 0 + 0 + \sum_{j=1}^{b+p-1} z_j^{\partial} \cdot C_j \cdot H^\alpha \text{ in } \mathbb{Z}[F/H^\alpha]. \qquad (8.1.4)$$

For each element x of the distinguished free basis of H, we have a specified free product decomposition $F = \langle x \rangle * Q_x$ for a well-defined group Q_x, and hence we have the Fox derivative $\frac{\partial}{x\partial} \colon F \to \mathbb{Z}[F]$, uniquely determined by the fact that it sends x to 1 and vanishes on Q_x. (This can be described in terms of a decomposition of the augmentation ideal of $\mathbb{Z}[F]$ as a direct sum with one of the summands being $(x-1)\mathbb{Z}[F] \simeq \mathbb{Z}[F]$. It can also be described in terms of paths in the Bass-Serre tree for the expression of F as an HNN extension with vertex group Q_x and trivial edge group.)

For $1 \leqslant j \leqslant b+p-1$, taking $\partial = \frac{\partial}{z_j\partial}$ in (8.1.2) and (8.1.4), we get

$$\prod_{j'=j+1}^{b+p-1} z_{j'} \cdot H^\alpha = C_j \cdot H^\alpha,$$

so C_j lies in $\prod_{j'=j+1}^{b+p-1} z_{j'} \cdot H^\alpha$. Since $z_j^{C_j} \in H^\alpha$, we can inductively show that C_j and z_j lie in H^α for $b+p-1 \geqslant j \geqslant 1$.

For $1 \leqslant i \leqslant g$, taking $\partial = \frac{\partial}{x_i \partial}$, resp. $\partial = \frac{\partial}{y_i \partial}$, in (8.1.2) and (8.1.4) we get

$$y_i \cdot (1 - \overline{y}_i^{x_i y_i}) \cdot \prod_{i'=i+1}^{g} [x_{i'}, y_{i'}] \cdot \prod_{j=1}^{b+p-1} z_j \cdot H^\alpha = 0,$$

resp.

$$(1 - x_i^{y_i}) \cdot \prod_{i'=i+1}^{g} [x_{i'}, y_{i'}] \cdot \prod_{j=1}^{b+p-1} z_j \cdot H^\alpha = 0.$$

Thus $\overline{y}_i^{x_i y_i}$ and $x_i^{y_i}$ $(= x_i^{x_i y_i})$, lie in $\prod_{i'=i+1}^{g}[x_{i'}, y_{i'}] \cdot H^\alpha$. By induction, $x_i y_i$, x_i and y_i lie in H^α for $g \geqslant i \geqslant 1$.

Thus $H^\alpha \geqslant H$. But H is a free factor of F, and hence of $H^\alpha \geqslant H$. Any free factor complementary to H in H^α can be collapsed to the trivial group, giving a surjective map $H^\alpha \xrightarrow{\text{coll}} H$. The composition $H \xrightarrow{\alpha} H^\alpha \xrightarrow{\text{coll}} H$ is a surjective endomorphism of the finitely generated free group H. By a theorem of Nielsen, this endomorphism is bijective. Hence, both (surjective) factors are injective. Thus α is injective on H, and $H^\alpha \xrightarrow{\text{coll}} H$ is injective. Hence the complementary free factor is trivial, and $H^\alpha = H$. Thus α acts as an automorphism on H.   □

We now obtain simplifications in the descriptions of both $\mathrm{Aut}_{g,b,p-1\perp\hat{\imath}}$ and $\mathrm{Aut}_{g,b-1\perp\hat{\imath},p}$.

8.2. Remarks. Recall from Notation 4.3 that, for $p \geqslant 1$,

$$\mathrm{Aut}_{g,b,p-1\perp\hat{\imath}} = \mathrm{Aut}(\Sigma_{g,b,p} * E_b, \Sigma_{g,b,p}, \{z_j^{\pm e_j}\}_{j=1}^{b}, \{[t_k]^{\pm 1}\}_{k=1}^{p-1}, t_p^{\pm 1}, \Omega_{g,b+p}).$$

Notice that the restraint that $\Sigma_{g,b,p}$ be mapped to itself is redundant where $t_p^{\pm 1}$ is fixed, by Theorem 8.1. Thus

$$\mathrm{Aut}_{g,b,p-1\perp\hat{\imath}} = \mathrm{Aut}(\Sigma_{g,b,p} * E_b, \{z_j^{\pm e_j}\}_{j=1}^{b}, \{[t_k]^{\pm 1}\}_{k=1}^{p-1}, t_p^{\pm 1}, \Omega_{g,b+p}).$$

Similarly, for $b \geqslant 1$, an analogous remark holds for $\mathrm{Aut}_{g,b-1\perp\hat{\imath},p}$, since this is isomorphic to $\mathrm{Aut}_{g,b-1,p\perp\hat{\imath}}$, by Proposition 7.1(iv).   □

9. Converting a boundary to a puncture

In this section we construct short exact sequences of groups which correspond to converting boundary components to punctures.

9.1. Definition. Let $b \geqslant 1$.

There is a natural embedding

$$\Sigma_{g,b-1,p+1} * E_{b-1} \to \Sigma_{g,b,p} * E_b$$

which sends t_{p+1} to z_b, and we may view it as an inclusion because we understand $t_{p+1} = z_b$. Even in the trivial cases, pullback along this inclusion, or restriction, induces a unique sign-preserving map $\mathrm{Aut}_{g,b-1\perp1,p} \to \mathrm{Aut}_{g,b-1,p\perp1}$. The latter map carries $\widetilde{\Sigma}_{g,b,p}$ to $\widetilde{\Sigma}_{g,b-1,p+1}$, and hence determines a map

$$\mathrm{elim}(e_b)\colon \mathrm{Out}_{g,b-1\perp1,p} \to \mathrm{Out}_{g,b-1,p\perp1}.$$

It corresponds to converting the bth boundary component into a new puncture, the $(p+1)$st puncture.

By Proposition 7.1 (iii)–(v), we have isomorphisms

$$\mathrm{Out}_{g,b-1\perp1,p} \simeq \mathrm{Aut}_{g,b-1\perp\hat{\imath},p} \simeq \mathrm{Aut}_{g,b-1,p\perp\hat{\imath}},$$

and quotient isomorphisms

$$\mathrm{Out}_{g,b-1\perp1,p}/\langle\breve{\sigma}_b\rangle \simeq \mathrm{Aut}_{g,b-1\perp\hat{\imath},p}/\langle\sigma_b\widetilde{z}_b^{-1}\rangle \simeq \mathrm{Aut}_{g,b-1,p\perp\hat{\imath}}/\langle\widetilde{t}_{p+1}\rangle.$$

The latter group is isomorphic to $\mathrm{Out}_{g,b-1,p\perp1}$, by Proposition 7.2. It is not difficult to see that we have a factorization of $\mathrm{elim}(e_b)$ as

$$\mathrm{Out}_{g,b-1\perp1,p} \xrightarrow{\sim} \mathrm{Aut}_{g,b-1\perp\hat{\imath},p}$$
$$\xrightarrow{\sim} \mathrm{Aut}_{g,b-1,p\perp\hat{\imath}}$$
$$\to \mathrm{Aut}_{g,b-1,p\perp\hat{\imath}}/\langle\widetilde{t}_{p+1}\rangle$$
$$\xrightarrow{\sim} \mathrm{Out}_{g,b-1,p\perp1}. \qquad \square$$

We make the resulting exact sequence explicit.

9.2. Proposition. *If $b \geqslant 1$, then there is an exact sequence*

$$1 \to \langle\breve{\sigma}_b\rangle \to \mathrm{Out}_{g,b-1\perp1,p} \xrightarrow{\mathrm{elim}(e_b)} \mathrm{Out}_{g,b-1,p\perp1} \to 1. \qquad \square$$

9.3. Notation. It is easy to check that the following hold in $\mathrm{Aut}_{g,b,p}$.

$$\begin{aligned}
\sigma_1 &= 1 && \text{if } (g,b,p) = (0,1,0); \\
\sigma_1 &= \widetilde{z}_1 && \text{if } (g,b,p) = (0,1,1); \\
\sigma_1 &= \widetilde{z}_1\sigma_2 && \text{if } (g,b,p) = (0,2,0).
\end{aligned}$$

Let $B_{g,b,p} = \langle\sigma_j \mid 1 \leqslant j \leqslant b\rangle \leqslant \mathrm{Aut}_{g,b,p}$, and

$$\breve{B}_{g,b,p} = \langle\breve{\sigma}_j \mid 1 \leqslant j \leqslant b\rangle \leqslant \mathrm{Out}_{g,b,p}.$$

It is then not difficult to show that $B_{g,b,p}$ is presented as the abelian group with the given generators together with the relation $\sigma_1 = 1$ if $(g,b,p) = (0,1,0)$. Also, $\breve{B}_{g,b,p}$ is presented as the abelian group with the given generators together with the following relations:

$$\begin{aligned}
\breve{\sigma}_1 &= 1 && \text{if } (g,b,p) = (0,1,0) \quad \text{or} \quad (0,1,1); \\
\breve{\sigma}_1 &= \breve{\sigma}_2 && \text{if } (g,b,p) = (0,2,0).
\end{aligned}$$

For $1 \leqslant j \leqslant b$, σ_j can be thought of as right multiplying e_j by $\overline{z}_j^{e_j}$, and it follows that σ_j commutes with any element of $\mathrm{Aut}_{g,b,p}$ which fixes $z_j^{e_j}$. Hence, it can be seen that $B_{g,b,p}$ is centralized by $\mathrm{Aut}^+_{g,1\perp b,p}$, which has index $2 \times b!$ in $\mathrm{Aut}_{g,b,p}$; see Notation 4.3. It is then straightforward to show that $B_{g,b,p}$ is normal in $\mathrm{Aut}_{g,b,p}$.

Similar statements hold when Aut is replaced with Out. □

Repeated applications of Proposition 9.2, and extending by Sym_b, give the following.

9.4. Proposition. *There is an exact sequence*

$$1 \to \breve{B}_{g,b,p} \to \mathrm{Out}_{g,b,p} \xrightarrow{\mathrm{elim}(e_1,..,e_b)} \mathrm{Out}_{g,0,p\perp b} \to 1. \qquad \square$$

9.5. Remark. It is not difficult to show that there is also an exact sequence

$$1 \to B_{g,b,p} \to \mathrm{Aut}_{g,b,p} \to \mathrm{Aut}_{g,0,p\perp b} \to 1.$$

We will see a related result in Remarks 11.3(iii). □

We next observe that topologists have shown that the algebraic mapping-class group agrees with the topological mapping-class group. See also Remarks 11.3(ii).

9.6. Theorem. *The natural map $\mathcal{MC}_{g,b,p} \to \mathrm{Out}_{g,b,p}$ is an isomorphism.*

Proof. We claim that the natural map $\mathcal{MC}_{g,0,b+p} \to \mathrm{Out}_{g,0,b+p}$ is an isomorphism. In the trivial cases, this holds by our artificial definitions. In the nontrivial cases, this isomorphism is a deep, classic result; see [17] or [13, Theorem 2.9.A].

The Dehn twists around the boundary components determine an exact sequence

$$1 \to \breve{B}_{g,b,p} \to \mathcal{MC}_{g,b,p} \to \mathcal{MC}_{g,0,p\perp b} \to 1;$$

see [23, Theorem 4.1]. Now the desired result follows from Proposition 9.4 and the five lemma. □

10. More isomorphisms

10.1. Theorem. *If $p \geqslant 1$, then $\mathrm{Aut}_{g,b,p-1\perp\hat{\imath}}/\langle\tilde{t}_p\rangle \simeq \mathrm{Aut}_{g,b,p-1}$.*

Proof. By Proposition 7.2, $\mathrm{Out}_{g,b,p-1\perp 1} \simeq \mathrm{Aut}_{g,b,p-1\perp\hat{\imath}}/\langle\tilde{t}_p\rangle$. Let us denote by $\mathrm{prelim}(t_p)$ the composite

$$\mathrm{Out}_{g,b,p-1\perp 1} \xrightarrow{\sim} \mathrm{Aut}_{g,b,p-1\perp\hat{\imath}}/\langle\tilde{t}_p\rangle \to \mathrm{Aut}_{g,b,p-1},$$

where it is understood that signs are preserved. It is known that if $b = 0$ then $\mathrm{prelim}(t_p)$ is an isomorphism; the history and references, together with an algebraic proof using [21], can be found in [9]. Hence

$$\mathrm{prelim}(t_{b+p}) \colon \mathrm{Out}_{g,0,b+p-1\perp 1} \to \mathrm{Aut}_{g,0,b+p-1}$$

is an isomorphism, and restricting gives another isomorphism

$$\mathrm{prelim}(t_{b+p}) \colon \mathrm{Out}_{g,0,b\perp p-1\perp 1} \xrightarrow{\sim} \mathrm{Aut}_{g,0,b\perp p-1}. \qquad (10.1.1)$$

Notice that $\mathrm{prelim}(t_p) \colon \mathrm{Out}_{g,b,p-1\perp 1} \to \mathrm{Aut}_{g,b,p-1}$ carries the subgroup $\breve{B}_{g,b,p}$ of $\mathrm{Out}_{g,b,p-1\perp 1}$ onto the subgroup $B_{g,b,p-1}$ of $\mathrm{Aut}_{g,b,p-1}$; we claim it does so bijectively. This is clear if $B_{g,b,p-1}$ has rank b, so it remains to consider the case

$$(g, b, p-1) = (0, 1, 0),$$

and here both groups are trivial, so the claim holds.

Now $\mathrm{prelim}(t_p)$ induces a map

$$\mathrm{Out}_{g,b,p-1\perp 1}/\check{B}_{g,b,p} \to \mathrm{Aut}_{g,b,p-1}/B_{g,b,p-1},$$

which is an isomorphism, because it is equivalent to the isomorphism (10.1.1) by Proposition 9.4 and Remark 9.5.

Hence $\mathrm{prelim}(t_p)$ is itself an isomorphism. □

By Proposition 7.2, $\mathrm{Out}_{g,b,p-1\perp 1} \simeq \mathrm{Aut}_{g,b,p-1\perp\hat{1}}/\langle\widetilde{t_p}\rangle$, so we have the following.

10.2. Corollary. *If $p \geqslant 1$, then $\mathrm{Out}_{g,b,p-1\perp 1} \simeq \mathrm{Aut}_{g,b,p-1}$.* □

10.3. Remark. Corollary 10.2 can be expressed as $\mathrm{Aut}_{g,b,p} \simeq \mathrm{Out}_{g,b,p\perp 1}$. We feel that the existence of this isomorphism justifies the artificial conventions that led to it. □

11. Eliminating a puncture

In this section, we construct a short exact sequences of groups which corresponds to eliminating a puncture, and obtain partial splittings in some cases.

11.1. Definitions. Suppose that $p \geqslant 1$.

View $\Sigma_{g,b,p-1} * E_b$ as the quotient of $\Sigma_{g,b,p} * E_b$ by the normal subgroup generated by t_p.

There is induced a map

$$\mathrm{mod}(t_p)\colon \mathrm{Aut}_{g,b,p-1\perp 1} \to \mathrm{Aut}_{g,b,p-1}$$

which assigns, to any $\alpha \in \mathrm{Aut}_{g,b,p-1\perp 1}$, the element with the same sign and the induced action on the quotient group $\Sigma_{g,b,p-1} * E_b$ of $\Sigma_{g,b,p} * E_b$.

Now $\mathrm{mod}(t_p)$ carries $\widetilde{\Sigma}_{g,b,p}$ to $\widetilde{\Sigma}_{g,b,p-1}$, so induces a map

$$\mathrm{elim}(t_p)\colon \mathrm{Out}_{g,b,p-1\perp 1} \to \mathrm{Out}_{g,b,p-1}.$$

It is not difficult to see that we have a factorization of $\mathrm{elim}(t_p)$ as

$$\mathrm{Out}_{g,b,p-1\perp 1} \xrightarrow{\sim} \mathrm{Aut}_{g,b,p-1\perp\hat{1}}/\langle\widetilde{t_p}\rangle \xrightarrow{\sim} \mathrm{Aut}_{g,b,p-1} \to \mathrm{Out}_{g,b,p-1}.$$
□

Since the kernel of $\mathrm{Aut}_{g,b,p-1} \to \mathrm{Out}_{g,b,p-1}$ is $\widetilde{\Sigma}_{g,b,p-1}$, we get a slight generalization of Maclachlan's form of the Bers-Birman sequence; see [9, Section 1].

11.2. Corollary. *If $p \geqslant 1$, then there is an exact sequence*

$$1 \to \widetilde{\Sigma}_{g,b,p-1} \to \mathrm{Out}_{g,b,p-1\perp 1} \xrightarrow{\mathrm{elim}(t_p)} \mathrm{Out}_{g,b,p-1} \to 1.$$
□

11.3. Remarks. (i) We shall describe the copy of $\widetilde{\Sigma}_{g,b,p-1}$ in $\mathrm{Out}_{g,b,p-1\perp 1}$ explicitly in Remarks 12.6(iii).

(ii) We leave it as an exercise to show that $\mathrm{Out}_{g,b,p}$ has a descending subnormal series in which the sequence of isomorphism classes of factor groups is given by

$$\mathrm{Sym}_{b\perp p}, \; \mathrm{Out}_{g,0,0}, \; \widetilde{\Sigma}_{g,0,0}, \; \widetilde{\Sigma}_{g,0,1}, \; \ldots, \; \widetilde{\Sigma}_{g,0,b+p-2}, \; \widetilde{\Sigma}_{g,0,b+p-1}, \; \breve{B}_{g,b,p}.$$

A corresponding result is known for $\mathcal{MC}_{g,b,p}$, and this gives another way to verify that $\mathcal{MC}_{g,b,p} \simeq \mathrm{Out}_{g,b,p}$.

(iii) By Proposition 7.1(iv) and Corollary 10.2, we can rephrase Proposition 9.2 to say that the sequence

$$1 \to \langle \widetilde{t}_p \rangle \to \mathrm{Aut}_{g,b,p-1\perp \hat{1}} \xrightarrow{\;\mathrm{mod}\;(t_p)\;} \mathrm{Aut}_{g,b,p-1} \to 1$$

is exact. The case where $(g,b) = (0,0)$ is a well-known result of Magnus [18]. □

By combining Proposition 7.1(ii)–(iii) and Corollary 10.2, we get a finite-index splitting for the exact sequence in Corollary 11.2, in some cases.

11.4. Corollary. *If $b \geqslant 1$ and $p \geqslant 1$, then $\Sigma_{g,b,p-1}$ is free of rank $2g + b + p - 2$, and $\mathrm{Out}_{g,b-1\perp 1,p-1\perp 1} \simeq \mathrm{Aut}_{g,b-1\perp 1,p-1}$*

$$= \mathrm{Aut}_{g,b-1\perp \hat{1},p-1} \ltimes \Sigma_{g,b,p-1}$$

$$\simeq \mathrm{Out}_{g,b-1\perp 1,p-1} \ltimes \Sigma_{g,b,p-1}.$$

□

We shall give another description of the splitting map

$$\mathrm{Out}_{g,b-1\perp 1,p-1} \to \mathrm{Out}_{g,b-1\perp 1,p-1\perp 1} \tag{11.4.1}$$

in Remarks 13.3.

The splitting can be iterated. We use Notation 4.3 and the convention that, in multiple semidirect products, left parentheses are understood to accumulate on the left; for example, $A \ltimes B \ltimes C \ltimes D$ denotes $((A \ltimes B) \ltimes C) \ltimes D$.

11.5. Corollary. *If $b \geqslant 1$, then*

$$\mathrm{Out}_{g,b-1\perp 1,1\perp p} \simeq \mathrm{Out}_{g,b-1\perp 1,0} \ltimes \Sigma_{g,b,0} \ltimes \Sigma_{g,b,1} \ltimes \cdots \ltimes \Sigma_{g,b,p-1};$$

$\Sigma_{g,b,i}$ is free of rank $2g + b + i - 1$, for $i = 0, \ldots, p-1$. □

12. Eliminating a boundary component

In this section, we construct a short exact sequence of groups which corresponds to eliminating a boundary component.

12.1. Definition. Suppose that $b \geqslant 1$.

In Definitions 9.1 and 11.1, we introduced the restriction and quotient maps

$$\mathrm{elim}(e_b)\colon \mathrm{Out}_{g,b-1\perp 1,p} \to \mathrm{Out}_{g,b-1,p\perp 1},$$

$$\mathrm{elim}(t_{p+1})\colon \mathrm{Out}_{g,b-1,p\perp 1} \to \mathrm{Out}_{g,b-1,p},$$

respectively. Here $t_{p+1} = z_b$, and we define

$$\mathrm{elim}(z_b, e_b) = \mathrm{elim}(e_b) \cdot \mathrm{elim}(t_{p+1}): \ \mathrm{Out}_{g,b-1\perp1,p} \to \mathrm{Out}_{g,b-1,p}. \qquad \square$$

This is a surjective map and we shall want to describe the kernel. Notice that $\mathrm{elim}(e_b)$ has kernel $\langle \breve{o}_b \rangle$, and $\mathrm{elim}(t_{p+1})$ has kernel $\widetilde{\Sigma}_{g,b-1,p}$, so the kernel of $\mathrm{elim}(z_b, e_b)$ is an extension of $\langle \breve{o}_b \rangle$ by $\widetilde{\Sigma}_{g,b-1,p}$, and we shall find that it is $\check{\Sigma}_{g,b-1,p}$. For convenience, we recall our factorizations of $\mathrm{elim}(e_b)$ and $\mathrm{elim}(t_{p+1})$ through automorphism groups, and deduce a corresponding factorization for $\mathrm{elim}(z_b, e_b)$.

$\mathrm{elim}(e_b):$

$$\begin{array}{llll}
\mathrm{Out}_{g,b-1\perp1,p} & \xrightarrow{\sim} & \mathrm{Aut}_{g,b-1\perp\hat{1},p} & \text{fix } e_b \\
& \xrightarrow{\sim} & \mathrm{Aut}_{g,b-1,p\perp\hat{1}} & \text{ignore } e_b, \text{ write } z_b = t_{p+1} \\
& \longrightarrow & \mathrm{Aut}_{g,b-1,p\perp\hat{1}} / \langle \widetilde{t}_{p+1} \rangle & \text{mod } \langle \widetilde{t}_{p+1} \rangle \\
& \xrightarrow{\sim} & \mathrm{Out}_{g,b-1,p\perp1} & \text{natural.}
\end{array}$$

$\mathrm{elim}(t_{p+1}):$

$$\begin{array}{llll}
\mathrm{Out}_{g,b-1,p\perp1} & \xrightarrow{\sim} & \mathrm{Aut}_{g,b-1,p\perp\hat{1}} / \langle \widetilde{t}_{p+1} \rangle & \text{fix } t_{p+1} \\
& \xrightarrow{\sim} & \mathrm{Aut}_{g,b-1,p} & \text{kill } t_{p+1} \\
& \longrightarrow & \mathrm{Aut}_{g,b-1,p} / \widetilde{\Sigma}_{g,b-1,p} & \text{mod } \widetilde{\Sigma}_{g,b-1,p}. \\
& \xrightarrow{\sim} & \mathrm{Out}_{g,b-1,p} & \text{natural.}
\end{array}$$

$\mathrm{elim}(z_b, e_b):$

$$\begin{array}{llll}
\mathrm{Out}_{g,b-1\perp1,p} & \xrightarrow{\sim} & \mathrm{Aut}_{g,b-1\perp\hat{1},p} & \text{fix } e_b \\
& \xrightarrow{\sim} & \mathrm{Aut}_{g,b-1,p\perp\hat{1}} & \text{ignore } e_b, \text{ write } z_b = t_{p+1} \\
& \longrightarrow & \mathrm{Aut}_{g,b-1,p\perp\hat{1}} / \langle \widetilde{t}_{p+1} \rangle & \text{mod } \langle \widetilde{t}_{p+1} \rangle \\
& \xrightarrow{\sim} & \mathrm{Aut}_{g,b-1,p} & \text{kill } t_{p+1} \\
& \xrightarrow{\sim} & \mathrm{Aut}_{g,b-1,p} / \widetilde{\Sigma}_{g,b-1,p} & \text{mod } \widetilde{\Sigma}_{g,b-1,p}. \\
& \xrightarrow{\sim} & \mathrm{Out}_{g,b-1,p} & \text{natural.}
\end{array}$$

We now describe elements of $\mathrm{Aut}_{g,b-1\perp1,p}$ which vanish in $\mathrm{Aut}_{g,b-1,p}$ if we ignore e_b and kill z_b. Moreover, if we use the above factorization, that is, first fixing e_b, we find these elements map to generators of $\widetilde{\Sigma}_{g,b-1,p}$ in $\mathrm{Aut}_{g,b-1,p}$.

12.2. Definitions. Suppose that $b \geqslant 1$.

We want to construct a homomorphism

$$\Sigma_{g,b,p} \to \mathrm{Aut}_{g,b,p}, \qquad x \mapsto \widehat{x}. \tag{12.2.1}$$

Recall that $\Sigma_{g,b,p}$ is free on $X_{g,b-1,p}$. We begin by defining, for each $x \in X_{g,b-1,p}$, an endomorphism $\widehat{x}$ of the free group on $X_{g,b,p} \cup E_b$, namely, for $p \geqslant k \geqslant 1$,

$1 \leqslant i \leqslant g$, $1 \leqslant j \leqslant b-1$, we define the following.

$$\hat{t}_k : \begin{cases} w & \mapsto & w^{[\bar{t}_k, \bar{z}_b]} & \text{for } w = t_p, ., t_{k+1}, \\ t_k & \mapsto & t_k^{\bar{z}_b}, \\ w & \mapsto & w & \text{for } w = t_{k-1}, ., t_1, x_1, ., y_g, z_1, e_1, ., z_{b-1}, e_{b-1}, \\ z_b & \mapsto & z_b^{t_k \bar{z}_b}, \\ e_b & \mapsto & z_b t_k e_b. \end{cases}$$

$$\hat{x}_i : \begin{cases} w & \mapsto & w^{[z_b, \bar{x}_i]} & \text{for } w = t_p, ., t_1, x_1, y_1, ., x_{i-1}, y_{i-1}, \\ x_i & \mapsto & x_i^{z_b \bar{x}_i}, \\ y_i & \mapsto & x_i \bar{z}_b \bar{x}_i y_i, \\ w & \mapsto & w & \text{for } w = x_{i+1}, y_{i+1}, ., x_g, y_g, z_1, e_1, ., z_{b-1}, e_{b-1}, \\ z_b & \mapsto & z_b^{\bar{x}_i}, \\ e_b & \mapsto & x_i \bar{z}_b e_b. \end{cases}$$

$$\hat{y}_i : \begin{cases} w & \mapsto & w^{[\bar{y}_i, \bar{z}_b]} & \text{for } w = t_p, ., t_1, x_1, y_1, ., x_{i-1}, y_{i-1}, \\ x_i & \mapsto & z_b x_i [\bar{y}_i, \bar{z}_b], \\ y_i & \mapsto & y_i^{\bar{z}_b}, \\ w & \mapsto & w & \text{for } w = x_{i+1}, y_{i+1}, ., x_g, y_g, z_1, e_1, ., z_{b-1}, e_{b-1}, \\ z_b & \mapsto & z_b^{\bar{y}_i \bar{z}_b}, \\ e_b & \mapsto & z_b y_i e_b. \end{cases}$$

$$\hat{z}_j : \begin{cases} w & \mapsto & w^{[\bar{z}_j, \bar{z}_b]} & \text{for } w = t_p, ., t_1, x_1, ., y_g, z_1, ., z_{j-1}, \\ w & \mapsto & [\bar{z}_b, \bar{z}_j] w & \text{for } w = e_1, ., e_{j-1}, \\ z_j & \mapsto & z_j^{\bar{z}_b}, \\ e_j & \mapsto & z_b e_j, \\ w & \mapsto & w & \text{for } w = z_{j+1}, e_{j+1}, ., z_{b-1}, e_{b-1}, \\ z_b & \mapsto & z_b^{\bar{z}_j \bar{z}_b}, \\ e_b & \mapsto & z_b z_j e_b. \end{cases}$$

To facilitate the definitions, we briefly postpone, to Lemma 12.3, the verification that, for each $x \in X_{g,b-1,p}$, the above $\hat{x}$ determines an element of $\mathrm{Aut}^+_{g,b,p}$, again denoted $\hat{x}$, and hence we have a homomorphism (12.2.1). There is then an induced homomorphism

$$\Sigma_{g,b,p} \to \mathrm{Out}_{g,b,p}, \quad x \mapsto \check{x}.$$

It is easy to check that the composite

$$\mathrm{Out}_{g,b-1 \perp 1,p} \xrightarrow{\sim} \mathrm{Aut}_{g,b-1 \perp \hat{1},p} \to \mathrm{Aut}_{g,b-1,p}$$

carries $\check{x}$ to $\tilde{x}$, for each $x \in X_{g,b-1,p}$ Here one considers the unique $s_x \in \Sigma_{g,b,p}$ such that $e_b^{\hat{x}} = s_x e_b$, and calculates the action of $\hat{x}\tilde{s}_x$ modulo the normal subgroup generated by z_b and e_b. □

12.3. Lemma. *Let $b \geqslant 1$ and $x \in X_{g,b-1,p}$. Then $\hat{x}$ induces an element of* $\mathrm{Aut}^+_{g,b-1 \perp 1,p}$.

Proof. We observe the following.

$$\hat{t}_k : \begin{cases} w & \mapsto & w & \text{for } w = t_{k-1}, ., t_1, x_1, ., y_g, z_1, ., z_{b-1}, \\ z_b & \mapsto & z_b^{\bar{t}_k \bar{z}_b} = z_b[\bar{t}_k, \bar{z}_b], \\ w & \mapsto & w^{[\bar{t}_k, \bar{z}_b]} & \text{for } w = t_p, ., t_{k+1}, \\ t_k & \mapsto & t_k^{\bar{z}_b} = [\bar{z}_b, \bar{t}_k]t_k. \end{cases}$$

$$\hat{x}_i : \begin{cases} w & \mapsto & w & \text{for } w = x_{i+1}, y_{i+1}, ., x_g, y_g, z_1, ., z_{b-1}, \\ z_b & \mapsto & z_b^{\bar{x}_i} = z_b[z_b, \bar{x}_i], \\ w & \mapsto & w^{[z_b, \bar{x}_i]} & \text{for } w = t_p, ., t_1, x_1, y_1, ., x_{i-1}, y_{i-1}, \\ [x_i, y_i] & \mapsto & [\bar{x}_i, z_b][x_i, y_i]. \end{cases}$$

$$\hat{y}_i : \begin{cases} w & \mapsto & w & \text{for } w = x_{i+1}, y_{i+1}, ., x_g, y_g, z_1, ., z_{b-1}, \\ z_b & \mapsto & z_b^{\bar{y}_i \bar{z}_b} = z_b[\bar{y}_i, \bar{z}_b], \\ w & \mapsto & w^{[\bar{y}_i, \bar{z}_b]} & \text{for } w = t_p, ., t_1, x_1, y_1, ., x_{i-1}, y_{i-1}, \\ [x_i, y_i] & \mapsto & [\bar{z}_b, \bar{y}_i][x_i, y_i]. \end{cases}$$

$$\hat{z}_j : \begin{cases} w & \mapsto & w & \text{for } w = z_{j+1}, ., z_{b-1}, \\ z_b & \mapsto & z_b^{\bar{z}_j \bar{z}_b} = z_b[\bar{z}_j, \bar{z}_b], \\ w & \mapsto & w^{[\bar{z}_j, \bar{z}_b]} & \text{for } w = t_p, ., t_1, x_1, ., y_g, z_1, ., z_{j-1}, \\ z_j & \mapsto & z_j^{\bar{z}_b} = [\bar{z}_b, \bar{z}_j]z_j. \end{cases}$$

It can now be seen that $\hat{x}$ fixes some $\Sigma_{g,b,p}$-conjugate of $w_{g,b,p}$. Hence $\hat{x}$ induces an endomorphism of $\Sigma_{g,b,p}$, and hence an endomorphism $\hat{x}$ of $\Sigma_{g,b,p} * E_b$. Moreover $\hat{x}$ fixes the $z_j^{e_j}$ and the $[t_k]$. One can check that this $\hat{x}$ is an automorphism either by straightforward calculation, or by applying Theorem 8.1. It is now clear that, in terms of Notation 4.3, we have an element $\hat{x}$ of $\mathrm{Aut}^+_{g,1 \perp b, 1 \perp p} \leqslant \mathrm{Aut}_{g,b-1 \perp 1,p}$. □

12.4. Lemma. *If $b \geqslant 1$, the kernel of* $\mathrm{elim}(z_b, e_b) : \mathrm{Out}_{g,b-1 \perp 1,p} \to \mathrm{Out}_{g,b-1,p}$ *is presented with the generating set* $\{\breve{o}_b, \check{x} \mid x \in X_{g,b-1,p}\} \subseteq \mathrm{Out}_{g,b,p}$ *and the relations of* $\check{\Sigma}_{g,b-1,p}$.

Proof. We have seen that $\mathrm{Ker}(\mathrm{elim}(z_b, e_b))$ is an extension of $\langle \breve{o}_b \rangle$ by $\tilde{\Sigma}_{g,b-1,p}$, and it is straightforward to see that the given set generates $\mathrm{Ker}(\mathrm{elim}(z_b, e_b))$.

Notice that the given generators lie in $\mathrm{Out}^+_{g,1 \perp b, 1 \perp p}$, and therefore commute with $\breve{o}_b$.

From its presentation, we see that $\check{\Sigma}_{g,b-1,p}$ too is an extension of $\langle \check{\sigma}_b \rangle$ by $\tilde{\Sigma}_{g,b-1,p}$. By the five lemma, it suffices to check that the given generators of $\mathrm{Ker}(\mathrm{elim}(z_b, e_b))$ satisfy all the relations of $\check{\Sigma}_{g,b-1,p}$.

We compute

$$\hat{x}_i^{-1}: \begin{cases} w & \mapsto & w^{[x_i, \overline{z}_b]} & \text{for } w = t_p, .., t_1, x_1, y_1, .., x_{i-1}, y_{i-1}, \\ x_i & \mapsto & x_i^{\overline{z}_b}, \\ y_i & \mapsto & z_b y_i, \\ w & \mapsto & w & \text{for } w = x_{i+1}, y_{i+1}, .., x_g, y_g, z_1, e_1, .., z_{b-1}, e_{b-1}, \\ z_b & \mapsto & z_b^{x_i \overline{z}_b}, \\ e_b & \mapsto & z_b \overline{x}_i e_b. \end{cases}$$

$$\hat{y}_i^{-1}: \begin{cases} w & \mapsto & w^{[z_b, y_i]} & \text{for } w = t_p, .., t_1, x_1, y_1, .., x_{i-1}, y_{i-1}, \\ x_i & \mapsto & \overline{y}_i z_b y_i x_i [z_b, y_i], \\ y_i & \mapsto & y_i^{z_b y_i}, \\ w & \mapsto & w & \text{for } w = x_{i+1}, y_{i+1}, .., x_g, y_g, z_1, e_1, .., z_{b-1}, e_{b-1}, \\ z_b & \mapsto & z_b^{y_i}, \\ e_b & \mapsto & \overline{y}_i \overline{z}_b e_b. \end{cases}$$

If $v = [x_i, y_i]$, then

$$\hat{v}: \begin{cases} w & \mapsto & w^{[\overline{v}, \overline{z}_b]} & \text{for } w = t_p, .., t_1, x_1, y_1, .., x_{i-1}, y_{i-1}, \\ w & \mapsto & w^{\overline{z}_b} & \text{for } w = x_i, y_i, \\ w & \mapsto & w & \text{for } w = x_{i+1}, y_{i+1}, .., x_g, y_g, z_1, e_1, .., z_{b-1}, e_{b-1}, \\ z_b & \mapsto & z_b^{\overline{v}\,\overline{z}_b}, \\ e_b & \mapsto & z_b v z_b e_b. \end{cases}$$

If $v = \coprod_{k'=p}^{k} t_{k'}$, then

$$\hat{v}: \begin{cases} w & \mapsto & w^{\overline{z}_b} & \text{for } w = t_p, .., t_k, \\ w & \mapsto & w & \text{for } w = t_{k-1}, .., t_1, x_1, y_1, .., x_g, y_g, z_1, e_1, .., z_{b-1}, e_{b-1}, \\ z_b & \mapsto & z_b^{\overline{v}\,\overline{z}_b}, \\ e_b & \mapsto & z_b v z_b^{p-k} e_b. \end{cases}$$

If $v = \coprod\limits_{k=p}^{1} t_k \cdot \prod\limits_{i'=1}^{i} [x_{i'}, y_{i'}]$, then

$$\hat{v}: \begin{cases} w & \mapsto & w^{\overline{z}_b} & \text{for } w = t_p, ., t_1, x_1, y_1, ., x_i, y_i, \\ w & \mapsto & w & \text{for } w = x_{i+1}, y_{i+1}, ., x_g, y_g, z_1, e_1, ., z_{b-1}, e_{b-1}, \\ z_b & \mapsto & z_b^{\overline{v}\,\overline{z}_b}, \\ e_b & \mapsto & z_b v z_b^{p+2i-1} e_b. \end{cases}$$

If $v = \coprod\limits_{k=p}^{1} t_k \cdot \prod\limits_{i=1}^{g} [x_i, y_i] \cdot \prod\limits_{j'=1}^{j} z_{j'}$, then

$$\hat{v}: \begin{cases} w & \mapsto & w^{\overline{z}_b} & \text{for } w = t_p, ., t_1, x_1, y_1, ., x_g, y_g, z_1, ., z_j, \\ w & \mapsto & z_b w & \text{for } w = e_1, ., e_j, \\ w & \mapsto & w & \text{for } w = z_{j+1}, e_{j+1}, ., z_{b-1}, e_{b-1}, \\ z_b & \mapsto & z_b^{\overline{v}\,\overline{z}_b}, \\ e_b & \mapsto & z_b v z_b^{p+2g+j-1} e_b. \end{cases}$$

If $v = \coprod\limits_{k=p}^{1} t_k \cdot \prod\limits_{i=1}^{g} [x_i, y_i] \cdot \prod\limits_{j=1}^{b-1} z_j$, then $[v] = [w_{g,b-1,p}]$, $[z_b v] = [w_{g,b,p}]$, and

$$\hat{v}: \begin{cases} w & \mapsto & w^{\overline{z}_b} & \text{for } w = t_p, ., t_1, x_1, y_1, ., x_g, y_g, z_1, ., z_{b-1}, \\ w & \mapsto & z_b w & \text{for } w = e_1, ., e_{b-1}, \\ z_b & \mapsto & z_b^{\overline{v}\,\overline{z}_b}, \\ e_b & \mapsto & z_b v z_b^{2g+b+p-2} e_b. \end{cases}$$

The latter gives rise to $\tilde{z}_b^{-1} \sigma_b^{3-2g-b-p}$ in $\mathrm{Aut}_{g,b,p}$, and hence to $\check{\sigma}_b^{3-2g-b-p}$ in $\mathrm{Out}_{g,b,p}$. Since $3 - 2g - b - p = \chi_{g,b-1,p}$, we see that we have verified one of the relations.

The other relations are straightforward to check. $\qquad\square$

12.5. Theorem. *If $b \geqslant 1$, then there is an exact sequence*

$$1 \to \check{\Sigma}_{g,b-1,p} \to \mathrm{Out}_{g,b-1\perp 1,p} \xrightarrow{\mathrm{elim}(z_b, e_b)} \mathrm{Out}_{g,b-1,p} \to 1. \quad\square \qquad (12.5.1)$$

The image of $\check{\Sigma}_{g,b-1,p}$ in $\mathrm{Out}_{g,b-1\perp 1,p}$ will again be denoted $\check{\Sigma}_{g,b-1,p}$.

12.6. Remarks. (i) In a standard way, (12.5.1) gives an action of $\mathrm{Out}_{g,b-1\perp 1,p}$ on $\check{\Sigma}_{g,b-1,p}$, and an outer action of $\mathrm{Out}_{g,b-1,p}$ on $\check{\Sigma}_{g,b-1,p}$, that is, (12.5.1) gives homomorphisms

$$\mathrm{Out}_{g,b-1\perp 1,p} \to \mathrm{Aut}(\check{\Sigma}_{g,b-1,p}) \quad \text{and} \quad \mathrm{Out}_{g,b-1,p} \to \mathrm{Out}(\check{\Sigma}_{g,b-1,p}).$$

We shall discuss the action of $\mathrm{Out}_{g,b-1\perp 1,p}$ on $\check{\Sigma}_{g,b-1,p}$ in Section 14.

(ii) Étienne Ghys has pointed out the following to us.

Suppose that $(b,p) = (1,0)$ and that $g \geqslant 2$. Give $S_{g,0,0}$ a hyperbolic metric, and let $TS_{g,0,0}$ denote the unit-tangent bundle over $S_{g,0,0}$.

Here, the exact sequence (12.5.1) corresponds to Mess's exact sequence

$$1 \to \pi_1(TS_{g,0,0}) \to \mathcal{MC}_{g,1,0} \to \mathcal{MC}_{g,0,0} \to 1,$$

described in [22] and [13, Section 6.3].

The resulting outer action of $\mathcal{MC}_{g,0,0}$ on $\pi_1(TS_{g,0,0})$ is related to classical geometric constructions, as follows. The universal covering space, $\tilde{S}_{g,0,0}$, has a circle at infinity, $\partial\tilde{S}_{g,0,0}$. The set of ordered triples of distinct points of $\partial\tilde{S}_{g,0,0}$, modulo the action of $\Sigma_{g,0,0}$, gives $TS_{g,0,0}$. Now, given an element of $\mathcal{MC}_{g,0,0}$, we can choose a representative self-homeomorphism of $S_{g,0,0}$, and lift this to a self-homeomorphism of $\tilde{S}_{g,0,0}$, and extend the latter continuously to get a self-homeomorphism of $\partial\tilde{S}_{g,0,0}$. Now we get a self-homeomorphism of the set of ordered triples of distinct points of $\partial\tilde{S}_{g,0,0}$, and, on dividing out by the action of $\Sigma_{g,0,0}$, we get a self-homeomorphism of $TS_{g,0,0}$, and hence an outer automorphism of $\pi_1(TS_{g,0,0})$. See [26, p. 31], for example. □

(iii) If $b \geqslant 1$, then, by Proposition 9.2 and Theorem 12.5, there is a copy of

$$\check{\Sigma}_{g,b-1,p}/\langle\check{\sigma}_b\rangle \quad \simeq \quad \tilde{\Sigma}_{g,b-1,p}$$

in $\mathrm{Out}_{g,b-1\perp1,p}/\langle\check{\sigma}_b\rangle \simeq \mathrm{Out}_{g,b-1,p\perp1}$. Consider any $x \in X_{g,b-1,p}$. To see the image of $\check{x}$ in $\mathrm{Out}_{g,b-1,p\perp1}$, we choose the representative of $\check{x}$ which fixes e_b, then ignore e_b and write z_b as t_{p+1}. We now have an explicit description of the generators of the copy of $\tilde{\Sigma}_{g,b-1,p}$ in $\mathrm{Out}_{g,b-1,p\perp1}$.

Replacing (b,p), $p \geqslant 0$, with $(b+1,p-1)$, $p \geqslant 1$, we get an explicit description of the copy of $\tilde{\Sigma}_{g,b,p-1}$ in $\mathrm{Out}_{g,b,p-1\perp1}$ given by Corollary 11.2. □

13. Another semidirect product decomposition

For $b \geqslant 2$, we shall define a map

$$\mathrm{pinch}(z_{b-1})\colon \mathrm{Out}_{g,b-2\perp1,p} \to \mathrm{Out}_{g,b-2\perp1\perp1,p}$$

which corresponds to identifying two disjoint closed subintervals of the $(b-1)$st boundary component, an identification which does not behave well with respect to the referential S^1.

13.1. Definitions. Suppose that $b \geqslant 2$.

There is a unique endomorphism of the free group on $X_{g,b,p} \cup E_b$ which is the identity on

$$(X_{g,b,p} \cup E_b) - \{z_{b-1}\}$$

and sends z_{b-1} to $z_{b-1}z_b$. Notice that this is an automorphism which sends $w_{g,b-1,p}$ to $w_{g,b,p}$, and hence induces an isomorphism

$$\Sigma_{g,b-1,p} * E_{b-1} * \langle z_b, e_b \mid \ \rangle \xrightarrow{\sim} \Sigma_{g,b,p} * E_b. \tag{13.1.1}$$

There is a natural sign-preserving map

$$\mathrm{mod}(z_b, e_b)\colon \mathrm{Aut}_{g,b-1\perp\hat{\imath},p} \to \mathrm{Aut}_{g,b-1,p}$$

corresponding to considering the action modulo the normal subgroup generated by e_b and z_b. This induces a map on index $b-1$ subgroups

$$\mathrm{Aut}_{g,b-2\perp1\perp\hat{\imath},p} \to \mathrm{Aut}_{g,b-2\perp1,p},$$

and, by abuse of notation, we also call this map $\mathrm{mod}(z_b, e_b)$.

We shall construct a map

$$\mathrm{Aut}_{g,b-2\perp\hat{\imath},p} \to \mathrm{Aut}_{g,b-2\perp1\perp\hat{\imath},p}, \quad \alpha \mapsto \alpha'',$$

such that the composite

$$\mathrm{Aut}_{g,b-2\perp\hat{\imath},p} \to \mathrm{Aut}_{g,b-2\perp1\perp\hat{\imath},p} \xrightarrow{\mathrm{mod}(z_b,e_b)} \mathrm{Aut}_{g,b-2\perp1,p}$$

is the inclusion map.

Consider any $\alpha \in \mathrm{Aut}_{g,b-2\perp\hat{\imath},p}$.

If α fixes z_{b-1}, we extend α to an automorphism α' of

$$\Sigma_{g,b-1,p} * E_{b-1} * \langle z_b, e_b \mid \ \rangle$$

that fixes z_b and e_b. Then, via the isomorphism (13.1.1), α' induces an automorphism α'' of $\Sigma_{g,b,p} * E_b$. Notice that α'' fixes $z_{b-1}z_b$, z_b, e_b, and hence also z_{b-1}, $z_{b-1}^{e_{b-1}}$. It is clear that $\mathrm{mod}(z_b, e_b)$ carries α'' to α.

If α inverts z_{b-1}, we restrict α to $\Sigma_{g,b-1,p} * E_{b-2}$ and extend this restricted map to a map α' which inverts z_b, fixes e_b, and sends e_{b-1} to $\bar{z}_b e_{b-1}$. Then, via the isomorphism (13.1.1), α' induces an automorphism α'' of $\Sigma_{g,b,p} * E_b$. Notice that α'' inverts $z_{b-1}z_b$ and z_b, fixes e_b, and sends e_{b-1} to $\bar{z}_b e_{b-1}$. Thus α'' sends z_{b-1} to $\bar{z}_{b-1}^{z_b}$, and inverts $z_{b-1}^{e_{b-1}}$. Again, it is clear that $\mathrm{mod}(z_b, e_b)$ carries α'' to α.

In all cases, α'' fixes

$$\{(z_{b-1}, e_{b-1}, z_b, e_b), (\bar{z}_{b-1}^{z_b}, \bar{z}_b e_{b-1}, \bar{z}_b, e_b)\}.$$

It is not difficult to show that $\alpha \mapsto \alpha''$ is a homomorphism.

Also, $\sigma_{b-1}\tilde{z}_{b-1}^{-1}$ is mapped to τ_{b-1}^2, which can be expressed as

$$\sigma_{b-1}\hat{z}_{b-1}^{-1}[\tilde{z}_{b-1}, \tilde{z}_b].$$

By Proposition 7.1(iii), $\mathrm{Aut}_{g,b-2\perp\hat{\imath},p} \simeq \mathrm{Out}_{g,b-2\perp1,p}$, and hence

$$\mathrm{Aut}_{g,b-2\perp1\perp\hat{\imath},p} \simeq \mathrm{Out}_{g,b-2\perp1\perp1,p}.$$

Thus we get a homomorphism

$$\mathrm{pinch}(z_{b-1})\colon \mathrm{Out}_{g,b-2\perp1,p} \to \mathrm{Out}_{g,b-2\perp1\perp1,p} \tag{13.1.2}$$

which is a left inverse of (the index $b-1$ restriction)

$$\mathrm{elim}(z_b, e_b)\colon \mathrm{Out}_{g,b-2\perp1\perp1,p} \to \mathrm{Out}_{g,b-2\perp1,p}. \tag{13.1.3}$$

Notice that $\mathrm{pinch}(z_{b-1})$ maps $\breve{\sigma}_{b-1}$ to $\breve{\tau}_{b-1}^2 = \breve{\sigma}_{b-1}\breve{z}_{b-1}^{-1}$, or, with extra notation, maps $\breve{\sigma}_{b-1}^{(g,b-1,p)}$ to $\breve{\sigma}_{b-1}^{(g,b,p)}\breve{z}_{b-1}^{-1}$. $\qquad\square$

We have constructed a finite-index splitting of the exact sequence in Theorem 12.5.

13.2. Theorem. *If $b \geqslant 2$, then there exists an isomorphism*

$$\mathrm{Out}_{g,b-2\perp1\perp1,p} \;\simeq\; \mathrm{Out}_{g,b-2\perp1,p} \;\ltimes\; \check{\Sigma}_{g,b-1,p},$$

such that

$\check{\sigma}_j$	$\leftrightarrow$	$\check{\sigma}_j$	$\ltimes$	1 *for* $j = 1,.,b-2$,
$\check{\sigma}_{b-1}^{(g,b,p)}$	$\leftrightarrow$	$\check{\sigma}_{b-1}^{(g,b-1,p)}$	$\ltimes$	$\check{z}_{b-1}$,
$\check{\sigma}_b$	$\leftrightarrow$	1	$\ltimes$	$\check{\sigma}_b$. $\square$

13.3. Remarks. (i) We shall describe the action of $\mathrm{Out}_{g,b-2\perp1\perp1,p}$ on $\check{\Sigma}_{g,b-1,p}$ in Section 14. The action of $\mathrm{Out}_{g,b-2\perp1,p}$ on $\check{\Sigma}_{g,b-1,p}$ is then by pullback along (13.1.2).

(ii) We have seen that (13.1.2), the map $\mathrm{pinch}(z_{b-1})$, is a left inverse of (13.1.3), (an index $b-1$ restriction of) the map $\mathrm{elim}(z_b, e_b)$. The factorization

$$\mathrm{elim}(z_b, e_b) = \mathrm{elim}(e_b) \cdot \mathrm{elim}(t_{p+1})$$

passes to the index $b-1$ subgroups. Thus $\mathrm{pinch}(z_{b-1})$ followed by (an index $b-1$ restriction)

$$\mathrm{elim}(e_b)\colon \mathrm{Out}_{g,b-2\perp1\perp1,p} \to \mathrm{Out}_{g,b-2\perp1,p\perp1},$$

is a left inverse of (an index $b-1$ restriction)

$$\mathrm{elim}(t_{p+1})\colon \mathrm{Out}_{g,b-2\perp1,p\perp1} \to \mathrm{Out}_{g,b-2\perp1,p}.$$

This corresponds to pinching the $(b-1)$st boundary component, and then eliminating the bth boundary component in two steps, first converting it to a puncture, and then eliminating the puncture, to get back the original surface.

If we replace p with $p-1$, and b with $b+1$, we get the map (11.4.1). $\square$

We can iterate Theorem 13.2.

13.4. Corollary. *If $b \geqslant 1$, then*

$$\mathrm{Out}_{g,1^{\perp b},p} \;\simeq\; \mathrm{Out}_{g,1,p} \ltimes \check{\Sigma}_{g,1,p} \ltimes \check{\Sigma}_{g,2,p} \cdots \ltimes \check{\Sigma}_{g,b-1,p}. \qquad \square$$

13.5. Corollary. *If $b \geqslant 1$, then*

$$\mathrm{Out}_{g,1^{\perp b},1^{\perp p}}$$
$$\simeq \mathrm{Out}_{g,1^{\perp b},0} \ltimes \Sigma_{g,b,0} \ltimes \Sigma_{g,b,1} \ltimes \cdots \ltimes \Sigma_{g,b,p-1}$$
$$\simeq \mathrm{Out}_{g,1,0} \ltimes \check{\Sigma}_{g,1,0} \ltimes \check{\Sigma}_{g,2,0} \ltimes \cdots \ltimes \check{\Sigma}_{g,b-1,0} \ltimes \Sigma_{g,b,0} \ltimes \Sigma_{g,b,1} \ltimes \cdots \ltimes \Sigma_{g,b,p-1}.$$

Proof. This follows from Corollary 11.5 and Corollary 13.4.

Alternatively, since $b + p \geqslant 1$,

$$\mathrm{Out}_{g,1^{\perp b+p},0} \simeq \mathrm{Out}_{g,1,0} \ltimes \check{\Sigma}_{g,1,0} \ltimes \check{\Sigma}_{g,2,0} \ltimes \cdots \ltimes \check{\Sigma}_{g,b+p-1,0}.$$

Since $\langle \breve{\sigma}_j \mid b+1 \leqslant j \leqslant b+p \rangle$ lies in the normal subgroup

$$\breve{\Sigma}_{g,b,0} \ltimes \breve{\Sigma}_{g,b+1,0} \ltimes \cdots \ltimes \breve{\Sigma}_{g,b+p-1,0},$$

we get the result. $\square$

14. Description of an action

14.1. Remarks. Let $b \geqslant 1$.

By Theorem 12.5, $\breve{\Sigma}_{g,b-1,p}$ is a normal subgroup of $\mathrm{Out}_{g,b-1\perp1,p}$; we want to describe the resulting action by conjugation.

Since $\breve{\sigma}_b$ is fixed by positive elements and inverted by negative elements, it remains to describe the action on the other generators of $\breve{\Sigma}_{g,b-1,p}$.

Recall that $\Sigma_{g,b,p}$ is free on $X_{g,b-1,p}$, and has a distinguished element, z_b.

By Proposition 7.1(iii), $\mathrm{Out}_{g,b-1\perp1,p} \simeq \mathrm{Aut}_{g,b-1\perp\hat{\imath},p}$, and the latter acts on $\Sigma_{g,b,p}$, fixing z_b.

Definitions 12.2 give a homomorphism $\Sigma_{g,b,p} \to \breve{\Sigma}_{g,b-1,p}$, $x \mapsto \breve{x}$; for example, $\breve{z}_b = \breve{\sigma}_b^{-X_{g,b-1,p}}$.

Consider any $v \in \Sigma_{g,b,p}$ and any $\alpha \in \mathrm{Aut}_{g,b-1\perp\hat{\imath},p}$. Thus v, $v^\alpha \in \Sigma_{g,b,p}$ and $\breve{v}$, $\breve{v^\alpha} \in \breve{\Sigma}_{g,b-1,p}$. Since $\breve{\alpha}$ normalizes $\breve{\Sigma}_{g,b-1,p}$, we see that $\breve{v}^{\breve{\alpha}}$ lies in $\breve{\Sigma}_{g,b-1,p}$. Moreover, under the composite

$$\mathrm{Out}_{g,b-1\perp1,p} \to \mathrm{Out}_{g,b-1,p\perp1} \xrightarrow{\sim} \mathrm{Aut}_{g,b-1,p},$$

both $\widetilde{v^\alpha}$ and $\breve{v}^{\breve{\alpha}}$ map to $\widetilde{v^\alpha} = \tilde{v}^\alpha$. Since the kernel of this composite is generated by $\breve{\sigma}_b$, we see that $\breve{v}^{\breve{\alpha}} = \breve{v^\alpha} \breve{\sigma}_b^n$ for some integer n. To describe the action, it remains to describe n in terms of v and α. $\square$

14.2. Definition. Suppose that $b \geqslant 1$.

We now define the *twisting-number map* tw$\colon \Sigma_{g,b,p} \to \mathbb{Z}$.

For each $v \in \Sigma_{g,b,p}$, define the *twisting number* of v, tw(v), as follows. Recall that $\Sigma_{g,b,p}$ is a free group, with basis $X_{g,b-1,p}$. When v is expressed as a reduced word in this basis, tw(v) sums the number of occurrences in v of subwords of the form

$$t_k, \quad \overline{x}_i, \quad y_i, \quad z_j, \quad x_i \cdots y_i, \quad y_i \cdots \overline{x}_i, \quad \overline{y}_i \cdots x_i, \quad \overline{x}_i \cdots \overline{y}_i,$$

and subtracts the number of occurrences in v of subwords of the form

$$\overline{t}_k, \quad x_i, \quad \overline{y}_i, \quad \overline{z}_j, \quad \overline{y}_i \cdots \overline{x}_i, \quad x_i \cdots \overline{y}_i, \quad \overline{x}_i \cdots y_i, \quad y_i \cdots x_i;$$

here $x_i \cdots y_i$ represents all reduced words that begin with x_i and end with y_i. For example,

$$\mathrm{tw}([x_i, y_i]) = \mathrm{tw}(\overline{x}_i \overline{y}_i x_i y_i) = 0 + 0 + 1 - 1 + 1 + 1 = 2$$

and

$$\mathrm{tw}(\overline{z}_b) = p + 2g + b - 1 = 2 - X_{g,b-1,p}.$$ $\square$

14.3. Remarks. (i) For a reduced word w in a given basis X of a free group F, let $f_w \colon F \to \mathbb{Z}$ be the map which assigns, to each reduced word v in $X \cup X^{-1}$, the number of occurrences of w as a subword of v, minus the number of occurrences of $\overline{w}$ as a subword of v. An interesting history of the use of such maps, especially in the study of $\Sigma_{g,0,1}$, can be found in [3, Section 1.1].

The twisting-number map is a sum of infinitely many such maps.

(ii) One can express the twisting-number map as the result of applying the endomorphism

$$\sum_{k=1}^{p} \frac{\partial}{t_k \partial} + \sum_{i=1}^{g}\left(\frac{\partial}{y_i \partial} - \frac{\partial}{x_i \partial} + \frac{\partial^2}{x_i \partial \cdot y_i \partial} - \frac{\partial^2}{y_i \partial \cdot x_i \partial} \right) + \sum_{j=1}^{b-1} \frac{\partial}{z_j \partial}$$

of $\mathbb{Z}[\Sigma_{g,b,p}]$, and then applying the augmentation map $\mathbb{Z}[\Sigma_{g,b,p}] \to \mathbb{Z}$. $\square$

We can now describe the action of $\mathrm{Out}_{g,b-1\perp 1,p}$ on $\check{\Sigma}_{g,b-1,p}$.

14.4. Theorem. *Suppose that* $b \geqslant 1$. *Let* $v \in \Sigma_{g,b,p}$ *and* $\alpha \in \mathrm{Aut}_{g,b-1\perp \hat{1},p}$. *Then*

$$\widehat{v}^{\alpha} = \widehat{v^{\alpha}}\, \sigma_b^{\mathrm{tw}(v^{\alpha}) - \mathrm{sign}(\alpha)\,\mathrm{tw}(v)} \tag{14.4.1}$$

in $\mathrm{Aut}_{g,b,p}$. *Hence* $\check{v}^{\check{\alpha}} = \widecheck{v^{\alpha}}\, \check{\sigma}_b^{\mathrm{tw}(v^{\alpha}) - \mathrm{sign}(\alpha)\,\mathrm{tw}(v)}$ *in* $\mathrm{Out}_{g,b,p}$.

Proof. Let $\beta = (\widehat{v}^{\alpha})^{-1}\, \widehat{v^{\alpha}}\, \sigma_b^{\mathrm{tw}(v^{\alpha}) - \mathrm{sign}(\alpha)\,\mathrm{tw}(v)}$. It follows from Remarks 14.1 that we can write $\beta = \sigma_b^m \widetilde{w}^{-1}$ for some $m \in \mathbb{Z}$ and some $w \in \Sigma_{g,b,p}$. We want to show that $m = 0$ and $w = 1$.

Now $e_b^{\beta} = w z_b^m e_b$ and, if $b \geqslant 2$, then $e_{b-1}^{\beta} = w e_{b-1}$.

In this paragraph we consider consequences of making z_b central in $\Sigma_{g,b,p} * E_b$. Notice that $\mathrm{Aut}_{g,b-1\perp 1,p}$ acts on this quotient group, and, in particular, so does $\widehat{x}$ for each $x \in X_{g,b-1,p}$. Explicitly, for $p \geqslant k \geqslant 1$, $1 \leqslant i \leqslant g$, $1 \leqslant j \leqslant b-1$, if z_b is made central then we can write the following.

$$\widehat{t}_k : \begin{cases} w \mapsto w & \text{for} \quad w \in X_{g,b,p} \cup E_{b-1}, \\ e_b \mapsto z_b t_k e_b. \end{cases}$$

$$\widehat{x}_i : \begin{cases} w \mapsto w & \text{for} \quad w \in E_{b-1} \cup X_{g,b,p} - \{y_i\}, \\ y_i \mapsto \overline{z}_b y_i, \\ e_b \mapsto x_i \overline{z}_b e_b. \end{cases}$$

$$\widehat{y}_i : \begin{cases} w \mapsto w & \text{for} \quad w \in E_{b-1} \cup X_{g,b,p} - \{x_i\}, \\ x_i \mapsto z_b x_i, \\ e_b \mapsto z_b y_i e_b. \end{cases}$$

$$\hat{z}_j : \begin{cases} w \mapsto w & \text{for} \quad w \in X_{g,b,p} \cup E_{b-1} - \{e_j\}, \\ e_j \mapsto z_b e_j, \\ e_b \mapsto z_b z_j e_b. \end{cases}$$

It is straightforward to check that $e_b^{\hat{v}} = v z_b^{\text{tw}(v)} e_b$. Hence $e_b^{\hat{v}^\alpha} = v^\alpha z_b^{\text{tw}(v^\alpha)} e_b$, and

$$e_b^{\overline{v}^\alpha} = e_b^{\widehat{\overline{\alpha} v \alpha}} = e_b^{\widehat{v}\alpha} = (v z_b^{\text{tw}(v)} e_b)^\alpha = v^\alpha z_b^{\text{sign}(\alpha)\,\text{tw}(v)} e_b.$$

It follows that β fixes e_b.

In summary, $w z_b^m$ is annihilated if z_b is made central.

Let us first consider the special case where $b \geqslant 2$, and α fixes

$$\{(z_{b-1}, e_{b-1}, z_b, e_b), (\overline{z}_{b-1}^{z_b}, \overline{z}_b e_{b-1}, \overline{z}_b, e_b)\},$$

and v does not involve z_{b-1}, that is, $v \in \langle X_{g,b-2,p}\rangle$. It follows that $e_{b-1}^\beta = e_{b-1}$. But $e_{b-1}^\beta = w e_{b-1}$, so $w = 1$. Thus, making z_b central annihilates z_b^m. Since $b \geqslant 2$, z_b is a primitive element of the free group $\Sigma_{g,b,p} * E_b$. It follows that $m = 0$. Thus this special case of the theorem holds.

If we now apply $\text{elim}(z_{b-1}, e_{b-1})$ to the foregoing special case, we get an $\alpha' \in \text{Aut}_{g,b-2\perp\hat{1},p}$ and a $v' \in \Sigma_{g,b-1,p}$ for which the conclusion of the theorem holds. By considering $\text{pinch}(z_b)\colon \text{Out}_{g,b-1\perp1,p} \to \text{Out}_{g,b-1\perp1\perp1,p}$, we see that the general case arises in this way. $\qquad\square$

14.5. Remark. It is not difficult to check that (14.4.1) can be used to define an action of $\text{Out}_{g,b-1\perp1,p}$ on $\Sigma_{g,b,p} \times \langle\sigma_b\rangle$. In checking a condition of the form $\widehat{vw}^\alpha = \hat{v}^\alpha\hat{w}^\alpha$, it is useful to know that $\langle v^\alpha, w^\alpha\rangle = \text{sign}(\alpha)\langle v, w\rangle$, where $\langle v, w\rangle$ denotes the result of applying the augmentation homomorphism to

$$\sum_{i=1}^{g}\left(\frac{v\partial}{x_i\partial}\frac{w\partial}{y_i\partial} - \frac{v\partial}{y_i\partial}\frac{w\partial}{x_i\partial}\right).$$

Notice that $\langle -, -\rangle$ can be given via the usual symplectic product on the abelianization of $\Sigma_{g,b,p}$.

It might be interesting to investigate the connection between the foregoing and the topological analysis of Humphries-Johnson [12]. $\qquad\square$

We can now calculate with some of the Dehn-twist automorphisms of Definitions 3.10.

14.6. Lemma. *If $b \geqslant 1$ and $g \geqslant 1$, then the following hold in $\mathrm{Aut}^{+}_{g,b-1\perp1,p}$.*

$$\widehat{x}_i^{\,\alpha_i} = \widehat{x_i^{\alpha_i}} = \widehat{y}_i^{\,-1}\widehat{x}_i, \ \text{if} \ 1 \leqslant i \leqslant g.$$

$$\widehat{y}_i^{\,\beta_i} = \widehat{y_i^{\beta_i}} = \widehat{x}_i\widehat{y}_i, \ \text{if} \ 1 \leqslant i \leqslant g.$$

$$\widehat{x}_1^{\,\gamma_0} = \widehat{x_1^{\gamma_0}}\,\sigma_b = \widehat{x}_1\,\widehat{t}_p\,\widehat{x}_1^{\,-1}\,\widehat{y}_1^{\,-1}\,\widehat{x}_1\,\sigma_b, \ \text{if} \ p \geqslant 1.$$

$$\widehat{x}_i^{\,\gamma_i} = \widehat{x_i^{\gamma_i}}\,\sigma_b^{-1} = \widehat{x}_{i+1}^{\,-1}\,\widehat{y}_{i+1}\,\widehat{x}_{i+1}\,\widehat{y}_i^{\,-1}\,\widehat{x}_i\,\sigma_b^{-1}, \ \text{if} \ 1 \leqslant i \leqslant g-1.$$

$$\widehat{y}_i^{\,\gamma_i} = \widehat{y_i^{\gamma_i}} = \widehat{x}_{i+1}^{\,-1}\,\widehat{y}_{i+1}\,\widehat{x}_{i+1}\,\widehat{y}_i\,\widehat{x}_{i+1}^{\,-1}\,\widehat{y}_{i+1}^{\,-1}\,\widehat{x}_{i+1}, \ \text{if} \ 1 \leqslant i \leqslant g-1.$$

$$\widehat{x}_{i+1}^{\,\gamma_i} = \widehat{x_{i+1}^{\gamma_i}}\,\sigma_b = \widehat{x}_{i+1}\,\widehat{y}_i\,\widehat{x}_{i+1}^{\,-1}\,\widehat{y}_{i+1}^{\,-1}\,\widehat{x}_{i+1}\sigma_b, \ \text{if} \ 1 \leqslant i \leqslant g-1.$$

$$\widehat{x}_g^{\,\gamma_g} = \widehat{x_g^{\gamma_g}}\,\sigma_b^{-1} = \widehat{z}_1^{\,-1}\,\widehat{y}_g^{\,-1}\,\widehat{x}_g\,\sigma_b^{-1}, \ \text{if} \ b \geqslant 2.$$

Proof. We compute

$$\mathrm{tw}(x_i^{\alpha_i}) = \mathrm{tw}(\overline{y}_i x_i) = -(1) + (-1) + 1 = -1,$$

$$\mathrm{tw}(y_i^{\beta_i}) = \mathrm{tw}(x_i y_i) = (-1) + (1) + 1 = 1,$$

$$\mathrm{tw}(x_1^{\gamma_0}) = \mathrm{tw}(x_1 t_p \overline{x}_1 \overline{y}_1 x_1),$$
$$= (-1) + 1 - (-1) - 1 + (-1) + 1 + (-1) + 1 = 0,$$

$$\mathrm{tw}(x_i^{\gamma_i}) = \mathrm{tw}(\overline{x}_{i+1} y_{i+1} x_{i+1} \overline{y}_i x_i)$$
$$= -(-1) + (1) + (-1) - (1) + (-1) - 1 - 1 + 1 = -2,$$

$$\mathrm{tw}(y_i^{\gamma_i}) = \mathrm{tw}(\overline{x}_{i+1} y_{i+1} x_{i+1} y_i \overline{x}_{i+1} \overline{y}_{i+1} x_{i+1}) = \mathrm{tw}(y_i),$$

$$\mathrm{tw}(x_{i+1}^{\gamma_i}) = \mathrm{tw}(x_{i+1} y_i \overline{x}_{i+1} \overline{y}_{i+1} x_{i+1})$$
$$= (-1) + (1) - (-1) - (1) + (-1) - 1 + 1 + 1 = 0.$$

$$\mathrm{tw}(x_g^{\gamma_g}) = \mathrm{tw}(\overline{z}_1 \overline{y}_g x_g) = -(1) - (1) + (-1) + 1 = -2,$$

Hence

$$\mathrm{tw}(x_i^{\alpha_i}) - \mathrm{tw}(x_i) = (-1) - (-1) = 0,$$

$$\mathrm{tw}(y_i^{\beta_i}) - \mathrm{tw}(y_i) = 1 - 1 = 0,$$

$$\mathrm{tw}(x_1^{\gamma_0}) - \mathrm{tw}(x_1) = (0) - (-1) = 1,$$

$$\mathrm{tw}(x_i^{\gamma_i}) - \mathrm{tw}(x_i) = (-2) - (-1) = -1,$$

$$\mathrm{tw}(y_i^{\gamma_i}) - \mathrm{tw}(y_i) = 0,$$

$$\mathrm{tw}(x_{i+1}^{\gamma_i}) - \mathrm{tw}(x_{i+1}) = (0) - (-1) = 1,$$

$$\mathrm{tw}(x_g^{\gamma_g}) - \mathrm{tw}(x_g) = (-2) - (-1) = -1.$$

The result now follows. □

The foregoing lemma is useful for calculating normal closures, and we now record one case.

14.7. Theorem (Wajnryb). *If $g \geqslant 2$ and $b \geqslant 1$, then, the normal closure of $\check{\gamma}_0 \, \check{\alpha}_1^{-1}$ in $\mathrm{Out}^{+}_{g,b-1\perp1,0}$ is the subgroup $\check{\Sigma}_{g,b-1,0}$ which occurs in the exact sequence*

$$1 \to \check{\Sigma}_{g,b-1,0} \to \mathrm{Out}^{+}_{g,b-1\perp1,0} \xrightarrow{\mathrm{elim}(z_b, e_b)} \mathrm{Out}^{+}_{g,b-1,0} \to 1.$$

Proof. It is not difficult to show that, in $\mathrm{Aut}_{g,b,0}$,

$$\gamma_0 \alpha_1^{-1} = \hat{x}_1^{-1} \, \hat{y}_1 \, \hat{x}_1 : \begin{cases} x_1 & \mapsto & y_1 x_1 z_b \overline{x}_1 \overline{y}_1 x_1, \\ e_b & \mapsto & \overline{x}_1 \overline{y}_1 x_1 z_b e_b, \\ z_b^{e_b} & \mapsto & z_b^{e_b}, \\ w & \mapsto & w \quad \text{for all } w \in X_{g,b,0} \cup E_b - \{x_1, z_b, e_b\}. \end{cases}$$

Hence, in $\mathrm{Out}^{+}_{g,b-1\perp1,0}$, we have $\check{\gamma}_0 \, \check{\alpha}_1^{-1} = \check{x}_1^{-1} \, \check{y}_1 \, \check{x}_1$. By applying Lemma 14.6, we deduce that the normal closure of this element in $\mathrm{Out}^{+}_{g,b-1\perp1,0}$ is all of $\check{\Sigma}_{g,b-1,0}$. □

15. Examples

In this section we describe some illustrative examples taken from topology, some of which do not have algebraic justifications at the time of writing.

15.1. Example. A sphere with p punctures, $S_{0,0,p}$.

By Notation 4.3, $\mathrm{Out}_{0,0,p}$ is an extension of $\mathrm{Out}_{0,0,1\perp p}$ by $C_2 \times \mathrm{Sym}_p$. By Corollary 11.2, $\mathrm{Out}_{0,0,1\perp p}$ has an ascending normal series with sequence of factor groups

$$\widetilde{\Sigma}_{0,0,p-1}, \widetilde{\Sigma}_{0,0,p-2}, \cdots, \widetilde{\Sigma}_{0,0,1}.$$

(Each $\widetilde{\Sigma}_{0,0,i}$ is free, so the normal series gives rise to a decomposition of the form

$$\mathrm{Out}_{0,0,1\perp p} = \widetilde{\Sigma}_{0,0,p-1} \rtimes \widetilde{\Sigma}_{0,0,p-2} \rtimes \cdots \rtimes \widetilde{\Sigma}_{0,0,0},$$

where left parentheses are understood to accumulate on the left. Such an expression is not as informative as the sequence of free factor groups of a normal series.)

If $0 \leqslant i \leqslant 2$, then $\widetilde{\Sigma}_{0,0,i} = 1$, and, if $i \geqslant 3$, then $\widetilde{\Sigma}_{0,0,i} = \Sigma_{0,0,i}$ is free of rank $i - 1$.

If $p \leqslant 3$, then $\mathrm{Out}_{0,0,p} = C_2 \times \mathrm{Sym}_p$.

The group $\mathrm{Out}_{0,0,4}$ is an extension of a free group of rank two by $C_2 \times \mathrm{Sym}_4$, and the latter is isomorphic to $\mathrm{PGL}_2(\mathbb{Z}/2\mathbb{Z}) \ltimes (\mathbb{Z}/2\mathbb{Z})^2$; in Remarks 15.8(v) and (viii), we shall see that $\mathrm{Out}_{0,0,4} \simeq \mathrm{PGL}_2(\mathbb{Z}) \ltimes (\mathbb{Z}/2\mathbb{Z})^2$.

The groups $\mathrm{Out}_{0,0,5}$ and $\mathrm{Out}_{0,0,6}$ have connections with Remarks 15.8(iv) and Example 15.5, respectively. □

15.2. Example. A closed disc with p punctures, $S_{0,1,p}$.

By Proposition 7.1(iv), $\mathrm{Out}_{0,1,p} \simeq \mathrm{Aut}_{0,0,p\perp\hat{\imath}}$, a group of automorphisms of a free group of rank p.

The subgroup $\mathrm{Out}^{+}_{0,1,p}$ is *the braid group on p strings.* Emil Artin showed that, in terms of Definitions 6.2 and Example 6.3,

$$\mathrm{Out}^{+}_{0,1,p} = \mathrm{Artin}[\check{\mu}_1 - \check{\mu}_2 - \check{\mu}_3 - \cdots - \check{\mu}_{p-2} - \check{\mu}_{p-1}]. \tag{15.2.1}$$

The subgroup $\text{Out}_{0,1,1^{\perp p}}^{+}$ is the *pure* braid group on p strings. By Corollary 11.5, we have a decomposition

$$\text{Out}_{0,1,1^{\perp p}}^{+} \quad \simeq \quad \Sigma_{0,1,1} \ltimes \Sigma_{0,1,2} \ltimes \cdots \ltimes \Sigma_{0,1,p-1}, \qquad (15.2.2)$$

with each $\Sigma_{0,1,i}$ free of rank i. This decomposition can be deduced directly from (15.2.1), and vice-versa.

The braid group on p strings, $\text{Out}_{0,1,p}^{+}$, modulo its center, $\langle \check{\sigma}_1 \rangle$, is isomorphic to $\text{Out}_{0,0,p\perp 1}^{+}$. This is a trivial group if $p \leqslant 2$, while $\text{Out}_{0,0,3\perp 1}^{+} \simeq \text{PSL}_2(\mathbb{Z})$, and $\text{Out}_{0,0,4\perp 1}^{+} \simeq \text{Aut}^{+}(F_2)$; see Remarks 15.8(viii) and (iv). □

15.3. Remarks. Let us mention some history concerning (15.2.1).

In 1925, Artin [1] proved (15.2.1) by an intuitive topological argument, and, in 1947, Artin [2] indicated that there were difficulties that could be corrected.

In 1934, Magnus [18] gave an algebraic proof of (15.2.1).

In 1945, Markov [19] gave a similar algebraic proof.

In 1947, Artin [2] and Bohnenblust [5] gave a similar algebraic proof, in parallel articles; Artin proved that the generators suffice, and Bohnenblust proved that the relations suffice. In 1948, Chow [7] simplified the latter proof.

All the algebraic proofs of the sufficiency of the relations that are cited above involve some form of the Reidemeister-Schreier rewriting process and the decomposition corrresponding to $\text{Out}_{0,1,p-1\perp 1}^{+} \simeq \text{Out}_{0,1,p-1}^{+} \ltimes \widetilde{\Sigma}_{0,1,p-1}$.

Larue [15] and Shpilrain [25] independently gave a new algebraic proof of the sufficiency of the relations, by using a result of Dehornoy [8]. □

15.4. Notation. Let $\Sigma_{g,b,p^{(2)}}$ denote the group obtained from $\Sigma_{g,b,p}$ by imposing the relations $t_p^2 = t_{p-1}^2 = \cdots = t_1^2 = 1$. This is the orbifold group of the surface $S_{g,b,p^{(2)}}$ with p double points, as opposed to punctures. There is little topological difference between a double point and a puncture.

There are natural sign-preserving maps

$$\text{Aut}_{g,b,p} \to \text{Aut}_{g,b,p^{(2)}} \quad \text{and} \quad \text{Out}_{g,b,p} \to \text{Out}_{g,b,p^{(2)}} .$$

The latter map can be shown to be an isomorphism by using topological arguments. We shall apply this fact only in this section, which is dedicated to examples. □

We now look at the hyperelliptic involution. This is an order-two, orientation-preserving homeomorphism of the surface of genus g induced by a $180°$-rotation of Euclidean three-space about the x-axis, where the surface is embedded so as to meet the axis in $2g + 2$ points, and to be invariant under the rotation.

15.5. Example. $S_{g,0,0}$ is a double branched cover of $S_{0,0,(2g+2)^{(2)}}$.

It is not difficult to show that there is a decomposition of the form

$$\Sigma_{0,0,(2g+2)^2} = \Sigma_{g,0,0} \rtimes C_2.$$

Here $\Sigma_{g,0,0}$ is the unique torsion-free, index-two subgroup of $\Sigma_{0,0,(2g+2)^{(2)}}$, and hence is a characteristic subgroup. Thus there is a sign-preserving restriction map

$$\mathrm{Aut}_{0,0,(2g+2)^{(2)}} \to \mathrm{Aut}_{g,0,0},$$

and an induced map

$$\mathrm{Aut}_{0,0,(2g+2)^{(2)}} / \Sigma_{g,0,0} \to \mathrm{Out}_{g,0,0}, \qquad (15.5.1)$$

where we understand that $\Sigma_{g,0,0}$ represents the image of the composite of natural maps

$$\Sigma_{g,0,0} \to \Sigma_{0,0,(2g+2)^{(2)}} \to \mathrm{Aut}_{0,0,(2g+2)^{(2)}} \,.$$

For $g = 0$, $\mathrm{Out}_{0,0,0} = \mathrm{Out}_{0,0,2} = C_2$, and (15.5.1) is bijective.

For $g \geqslant 1$, the domain of (15.5.1) is an extension of C_2 by $\mathrm{Out}_{0,0,2g+2}$, since there is a short exact sequence

$$1 \to C_2 \to \mathrm{Aut}_{0,0,(2g+2)^{(2)}} / \Sigma_{g,0,0} \to \mathrm{Out}_{0,0,(2g+2)^{(2)}} \to 1.$$

If $g = 1$, then (15.5.1) has kernel of order four, and is split surjective; see Remarks 15.8(v) and (viii).

For $g = 2$, Birman-Hilden [4] showed that (15.5.1) is an isomorphism, and hence

$$\mathrm{Aut}_{0,0,6^{(2)}} \simeq \mathrm{Aut}_{2,0,0} \quad \text{and} \quad \mathrm{Out}_{0,0,6} \simeq \mathrm{Out}_{2,0,0} / C_2.$$

If $g \geqslant 3$, then (15.5.1) is injective, but not surjective.   □

We now remove an invariant disc around one of the skewering points. We note that the hyperelliptic involution is not admissible here, so does not give rise to an element of order 2 in the corresponding mapping-class group.

15.6. Example. $S_{g,1,0}$ is a double branched cover of $S_{0,1,(2g+1)^{(2)}}$.

Again, the group

$$\Sigma_{0,1,(2g+1)^{(2)}} = \langle z_k \; (1 \leqslant k \leqslant 2g+2) \mid \prod_{k=1}^{2g+2} z_k = 1, z_k^2 = 1 \; (2 \leqslant k \leqslant 2g+2) \rangle$$

has a decomposition of the form $\Sigma_{0,1,(2g+1)^{(2)}} = \Sigma_{g,1,0} \rtimes C_2$. Explicitly, there is a homomorphism

$$\Sigma_{g,1,0} * E_1 \qquad \to \qquad \Sigma_{0,1,(2g+1)^{(2)}} * E_1$$

given by

$$x_i \qquad \mapsto \qquad z_{2i+2} z_{2i+1} \qquad \text{for } 1 \leqslant i \leqslant g,$$

$$y_i \qquad \mapsto \qquad z_{2i+2} \cdot \prod_{k=2}^{2i+2} z_k \qquad \text{for } 1 \leqslant i \leqslant g,$$

$$z_1 \qquad \mapsto \qquad z_1^2,$$

$$e_1 \qquad \mapsto \qquad e_1.$$

Here

$$[x_i, y_i] \qquad \longmapsto \qquad \coprod_{k=2i}^{2} z_k \cdot \coprod_{k=2i+1}^{2i+2} z_k \cdot \coprod_{k=2}^{2i+2} z_k.$$

It follows that, for $1 \leqslant i \leqslant g$,

$$\prod_{i'=1}^{i} [x_{i'}, y_{i'}] \qquad \longmapsto \qquad \coprod_{k=2}^{2i+2} z_k \cdot \coprod_{k=2}^{2i+2} z_k = (\coprod_{k=2}^{2i+2} z_k)^2,$$

and we see that the map is well-defined. It is not difficult to show that it is injective, and identifies $\Sigma_{g,1,0}$ with a characteristic, index-two subgroup of $\Sigma_{0,1,(2g+1)^{(2)}}$. It can be shown that the u_i used in Remarks 5.1 maps to $z_{i+1} z_{i+2}$.

The homomorphism $\Sigma_{g,1,0} * E_1 \to \Sigma_{0,1,(2g+1)^{(2)}} * E_1$ determines a sign-preserving homomorphism $\mathrm{Aut}_{0,\hat{1},2g+1} \to \mathrm{Aut}_{g,\hat{1},0}$, and this in turn is equivalent to a homomorphism

$$\mathrm{Out}_{0,1,2g+1} \to \mathrm{Out}_{g,1,0}. \tag{15.6.1}$$

It can be shown that (15.6.1) is injective. For $g \leqslant 1$, (15.6.1) is bijective; see Remarks 15.8(vi) and (viii).

In (15.6.1), $\breve{\sigma}_1^2 \ (= \sigma_1^2 \breve{z}_1^{-2})$ is mapped to $\breve{\sigma}_1^{-1} \ (= \sigma_1^{-1} \breve{z}_1)$, and we get an injective quotient map

$$\mathrm{Out}_{0,1,2g+1} / \langle \breve{\sigma}_1^2 \rangle \to \mathrm{Out}_{g,1,0} / \langle \breve{\sigma}_1 \rangle.$$

For $g \geqslant 1$, this says that an extension of C_2 by $\mathrm{Out}_{0,0,2g+1 \perp 1}$ embeds in $\mathrm{Out}_{g,0,1}$. It follows, from Example 15.5, that this embedding remains injective even after composition with

$$\mathrm{elim}(t_1) \colon \mathrm{Out}_{g,0,1} \to \mathrm{Out}_{g,0,0}.$$

In general, we have the braid group on $2g+1$ strings, $\mathrm{Out}_{0,1,2g+1}^{+}$, embedded in $\mathrm{Out}_{g,1,0}^{+}$; with reference to Example 6.3, the corresponding map of Artin diagrams is as follows.

$$
\begin{array}{ccccccccc}
\breve{\mu}_1 & - & \breve{\mu}_2 & - & \breve{\mu}_3 & - & \breve{\mu}_4 & - & \breve{\mu}_5 & - & \cdots & - & \breve{\mu}_{2g-1} & - & \breve{\mu}_{2g} \\
\downarrow & & \downarrow & & \downarrow & & \downarrow & & \downarrow & & & & \downarrow & & \downarrow \\
\breve{\alpha}_1 & - & \breve{\beta}_1 & - & \breve{\gamma}_1 & - & \breve{\beta}_2 & - & \breve{\gamma}_2 & - & \cdots & - & \breve{\gamma}_{g-1} & - & \breve{\beta}_g.
\end{array}
\qquad \square
$$

Rather than remove one invariant disc, we can remove a pair of disjoint discs which are interchanged by the rotation.

15.7. Example. $S_{g,2,0}$ is a double branched cover of $S_{0,1,(2g+2)^{(2)}}$. Here there is a homomorphism

$$\Sigma_{g,2,0} * E_2 \qquad \longrightarrow \qquad \Sigma_{0,1,(2g+2)^{(2)}} * E_1$$

given by

$$x_i \quad \mapsto \quad z_{2i+2}z_{2i+1} \quad \text{for } 1 \leqslant i \leqslant g,$$

$$y_i \quad \mapsto \quad z_{2i+2} \cdot \prod_{k=2}^{2i+2} z_k \quad \text{for } 1 \leqslant i \leqslant g,$$

$$z_1 \quad \mapsto \quad z_1^{z_{2g+3}},$$

$$e_1 \quad \mapsto \quad z_{2g+3}e_1,$$

$$z_2 \quad \mapsto \quad z_1,$$

$$e_2 \quad \mapsto \quad e_1.$$

This map is well defined, since

$$\prod_{i=1}^{g}[x_i, y_i]z_1z_2 \quad \mapsto \quad (\prod_{k=2}^{2g+2} z_k)^2(z_{2g+3}z_1)^2 = 1,$$

and it identifies $\Sigma_{g,2,0}$ with a characteristic, index-two subgroup of $\Sigma_{0,1,(2g+2)^{(2)}}$. Notice that $\Sigma_{0,1,(2g+2)^{(2)}}e_1$ has a corresponding partition in two. Here we get a homomorphism $\mathrm{Aut}_{0,1,2g+2} \rightarrow \mathrm{Aut}_{g,2,0}$. For example, the image of σ_1 is $\sigma_1\sigma_2$, and the image of $\tilde{z}_{2g+3}$ interchanges e_1 and e_2. Now $\mathrm{Aut}_{0,\hat{1},2g+2}$ is mapped to $\mathrm{Aut}_{g,1\perp\hat{1},0}$, so we get a homomorphism

$$\mathrm{Out}_{0,1,2g+2} \rightarrow \mathrm{Out}_{g,1\perp1,0}. \tag{15.7.1}$$

Topological arguments can be used to show the injectivity of (15.7.1); see [24].

For $g = 0$, (15.7.1) is bijective.

For $g \geqslant 1$, since (15.7.1) carries $\breve{\sigma}_1 = \sigma_1\tilde{z}_1^{-1}$ to $\breve{\sigma}_1\breve{\sigma}_2 = \sigma_1\sigma_2\tilde{z}_2^{-1}$, composing (15.7.1) with $\mathrm{elim}(e_2)$ and with $\mathrm{elim}(e_1, e_2)$ gives embeddings

$$\mathrm{Out}_{0,1,2g+2} \rightarrow \mathrm{Out}_{g,1,1} \quad \text{and} \quad \mathrm{Out}_{0,0,2g+2\perp1} \rightarrow \mathrm{Out}_{g,0,1\perp1},$$

respectively. For $g = 1$, these are bijective; see Remarks 15.8(ii) and (iv).

For $g \geqslant 1$, the braid group $\mathrm{Out}^+_{0,1,2g+2}$ embeds in $\mathrm{Out}^+_{g,1,1}$, and the corresponding map of Artin diagrams is as follows.

$$\breve{\mu}_1 - \breve{\mu}_2 - \breve{\mu}_3 - \breve{\mu}_4 - \breve{\mu}_5 - \cdots - \breve{\mu}_{2g-1} - \breve{\mu}_{2g} - \breve{\mu}_{2g+1}$$
$$\downarrow \quad\quad \downarrow \quad\quad \downarrow \quad\quad \downarrow \quad\quad \downarrow \quad\quad\quad\quad \downarrow \quad\quad\quad \downarrow \quad\quad\quad \downarrow$$
$$\breve{\alpha}_1 - \breve{\beta}_1 - \breve{\gamma}_1 - \breve{\beta}_2 - \breve{\gamma}_2 - \cdots - \breve{\gamma}_{g-1} - \breve{\beta}_g - \breve{\gamma}_g. \qquad \square$$

15.8. Remarks. Let us collect together some of the results related to the sphere and the torus, as follows. Let V denote $C_2 \times C_2$. Let F_2 denote a free group of rank two, $F_2 \simeq \Sigma_{1,1} \simeq \Sigma_{0,3}$. The following hold.

(i) $\mathrm{Aut}_{1,0,2} \simeq \mathrm{Aut}_{0,0,4^{(2)}\perp1}$.

(ii) $\mathrm{Out}_{1,1,1} \simeq \mathrm{Out}_{0,1,4}$
$\simeq \mathrm{Aut}_{1,1,0} \simeq \mathrm{Aut}_{1,0,\hat{1}} \ltimes F_2 \simeq \mathrm{Aut}_{0,0,4\perp\hat{1}} \simeq \mathrm{Aut}_{1,0,1\perp\hat{1}}$.

(iii) $\mathrm{Out}_{1,0,2} = \mathrm{Out}_{1,0,1\perp1} \times C_2$.

(iv) $\mathrm{Out}_{1,0,2}/C_2 \simeq \mathrm{Out}_{1,0,1\perp 1} \simeq \mathrm{Out}_{0,0,4\perp 1}$
$\simeq \mathrm{Aut}_{0,0,4} \simeq \mathrm{Aut}_{1,0,1} \simeq \mathrm{Aut}(F_2) \simeq \mathrm{Aut}_{0,0,3^{(2)}\perp 1}$.

(v) $\mathrm{Out}_{0,0,4} = \mathrm{Out}_{0,0,3\perp 1} \ltimes V$.

(vi) $\mathrm{Out}_{1,1,0} \simeq \mathrm{Out}_{0,1,3}$
$\simeq \mathrm{Aut}_{1,0,\hat{\imath}} \simeq \mathrm{Aut}_{0,0,3\perp \hat{\imath}}$.

(vii) $\mathrm{Out}_{1,0,1} \simeq \mathrm{Out}_{1,0,0}$
$\simeq \mathrm{Aut}_{1,0,0} \simeq \mathrm{GL}_2(\mathbb{Z})$.

(viii) $\mathrm{Out}_{0,0,4}/V \simeq \mathrm{Out}_{1,0,1}/C_2 \simeq \mathrm{Out}_{0,0,3\perp 1}$
$\simeq \mathrm{PGL}_2(\mathbb{Z}) \simeq \mathrm{Aut}_{0,0,3}$.

The isomorphisms of the positive subgroups are of interest since $\mathrm{Out}_{0,1,4}^+$ (see (ii)) is the braid group on four strings, and passing modulo the center gives $\mathrm{Out}_{0,0,4\perp 1}^+$ (see (iv)). Similarly, $\mathrm{Out}_{0,1,3}^+$ (see (vi)) is the braid group on three strings, and passing modulo the center gives $\mathrm{Out}_{0,0,3\perp 1}^+$ (see (viii)).

Many of the interconnections arise from an action of $V \times C_2$ on $S_{1,0,4}$. Consider the affine action of $\mathbb{Z}^2 \rtimes \mathrm{GL}_2(\mathbb{Z})$ on the plane $\mathbb{R}^2$. This induces an action on the punctured plane $\mathbb{R}^2 - \mathbb{Z}^2$. Modulo the action of $(2\mathbb{Z})^2$, this gives an action of $V \rtimes \mathrm{GL}_2(\mathbb{Z})$ on $(\mathbb{R}^2 - \mathbb{Z}^2)/(2\mathbb{Z})^2 = S_{1,0,4}$. This in turn gives a subgroup $V \rtimes \mathrm{GL}_2(\mathbb{Z})$ of $\mathrm{Out}_{1,0,4}$, which we denote by $\mathrm{Out}_{1,0,4}^\dagger$ for the purposes of this digression. The subgroup $V \times C_2$ acts on $S_{1,0,4}$, with the generator of C_2 acting as the hyperelliptic involution fixing each puncture. We find $S_{1,0,4}/C_2 = S_{0,0,4}$, $S_{1,0,4}/V = S_{1,0,1}$ and $S_{1,0,4}/(V \times C_2) = S_{0,0,3^{(2)}}$. The resulting action of V on $S_{1,0,4}/C_2 = S_{0,0,4}$ corresponds to taking an appropriate group of symmetries of the two-skeleton of a regular tetrahedron, and deleting the (four) vertices.

The abelianization of $\Sigma_{0,0,3^{(2)}\perp 1}$ is $V \times C_2$, and we find that $\Sigma_{0,0,3^{(2)}\perp 1}$ is an extension of $\Sigma_{1,0,1}$ by C_2, of $\Sigma_{0,0,4}$ by V, and of $\Sigma_{1,0,4}$ by $V \times C_2$. The subgroups $\Sigma_{1,0,4}$, $\Sigma_{1,0,1}$, and $\Sigma_{0,0,4}$ of $\Sigma_{0,0,3^{(2)}\perp 1}$ are characteristic, or $\mathrm{Aut}_{0,0,3^{(2)}\perp 1}$-invariant, and it can be shown that there are natural identifications

$$\mathrm{Aut}_{0,0,3^{(2)}\perp 1} = \mathrm{Aut}_{1,0,1} = \mathrm{Aut}_{0,0,4} = \mathrm{Aut}_{1,0,4}^\dagger,$$

where $\mathrm{Aut}_{1,0,4}^\dagger$ denotes the image of the restriction map $\mathrm{Aut}_{0,0,3^{(2)}\perp 1} \to \mathrm{Aut}_{1,0,4}$. This gives identifications

$$\mathrm{Out}_{0,0,3^{(2)}\perp 1} = \mathrm{Out}_{1,0,1}/C_2 = \mathrm{Out}_{0,0,4}/V = \mathrm{Out}_{1,0,4}^\dagger/(V \times C_2).$$

Let us say a brief word about the proofs of (i)-(viii).

We first consider (iv). Notice that $\Sigma_{1,0,1}$ is free on $\{x_1, y_1\}$, that $\mathrm{Aut}_{1,0,\hat{\imath}}$ is the group of all automorphisms of $\Sigma_{1,0,1}$ which fix or invert the commutator $[x_1, y_1]$, and that $\mathrm{Aut}_{1,0,1}$ is the group of automorphisms of $\Sigma_{1,0,1}$ which carry the commutator to a conjugate of itself or its inverse. Now, by a result of Nielsen, $\mathrm{Aut}_{1,0,1}$ is all of $\mathrm{Aut}(\Sigma_{1,0,1}) \simeq \mathrm{Aut}(F_2)$. The isomorphism $\mathrm{Out}_{0,0,4\perp 1}^+ \to \mathrm{Aut}_{1,0,1}^+$ is described explicitly in [10], using presentations.

It is a simple matter to verify (i), (ii) and (iii) using (iv).

In (v), V is generated by $\breve{\mu}_1 \breve{\mu}_3^{-1}$, $\breve{\mu}_2 \breve{\mu}_1 \breve{\mu}_3^{-1} \breve{\mu}_2^{-1}$. We leave the proof of (v) as an exercise. Notice the similarity to the classic decomposition $\mathrm{Sym}_4 = \mathrm{Sym}_{3\perp 1} \ltimes V$.

Now (vi), (vii) and (viii) are straightforward.   □

16. Wajnryb's A-relator

As in Example 6.3,

$$
\begin{array}{ccccc}
\breve{\sigma}_1 & \breve{\alpha}_1 & \breve{\alpha}_2 & \breve{\alpha}_{g-1} & \breve{\alpha}_g \\
| & | & | & | & | \\
\breve{\gamma}_0- & \breve{\beta}_1-\breve{\gamma}_1- & \breve{\beta}_2- \cdots - & \breve{\beta}_{g-1}-\breve{\gamma}_{g-1}- & \breve{\beta}_g-\breve{\gamma}_g
\end{array}
$$

is an Artin diagram in $\mathrm{Out}_{g,1,0}$. We want to find an interesting expression in $\mathrm{Out}^+_{g,1,0}$ for $\breve{\sigma}_1$ in terms of the DLH generators (5.2.1). We will show that the two elements of

$$
\text{Z-Artin}\left[\begin{array}{cc} \breve{\sigma}_1 & \breve{\alpha}_1 \\ & | \\ & \breve{\beta}_1 - \breve{\gamma}_1 - \breve{\beta}_2 - \breve{\gamma}_2 - \breve{\beta}_3 - \cdots - \breve{\gamma}_{g-1} - \breve{\beta}_g \end{array}\right]
$$

in $\mathrm{Out}_{g,1,0}$ are subject to the relation

$$
\text{Z-Artin}\left[\begin{array}{c} \breve{\sigma}_1 \end{array}\right] = \text{Z-Artin}^2\left[\begin{array}{c} \breve{\alpha}_1 \\ | \\ \breve{\beta}_1-\breve{\gamma}_1-\breve{\beta}_2-\breve{\gamma}_2-\breve{\beta}_3- \cdots -\breve{\gamma}_{g-1}-\breve{\beta}_g \end{array}\right]
$$

in $\mathrm{Out}_{g,1,0}$, that is,

$$
(\breve{\alpha}_1 \cdot \prod_{i=1}^{g-1}(\breve{\beta}_i\breve{\gamma}_i) \cdot \breve{\beta}_g)^{4g+2} = \breve{\sigma}_1 \quad \text{in } \mathrm{Out}_{g,1,0}. \tag{16.0.1}
$$

This will follow from the next result.

16.1. Lemma. In $\mathrm{Aut}_{g,1,0}$, $\quad (\alpha_1 \cdot \prod_{i=1}^{g-1}(\beta_i\gamma_i) \cdot \beta_g)^{4g+2} = \sigma_1\tilde{z}_1^{-1}.$

Proof. Consider first $\mathrm{Aut}_{0,1,(2g+1)(2)}$, as in Example 15.6. For $1 \leqslant k \leqslant 2g$, the action of μ_k on $\Sigma_{0,1,(2g+1)(2)}$ is given by

$$
\mu_k : \begin{cases} z_{k'} & \mapsto & z_{k'} & \text{for } k' = 1,.,k, \\ z_{k+1} & \mapsto & z_{k+2}, \\ z_{k+2} & \mapsto & z_{k+2}z_{k+1}z_{k+2}, \\ z_{k'} & \mapsto & z_{k'} & \text{for } k' = k+3,.,2g+2. \end{cases}
$$

It is then not difficult to show that the action of $\prod_{k=1}^{2g}\mu_k$ on $\Sigma_{0,1,(2g+1)(2)} * E_1$ is given by

$$
\prod_{k=1}^{2g}\mu_k : \begin{cases} e_1 & \mapsto & e_1, & z_1^{e_1} \mapsto z_1^{e_1}, \\ z_2 & \mapsto & z_{2g+2}, \\ z_k & \mapsto & z_{2g+2}z_{k-1}z_{2g+2} & \text{for } k = 3,.,2g+2. \end{cases}
$$

Hence

$$\prod_{k=1}^{2g+1} \mu_k \cdot \widetilde{z}_{2g+2}: \begin{cases} e_1 & \mapsto & z_{2g+2}e_1, \quad z_1^{e_1} \mapsto z_1^{e_1}, \\ z_2 & \mapsto & z_{2g+2}, \\ z_k & \mapsto & z_{k-1} \qquad \text{for } k = 3, ., 2g+2. \end{cases}$$

Thus

$$(\prod_{k=1}^{2g+1} \mu_k \cdot \widetilde{z}_{2g+2})^{2g+1}: \begin{cases} e_1 & \mapsto & \prod_{k=2}^{2g+2} z_k \cdot e_1, \quad z_1^{e_1} \mapsto z_1^{e_1}, \\ z_k & \mapsto & z_k \qquad \text{for } k = 2, ., 2g+2. \end{cases}$$

This map is σ_1, since $\prod_{k=2}^{2g+2} z_k = z_1^{-1}$. Thus we can write

$$\sigma_1 = (\prod_{k=1}^{2g+1} \mu_k \cdot \widetilde{z}_{2g+2})^{2g+1} = (\prod_{k=1}^{2g+1} \mu_k)^{2g+1} \cdot \prod_{k=2g+2}^{2} \widetilde{z}_k = (\prod_{k=1}^{2g+1} \mu_k)^{2g+1} \cdot \widetilde{z}_1.$$

Hence $\sigma_1^2 \widetilde{z}_1^{-2} = (\prod_{k=1}^{2g+1} \mu_k)^{4g+2}$ in $\text{Aut}_{0,\hat{1},(2g+1)^{(2)}}$.

Now applying the map $\text{Aut}_{0,\hat{1},2g+1} \rightarrow \text{Aut}_{g,\hat{1},0}$ of Example 15.6 gives the desired result. □

We can now verify Examples 6.5.

16.2. Theorem (Wajnryb). *Any presentation of* $\text{Out}_{g,1,0}^+$ *on the DLH generators*

$$\{\breve{\alpha}_1, \breve{\alpha}_2, \dots, \breve{\alpha}_{\min\{2,g\}}, \breve{\beta}_2, \breve{\beta}_3, \dots, \breve{\beta}_g, \breve{\gamma}_1, \breve{\gamma}_2, \dots, \breve{\gamma}_{g-1}\}$$

together with the relator $(\breve{\alpha}_1 \cdot \prod_{i=1}^{g-1}(\breve{\beta}_i \breve{\gamma}_i) \cdot \breve{\beta}_g)^{4g+2}$ *gives a presentation of* $\text{Out}_{g,0,1}^+$.

Proof. By Lemma 16.1, (16.0.1) holds in $\text{Out}_{g,1,0}^+$. By Proposition 9.2, $\text{Out}_{g,0,1}^+$ is naturally isomorphic to $\text{Out}_{g,1,0}^+$ modulo the (normal) subgroup generated by $\breve{\sigma}_1$. □

17. Matsumoto's A-D-relator

Suppose that $g \geq 2$.

We want to find an interesting expression in $\text{Out}_{g,1,0}^+$ for $\breve{\gamma}_0$ in terms of the DLH generators (5.2.1).

We will find that the three elements of

$$\text{Z-Artin} \begin{bmatrix} & \breve{\alpha}_1 & \breve{\alpha}_2 & \\ & & | & \\ \breve{\gamma}_0 & \breve{\gamma}_1 - & \breve{\beta}_2 - \breve{\gamma}_2 - \breve{\beta}_3 - \cdots - \breve{\gamma}_{g-1} - \breve{\beta}_g \end{bmatrix}$$

in $\text{Out}_{g,1,0}$ are subject to the relation

$$\text{Z-Artin}\begin{bmatrix} \\ \breve{\gamma}_0 \\ \end{bmatrix} \cdot \text{Z-Artin}^{2g-3}\begin{bmatrix} \breve{\alpha}_1 \\ \\ \end{bmatrix}$$

$$= \text{Z-Artin}\begin{bmatrix} & \breve{\alpha}_2 & \\ & | & \\ \breve{\gamma}_1 - \breve{\beta}_2 - \breve{\gamma}_2 - \breve{\beta}_3 - \cdots - \breve{\gamma}_{g-1} - \breve{\beta}_g \end{bmatrix}$$

in $\text{Out}_{g,1,0}$; that is, $\breve{\gamma}_0 \breve{\alpha}_1^{2g-3} = (\breve{\alpha}_2 \cdot \prod_{i=2}^{g}(\breve{\gamma}_{i-1}\breve{\beta}_i))^{4g-4}$ in $\text{Out}_{g,1,0}$, a fact which can be seen from the next result.

17.1. Lemma. *If $g \geqslant 2$, then, in $\text{Aut}_{g,1,0}$,*

$$(\overline{\alpha}_1 \cdot (\alpha_2 \cdot \prod_{i=2}^{g}(\gamma_{i-1}\beta_i))^2 \cdot \tilde{x}_g)^{2g-2} \cdot \tilde{y}_1^{-1}\alpha_1 = \gamma_0.$$

Proof. We simplify the exposition by viewing $\Sigma_{g,1,0}$ as a subgroup of index two in $\Sigma_{0,1,(2g+1)^{(2)}}$, as in Example 15.6. Thus we identify

$$x_i \quad = \quad z_{2i+2}z_{2i+1} \qquad\qquad \text{for } i = 1,.,g,$$

$$y_i \quad = \quad z_{2i+2} \cdot \prod_{k=2}^{2i+2} z_k \qquad \text{for } i = 1,.,g.$$

Hence

$$y_1 \quad = \quad z_4 z_2 z_3 z_4,$$

$$\overline{x}_1 \quad = \quad z_3 z_4,$$

$$y_1 \cdot \prod_{i'=2}^{i} \overline{x}_{i'} \cdot \overline{y}_i x_i \quad = \quad z_4 z_{2i+1} \qquad \text{for } i = 2,.,g,$$

$$y_1 \cdot \prod_{i'=2}^{i} \overline{x}_{i'} \cdot \overline{y}_i \quad = \quad z_4 z_{2i+2} \qquad \text{for } i = 2,.,g.$$

Notice that both

$$\{z_4 z_k \mid 2 \leqslant k \leqslant 2g+2, k \neq 4\} \quad \text{and} \quad \{z_4 z_2 z_3 z_4, z_3 z_4, z_4 z_k \mid 5 \leqslant k \leqslant 2g+2\}$$

are free generating sets of $\Sigma_{g,1,0}$.

It is not difficult to show that the action of $\prod_{k=3}^{2g}\mu_k$ on $\Sigma_{0,1,(2g+1)^{(2)}}$ is given by

$$\prod_{k=3}^{2g}\mu_k : \begin{cases} z_{k'} & \mapsto \quad z_{k'} \qquad\qquad\qquad \text{for } k' = 1,2,3, \\ z_4 & \mapsto \quad z_{2g+2}, \\ z_{k'} & \mapsto \quad z_{2g+2}z_{k'-1}z_{2g+2} \quad \text{for } k' = 5,.,2g+2. \end{cases}$$

On restricting to $\Sigma_{g,1,0}$, we find that this says that the action of $\prod_{i=2}^{g}(\gamma_{i-1}\beta_i)$ on $\Sigma_{g,1,0} * E_1$ is given by

$$\prod_{i=2}^{g}(\gamma_{i-1}\beta_i) : \begin{cases} e_1 & \mapsto & e_1, \\ z_2 z_4 & \mapsto & z_2 z_{2g+2}, \\ z_3 z_4 & \mapsto & z_3 z_{2g+2}, \\ z_4 z_k & \mapsto & z_{k-1} z_{2g+2} \quad \text{for } k = 5, ., 2g+2. \end{cases}$$

Here, and throughout the remainder of the proof, it is tacitly understood that the image of z_1 is to be determined by the facts that $z_1^{e_1}$ is fixed and $z_1 = e_1 \cdot z_1^{e_1} \cdot \bar{e}_1$.

It is not difficult to check that the action of α_2 on $\Sigma_{g,1,0} * E_1$ is given by

$$\alpha_2 : \begin{cases} e_1 & \mapsto & e_1, \\ z_2 z_4 & \mapsto & z_2 z_4, \\ z_3 z_4 & \mapsto & z_3 z_4, \\ z_4 z_5 & \mapsto & z_4 z_5, \\ z_4 z_k & \mapsto & z_4 z_2 z_3 z_4 z_5 z_k \quad \text{for } k = 6, ., 2g+2. \end{cases}$$

Hence

$$\alpha_2 \cdot \prod_{i=2}^{g}(\gamma_{i-1}\beta_i) : \begin{cases} e_1 & \mapsto & e_1, \\ z_2 z_4 & \mapsto & z_2 z_{2g+2}, \\ z_3 z_4 & \mapsto & z_3 z_{2g+2}, \\ z_4 z_5 & \mapsto & z_4 z_{2g+2}, \\ z_4 z_k & \mapsto & z_{2g+2} z_2 z_3 z_4 z_{k-1} z_{2g+2} \quad \text{for } k = 6, ., 2g+2, \end{cases}$$

and

$$(\alpha_2 \cdot \prod_{i=2}^{g}(\gamma_{i-1}\beta_i))^2 : \begin{cases} e_1 & \mapsto & e_1, \\ z_2 z_4 & \mapsto & z_3 z_4 z_{2g+1} z_{2g+2}, \\ z_3 z_4 & \mapsto & z_3 z_2 z_3 z_4 z_{2g+1} z_{2g+2}, \\ z_4 z_5 & \mapsto & z_{2g+2} z_2 z_3 z_4 z_{2g+1} z_{2g+2}, \\ z_4 z_6 & \mapsto & z_{2g+2} z_{2g+1} z_{2g+2} z_2 z_3 z_4 z_{2g+1} z_{2g+2}, \\ z_4 z_k & \mapsto & z_{2g+2} z_{2g+1} z_4 z_{k-2} z_{2g+1} z_{2g+2} \quad \text{for } k=7, ., 2g+2. \end{cases}$$

Now $x_g = z_{2g+2} z_{2g+1}$, and therefore, even with $g = 2$,

$$(\alpha_2 \cdot \prod_{i=2}^{g}(\gamma_{i-1}\beta_i))^2 \cdot \tilde{x}_g : \begin{cases} e_1 & \mapsto & z_{2g+1} z_{2g+2} e_1, \\ z_2 z_4 & \mapsto & z_{2g+1} z_{2g+2} z_3 z_4, \\ z_3 z_4 & \mapsto & z_{2g+1} z_{2g+2} z_3 z_2 z_3 z_4, \\ z_4 z_5 & \mapsto & z_{2g+1} z_2 z_3 z_4, \\ z_4 z_6 & \mapsto & z_{2g+2} z_2 z_3 z_4, \\ z_4 z_k & \mapsto & z_4 z_{k-2} \quad \text{for } k = 7, ., 2g+2. \end{cases}$$

Notice that this map fixes $z_4 z_2 z_3 z_4$. Also, $\overline{\alpha}_1$ acts like the restriction of $\overline{\mu}_1$, and $\overline{\mu}_1$ sends z_2 to $z_2 z_3 z_2$, and sends z_3 to z_2, and fixes all other z_k. In particular it also fixes $z_4 z_2 z_3 z_4$. Thus we have

$$\overline{\alpha}_1 \cdot (\alpha_2 \cdot \prod_{i=2}^{g}(\gamma_{i-1}\beta_i))^2 \cdot \tilde{x}_g : \begin{cases} e_1 & \mapsto & z_{2g+1}z_{2g+2}e_1, \\ z_4 z_2 z_3 z_4 & \mapsto & z_4 z_2 z_3 z_4, \\ z_3 z_4 & \mapsto & z_{2g+1}z_{2g+2}z_3 z_4, \\ z_4 z_5 & \mapsto & z_{2g+1}z_2 z_3 z_4, \\ z_4 z_6 & \mapsto & z_{2g+2}z_2 z_3 z_4, \\ z_4 z_k & \mapsto & z_4 z_{k-2} \quad \text{for } k = 7, ., 2g+2. \end{cases}$$

Let $\phi = \overline{\alpha}_1 \cdot (\alpha_2 \cdot \prod_{i=2}^{g}(\gamma_{i-1}\beta_i))^2 \cdot \tilde{x}_g$.

It can be shown by induction that, for $1 \leqslant n \leqslant g-1$,

$$\phi^n : \begin{cases} e_1 & \mapsto & \prod_{k=2g+3-2n}^{2g+2} z_k \cdot e_1, \\ z_4 z_2 z_3 z_4 & \mapsto & z_4 z_2 z_3 z_4, \\ z_3 z_4 & \mapsto & \prod_{k=2g+3-2n}^{2g+2} z_k \cdot z_3 z_4, \\ z_4 z_k & \mapsto & z_{2g-2n+k-2}z_2 z_3 z_4 \quad \text{for } k = 5, ., 2n+4, \\ z_4 z_k & \mapsto & z_4 z_{k-2n} \quad \text{for } k = 2n+5, ., 2g+2. \end{cases}$$

In particular,

$$\phi^{g-1} : \begin{cases} e_1 & \mapsto & \prod_{k=5}^{2g+2} z_k \cdot e_1, \\ z_4 z_2 z_3 z_4 & \mapsto & z_4 z_2 z_3 z_4, \\ z_3 z_4 & \mapsto & \prod_{k=5}^{2g+2} z_k \cdot z_3 z_4, \\ z_4 z_k & \mapsto & z_k z_2 z_3 z_4 \quad \text{for } k = 5, ., 2g+2. \end{cases}$$

Squaring, we find

$$\phi^{2g-2} : \begin{cases} e_1 & \mapsto & z_4 z_3 z_2 \cdot \prod_{k=5}^{2g+2} z_k \cdot z_2 z_3 z_4 \cdot \prod_{k=5}^{2g+2} z_k \cdot e_1, \\ z_4 z_2 z_3 z_4 & \mapsto & z_4 z_2 z_3 z_4, \\ z_3 z_4 & \mapsto & z_4 z_3 z_2 \cdot \prod_{k=5}^{2g+2} z_k \cdot z_2 z_3 z_4 \cdot \prod_{k=5}^{2g+2} z_k \cdot z_3 z_4, \\ z_4 z_k & \mapsto & z_4 z_3 z_2 z_k z_4 z_2 z_3 z_4 \quad \text{for } k = 5, ., 2g+2. \end{cases}$$

Since $\bar{z}_1 = (\prod_{k=2}^{2g+2} z_k)^2$, this is

$$\phi^{2g-2} : \begin{cases} e_1 & \mapsto & z_4 z_3 z_2 z_4 z_3 z_2 \bar{z}_1 e_1, \\ z_4 z_2 z_3 z_4 & \mapsto & z_4 z_2 z_3 z_4, \\ z_3 z_4 & \mapsto & z_4 z_3 z_2 z_4 z_3 z_2 \bar{z}_1 z_3 z_2, \\ z_4 z_k & \mapsto & z_4 z_3 z_2 z_k z_4 z_2 z_3 z_4 \quad \text{for } k = 5, ., 2g+2. \end{cases}$$

Since $\bar{y}_1 = z_4 z_3 z_2 z_4$, we have

$$\phi^{2g-2} \cdot \widetilde{y}_1^{-1} : \begin{cases} e_1 & \mapsto & z_3 z_2 \bar{z}_1 e_1, \\ z_4 z_2 z_3 z_4 & \mapsto & z_4 z_2 z_3 z_4, \\ z_3 z_4 & \mapsto & z_3 z_2 \bar{z}_1 z_2 z_4, \\ z_4 z_k & \mapsto & z_4 z_k \quad \text{for } k = 5, ., 2g+2. \end{cases}$$

Recall that α_1 acts as the restriction of μ_1, and that μ_1 sends z_2 to z_3, sends z_3 to $z_3 z_2 z_3$, and fixes all the other z_k. Thus

$$\phi^{2g-2} \cdot \widetilde{y}_1^{-1} \alpha_1 : \begin{cases} e_1 & \mapsto & z_3 z_2 \bar{z}_1 e_1, \\ z_4 z_2 z_3 z_4 & \mapsto & z_4 z_2 z_3 z_4, \\ z_3 z_4 & \mapsto & z_3 z_2 \bar{z}_1 z_3 z_4, \\ z_4 z_k & \mapsto & z_4 z_k \quad \text{for } k = 5, ., 2g+2. \end{cases}$$

It is straightforward to show that this is γ_0, as claimed. $\qquad\square$

We can now verify Examples 6.6(ii).

17.2. Theorem (Matsumoto). *If $g \geqslant 2$, then any presentation of $\mathrm{Out}^+_{g,1,0}$ on the DLH generators $\{\breve{\alpha}_1, \breve{\alpha}_2, \breve{\beta}_2, \breve{\beta}_3, \dots, \breve{\beta}_g, \breve{\gamma}_1, \breve{\gamma}_2, \dots, \breve{\gamma}_{g-1}\}$ together with the relator*
$$\breve{\alpha}_1^{2-2g} \cdot (\breve{\alpha}_2 \cdot \prod_{i=1}^{g-1} (\breve{\gamma}_i \breve{\beta}_{i+1}))^{4g-4}$$
gives a presentation of $\mathrm{Out}^+_{g,0,0}$.

Proof. It follows from Lemma 17.1 that, in $\mathrm{Out}^+_{g,1,0}$,

$$\breve{\alpha}_1^{2-2g} \cdot (\breve{\alpha}_2 \cdot \prod_{i=1}^{g-1} (\breve{\gamma}_i \breve{\beta}_{i+1}))^{4g-4} = \breve{\gamma}_0 \, \breve{\alpha}_1^{-1}.$$

By Theorem 14.7, $\mathrm{Out}^+_{g,0,0}$ is isomorphic to $\mathrm{Out}^+_{g,1,0}$ modulo the normal subgroup generated by $\breve{\gamma}_0 \breve{\alpha}_1^{-1}$, as desired. $\qquad\square$

18. Matsumoto's A-A and E-E relators

The interested reader can verify all of the following remarks, and deduce that the A-A-relation (6.4.1) holds in $\mathrm{Out}^+_{g,1,0}$ for $g \geqslant 2$, and the E-E-relation (6.4.2) holds in $\mathrm{Out}^+_{g,1,0}$ for $g \geqslant 3$.

18.1. Remarks. (i) We have the Artin diagram

$$
\begin{array}{ccccc}
\breve{\sigma}_1 & & \breve{\alpha}_1 & & \breve{\alpha}_2 \\
& & | & & | \\
\breve{\gamma}_0- & \breve{\beta}_1-\breve{\gamma}_1- & \breve{\beta}_2-\breve{\gamma}_2
\end{array}
$$

in $\mathrm{Out}_{2,1,0}$. Tedious calculations, similar to those in the proof of Lemma 17.1, show that

$$
\text{Z-Artin}\left[\begin{array}{c}\breve{\sigma}_1 \\ \\ \\ \end{array}\right] = \text{Z-Artin}\left[\begin{array}{ccc}\breve{\alpha}_1 & & \breve{\alpha}_2 \\ | & & | \\ \breve{\beta}_1-\breve{\gamma}_1- & \breve{\beta}_2\end{array}\right]
$$

and

$$
\text{Z-Artin}\left[\begin{array}{c}\breve{\sigma}_1 \\ \\ \\ \end{array}\right] = \text{Z-Artin}^2\left[\begin{array}{c}\breve{\alpha}_2 \\ | \\ \breve{\beta}_1-\breve{\gamma}_1- \breve{\beta}_2\end{array}\right]
$$

in $\mathrm{Out}_{2,1,0}$, that is, $\breve{\sigma}_1 = (\breve{\alpha}_1\breve{\beta}_1\breve{\gamma}_1\breve{\beta}_2\breve{\alpha}_2)^6$ and $\breve{\sigma}_1 = (\breve{\beta}_1\breve{\gamma}_1\breve{\beta}_2\breve{\alpha}_2)^{10}$ in $\mathrm{Out}_{2,\hat{1},0}$. The latter is essentially the same as the $g = 2$ case of (16.0.1).

(ii) Suppose that $g \geqslant 2$. By (i), the equation

$$
(\alpha_1\beta_1\gamma_1\beta_2\alpha_2)^6 = (\beta_1\gamma_1\beta_2\alpha_2)^{10}
$$

holds in $\mathrm{Aut}^+_{2,\hat{1},0} \simeq \mathrm{Out}^+_{2,1,0}$ and, hence, in $\mathrm{Aut}^+_{g,\hat{1},0} \simeq \mathrm{Out}^+_{g,1,0}$. Thus the A-A relation (6.4.1) holds in $\mathrm{Out}^+_{g,1,0}$. $\quad\square$

18.2. Remarks. (i) We have the Artin diagram

$$
\begin{array}{ccccccc}
\breve{\sigma}_1 & & \breve{\alpha}_1 & & \breve{\alpha}_2 & & \breve{\alpha}_3 \\
& & | & & | & & | \\
\breve{\gamma}_0- & \breve{\beta}_1-\breve{\gamma}_1- & \breve{\beta}_2- & \breve{\gamma}_2- & \breve{\beta}_3-\breve{\gamma}_3
\end{array}
$$

in $\mathrm{Out}_{3,1,0}$. It can be shown that

$$
\text{Z-Artin}\left[\begin{array}{c}\breve{\sigma}_1 \\ \\ \\ \end{array}\right] = \text{Z-Artin}\left[\begin{array}{ccc}\breve{\alpha}_1 & & \breve{\alpha}_2 \\ | & & | \\ \breve{\beta}_1-\breve{\gamma}_1- & \breve{\beta}_2-\breve{\gamma}_2- \breve{\beta}_3\end{array}\right]
$$

and

$$
\text{Z-Artin}\left[\begin{array}{c}\breve{\sigma}_1 \\ \\ \\ \end{array}\right] = \text{Z-Artin}\left[\begin{array}{c}\breve{\alpha}_2 \\ | \\ \breve{\beta}_1-\breve{\gamma}_1- \breve{\beta}_2-\breve{\gamma}_2- \breve{\beta}_3\end{array}\right]
$$

in $\mathrm{Out}_{3,1,0}$, that is, $\breve{\sigma}_1 = (\breve{\alpha}_2\breve{\alpha}_1\breve{\beta}_1\breve{\gamma}_1\breve{\beta}_2\breve{\gamma}_2\breve{\beta}_3)^9$ and $\breve{\sigma}_1 = (\breve{\alpha}_2\breve{\beta}_1\breve{\gamma}_1\breve{\beta}_2\breve{\gamma}_2\breve{\beta}_3)^{12}$ in $\mathrm{Out}_{3,1,0}$.

(ii) Suppose that $g \geqslant 3$. By (i), the equation

$$
(\alpha_2\alpha_1\beta_1\gamma_1\beta_2\gamma_2\beta_3)^9 = (\alpha_2\beta_1\gamma_1\beta_2\gamma_2\beta_3)^{12}
$$

holds in $\mathrm{Aut}^+_{3,\hat{1},0} \simeq \mathrm{Out}^+_{3,1,0}$, and, hence, in $\mathrm{Aut}^+_{g,\hat{1},0} \simeq \mathrm{Out}^+_{g,1,0}$. Thus the E-E relation (6.4.2) holds in $\mathrm{Out}^+_{g,1,0}$. $\qquad\square$

19. Four copies of $\mathrm{Artin}[D_{2g+1}]$

19.1. Remarks. Let $g \geqslant 1$.

We consider the following four Artin diagrams.

$$
\begin{array}{cc}
\check{\sigma}_1 & \check{\alpha}_1 \\
 & | \\
\check{\gamma}_0 - \check{\beta}_1 - \check{\gamma}_1 - \cdots - \check{\gamma}_{g-1} - \check{\beta}_g & \text{in } \mathrm{Out}_{g,1,1}
\end{array}
\tag{19.1.1}
$$

$$
\begin{array}{ccc}
& \check{\sigma}_2 & \check{\alpha}_1 \\
& & | \\
\check{\sigma}_1 & \check{\gamma}_0 - \check{\beta}_1 - \check{\gamma}_1 - \cdots - \check{\gamma}_{g-1} - \check{\beta}_g & \text{in } \mathrm{Out}_{g,2,0}
\end{array}
\tag{19.1.2}
$$

$$
\begin{array}{ccc}
& \check{\sigma}_3 & \check{\alpha}_1 \\
& & | \\
\check{\tau}_1^2 & \check{\gamma}_0 - \check{\beta}_1 - \check{\gamma}_1 - \cdots - \check{\gamma}_{g-1} - \check{\beta}_g & \text{in } \mathrm{Out}_{g,3,0}
\end{array}
\tag{19.1.3}
$$

$$
\begin{array}{ccc}
& \check{\alpha}_1 & \check{\alpha}_2 \\
& & | \\
\check{\gamma}_0 & \check{\gamma}_1 - \check{\beta}_2 - \check{\gamma}_2 - \cdots - \check{\gamma}_g - \check{\beta}_{g+1} & \text{in } \mathrm{Out}_{g+1,1,0}
\end{array}
\tag{19.1.4}
$$

In Remarks 19.2, 19.3, and 19.4, we will discuss partial isomorphisms

$$
\mathrm{Out}_{g,1,1} \longleftrightarrow \mathrm{Out}_{g,2,0} \longleftrightarrow \mathrm{Out}_{g,3,0} \longleftrightarrow \mathrm{Out}_{g+1,1,0} \, .
\tag{19.1.5}
$$

We will see that each of the Artin diagrams (19.1.1)–(19.1.4) gives an embedding of $\mathrm{Artin}[D_{2g+1}]$ in the corresponding group in (19.1.5). This was proved in [24] for $\mathrm{Out}_{g,1,1}$, and will follow for the others, since the partial isomorphisms (19.1.5) respect the Artin diagrams.

We will see that the partial isomorphisms (19.1.5) behave on the complementary part of the Artin diagrams according to the following:

$\mathrm{Out}_{g,1,1}$		$\mathrm{Out}_{g,2,0}$		$\mathrm{Out}_{g,3,0}$		$\mathrm{Out}_{g+1,1,0}$
	$\longleftrightarrow$		$\longleftrightarrow$		$\longleftrightarrow$	
		$\check{\sigma}_1$	$\leftrightarrow$	$\check{\tau}_1^2$	$\leftrightarrow$	$\check{\gamma}_0$
		$\check{\sigma}_2$	$\leftrightarrow$	$\check{\sigma}_3$	$\leftrightarrow$	$\check{\alpha}_1$
$\check{\sigma}_1$	$\leftrightarrow$	$\check{\sigma}_1 \check{z}_2 = \check{\sigma}_1 \check{\sigma}_2^{2g-1}$	$\leftrightarrow$	$\check{\tau}_1^2 \check{\sigma}_3^{2g-1}$	$\leftrightarrow$	$\check{\gamma}_0 \check{\alpha}_1^{2g-1}$

Hence Z-Artin$[D_{2g+1}]$ gets identified with

$$\breve{\sigma}_1 \qquad\qquad \text{in } \mathrm{Out}_{g,1,1},$$
$$\breve{\sigma}_1\breve{\sigma}_2^{2g-1} \qquad\qquad \text{in } \mathrm{Out}_{g,2,0},$$
$$\breve{\tau}_1^2\breve{\sigma}_3^{2g-1} \qquad\qquad \text{in } \mathrm{Out}_{g,3,0},$$
$$\breve{\gamma}_0\breve{\alpha}_1^{2g-1} \qquad\qquad \text{in } \mathrm{Out}_{g+1,1,0};$$

the latter was seen in Lemma 17.1 and the comment preceding it, and the others will follow from the partial isomorphisms (19.1.5).

We find it interesting that the power of $\breve{\alpha}_1$ in Matsumoto's A-D relator comes from $\breve{z}_2$. $\qquad\square$

19.2. Remarks. In Section 20, we shall construct a strange partial isomorphism

$$\mathrm{unelim}(b_2)\colon \ \mathrm{Out}_{g,1,1} \rightsquigarrow \mathrm{Out}_{g,2,0}\,.$$

It corresponds to distinguishing, or removing, an open disk centered at the puncture. As we asserted in Remarks 19.1, it respects the copies of Artin$[D_{2g+1}]$ and carries $\breve{\sigma}_1$ to $\breve{\sigma}_1\breve{z}_2 = \breve{\sigma}_1\breve{\sigma}_2^{2g-1}$.

This is a partial left inverse of the map $\mathrm{elim}(b_2)\colon \mathrm{Out}_{g,1\perp1,0} \to \mathrm{Out}_{g,1,1}$ which respects the copies of Artin$[D_{2g+1}]$, and carries $\breve{\sigma}_1$ to $\breve{\sigma}_1$, and $\breve{z}_2$ to 1. $\qquad\square$

19.3. Remarks. Essentially as in Definitions 13.1, we can construct a map

$$\mathrm{pinch}(z_1)\colon \ \mathrm{Out}_{g,1\perp1,0} \to \mathrm{Out}_{g,1\perp1\perp1,0}$$

corresponding to pinching the first boundary component to get the first two boundary components, and renumbering the second boundary component as the third boundary component. It is a left inverse of the map

$$\mathrm{elim}(z_2, e_2)\colon \ \mathrm{Out}_{g,1\perp1\perp1,0} \to \mathrm{Out}_{g,1\perp1,0}$$

which corresponds to eliminating the second boundary component and renumbering the third boundary component as the second boundary component. Hence we have a partial isomorphism

$$\mathrm{pinch}(z_1)\colon \ \mathrm{Out}_{g,2,0} \rightsquigarrow \mathrm{Out}_{g,3,0}\,.$$

As we asserted in Remarks 19.1, it respects the copies of Artin$[D_{2g+1}]$, and carries $\breve{\sigma}_1$ to $\breve{\tau}_1^2$, and $\breve{\sigma}_2$ to $\breve{\sigma}_3$. $\qquad\square$

If $b \geqslant 2$ and we identify two boundary components of $S_{g,b,p}$, respecting the referential S^1, then we get a surface $S_{g+1,b-2,p}$. This operation behaves moderately well with respect to homotopies. We consider a special case.

19.4. Definition. There is a map

$$\Sigma_{g,3,0} * E_3 \to \Sigma_{g+1,1,0} * E_1$$

given by

$$
\begin{cases}
z_1 & \mapsto & z_1, & e_1 & \mapsto & e_1. \\
z_2 & \mapsto & \overline{x}_1\overline{y}_1 x_1, & e_2 & \mapsto & 1, \\
z_3 & \mapsto & y_1, & e_3 & \mapsto & x_1, \\
x_i & \mapsto & x_{i+1}, & y_i & \mapsto & y_{i+1}, & \text{for } i = 1, ., g.
\end{cases}
$$

There is then an induced map

$$\text{Aut}_{g,\hat{1}\perp\hat{1}\perp1,0} \qquad \to \qquad \text{Aut}_{g+1,\hat{1},0}, \tag{19.4.1}$$

$$
\begin{aligned}
\alpha_i & \mapsto & \alpha_{i+1} & \quad \text{for } i = 1, ., g, \\
\beta_i & \mapsto & \beta_{i+1} & \quad \text{for } i = 1, ., g, \\
\gamma_i & \mapsto & \gamma_{i+1} & \quad \text{for } i = 0, ., g-1, \\
\sigma_3 & \mapsto & \alpha_1, \\
\tau_1^2 \tilde{z}_2^{-1} \tilde{z}_1^{-1} & \mapsto & \gamma_0 \tilde{x}_1^{-1} \tilde{y}_1 \tilde{x}_1 \tilde{z}_1^{-1}.
\end{aligned}
$$

Here

$$
\gamma_0 \tilde{x}_1^{-1} \tilde{y}_1 \tilde{x}_1 \tilde{z}_1^{-1} :
\begin{cases}
x_1 & \mapsto & z_1 \overline{x}_1 \overline{y}_1 x_1 x_1, \\
w & \mapsto & w^{\overline{x}_1 y_1 x_1 \overline{z}_1} & \text{for all } w \in X_{g,1,0} - \{x_1, z_1\}, \\
z_1 & \mapsto & z_1, \\
e_1 & \mapsto & e_1.
\end{cases}
$$

It is easily seen that (19.4.1) is injective.

By Proposition 7.1(iii), (19.4.1) can be viewed as an injective partial map

$$\text{glue}(z_2^{e_2}, z_3^{-e_3}): \text{Out}_{g,3,0} \rightsquigarrow \text{Out}_{g+1,1,0}.$$

As we asserted in Remarks 19.1, it respects the copies of $\text{Artin}[D_{2g+1}]$, and carries $\breve{\sigma}_3$ to $\breve{\alpha}_1$ and $\breve{\tau}_1^2$ to $\breve{\gamma}_0$.   □

20. Converting a puncture to a boundary

We can convert a puncture into a boundary component by removing, or distinguishing, an open disk centered at the puncture. This gives an inclusion of surfaces which behaves badly with respect to homotopies, but we get partial homomorphisms of mapping-class groups.

The following is a consequence of Lemma 14.6.

20.1. Lemma. *If $b \geqslant 1$ and $g \geqslant 1$, then the following hold in* $\mathrm{Aut}_{g,b,p}$.

$$\widehat{x}_i^{\alpha_i} = (\widehat{y}_i \sigma_b^{1-i})^{-1} \, \widehat{x}_i \, \sigma_b^{i-1} \; \text{for } i = 1, ., g.$$

$$(\widehat{y}_i \, \sigma_b^{1-i})^{\beta_i} = \widehat{x}_i \, (\widehat{y}_i \, \sigma_b^{1-i}) \; \text{for } i = 1, ., g.$$

$$\widehat{x}_i^{\gamma_i} = \widehat{x}_{i+1}^{-1} \, (\widehat{y}_{i+1} \sigma_b^{-i}) \, \widehat{x}_{i+1} \, (\widehat{y}_i \sigma_b^{1-i})^{-1} \, \widehat{x}_i \; \text{for } i = 1, ., g - 1.$$

$$(\widehat{y}_i \, \sigma_b^{1-i})^{\gamma_i} = \widehat{x}_{i+1}^{-1} \, (\widehat{y}_{i+1} \, \sigma_b^{-i}) \, \widehat{x}_{i+1} \, (\widehat{y}_i \, \sigma_b^{1-i}) \, \widehat{x}_{i+1}^{-1} \, (\widehat{y}_{i+1} \, \sigma_b^{-i})^{-1} \, \widehat{x}_{i+1},$$
$$\text{for } i = 1, ., g - 1.$$

$$\widehat{x}_{i+1}^{\gamma_i} = \widehat{x}_{i+1} \, (\widehat{y}_i \, \sigma_b^{1-i}) \, \widehat{x}_{i+1}^{-1} \, (\widehat{y}_{i+1} \, \sigma_b^{-i})^{-1} \, \widehat{x}_{i+1}, \; \text{for } i = 1, ., g - 1. \qquad \square$$

20.2. Definition. Suppose that $b \geqslant 2$.

We shall construct a partial splitting of the exact sequence in Proposition 9.2. To be more precise, we shall construct a partial splitting of the index $b - 1$ exact subsequence

$$1 \to \langle \breve{\sigma}_b \rangle \to \mathrm{Out}_{g,b-2\perp1\perp1,p} \to \mathrm{Out}_{g,b-2\perp1,p\perp1} \to 1. \qquad (20.2.1)$$

By Theorem 13.2 and Corollary 11.4, we have semidirect product decompositions

$$\mathrm{Out}_{g,b-2\perp1\perp1,p} \simeq \mathrm{Out}_{g,b-2\perp1,p} \ltimes \breve{\Sigma}_{g,b-1,p}, \qquad (20.2.2)$$

$$\mathrm{Out}_{g,b-2\perp1,p\perp1} \simeq \mathrm{Out}_{g,b-2\perp1,p} \ltimes \breve{\Sigma}_{g,b-1,p}. \qquad (20.2.3)$$

Recall that an element of $\mathrm{Out}_{g,b-2\perp1,p}$ lifts back to a unique representative in $\mathrm{Aut}_{g,b-2\perp\hat{1},p}$, and this acts on $\breve{\Sigma}_{g,b-1,p}$. The representative in turn lifts back to a well-defined second representative in $\mathrm{Aut}_{g,b-2\perp1\perp\hat{1},p}$, where $z_{b-1}z_b$ replaces z_{b-1}. This second representative acts on $\Sigma_{g,b,p}$, and hence acts on $\breve{\Sigma}_{g,b-1,p}$ by the formula in Theorem 14.4.

Moreover, (20.2.1) arises from (20.2.2) and (20.2.3) using the identification

$$\breve{\Sigma}_{g,b-1,p}/\langle \breve{\sigma}_b \rangle = \widetilde{\Sigma}_{g,b-1,p}.$$

Since $\widetilde{\Sigma}_{g,b-1,p}$ is free (of rank $2g+b+p-2$), we can construct an isomorphism

$$\phi: \quad \widetilde{\Sigma}_{g,b-1,p} \quad \times \quad \langle \breve{\sigma}_b \rangle \quad \xrightarrow{\;\sim\;} \quad \breve{\Sigma}_{g,b-1,p}$$

$$\begin{aligned}
\widetilde{x}_i \quad &\mapsto \quad \breve{x}_i && \text{for } i = 1, ., g, \\
\widetilde{y}_i \quad &\mapsto \quad \breve{y}_i \, \breve{\sigma}_b^{1-i} && \text{for } i = 1, ., g, \\
\widetilde{z}_j \quad &\mapsto \quad \breve{z}_j && \text{for } j = 1, ., b - 2, \\
\widetilde{t}_k \quad &\mapsto \quad \breve{t}_k && \text{for } k = 1, ., p, \\
\breve{\sigma}_b \quad &\mapsto \quad \breve{\sigma}_b.
\end{aligned}$$

Notice that

$$\widetilde{z}_{b-1} = \coprod_{j=b-2}^{1} \widetilde{z}_j^{-1} \cdot \coprod_{i=g}^{1} [\widetilde{y}_i, \widetilde{x}_i] \cdot \prod_{k=1}^{p} \widetilde{t}_k^{-1}$$

is mapped to

$$\coprod_{j=b-2}^{1} \breve{z}_j^{-1} \cdot \coprod_{i=g}^{1} [\breve{y}_i, \breve{x}_i] \cdot \prod_{k=1}^{p} \breve{t}_k^{-1} = \breve{z}_{b-1}\breve{z}_b = \breve{z}_{b-1}\breve{\sigma}_b^{1-2g-b-p}.$$

Here $\mathrm{Out}_{g,b-2\perp1,p}$ acts on the domain and codomain of ϕ; let N denote the subgroup of $\mathrm{Out}_{g,b-2\perp1,p}$ consisting of those elements which commute with ϕ, so ϕ is a map of N-groups. Then ϕ lifts to an isomorphism

$$(N \ltimes \widetilde{\Sigma}_{g,b-1,p}) \times \langle \breve{\sigma}_b \rangle \quad \xrightarrow{\sim} \quad N \ltimes \widetilde{\Sigma}_{g,b-1,p}$$

We represent the restriction

$$N \ltimes \widetilde{\Sigma}_{g,b-1,p} \to N \ltimes \widetilde{\Sigma}_{g,b-1,p}$$

as a partial map

$$\mathrm{unelim}(e_b)\colon \mathrm{Out}_{g,b-2\perp1,p\perp1} \rightsquigarrow \mathrm{Out}_{g,b-2\perp1\perp1,p}.$$

It is a partial splitting of (20.2.1), with domain $N \ltimes \widetilde{\Sigma}_{g,b-1,p}$.

It follows from Lemma 20.1 that N contains $\{\breve{\alpha}_1\} \cup \{\breve{\beta}_i\}_{i=1}^{g} \cup \{\breve{\gamma}_i\}_{i=1}^{g-1}$.

We claim that N also contains $\breve{\sigma}_{b-1}^{(g,b-1,p)}$. The representative of $\breve{\sigma}_{b-1}^{(g,b-1,p)}$ in $\mathrm{Aut}_{g,b-2\perp\hat{i},p}$ is $\sigma_{b-1}\widetilde{z}_{b-1}^{-1}$, and the second representative, in $\mathrm{Aut}_{g,b-2\perp1\perp\hat{i},p}$, is

$$\tau_{b-1}^2 \widetilde{z}_b^{-1} \widetilde{z}_{b-1}^{-1}.$$

The latter acts on each element of $\{x_i, y_i\}_{i=1}^{g} \cup \{z_j\}_{j=1}^{b-2} \cup \{t_k\}_{k=1}^{p}$ as conjugation by

$$\coprod_{k=p}^{1} t_k \cdot \prod_{i=1}^{g} [x_i, y_i] \cdot \prod_{j=1}^{b-2} z_j.$$

It is not difficult to show that this commutes with ϕ, so N contains $\breve{\sigma}_{b-1}^{(g,b-1,p)}$.

Hence $N \ltimes \widetilde{\Sigma}_{g,b-1,p}$, the domain of $\mathrm{unelim}(e_b)$, contains

$$\{\breve{\alpha}_1\} \cup \{\breve{\beta}_i\}_{i=1}^{g} \cup \{\breve{\gamma}_i\}_{i=1}^{g-1},$$

and also contains $\breve{\gamma}_0 = \breve{\alpha}_1 \ltimes \widetilde{x}_1^{-1}\widetilde{y}_1\widetilde{x}_1$, and $\breve{\sigma}_{b-1}^{(g,b,p)} = \breve{\sigma}_{b-1}^{(g,b-1,p)} \ltimes \widetilde{z}_{b-1}$. Now $\breve{\gamma}_0 = \breve{\alpha}_1 \ltimes \widetilde{x}_1^{-1}\widetilde{y}_1\widetilde{x}_1$ is mapped to $\breve{\alpha}_1\breve{x}_1^{-1}\breve{y}_1\breve{x}_1 = \breve{\gamma}_0$; and, $\breve{\sigma}_{b-1}^{(g,b,p)} = \breve{\sigma}_{b-1}^{(g,b-1,p)} \ltimes \widetilde{z}_{b-1}$ is mapped to $\breve{\sigma}_{b-1}^{(g,b-1,p)} \ltimes \breve{z}_{b-1}\breve{z}_b = \breve{\sigma}_{b-1}^{(g,b,p)}\breve{z}_b$.

If $g \geqslant 2$, then N does not contain $\breve{\alpha}_2$, nor any proper power of $\breve{\alpha}_2$, and hence N has infinite index in $\mathrm{Out}_{g,b-2\perp1,p}$. $\qquad\square$

20.3. Example. $g \geqslant 2$, $b = 2$, $p = 0$.

As we asserted in Remarks 19.1, the injective partial map

$$\mathrm{unelim}(e_2)\colon \mathrm{Out}_{g,1,1} \rightsquigarrow \mathrm{Out}_{g,2,0}$$

respects the copies of $\mathrm{Artin}[D_{2g+1}]$, and carries $\breve{\sigma}_1$ to $\breve{\sigma}_1\breve{z}_2 = \breve{\sigma}_1\breve{\sigma}_2^{2g-1}$. $\qquad\square$

Acknowledgements. We thank Mladen Bestvina, Brian Bowditch, Étienne Ghys, Joan Porti and Bronek Wajnryb for many informative, interesting comments.

The research of the first-named author was funded by the DGI (Spain) through projects BFM2000-0354 and BFM2003-06613. The research of the second-named author was partially funded by the NSA.

References

[1] E. Artin, *Theorie der Zöpfe*, Abh. Math. Sem. Univ. Hamburg **4** (1925), 47–72.

[2] E. Artin, *Theory of braids*, Ann. of Math. **48** (1947), 101–126.

[3] Christophe Bavard, *Longueur stable des commutateurs*, Enseign. Math. (2) **37** (1991), 109–150.

[4] Joan S. Birman and Hugh M. Hilden, *On the mapping class groups of closed surfaces as covering spaces*, pp. 81–115 in Advances in the theory of Riemann surfaces (Proc. Conf., Stony Brook, N.Y., 1969), Ann. of Math. Studies, No. **66**, Princeton Univ. Press, Princeton, N.J., 1971.

[5] F. Bohnenblust, *The algebraic braid groups*, Ann. of Math. **48** (1947), 127–136.

[6] E. Brieskorn and K. Saito, *Artin-Gruppen und Coxeter-Gruppen*, Invent. Math. **17** (1972), 245–271.

[7] Wei-Liang Chow, *On the algebraical braid group*, Ann. of Math. **49** (1948), 654–658.

[8] P. Dehornoy, *Braid groups and left distributive operations*, Trans. Amer. Math. Soc. **345** (1994), 115–150.

[9] Warren Dicks and Edward Formanek, *Automorphism subgroups of finite index in algebraic mapping class groups*, J. Algebra **189** (1997), 58–89.

[10] Joan L. Dyer, Edward Formanek and Edna K. Grossman, *On the linearity of automorphism groups of free groups*, Archiv der Math. **38** (1982), 404–409.

[11] Stephen P. Humphries, *Generators for the mapping class group*, pp. 44–47 in Topology of low-dimensional manifolds (Proc. Second Sussex Conf., Chelwood Gate, 1977), Lecture Notes in Math., **722**, Springer, Berlin, 1979.

[12] Stephen P. Humphries and Dennis Johnson, *A generalization of winding number functions on surfaces*, Proc. London Math. Soc. **58** (1989), 366–386.

[13] Nikolai V. Ivanov, *Mapping class groups*, pp. 523–633 in Handbook of geometric topology, North-Holland, Amsterdam, 2002.

[14] Catherine Labruère and Luis Paris, *Presentations for the punctured mapping class groups in terms of Artin groups*, Geom. Topol. **46** (2001), 73–114.

[15] D. M. Larue, *On braid words and irreflexivity*, Algebra Univ. **31** (1994), 104–112.

[16] G. Levitt, *Automorphisms of hyperbolic groups and graphs of groups*, Geometriae Dedicata **114** (2005), 49–70.

[17] C. Maclachlan and W. J. Harvey, *On mapping class groups and Teichmüller spaces*, Proc. London Math. Soc. (3) **30** (1975), 496–512.

[18] W. Magnus, *Über Automorphismen von Fundamentalgruppen beranderter Flächen*, Math. Ann. **109** (1934), 617–646.

[19] A. Markov, *Foundations of the algebraic theory of braids* (Russian), Trav. Inst. Math. Steklov **16** (1945), 53 pages.

[20] Makoto Matsumoto, *A presentation of mapping class groups in terms of Artin groups and geometric monodromy of singularities*, Math. Ann. **316** (2000), 401–418.

[21] J. McCool, *Generators of the mapping class group (An algebraic approach)*, Publ. Math. **40** (1996), 457–468.

[22] G. Mess, *Unit tangent bundle subgroups of the mapping class groups*, Preprint IHES/M/90/30 (1990), 15.

[23] Luis Paris and Dale Rolfsen, *Geometric subgroups of mapping class groups*, J. Reine Angew. Math. **521** (2000), 47–83.

[24] B. Perron and J. P. Vannier, *Groupe de monodromie géométrique des singularités simples*, Math. Ann. **306** (1996), 231–245.

[25] Vladimir Shpilrain, *Representing braids by automorphisms*, Internat. J. Algebra Comput. **11** (2001), 773–777.

[26] Pekka Tukia, *Homeomorphic conjugates of Fuchsian groups*, J. Reine Angew. Math. **391** (1988), 1–54.

[27] Bronislaw Wajnryb, *A simple presentation for the mapping class group of an orientable surface*, Israel J. Math. **45** (1983), 157–174; Errata, **88** (1994), 425–427.

[28] Bronislaw Wajnryb, *An elementary approach to the mapping class group of a surface*, Geom. Topol. **3** (1999),405-466.

Warren Dicks
Departament de Matemàtiques
Universitat Autònoma de Barcelona
08193 Bellaterra (Barcelona)
Spain
e-mail: dicks@mat.uab.es
URL: http://mat.uab.es/~dicks/

Edward Formanek
Department of Mathematics
The Pennsylvania State University
University Park, PA 16802
USA
e-mail: formanek@math.psu.edu

Progress in Mathematics, Vol. 248, 117–218
© 2005 Birkhäuser Verlag Basel/Switzerland

Solved and Unsolved Problems Around One Group

Rostislav Grigorchuk

Abstract. This is a survey paper on various topics concerning self-similar groups and branch groups with a focus on those notions and problems that are related to a 3-generated torsion 2 group of intermediate growth G, constructed by the author in 1980, and its generalizations G_ω, $\omega \in \{0, 1, 2\}^{\mathbb{N}}$.

Among the topics touched in the paper are: actions on rooted trees and groups generated by finite automata, self-similarity, contracting properties, various notions of length, topology in the space of finitely generated groups, Kolmogorov complexity in relation to algorithmic problems, algorithmic problems such as word, conjugacy, and isomorphism problem, L-presentations of self-similar groups by generators and relators, branch groups and their subgroup structure, maximal and weakly maximal subgroups and the congruence subgroup property for groups acting on rooted trees, groups of finite type, intermediate growth and growth functions, amenability, Schreier graphs associated to groups acting on rooted trees and their asymptotic characteristics, C^*-algebras associated to self-similar groups and spectral properties of Markov and Hecke type operators, multidimensional rational maps which arise in the study of the spectral problem, etc.

The paper contains many open problems, some of which have long history but some are completely new.

Mathematics Subject Classification (2000). 20F69, 20F65, 20F50, 20F38, 20F19, 20F38, 20E08, 20F10, 20F05, 20E18, 20E15, 20E22, 20E28, 20E34, 20E07, 20E25.

Keywords. Rooted trees, self-similar groups, contracting groups, intermediate growth, torsion groups, branch groups, amenability, topology in the space of groups, Kolmogorov complexity, groups of finite type, maximal subgroups, weakly maximal subgroups, Schreier graphs, C^*-algebra, trace, Markov operator, Hecke type operator, spectrum, rational maps.

The author was supported by NSF grants DMS-0308985 and DMS-0456185 and by Swiss National Foundation for Scientific Research.

Contents

0. Introduction

Back in 1979, interested in growth, amenability and torsion property of finitely generated groups, I discovered a construction which produced several groups with very similar basic properties. Two of them, G and H, were described in the note [Gri80]. Both groups are infinite finitely generated torsion groups and thus belong to the class of Burnside groups [Adi79, GL02, Gup89].

In 1982 I proved that the first of these groups, G, has growth between polynomial and exponential, hence is amenable but not elementary amenable [Gri84a]. This group therefore provided simultaneous answers to the question of J. Milnor [Mil68] on existence of groups of intermediate growth, and to the question of M. Day [Day57]) on existence of amenable but not elementary amenable groups. Main structural properties of the groups G and H were described already in the note [Gri80], and later it started to become clear that these groups are examples of groups in two classes that we now call self-similar groups (or automaton groups) and branch groups.

Automaton groups are groups generated by the states of a finite invertible Mealy automaton. They constitute a subclass of the class of groups of finite automata which was defined in 1963 [Hoř63] and studied during the next two decades, mostly by Ukrainian and Russian mathematicians, with the summary of the results obtained in the first decade published in the book [GP72]. It was already known at that time that they are closely related to subgroups of iterated wreath products introduced by L. Kaloujnin and studied by P. Hall and others. One of the main tools of working with such groups was the method of tables developed by L. Kaloujnin [Kal45, Kal48] and his students V.I. Sushchanskii, Yu.V. Bodnarchuk, and others. Classical methods from computer science (mostly theory of finite automata) were also useful in some investigations (e.g. [Ale72]). The study of the class of automaton groups intensified in the beginning of the 1980-ies after examples appeared in [Gri80] and [GS83], together with new techniques featuring extensive use of self-similarity, projections and contracting.

Originally the groups G and H were defined as groups acting on the interval and on the square, respectively, by measure preserving transformations, but it was also clear that they act on a rooted tree. The construction which appeared in [GS83] used this model, and this point of view now dominates the field. The discovery of Gupta-Sidki p-groups played an extremely important role in the development of the theory of self-similar groups and branch groups; many fundamental properties were established for them.

It was observed already in the pioneering papers [Gri80, GS83] that such groups have unusual subgroup structure which is closely related to the structure of the tree on which the groups act. It was however not until 1997 that this was formalized in the form of a definition, thus bringing us to the notion of a branch group [Gri00a]. The importance of the class of branch groups lies in the fact that they constitute one of three classes into which the class of just infinite groups (i.e., infinite groups with finite proper quotients) naturally splits [Gri00a].

Every infinite finitely generated group has a just infinite quotient. Thus if a group property $\mathcal{P}$ is preserved under homomorphisms (such as torsion, subexponential growth, amenability, finite width, bounded generation, etc.) and there exists an infinite finitely generated group with the property $\mathcal{P}$ then there exists a a just infinite group with the same property. This, in part, explains why, during the last two decades, many examples with remarkable properties were found in the class of groups of branch type.

It was clear from the beginning that the self-similarity phenomenon observed in groups has relations to the self-similarity phenomena arising in other areas of mathematics.

Recent investigations perfectly confirm this. For example, it was observed that transitive actions on compact ultrametric spaces are isomorphic to actions on boundaries of spherically homogeneous rooted trees [GNS00]. This observation provides links to cyclic renormalization and attractors in dynamics [BOERT96], to holomorphic dynamics [Nek03, BGN03, Nek05] , to Grothendick dessins [Pil00], to operator algebras [BG00a, Nek04], to fractal geometry [BGN03] and to other areas of mathematics.

New ideas came with the new generation of mathematicians. The idea of self-similarity was developed wonderfully in the work of V. Nekrashevych whose results are collected in the book [Nek05]. Different spaces such as limit space, solenoid, hyperbolicity complex are associated with a self-similar group, which lead naturally to appearance of dynamical systems related to the group. An iterated monodromy group, denoted $IMG(f)$, corresponds to a self-covering map f of a topological space (under certain conditions). In many cases, when a rational map of $\mathbb{C}$ is iterated, the corresponding group is self-similar and the asymptotics of different combinatorial objects that can be associated with the group (first of all, of the associated Schreier graphs) is related to the geometry of the Julia set and other objects arising in dynamics [Nek05]. L. Bathtoldi developed a theory of algebras of branch type and applied it in the study of different types of growth. E. Pervova proved that maximal subgroups in many self-similar groups of branch type have finite index and that there are branch p-groups without the congruence subgroup property. A. Erschler developed a boundary theory of self-similar groups and obtained fascinating results concerning the growth and amenability. B. Virag converted the concept of self-similarity in groups into the self-similarity of random walks, thus opening a new page in theory of random walks on groups. Z. Sunik found strong links of self-similar groups to problems in combinatorics. For example he showed that some of the famous Hanoi towers problems can be modelled by branch groups.

Surprisingly, dynamical systems appear in different ways in self-similar and in branch groups. For instance, in [BG00b] they appear in connection with the study of the spectral problem for the discrete Laplace operator, while in [Nek03, BGN03, Nek05] they come from holomorphic dynamics. Also, the work [Lys85] and the geometry of Schreier graphs considered in [BG00a, BGN03] naturally led to consideration of substitutional dynamical systems related to self-similar groups.

The ideas of self-similarity and branching penetrated also into the theory of profinite groups [Gri00a]. There are indications that they may be useful in number theory [Bos00] and in Galois theory [AHM].

The number of results in the field and new connections to other areas of mathematics is growing so fast that a few books are already necessary to cover all the material. The number of open questions grows equally fast.

In this paper I discuss some topics related to the group G and both to the class of self-similar (or automaton) groups and the class of of branch groups. I recommend the paper [CSMS01] and Chapter VIII of the beautiful book [dlH00] by Pierre de la Harpe as sources for easy and quick introduction to the group G. The group H happened to be less lucky and there are very few publications related to it, one of which is [Vov00].

Many relevant topics are not included in this article, such as Burnside groups [Adi79, Gup89, Zel90, Zel91], associated algebras, random walks, zeta functions, to name but a few. I recommend [Sid98, Gri00a, CSMS01, GNS00, dlH00, BGŠ03, Nek05] as publications where much of additional material can be found.

The structure of the paper is as follows.

We start with the original definition of the group G followed by the definition via its action on the binary tree, and finally we present the group as a group generated by states of a five state automaton. We also define the class of self-similar groups and discuss its basic properties. One of important notions of the theory of groups acting on rooted trees is the notion of a portrait which we define in the first Section and which appears later, with modifications, in Section 4.

In Section 2 we define the family G_ω which was constructed in [Gri84a] as a generalization of the main example. The construction was created to show that the cardinality of the rates of growth of groups of intermediate growth is uncountable. As a byproduct it suggested introducing a topology on the space of finitely generated groups which we discuss in this Section as well.

In Section 3 we discuss properties of the word length function defined via canonical set of generators, the main of which is the contracting property. Self-similar contracting groups constitute very important subclass and often appear in applications.

In Section 4 and Section 5 we discuss algorithmic problems, especially the word problem and the conjugacy problem. The algorithms presenting solutions of these problems are of branch type and are quite elegant. But in contrast to the algorithm for solution of the word problem the algorithm solving the conjugacy problem is hard for practical realization. Surprisingly, the last fact is very useful from the point of view of modern commercial cryptography, which is looking for groups with easy word problem but difficult conjugacy problem.

At some point L-presentations come into the picture and we discuss a method of finding them. Also, based on the construction G_ω we discuss how to estimate the measure of complexity of the word problem in terms of Kolmogorov complexity.

There are very few results in direction of solving the isomorphism problem, and we mention some of them at the end of Section 5.

Section 6 deals with subgroups of G and features a brief introduction to branch groups. It contains comprehensive information about subgroups of G and touches a little bit on topics such as the congruence subgroup property and maximal subgroups.

Section 7 is about the sequence of finite quotients of G arising as restrictions of the action on levels of the tree. This sequence contains not only a lot of information about the group but it also contains all the information about the profinite closure of G. Topological groups appear here for the first time and don't stay for long. However, branch profinite groups and self-similar profinite groups are playing more and more important role in the latest investigations. Analysis of the structure of portraits of elements of G leads us to the notion of profinite group of finite type, the portraits of elements of which can be described by a finite set of forbidden configurations. This bears analogy with sub-shifts of finite type [Kit98]. Here group theory meets ergodic theory once again.

In Sections 8 and 9 we discuss some topics related to growth and amenability. Though it has been known for more than two decades that G has intermediate growth and is amenable, very little is known about the asymptotics of its growth, about growth of the associated Fölner function, and about the structure of the Cayley graph of the group. We recall some known facts, formulate some recent results and propose a number of open questions.

Some topics related to presentations of G and other groups acting on rooted trees are discussed in Section 10. Most of the attention is paid to the representation determined by the action of the group on the boundary of a tree with the uniform measure which is invariant under the whole group of tree automorphisms. Also we touch on Kaplansky Conjecture on Jacobson radical and explain why G is a candidate to solve it.

In next section we introduce two C^*-algebras which can naturally be attached to any group acting on a rooted tree and which are closures in norm topology of unitary representations considered in the previous section. The idea to use such algebras in the study of self-similar groups appeared in [BG00a], and further progress was made in [GŻ01] and [Nek04]. The C^*-algebras arising in this way also have self-similar properties, and this is discussed in detail in [Nek05]. We believe that these algebras may open a new page in the theory of operator algebras while the information about their properties will be a powerful tool in the study of self-similar groups.

Schreier graphs are the topic of Section 12. Although Schreier's construction is classics, Schreier graphs have not played a large role in group theory until recently.

The works [BG00a, BGN03] and [Nek05] show that in some problems Schreier graphs are more important than the Cayley graphs, and many questions arise about combinatorics, geometry and analysis on such graphs. Schreier graphs related to self-similar groups are often substitutional graphs which are limits of sequences of finite graphs obtained by iterating a substitutional rule.

Topics discussed in Section 12 and Section 13 include spectral properties, amenability, growth, growth of diameters, and expanding properties of Schreier graphs.

In the last Section we consider the spectral problem for discrete Laplace operator on Schreier graphs, for the group G, for the lamplighter group and for the Basilica group. Our wish is to have a solution of the spectral problem for all self-similar groups. Unfortunately there are only very few cases so far where such a solution has been obtained. We describe roughly the method which was used in all of the solved cases. It relates the spectral problem to the problem of description of invariant subsets for multidimensional rational maps, and to the question of weak form integrability of such maps.

An interesting object that appears in the spectral problem is the so called KNS-spectral measure introduced in [BG00a]. Developing new methods for solution of the spectral problem for other self-similar groups is a matter of extreme importance, and we formulate a number of questions in this direction.

The article contains a large number of open problems. Most of them are formulated rigorously, but sometimes we choose rather to express a wish for certain directions to be pursued or for certain circles of problems to be taken into consideration. Some of them are original but most already appeared in [GNS00, BGŠ03, BGN03, dlH00, dlH04, kou02] and other articles and books.

We hope the reader will not be disappointed.

0.1. Acknowledgments

The author would like to express his gratitude to Tatiana Smirnova-Nagnibeda and Zoran Sunik for for their unselfish contribution and fruitful discussions on the topics touched in the text. In addition, thanks to V. Nekrashevych, E. Pervova and L. Bartholdi for their valuable input improving the quality, correctness and the readability of the text.

0.2. Notation

Throughout the paper we will fix G as the notation for the first group from [Gri80], while the letter H and other letters will be used for different purposes in different parts of the article.

0.3. Terminology

The subject of self-similar groups is young and the terminology is not yet fixed. In this article, by a self-similar group I mean the group generated by the states of a finite invertible automaton. This is different from [Nek05] where finiteness of the automaton is not assumed. There are also some other, less important, differences in terminology.

1. The group G and self-similarity

We start with different definitions of the main object of this article, the group G. The first definition that we will give here is the original one which appeared in [Gri80]. Denote by Δ the set obtained from $[0, 1]$ by removing the set R of dyadic rational points $\frac{k}{2^m}$, $k, m \in \mathbb{Z}$. (All intervals in this section are of the form $(\alpha, \beta) \backslash R$.) The group G acts on Δ by Lebesgue measure preserving transformations (we consider the right action). Let us agree that the letter I written above an interval (α, β) denotes the identity transformation on (α, β) while the letter $\mathbf{P}$ denotes the permutation of two halves:

$$x^{\mathbf{P}} = \begin{cases} x + \dfrac{\beta - \alpha}{2} & \text{if } 0 < x < \alpha + \dfrac{\beta - \alpha}{2} \\ x - \dfrac{\beta - \alpha}{2} & \text{if } \alpha + \dfrac{\beta - \alpha}{2} < x < \beta \end{cases}$$

The group G is the group generated by four transformations a, b, c, d of Δ defined as with the operation of composition of transformations. The transformations b, c, d are defined by infinite periodic sequences

$$\begin{matrix} \mathbf{P} & \mathbf{P} & \mathbf{I} & \mathbf{P} & \mathbf{P} & \mathbf{I} & \cdots \\ \mathbf{P} & \mathbf{I} & \mathbf{P} & \mathbf{P} & \mathbf{I} & \mathbf{P} & \cdots \\ \mathbf{I} & \mathbf{P} & \mathbf{P} & \mathbf{I} & \mathbf{P} & \mathbf{P} & \cdots \end{matrix} \qquad (1.1)$$

which we can abbreviate as $(\mathbf{PPI})^\infty$, $(\mathbf{PIP})^\infty$, $(\mathbf{IPP})^\infty$, and the endpoints of the intervals are of the form $1 - \frac{1}{2^n}$, $n = 1, 2, \ldots$.

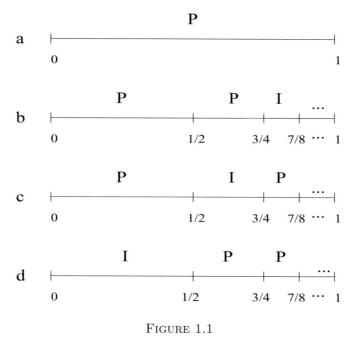

FIGURE 1.1

It is obvious that a, b, c, d satisfy the following relations

$$a^2 = b^2 = c^2 = d^2 = 1$$
$$bc = cb = d$$
$$bd = db = c \qquad\qquad (1.2)$$
$$cd = dc = b$$

where $1 \in G$ is the identity element, so G is indeed a 3-generated group. But for most of our considerations we will use all four generators. The group G has many other relations, for instance,

$$1 = (ad)^4 = (ac)^8 = (ab)^{16} = (adacac)^4 = \cdots$$

and is not a finitely presented group. Later we will discuss in more details the set of relators for G.

We now proceed to give another definition of the group G. To do so, let us identify the points $x \in \Delta$ with infinite binary sequences

$$x \leftrightarrow x_0 x_1 \ldots x_n \ldots,$$

$x_i \in \{0, 1\}$, $i = 0, 1, \ldots$, where $0.x_0x_1 \ldots$ is the binary expansion of the number $x \in \Delta$. The space $\{0, 1\}^{\mathbb{N}}$ of such sequences equipped with the Tychonoff topology (the topology of coordinate-wise convergence) is homeomorphic to the Cantor set.

The action of G on Δ transforms into the action by homeomorphisms on $\{0,1\}^{\mathbb{N}}$ defined recursively as

$$
\begin{aligned}
(xw)^a &= \bar{x}w \\
(0w)^b &= 0w^a &\qquad (1w)^b &= 1w^c \\
(0w)^c &= 0w^a &\qquad (1w)^c &= 1w^d \\
(0w)^d &= 0w &\qquad (1w)^d &= 1w^b
\end{aligned}
\tag{1.3}
$$

where $x \to \bar{x}$ is the permutation on $\{0,1\}$

$$
\varepsilon: \begin{cases} 0 & \to & 1 \\ 1 & \to & 0 \end{cases}
$$

$w \in \{0,1\}^{\mathbb{N}}$ and xw stands for concatenation of x and w.

This action is self-similar in the sense that for any $g \in G$ and $x \in \{0,1\}$ there are $h \in G$ and $y \in \{0,1\}$ such that for any $w \in \{0,1\}^{\mathbb{N}}$

$$
(xw)^g = yw^h.
\tag{1.4}
$$

The relations (1.3) also define an action of G on the set $\{0,1\}^*$ of finite words over the alphabet $\{0,1\}$. We will view $\{0,1\}^*$ as a free monoid generated by the symbols 0 and 1 (the empty word corresponds to the identity element). This monoid is in natural bijection with the vertices of the binary rooted tree T_2 embedded in the plane as shown in Figure 1.2:

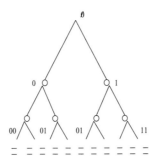

FIGURE 1.2

The set V of vertices splits into a disjoint union

$$
V = \bigsqcup_{n=0}^{\infty} V_n,
$$

where the root vertex corresponds to the empty set; and for $n \geq 1$, V_n is the set of vertices at the n-th level (i.e., the set of vertices at combinatorial distance n from the root vertex). The vertices of the the n-th level are ordered from left to right in the lexicographic order on $\{0,1\}^*$ (with the standard agreement that $0 < 1$).

Any element $g \in G$ induces a bijection on V. Moreover, this bijection preserves the structure of the tree (this can been easily seen if one considers G as a group generated by the states of a finite automaton, as below, or it can be seen directly from (1.3)). In this way we get an action of G by automorphisms of the rooted binary tree. The root vertex is the fixed point and the levels V_n are the invariant sets for the action of G.

Let now $T = T_d$, $d \geq 2$, be the d-regular rooted tree whose vertices can be identified with the elements of the free monoid generated by any set X of cardinality d. Let $X = \{x_1, \ldots, x_d\}$ and let $\mathcal{G}$ be the full group of automorphisms of the tree (we consider the right action on the tree). There is a natural embedding (in fact isomorphism)

$$\psi : \mathcal{G} \to \mathcal{G} \wr_X S_d \tag{1.5}$$

with $\wr_X$ denoting the permutational wreath product, so that the right-hand side in (1.5) is the semidirect product

$$\underbrace{(\mathcal{G} \times \cdots \times \mathcal{G})}_{d} \rtimes S_d$$

where S_d acts on the direct product by permuting the factors. Also there is a natural topology converting $\mathcal{G}$ into a profinite group. If $g \in \mathcal{G}$ and

$$\psi(g) = (g_1, \ldots, g_d)\alpha \tag{1.6}$$

$g_1, \ldots, g_d \in \mathcal{G}, \alpha \in S_d$, then α is the permutation of X induced by the action of g on the first level of the tree, and $g_1, \ldots, g_d$ are the projections of g on subtrees with roots at the first level. Using the canonical identification of any of these subtrees with T allows to view the projections g_i as elements of $\mathcal{G}$.

For instance, the map ψ acts on the generators of G by

$$\psi : \quad \begin{cases} a \to (1,1)\epsilon \\ b \to (a,c)e \\ c \to (a,d)e \\ d \to (1,b)e \end{cases}$$

or in simpler notation

$$\psi : \quad \begin{cases} a \to \varepsilon \\ b \to (a,c) \\ c \to (a,d) \\ d \to (1,b) \end{cases} \tag{1.7}$$

where e and ε are the trivial and the nontrivial element, respectively, in the symmetric group S_2.

The embedding (1.5) leads to the notion of a portrait $P(g)$ of an automorphism $g \in \mathcal{G}$. By a portrait we mean a labelling of the vertices of the tree by elements of the symmetric group S_d, defined as follows.

Using the decomposition (1.5) we label the root vertex by the element α. For each projection g_i we do the same. Namely we decompose g_i as

$$\psi(g_i) = (g_{i1}, \ldots, g_{id})\alpha_i$$

and label the i-th vertex of the first level by α_i. Then we repeat this for projections g_{ij} for the second level etc. There is a bijection between the elements of Aut T and portraits and the set of portraits is the set S_d^V of functions $V \to S_d$ (or labellings), where $V = V(T)$ is the set of vertices. The elements of S_d^V will also be called configurations. For more on portraits see [BGŠ03]. The topology on $\mathcal{G}$ is induced by Tychonoff topology on S_d^V.

Using the relations (1.3) we get the following portraits for generators of G (e and ε are elements of S_2). In each of the portraits $P(b), P(c), P(d)$, labels σ

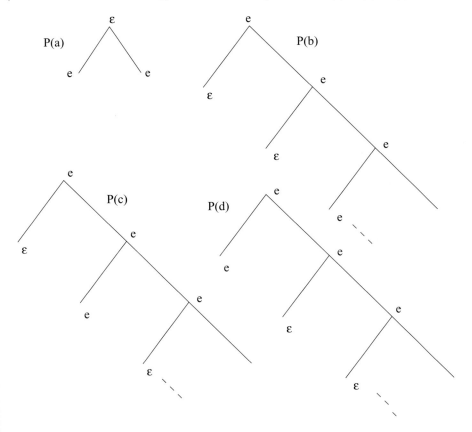

FIGURE 1.3

only appear at distance 1 from the rightmost ray of the tree. Labels at distance 1 from the rightmost ray form periodic sequences of period three, as in (1.1). Other vertices of T_2 that do not appear in Figure 1.3 are labelled by the identity element.

Presentation of elements by portraits allows us to visualize them and is useful in many situations.

Quite often we will rewrite (1.6) in the form

$$g = (g_1, \ldots, g_d)\alpha \tag{1.8}$$

and call this the decomposition of g.

The self-similarity of G that was mentioned above (see the relation (1.4)) can be interpreted here in the following way.

Let $g \in G$ and let $P = P(g)$ be the portrait of g.

For any vertex $u \in V(T)$ let P_u be the restriction of P on the subtree T_u with root u; and let g_u be the automorphism of T_u given by P_u. If we identity T_u with T using the canonical isomorphism, then g_u is again an element of G called the projection (or the slice) of g at vertex u. For the generators this is obvious and for the elements of G it follows easily from the formulas

$$(gf)_u = g_u f_{u^g},$$
$$(g^{-1})_u = (g_{u^{g-1}})^{-1}.$$

In the same way, self-similarity of any group L acting on a regular rooted tree essentially means that any projection of any element $g \in L$ at any vertex u is again an element of L. This remark will be refined below in Definition 1.1 to allow a distinction between weakly self-similar groups and self-similar groups.

The data contained in Figure 1.3 can be described by the diagram (a) given in Figure 1.4. If $q \in \{a, b, c, d, id\}$ is any node of the diagram, $u \in \{0, 1\}^*$ is a word,

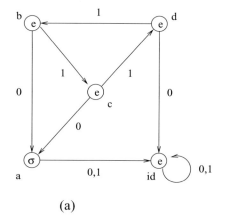
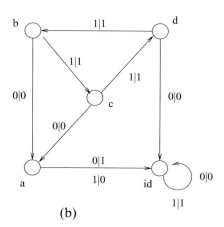

(a)   (b)

FIGURE 1.4

and s is the final node of the path ℓ_u in the diagram starting at q and determined by the word u, then the label (e or ε) of the node s coincides with the label of the vertex u in the portrait of the element q. The node id corresponds to the identity element of G given by the portrait where all the labels are the identity. We will call the subset $\{id, a, b, c, d\} \subset G$ the core (or nucleus) of G.

The diagram (a) in Figure 1.4 determines a noninitial Mealey automaton A over the alphabet $\{0, 1\}$ with five states a, b, c, d, id, with transition function given by the oriented edges and with exit function given by the labelling of the vertices by elements of the symmetric group S_2. Specifying which state q is initial we get an initial automaton A_q which can be viewed as a self-mapping $\tilde{A}_q$ of the set $\{0,1\}^* \cup \{0,1\}^{\mathbb{N}}$ consisting of finite and infinite strings

$$x_0 x_1 \cdots x_n \longrightarrow \boxed{A_q} \longrightarrow y_0 y_1 \cdots y_n$$

Given a sequence $x_0 x_1 \ldots x_n \ldots$ the automaton acts on the first symbol by its label $\sigma_q \in S_2$ and changes its initial state to the state given by the end of the arrow going out of q and labelled by x_0. Now the automaton is in a new state, reads the next symbol x_1, transforms it and moves to the next state according to the described rule; and continues to operate in this way. Composition of two maps given by two finite initial automata is again a transformation given by a finite automaton, and the inverse map can also be given by a finite automaton. Thus finite initial (invertible) automata constitute a group with a simple rule for composition and inversion. We recommend [GNS00] as a source of information about groups of finite automata (or automaton groups). An equivalent description of the automaton A is given by Figure 1.4 (b) which is usually called the Moore diagram.

We have seen that the group G can be considered as a group generated by the states of the automaton from Figure 1.4 (the state id corresponds to the identity element and can be deleted from the generating set). Similarly, any group generated by a finite automaton acts on a regular rooted tree and this action is self-similar because any slice of any generator is again a generator. And conversely, any group L acting on a rooted tree in a self-similar way can be generated by the states of some automaton (not necessary finite). Namely, having a generating set S of L one can consider the set of projections of the elements of S as the set of states of the automaton and construct the diagram of the automaton using the rewriting rules

$$s = (s_1, \ldots, s_m)\alpha, \tag{1.9}$$

$s \in S$, analogous to 1.8.

Definition 1.1. (i) *A group L is called self-similar if it is isomorphic to a group generated by all states of a finite invertible Mealey automaton.*

(ii) *A group L is called weakly self-similar if it is isomorphic to a group generated by the states of an invertible Mealey automaton.*

(iii) *A closed subgroup L in Aut T, T a regular rooted tree, is called self-similar if it is a closure of a self-similar subgroup of Aut T.*

A self-similar group L defined by an automaton over the alphabet X of cardinality d embeds into the wreath product $L \wr_X S_d$ through a map induced by the map ψ in (1.5). This embedding is completely determined by the Moore diagram of the corresponding automaton. In case of G the embedding is described in (1.7).

The groups G and H happen to be the first examples considered in the literature of groups generated by all the states of an automaton, so they are the first (nontrivial) examples of self-similar groups.

Groups generated by the states of an automaton form an extremely interesting and important class of groups related to many topics (see for instance [GNS00]), and many open questions about them await solution.

Problem 1.1.   (i) *What groups are weakly self-similar ?*
  (ii) *What groups are self-similar?*

As an example let us mention that it was shown only recently that a free group of finite rank can be generated by the states of a finite automaton [GM03], [VV05]. For instance, Mariya Vorobets and Yaroslav Vorobets show in [VV05] that the automaton given in Figure 1.5 generates a free group of rank 3 (thus answering a question of S. Sidki).

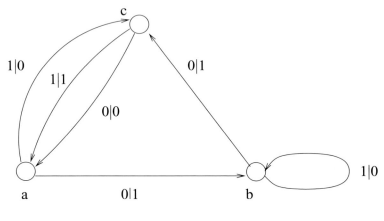

FIGURE 1.5

Perhaps the simplest example of an interesting group generated by a finite automaton is the lamplighter group $\mathbb{Z}/2\mathbb{Z} \wr \mathbb{Z}$ which is generated by the automaton in Figure 1.6.

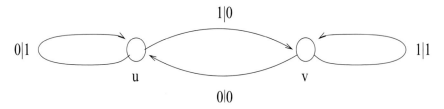

FIGURE 1.6

One more interesting example of an automaton group (the so-called Basilica group) will be discussed in Section 9.

There are two obvious restrictions which self-similar groups have to conform to. First, they have to be residually finite as the full group of automorphisms of a rooted tree is residually finite (the approximating sequence of finite groups is Aut $T/st(n)$, where $st(n)$ is the stabilizer of the level n of the tree). Also, a group generated by a finite automaton has solvable word problem (this follows from the minimization algorithm for finite automata, see Section 4).

2. Groups G_ω and the topology in the space of groups

The following construction generalizes the main example. Let $\Omega = \{0,1,2\}^{\mathbb{N}}$ be the space of sequences over the alphabet $\{0,1,2\}$. The bijection

$$0 \longleftrightarrow \begin{pmatrix} \mathbf{P} \\ \mathbf{P} \\ \mathbf{I} \end{pmatrix}$$

$$1 \longleftrightarrow \begin{pmatrix} \mathbf{P} \\ \mathbf{I} \\ \mathbf{P} \end{pmatrix}$$

$$2 \longleftrightarrow \begin{pmatrix} \mathbf{I} \\ \mathbf{P} \\ \mathbf{P} \end{pmatrix}$$

between symbols and columns will be used in the construction. Namely, for a sequence $\omega = \omega_0\omega_1\ldots \in \Omega$ replace each ω_i, $i = 0,1,\ldots$, by the corresponding vector and get a vector

$$\begin{pmatrix} U_\omega \\ V_\omega \\ W_\omega \end{pmatrix}$$

consisting of three words $U_\omega, V_\omega, W_\omega$ over the alphabet $\{\mathbf{I}, \mathbf{P}\}$. For instance, the triple corresponding to the sequence $\xi = 012\,012\ldots$ is

$$U_\xi = \mathbf{PPI}\ \mathbf{PPI}\ldots$$
$$V_\xi = \mathbf{PIP}\ \mathbf{PIP}\ldots$$
$$W_\xi = \mathbf{IPP}\ \mathbf{IPP}\ldots$$

Similarly, the triple corresponding to the sequence $\eta = 01\,01\ldots$ is

$$U_\eta = \mathbf{PPP}\ \mathbf{PPP}\ldots$$
$$V_\eta = \mathbf{PIP}\ \mathbf{IPI}\ldots$$
$$W_\eta = \mathbf{IPI}\ \mathbf{PIP}\ldots$$

Using the words $U_\omega, V_\omega, W_\omega$ construct the transformations $b_\omega, c_\omega, d_\omega$ of the interval Δ defined in a way analogous to the case of the sequence ξ (see Figure 1.1). Define the group G_ω as the group generated by the transformation a from 1.1 and by the transformations $b_\omega, c_\omega, d_\omega$, i.e. $G_\omega = \langle a, b_\omega, c_\omega, d_\omega \rangle$.

The following relations hold

$$a^2 = b_\omega^2 = c_\omega^2 = d_\omega^2 = 1$$
$$b_\omega c_\omega = c_\omega b_\omega = d_\omega$$
$$b_\omega d_\omega = d_\omega b_\omega = c_\omega$$
$$c_\omega d_\omega = d_\omega c_\omega = b_\omega$$

so all groups G_ω are 3-generated (in some degenerate cases even 2-generated, such as in the case of the sequence $0\,0\ldots0\ldots$), but we prefer to consider them as 4-generated groups with the system $\{a, b_\omega, c_\omega, d_\omega\}$ assumed as the canonical system of generators.

Let

$$\Lambda = \{\omega \in \Omega: \quad \text{each symbol } 0,1,2 \text{ occurs in } \omega \text{ infinitely many times}\}$$

$$\Xi = \{\omega \in \Omega: \text{ at least two symbols occur in } \omega \text{ infinitely many times}\}.$$

Then $\Lambda \subset \Xi \subset \Omega$. The groups G_ω are infinite groups and there is uncountably many pairwise non-isomorphic groups among them, since for each $\omega \in \Omega$ there is at most a countable set of sequences ζ such that $G_\omega \simeq G_\zeta$ [Gri84a]. Indeed there is not more than 6 such sequences [Nek05]. If $\omega \in \Omega - \Xi$ then G_ω is virtually abelian, if $\omega \in \Xi$ the group G_ω has degree of growth that is intermediate between polynomial and exponential (for more on growth see Section 8), and if $\omega \in \Lambda$ then G_ω is a torsion 2-group.

If ω is a periodic sequence then G_ω embeds in the group of finite automata. For instance for the sequence $\eta = (01)^\infty$ the group G_η is the group generated by the states of the automaton E given by Figure 2.1. Let $\tau\colon \Omega \to \Omega$ be the shift

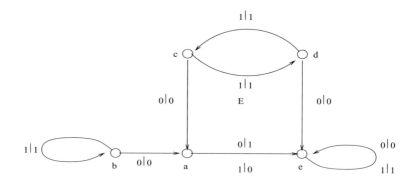

FIGURE 2.1. Automaton E

map given by $(\tau\omega)_n = \omega_{n+1}$. The sequence $\{G_{\tau^n\omega}\}_{n=0}^\infty$ is called the linking class of G_ω. In case ω is periodic it consists of finitely many groups while for non-periodic ω it consists of infinitely many non-isomorphic groups, which follows from the solution to the isomorphism problem by Nekrashevych [Nek05] (see Section 5).

The groups from one linking class share many properties and the study of a group G_ω is usually impossible without studying the whole linking class.

The study of the groups G_ω, $\omega \in \Omega$ inspired the invention of a topology on the set X_k of k-generated groups, $k \geq 2$, in [Gri84a]. More precisely, the elements of the space X_k are k-generated marked groups, i.e. pairs (M, S_M) consisting of a group M and a generating system $S_M = \{s_1, \ldots, s_k\}$ of k-elements with a given order (the change of the order changes the point). A (metrizable) topology in the space X_k is generated by a system of open sets $U(M, S, r)$, $r = 1, 2, \ldots$, consisting of those points $(L, S_L) \in X_k$ for which the Cayley graphs $\Gamma(M, S_M)$ and $\Gamma(L, S_L)$ of the groups M and L constructed with respect to the system of generators S_M and S_L are isomorphic in the neighborhoods of the identity element of radius r (here by isomorphism of graphs we mean a graph isomorphism which also preserves the labelling of the edges by generators and their inverses). Other interpretation of the elements of the space X_k can be obtained by replacing (M, S_M) by the corresponding Cayley graph $\Gamma(M, S_M)$ or by a pair (F_k, N) where F_k is a fixed free group with basis $s_1, \ldots, s_k$ and N is a normal subgroup such that F_k/N is isomorphic to M under the isomorphism that sends the generators of F_k to the corresponding generators of M.

For each $k \geq 2$, the space X_k is a totally disconnected compact metric space. The space X_k naturally embeds in X_{k+1}, simply by adding one more generator to the generating set. Thus each "point" (F_k, N) in X_k is presented by the pair $(F_{k+1}, \tilde{N})$ where $\tilde{N}$ is the normal subgroup in F_{k+1} generated by N and s_{k+1}. We can now consider the space $X = \bigcup_{k=2}^{\infty} X_k$ with a natural topology that has the property that its restriction to each X_k gives back the original topology of this space. Thus X is a locally compact totally disconnected space. It can be identified with the space of normal subgroups of the free group F_∞ of infinite rank with a basis $s_1, s_2, \ldots,$, which contains all generators a_n, $n \geq n_0$ for some n_0, with Chabauty topology.

The group $\mathcal{A}_k$ of automorphisms of F_k is generated by the elementary Nielsen transformations (interchange of generators $s_1, \ldots, s_k$, replacement of one generator s_i by its inverse s_i^{-1} and replacement of one generator s_i by $s_i s_j$, $j \neq i$) and naturally acts on the space X_k: if $\varphi \in \mathcal{A}_n$ then $(F_m, N)^\varphi = (F_m, N^\varphi)$. The group F_m/N is isomorphic to F_m/N^φ but the action of $\mathcal{A}_n$ on X_n is not transitive on classes of isomorphic groups. Nevertheless simple arguments [Cha91] based on use of Tietze transformations show that if two points (M, S_M), $(L, S_L) \in X_k$ correspond to isomorphic groups (i.e. $M \cong L$) then there is $\varphi \in \mathcal{A}_{2k}$ s.t. $(M, S_M)^\varphi = (L, S_L)$ if (M, S_M), (L, S_L) are viewed as elements of the space X_{2k}.

Therefore the group $\mathcal{A}_\infty$ of Nielsen automorphisms of the free group F_∞ of infinite rank with basis $s_1, \ldots, s_n, \ldots$ ($\mathcal{A}_\infty$ is generated by elementary Nielsen transformations) acts on X in such a way that this action is transitive on classes of isomorphic groups (that is if (M, S_M), $(L, S_L) \in X$ and $M \cong L$ then there is $\varphi \in \mathcal{A}_\infty$ s.t. $(M, S_M)^\varphi = (L, S_L)$).

Thus the space X encompasses the whole world of finitely generated groups and has many symmetries given by the action of $\mathcal{A}_\infty$. A number of questions arise about X.

Problem 2.1. *What is the topological type of the space X_k, $k = 2, 3, \ldots$?*

Recall that there is a complete system of invariants for metrizable totally disconnected compact spaces. Up to homeomorphism there is only one such space without isolated points, namely the Cantor set. If there are i_1 isolated points $((1 \leq i_1 \leq \infty))$ delete them. The space that is left is still compact and totally disconnected. If it has no isolated points, i.e. it is a Cantor set, we stop. Otherwise the we record the number i_2 of isolated points, delete them, and so on. Continuing this procedure one gets a sequence $i_1, i_2, \ldots$ in which each i_j is a positive integer or ∞).

The sequence $\bar{i} = \{i_j\}$, which can be empty (in case of the Cantor set), finite, or infinite is a complete invariant of totally disconnected compact spaces. The problem above asks what is this invariant for the set X_k and in particular what is the cardinality d of the sequence $\{i_j\}$ (which is called the Cantor-Bendixson rank [Kec95]).

A point of a totally disconnected set is called isolated point of rank j if it appears as isolated point in the j-th step of the above procedure of deletion of isolated points (so the isolated points of rank 1 are just ordinary isolated points).

Problem 2.2. *Describe the set of isolated points of rank j, $j = 1, 2, \ldots, d$ of the set X_k.*

Let us have a look at the isolated points of rank one. Let (M, S_M) be an isolated point of the space X_k presented by a pair (F_k, N) so $M \cong F_k/N$. Then M is finitely presented group as otherwise N would be a union $\bigcup_{i=1}^{\infty} N_i$ of a strictly increasing sequence of normal subgroups and therefore the point (F_k, N) would be the accumulating point of the sequence (F_k, N_i).

Let us say that a group M has the finite approximation property if for any finite subset $E \subset M$ there is a nontrivial normal subgroup $H \lhd M$ such that the image $\overline{E}$ of E in quotient group M/H has the same cardinality as E.

Theorem 2.1. *A point (M, S_M) is a isolated point in X_k if and only if M is a finitely presented group without the finite approximation property.*

Proof. Assume M is finitely presented, let (M, S_M) be identified with the pair (F_k, N), and let $\{(M_n, S_{M_n})\}_{n=1}^{\infty}$ converge to (M, S_M). Let $m = \max_{i \in I} |R_i|$ be the maximum length of a relator in some fixed finite presentation of M

$$M = \langle s_1, \ldots, s_k \mid R_i, i \in I \rangle.$$

As the Cayley graphs $\Gamma(M, S_M)$ and $\Gamma(M_n, S_{M_n})$ have the same neighborhoods of radius m for all $n \geq N_0$ (N_0 sufficiently large number) the relations $R_i = 1$ also hold in the groups $M_n, n \geq N_0$ and therefore all M_n with $n \geq N_0$ are quotients of M.

Assume M is not isolated and $\{(M_n, S_n)\}_{n=1}^{\infty}$ is an approximating sequence consisting of proper homomorphic images M_n of M. Thus $M_n = M/H_n$ with $H_n \neq 1$, for all n. Let $B_M(r)$ be the ball of radius r centered at the identity element in $\Gamma(M, S_M)$. For each r there is n_0 such that $B_M(r) \cong B_{M_n}(r)$ for $n \geq n_0$. Any finite set $E \subset M$ can be included in $B_M(r)$ for sufficiently large r. This implies the finite approximation property for M as $H_n \neq \{1\}$, for all n. By contraposition, if M does not have the finite approximation property than it is an isolated point.

On the other hand, if M has the finite approximation property then (M, S_M) is accumulating point of the set $\{(M/H, \overline{S}_M)\}$ consisting of all proper homomorphic images of M and corresponding images of systems of generators because using as E the set $B_M(r)$ and finding the corresponding normal subgroup $H \triangleleft M$ we get an isomorphism $B_M(r) \simeq B_{M/H}(r)$. $\square$

Examples of isolated points are finite groups, finitely presented simple groups and, more generally, finitely presented groups with nontrivial intersection of all nontrivial normal subgroups (also known as monolithic groups or subdirectly irreducible groups).

Problem 2.3. *Is there a description of finitely presented groups without the finite approximation property in terms of simple groups?*

Examples of isolated points of rank 2 are the finitely presented residually finite just infinite groups (recall that a group is just infinite if it is infinite but every proper quotient is finite). In [Gri00b] the class of just infinite groups is split in three classes: branch groups (defined in Section 6), almost hereditarily just infinite groups, and almost simple groups. Hereditarily just infinite group is a residually finite group in which every normal subgroup of finite index is just infinite (for instance $SL(n, \mathbb{Z}), n \geq 3$). "Almost" hereditarily just infinite (simple) means that there is a subgroup of finite index which is a direct power of several copies of some group which is hereditarily just infinite (simple). All finitely presented almost simple groups are isolated points of rank 1.

At the moment there are no known examples of finitely presented branch groups and no known examples of infinitely presented hereditarily just infinite groups.

One of the challenges related to the introduced spaces X_k is in finding closed subsets consisting of groups with specific properties. This was our original idea realized in [Gri84a] based on the construction of the groups G_ω. Namely, two groups G_ω and G_ζ have isomorphic neighborhoods of identity of radius $r = 2^t$ in the Cayley graphs with respect to the canonical systems of generators, where t is the coordinate where the sequences ω and ζ become different. This rule works for sequences from the subset $\Xi \subset \Omega$ and follows from the solution of the word problem in the groups G_ω (for more on this see Section 4). The groups $G_\omega, \omega \in \Omega - \Xi$ are virtually abelian and for them the above rule does not work. Let us modify our construction replacing the groups G_ω, $\omega \in \Omega - \Xi$, by the limits

$$\widetilde{G}_\omega = \lim_{n \to \infty} G_{\zeta_n} \tag{2.1}$$

where $\{\zeta_n\}_{n=1}^\infty$ is any sequence of points $\eta_n \in \Xi$ converging to ω. The limit (2.1) doesn't depend on the choice of sequence $\{\zeta_n\}$. The limit groups $\widetilde{G}_\omega$, $\omega \in \Omega - \Xi$, are virtually metabelian of exponential growth [Gri84a]. After this modification let us return to the notation G_ω and omit writing the tilda. The set $\{G_\omega : \omega \in \Omega\}$ is a closed subset of X_4 homeomorphic to a Cantor set. It consists of infinitely presented amenable groups only countably many of which (namely G_ω, $\omega \in \Omega - \Xi$) are elementary amenable, (more about amenability and elementary amenable groups is written in Section 9). The groups G_ω, $\omega \in \Xi$ have intermediate growth and the groups G_ω, $\omega \in \Lambda$ are torsion 2-groups.

Problem 2.4. *Find other interesting closed subsets in X_k consisting of groups with some particular (interesting) properties.*

The set defined as the closure of the set of torsion free Gromov's hyperbolic groups is studied in [Cha00]. Some interesting facts about the structure of the space X are summarized in [CG00]. A fact proven in [CG00] is that the class of groups known as fully residually free groups [KM98] and limit groups of Sela [Sel01] consists of accumulating points of pairs (M, S_M) where M is isomorphic to a free group of finite rank. Another fact is the result of Y. Shalom claiming that groups with Kazhdan property (T) constitute a closed subset in X [Sha00].

Given the action of the shift $\tau : \Omega \to \Omega$ on the set $Z = \{G_\omega : \omega \in \Omega\}$ one can study the typical properties of groups from Z using a τ-invariant measure. The most natural measure in our situation is the Bernoulli $\{\frac{1}{3}, \frac{1}{3}, \frac{1}{3}\}$ measure. Any group property that is preserved in subgroups of finite index and homomorphic images holds for the groups in Z with either probability one or probability zero. For instance, with probability 1 the groups in Z are torsion groups. It is also known that with probability 1 the groups in Z are of intermediate growth, and are branch groups. It would be interesting to find a τ-invariant property (i.e. a property common for all groups from the same linking class) for which it is difficult to say whether the measure is 0 or 1.

An interesting question raised by E. Ghys is the question on typical properties (those that occur with probability 1) of groups viewed as elements of the space X supplied with a probability measure μ. It is unclear what measure μ on X would be natural to consider. For the space X_k, it is natural to consider the uniform measure: the cylinder set of pairs (M, S_M) in X_k having a given neighborhood of radius r has to have a measure $\frac{1}{b(r)}$ where $b(r)$ is the number of possible different neighborhoods of radius r in X_k (by the way, it would be interesting to know the precise asymptotic of $b(r)$ when $r \to \infty$). It is unclear what we should do with the isolated points and how their deletion affects the type of the measure.

We have the action of the group $\mathcal{A}_\infty$ on X and the orbits consist of representatives of the same group. As we want to study statistical properties of finitely generated groups the measure μ has to be quasiinvariant (there is a very little chance to have invariant measure). Our discussion leads to the following question.

Problem 2.5.   a) *Is there a $\mathcal{A}_\infty$ quasiinvariant measure defined on the space X?*

    b) *If such a measure μ exists, which group properties are typical with respect to μ?*

    c) *Which group properties are typical with respect to the uniform measure on X_k?*

It is easy to see that the action of A_∞ on X is not totally disconnected so there is no good factorization X/A_∞ in topological sense. However, the partition into orbits is measurable and hence the factor space X/A_∞ with the induced σ-algebra of sets can be defined [Cha00].

3. Length functions

In this section we discuss the length properties of the elements of the group G, the most important of which is the contracting property. Some other length functions, such as for instance the depth function, are also discussed here. For details we recommend [Gri80, Gri84a, Gri98, CSMS01, dlH00]

As was already mentioned G is a 3-generated group but most of our considerations will be given with respect to the system of generators

$$\{a, b, c, d\}$$

which we will be considered canonical. The set

$$\{1, a, b, c, d\} \tag{3.1}$$

which represents the states of the automaton A from Figure 1.4 is called the core and the reason will be clear later.

We denote by $|g|$ the length of g with respect to the canonical systems of generators (i.e. the length of the shortest presentation of g as a product of generators). As the generators have order 2 we do not use negative powers and because of relations (1.2) the shortest representative of an element $g \in G$ has the form

$$\flat a * a \cdots * a * a \flat \tag{3.2}$$

where $*$ represents an element from the set $\{b, c, d\}$ and $\flat$ represents an element from the set $\{\emptyset, b, c, d\}$ ($\emptyset$ – empty symbol).

Let

$$\Gamma = \langle a, b, c, d \mid a^2 = b^2 = c^2 = d^2 = bcd = 1 \rangle. \tag{3.3}$$

Then $\Gamma \simeq C_2 * (C_2 \times C_2)$ (where the first factor is generated by a and $C_2 \times C_2 = \{1, b, c, d\}$ is the Klein group). Because of the relations (1.2) the group Γ naturally covers the group G in the sense that there is a canonical epimorphism $\Gamma \to G$. It is also clear that (3.2) is the normal form for the elements in the group Γ viewed as a free product.

The embedding (1.5) induces the embedding (also denoted ψ)

$$\psi\colon G \to G \wr S_2 \qquad (3.4)$$

(the permutational wreath product sign can be replaced by the standard wreath product sign in the case $d = 2$) where ψ is determined by (1.7). We have already agreed to write $g = (g_0, g_1)\alpha$ instead of $\psi(g) = (g_0, g_1)\alpha$, for $g_i \in G$, $\alpha \in S_2$. The following Lemma is a statement about length reduction with respect to the canonical system of generators.

Lemma 3.1. [Gri80] *Let* $g \in G, g = (g_0, g_1)\alpha$. *Then*

$$|g_i| \leq \frac{|g| + 1}{2}. \qquad (3.5)$$

If a shortest word of the form (3.2) *representing* g *has one of two symbols* $\flat$ *empty then*

$$|g_i| \leq \frac{|g|}{2}. \qquad (3.6)$$

Corollary 3.1.

$$|g_0| + |g_1| \leq |g| + 1. \qquad (3.7)$$

On the other hand we have

Lemma 3.2.

$$|g| \leq 2(|g_0| + |g_1|) \qquad (3.8)$$

The estimate follows from the fact that the image Im ψ is the group $(B \times B) \rtimes \tilde{D}$ where $B = \langle b \rangle^G$ has index 8 in G and $\tilde{D} = \langle (a, d), (d, a) \rangle$. For more details see [Gri84a]. The inequalities of the type (3.8) are important for the solution of the conjugacy problem (see Section 5).

Problem 3.1. *Let* L *be a self-similar group acting on a tree* T_d. *Is there a constant* C *s.t.*

$$|g| \leq C \sum_{i=1}^{d} |g_i| \qquad (3.9)$$

for the decomposition $g = (g_1, \ldots, g_d)\alpha$, *of any element* $g \in L$?

The property of the length given by the inequality (3.5) reflects the contracting property of the group (definition is provided in the next paragraph). We already mentioned that an embedding of type (3.4) holds for any self-similar group L and has the form

$$\psi\colon L \to L \wr_X S_d \qquad (3.10)$$

where d is the arity of the rooted tree on which the group acts and $X = \{x_1, \ldots, x_d\}$ is the corresponding alphabet on which the symmetric group S_d acts. We denote by $|g|$ the length of $g \in L$ with respect to the system of generators given by the states of the corresponding automaton.

Definition 3.1. *A self-similar group* L *whose action induces an embedding* (3.10) *is called contracting if there are constants* $\lambda < 1$ *and* C *such that for any element* $g = (g_1, \ldots, g_d)\alpha \in L$,

$$|g_i| \leq \lambda |g| + C, \qquad i = 1, \ldots, d. \tag{3.11}$$

The smaller the constant λ, the better contraction properties. The definition does not depend on the choice of generating system.

Problem 3.2. *What is the class of contracting self-similar groups? Can it be characterized in algebraic terms?*

Problem 3.3. *Is every contracting self-similar group amenable?*

Recall that a (discrete) group L is amenable if there is an invariant mean on the Banach space $B^\infty(L)$ of bounded functions [Gre69]. All known examples of contracting groups are amenable. We will discuss amenability in more detail in Section 9.

Estimates of the contracting coefficient λ from Definition 3.1 are important in the study of growth of groups, which is discussed in Section 8. While many contracting groups (in particular G) have subexponential growth, there are contracting groups of exponential growth (for instance Basilica groups defined in Section 9).

In the Definition 3.1 and last problem we consider the length of elements with respect to a system of generators (that is given by a set of states of corresponding automaton). One can consider a more general type of length functions on a group G and other (self-similar) groups, namely the functions $\ell\colon L \to \mathbb{R}$ satisfying the conditions

$$\begin{cases} \ell(g) = 0 \Leftrightarrow g = 1 & (3.12) \\ \ell(g) = \ell(g^{-1}) & (3.13) \\ \ell(gh) \leq \ell(g) + \ell(h) & (3.14) \end{cases}$$

and study its properties. One extra requirement for reasonable length function is to insist on the finiteness of the balls

$$B_1^\ell(n) = \{g \in L \colon \ell(g) \leq n\}$$

and hence the existence of a growth function

$$\gamma_L^\ell(n) = \#(B_1^\ell(n)). \tag{3.15}$$

where $\#$ denotes cardinality. The last condition should sometimes be strengthened by requiring that $\gamma_L^\ell(n)$ grows no faster than an exponential function.

Growth functions with respect to the word length induced by a generating set are discussed further in Section 8.

Length functions are closely related to weights [Gri96]. We are going to define now another type of length function called depth which first appeared in the work of S. Sidki [Sid87a].

Applying the embedding ψ n times to an element $g \in G$ gives the embedding

$$\psi_n\colon \ G \to G \wr_{X^n} Q_n$$

where the group $Q_n = \underbrace{C_2 \wr \cdots \wr C_2}_{2^n}$ is the group acting on the n-th level of the tree

(which can be identified with the set X^n, $X = \{0,1\}$). The group Q_n can also be viewed as a group acting on the part of the tree T up to level n and the action is given by restriction of the action of G. If

$$\psi_n(g) = (g_{0\cdots 0} \ldots, g_{i_1 \cdots i_n}, \ldots, g_{1\cdots 1})\alpha$$

$g_{i_1 \ldots i_n} \in G$, $\alpha \in Q_n$ then the element g can be represented by a diagram of the form given in Figure 3.1 where the vertices up to level $n-1$ are labeled by elements

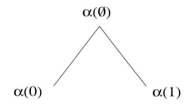

FIGURE 3.1

from S_2 according to the portrait of α. By Lemma 3.1

$$|g_{i_1 \ldots i_n}| \le \frac{|g|}{2^n} + 1$$

$\forall i_1, \ldots, i_n \in \{0,1\}$.

Thus there is $n \le \log_2 |g| + 1$ s.t. that all elements $g_{i_1 \ldots i_n}$ belong to the core $\{1, a, b, c, d\}$. The smallest such n is called the depth of the element g and is denoted by $d(g)$. We have

$$d(g) \le \lceil \log_2 |g| \rceil + 1. \tag{3.16}$$

Using the incquality (3.8) it is easy to get the estimate

$$\frac{1}{2} \log_2 |g| \le d(g).$$

Problem 3.4.   (i) *Are there constants $\lambda < 1$ and c such that*

$$d(g) \le \lambda \log_2 |g| + c \tag{3.17}$$

holds for any $g \in G$? What is the minimal value of λ?

(ii) *Are there constants $\mu > 1/2$ and C such that*

$$d(g) \geq \mu \log_2 |g| + C?$$

What is the supremum of values of μ?

(iii) *Is there a good description of the portraits of the elements of fixed length in the group G?*

There are five elements of depth 0 (namely $1, a, b, c, d$) and

$$d(gh) \leq \max(d(g), d(h)) + 1 \leq d(g) + d(h) + 1.$$

So $\ell(g) = d(g) + 1$ satisfies the inequality (3.14). The balls

$$\{g \in G\colon\ d(g) \leq n\} \tag{3.18}$$

are finite and have the cardinality

$$\leq 5^{1+2+\cdots+2^n} = 5^{2^{n+1}-1}.$$

Problem 3.5. *What is the precise growth function of G with respect to the depth?*

For groups with the contracting property naturally the notion of a core (or nucleus [BGN03], [Nek05]) arises. Namely the inequality (3.11) implies that if $|g| > \frac{C}{1-\lambda}$ then the projections g_i, $i = 1, \ldots, d$ have smaller length. Thus the set of elements for which there is no shortening of the lengths in the projections is contained in the ball

$$\left\{g \in L\colon\ |g| \leq \frac{C}{1-\lambda}\right\}$$

and hence is finite. It is called the core of the group and denoted by $\mathrm{core}(L)$.

The core depends on the system of generators. For the group G and the canonical system of generators

$$\mathrm{core}(G) = \{1, a, b, c, d\}.$$

The notion of the core leads to the notion of the core portrait $P_{\mathrm{core}}(g)$, $g \in L$. Namely when defining $P_{\mathrm{core}}(g)$ start with the same procedure as in the definition of the portrait $P(g)$ but stop the development of the portrait at those vertices u of the tree for which the corresponding projection g_u of the element g belongs to $\mathrm{core}(L)$. Attach to such a vertex u the label g_u. This procedure leads to a finite rooted tree (with degree of vertices $\leq d$) whose internal vertices are labeled by elements of the symmetric group S_d (according to the portrait $P(g)$) while the leaves are labeled by corresponding elements of the core. For instance the core portrait of the element $adacac$ in G is given in Figure 3.2.

The core portraits are useful in many situations, for instance in solving the conjugacy problem.

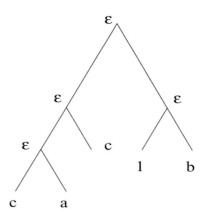

FIGURE 3.2

4. The word problem and L-presentations

The group G and the whole family G_ω have nice and interesting algorithmic properties and we start our discussion with the word problem (WP), whose solution is given by a new (at least in group theory) type of algorithm. We call it branch algorithm because of its branch structure. Let us mention first of all that the word problem for G can be solved by the algorithm which works for all groups generated by finite automata (and in particular for all self-similar groups) as G is one of the representatives of this class (as was described in Section 1). This algorithm works as follows. Given a word in the generating set one multiplies the corresponding automata using the composition rule and then to minimizes the product using the classical algorithm of minimization [Eil74].

    The word represents the identity element if and only if the minimal automaton is 1-state automaton with trivial exit function (that doesn't changes the input symbols). The rough upper estimate of complexity of such an algorithm (both in time and in space) is exponential as the number of states of composition of automata is equal to the product of numbers of states of factors. But indeed we don't know any example of an automaton group with exponential complexity of word problem. We even don't have an examples with complexity higher than polynomial. All known examples of automaton groups are either solvable groups of special type, or groups of matrices over $\mathbb{Z}$ or groups of branch type with good contracting properties. For all of them the WP has polynomial complexity. Thus we formulate

Problem 4.1.   a) *What is the maximal possible complexity of WP for self-similar groups?*

    b) *In particular, is there a self-similar group with superpolynomial complexity of WP? Is there a self-similar group with exponential complexity?*

The WP can be formulated for any finitely generated group (and even for infinitely generated if we do this carefully). The next theorem gives an information about a solvability of WP for the whole class G_ω.

Theorem 4.1 ([Gri83, Gri84a]). *The word problem is solvable in the group G_ω if and only if the sequence ω is recursive.*

We will describe the algorithm for the case of the group G modifying the exposition done in [Gri99]. As was mentioned in Section 3 a word $w = w(a, b, c, d)$ of the form (3.2) represents the normal form of elements in the group Γ given by the representation (3.3) For any word $u = u(a, b, c, d)$ we denote by $r(u)$ the reduction of u in Γ which can be done by rewriting u using the following rules

$$\begin{aligned} &\text{(i) } x^2 \to 1 \\ &\text{(ii) } xy \to z \end{aligned} \tag{4.1}$$

where $x, y, z \in \{b, c, d\}$ and for the rule of the second type x, y, z have to be different elements.

Now we are going to describe two rewriting processes ϕ_0, ϕ_1. Namely $\phi_i(w)$, $i \in 0, 1$ is the word obtained from w by associating with each of the letters b, c, d occurring in w a symbol in an accordance with the following rule:

$$\varphi_i: \begin{cases} b \to a \\ c \to a \\ d \to 1 \end{cases} \tag{4.2}$$

if the number of occurrences of a in w preceding the present occurrence of the symbol in question is even for $i = 0$ or odd for $i = 1$. In a similar way

$$\varphi_i: \begin{cases} b \to c \\ c \to d \\ d \to b \end{cases} \tag{4.3}$$

if the number of occurrences of a in w proceeding the present occurrence of the symbol in question is odd for $i = 0$ or even for $i = 1$.

ALGORITHM. To verify the relation $w(a, b, c, d) = 1$ proceed as follows.

(1) Calculate $|w|_a$ the number of occurrences of a in w. If $|w|_a$ is odd, then $w \neq 1$. If $|w|_a$ is even, then reduce w using the rules (4.1) (that is, calculate $r(w)$). If $r(w)$ is empty word then $w = 1$, if $r(w)$ is nonempty, then pass to step (2).
(2) Calculate $w_i = \phi_i(r(w)), i = 0, 1$, and return to step (1), but now verify two relations in G, namely, $w_i = 1, i = 0, 1$ keeping in mind that

$$(w = 1) \Leftrightarrow (w_i = 1, i = 0, 1)$$

One can visualize the algorithm as a branching rewriting process as in Figure 4.1, which stops at some level n if either some of the words $w_{i_1, i_2, \ldots, i_n}$, $i_1, \ldots, i_n = 0, 1$ have odd number of occurrences of a (and then $w \neq 1$) or all these words are empty (and then $w = 1$).

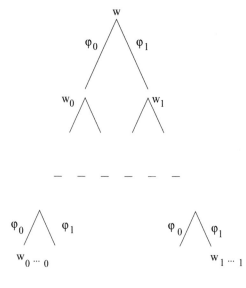

FIGURE 4.1

The process terminates at most at level $[\log_2 |w|] + 1$ because of the reduction of the length

$$|w_i| \leq \frac{|w| + 1}{2} \tag{4.4}$$

(this inequalities are analogous to (3.5) in case w is reduced in Γ). We see that when we apply parallel computations this process takes a logarithmic time to solve WP. Rough estimates give $n \log(n)$ for the time and cn estimate for the space needed, where n is the length of the input (the length of the word w) and c is some constant.

Problem 4.2.   a) *What is the actual time needed for the solution of WP for G?*
b) *Is it linear (when using a Turing machine with one tape)?*

For the group G_ω the word problem is solvable by similar branch algorithm, only now the rewriting rules $\varphi_i^{(k)}$, $i = 0, 1$ used for k-th level of the tree depend on the k-th digit of oracle ω and the (Turing) degree of unsolvability of WP for G_ω is the same as the degree of unrecursiveness of oracle ω [Gri83, Gri84a, Gri85].

In modern combinatorial group theory an important role is played by formal languages. Let us indicate a few interesting questions which arise around the group G.

A group M is better understood if there is a good normal form for its elements. By this we mean the one-to-one correspondence $g \rightarrow W_g$ between the elements of M and words over an alphabet $S \cup S^{-1}$ where S is a generating set. "Good" means that the subset $\mathcal{L} = \{W_g \colon g \in M\}$ of the set $\{S \cup S^{-1}\}^*$ of all words viewed as a formal language should belong to the class of well understood languages, for instance to be regular, context free, indicable or at least context sensitive language [Rev91, HU79]. In view of problems of Section 8 (like Problems 8.2, 8.3) it is better to deal with a geodesic normal forms, that is when the length of element with respect to generating system S coincides with the length of corresponding word:

$$|g|_S = |W_g|, \quad g \in M.$$

In this case the growth series $\Gamma_M^S(z) = \sum_{n=0}^{\infty} \gamma_M^S(n)z^n$ of the group M (see Section 8) coincides with the growth series $\Gamma_{\mathcal{L}}(z) = \sum_{n=0}^{\infty} \ell(n)z^n$ of the language $\mathcal{L}$ (here $\ell(n)$ is the number of words of length $\leq n$ in $\mathcal{L}$).

There are many examples of groups with regular geodesic normal form (for instance Coxeter groups and hyperbolic groups). But the geodesic normal form of a group of intermediate growth can not belong to this class and even cannot be a context free language as context free languages have either polynomial or exponential growth [BG02b, Inc01]. It was shown in [GM99] that there are indexed languages of intermediate growth.

Problem 4.3.   a) *Is there an indexed geodesic normal form for the elements of G?*

b) *What is the simplest class of languages that has a representative that can figurate as a geodesic normal form for the elements in G or in any other group of intermediate growth?*

Another language that naturally arises when studying a group by geometric means is the language $\mathcal{L}_{\text{geod}}$ consisting of words in $(S \cup S^{-1})^*$ determining a geodesic path beginning at the identity element in a Cayley graph. For instance, for hyperbolic groups $\mathcal{L}_{\text{geod}}$ is a regular language, hence has exponential growth (in the non-elementary case) and the set of all words in $\mathcal{L}_{\text{geod}}$ of length $\leq n$ can be listed by an algorithm in exponential time.

The growth of $\mathcal{L}_{\text{geod}}$ can be very different from growth of the language of geodesic normal form (i.e. of growth of the group). For instance, for $\mathbb{Z}^d$, $d \geq 2$ the growth is polynomial while the language of geodesics grows exponentially.

Problem 4.4.   a) *What is the rate of exponential growth of $\mathcal{L}_{\text{geod}}^G$?*

b) *To which class of functions belongs growth series of $\mathcal{L}_{\text{geod}}^G$?*

c) *Is there a group with intermediate (between polynomial and exponential) growth of the language of geodesics?*

Problem 4.5.   a) *To which class of languages does $\mathcal{L}_{\text{geod}}^G$ belong?*

b) *What is the minimal complexity of an algorithm which for given n produces all words of length $\leq n$ in $\mathcal{L}_{\text{geod}}^G$?*

One more natural language that can be associated to any finitely generated group is the word problem language $\mathcal{L}_{WP}$ consisting of all words from $\{S \cup S^{-1}\}^*$

representing the identity element. $\mathcal{L}_{WP}$ is a recursive set if and only if the word problem is solvable. The "nicer" the language $\mathcal{L}_{WP}$ the "better" the word problem.

Problem 4.6. *Is $\mathcal{L}_{WP}^G$ an indexed language?*

This is a longstanding open question which has a chance to have the positive solution because of the result of D. Holt and C. Röver [HR03] stating that the co-word problem language (i.e. the complement of $\mathcal{L}_{WP}$ in $(s \cup s^{-1})^*$) is indeed the indexed language. Therefore the study of growth series of indexed languages initiated in [GM99] looks quite relevant.

Problem 4.7.   a) *To which class of functions does the growth series of $\mathcal{L}_{WP}^G$ belong?*
    b) *What is the second term of the asymptotic of the growth of $\mathcal{L}_{WP}^G$?*

The latter question is equivalent to the question about the asymptotic of decay of probabilities $P_{11}^{(n)}$ of return for simple random walk on G. The first term is 7^n, which follows from the amenability of G and Kesten' criterion [Kes59].

We do not require in the last problems that the system of generators for G is the canonical one. The same type of questions can be attributed to any other self-similar group of intermediate growth or of branch type.

There are different ways to measure the complexity of algorithmic problems [LV97]. The one of them is based on use of Kolmogorov complexity was applied in [Gri85] to groups G_ω. The idea is as follows. Let M be an arbitrary group with system of generators $\{a_1, \ldots, a_m\}$ and let $A = \{a_1, \ldots, a_m, a_1^{-1}, \ldots a_m^{-1}\}$ be the set of generators and its inverses considered as an alphabet. Using the order $a_1 < a_2 < \cdots < a_m < a_1^{-1} < \cdots < a_m^{-1}$ extend it lexicographically to the order on the set A^* of all words over A. Let $\{w_n\}_{n=0}^\infty$ be enumeration of elements of A^* in this ordering with $w_0 = \emptyset$. Let $\mathcal{L} \subset A^*$ be a subset consisting of words representing the identity element (i.e. $\mathcal{L} = \mathcal{L}_{WP}^M$) and let

$$\xi = \xi_{\mathcal{L}} = \xi_0 \xi_1 \ldots \xi_{n-1} \ldots$$

be a characteristic sequence of $\mathcal{L}$ i.e.

$$\xi_n = \begin{cases} 1 & \text{if } w_n \in \mathcal{L} \\ 0 & \text{otherwise.} \end{cases}$$

Let $\xi^{(n)} = \xi_0 \xi_1 \ldots \xi_{n-1}$ be the prefix of ξ of length n. Then the Kolmogorov complexity $K(\xi^{(n)})$ as a function of n is a quantitative measure of complexity of word problems. The growth of $K(n) = K(\xi^{(n)})$ when $n \to \infty$ is not faster than linear. For a typical sequence η (with respect to uniform Bernoulli measure on $\{0,1\}^{\mathbb{N}}$)

$$\lim_{n \to \infty} \frac{1}{n} K(\eta^{(n)}) = 1.$$

but for the sequences $\xi^{(n)}$ constructed above one has

$$\lim_{n \to \infty} \frac{1}{n}(\xi^{(n)}) = 0$$

if the group M is infinite [Gri85].

In the study of algorithmic problems it is more natural to use a version of Kolmogorov complexity, denoted KR, and called the complexity of resolution (or length conditional Kolmogorov complexity [LV97]).

The function

$$R(n) = KR(\xi^{(n)})$$

is the quantitative measure of undecidability of word problems and the WP is decidable if and only if $R(n)$ is bounded.

It is more natural to consider values of $R(n)$ at points of the form

$$N_n = \sum_{i=0}^{n} (2m)^i = \frac{(2m)^{n+1} - 1}{2m - 1}$$

which counts the number of all words in A^* of length $\leq n$. The value

$$r(n) = r_M^A(n) = R(N_n)$$

represents the amount of information needed to solve the word problem for elements of length $\leq n$.

The definition of Kolmogorov complexity or of the complexity of resolution relies on the notion of a universal Turing machine. The difference between complexities defined by different universal (or optimal) machines is uniformly bounded. Therefore it is natural to study functions $K(n)$ and $R(n)$ up to equivalence which ignores bounded functions. For study of growth of functions $r_G^A(n)$ (which we call WP growth function) more natural is to use Milnor equivalence [Mil68]:

$$f_1(n) \sim f_2(n) \Leftrightarrow \exists c \quad \text{s.t.} \quad f_1(n) \leq cf_2(cn),$$
$$f_2(n) \leq cf_1(cn), \quad n = 1, 2, \ldots \ .$$

Then the class of equivalence $[r_M^A(n)]$ does not depend on the choice of generating system and is bounded by the class $[2^n]$ of exponential function. Moreover, in case M is recursively presented (for instance, finitely presented) $[r_M(n)] \prec [n]$ (linear bound). There are finitely generated groups with $r(n) \sim 2^n$ and there are finitely presented groups with $r(n) \sim n$.

For the groups G_ω

$$r_{G_\omega}(n) \sim R(\omega^{[\log_2 n]}) \tag{4.5}$$

so the complexity of word problems depend on a complexity of the oracle ω determining the group [Gri85].

Relation (4.5) shows that there is infinitely many degrees of growth of complexity of word problem in finitely generated groups and that there are groups with incomparable growth of complexity.

A number of open questions can be formulated about $K(n)$, $R(n)$ and $r(n)$ and further investigations in this direction are desirable. Let us recall an open question from [Gri85]. Let

$$\beta = \varlimsup_{n \to \infty} \sqrt[n]{|\mathcal{L}_n|}$$

where $\mathcal{L}_n$ is the set of words of length n in $\mathcal{L}$. Then

$$R(n) \le n^\delta \log n$$

where $\delta = \frac{\log \beta}{\log 2m}$. From Kesten's criterion of amenability [Kes59] follows that M is amenable if and only if $\delta < 1$. Thus for nonamenable groups the function $R(n)$ grows slower than n^ρ for some $\rho < 1$.

Problem 4.8. *Is there a finitely generated group for which the function $R(n)$ can not be estimated from above by a function of the form n^ρ where $\rho < 1$?*

If such a group exists it must be amenable. These and other considerations show that Kolmogorov complexity has interesting links to the subject of amenability and growth. It looks like the groups with most difficult word problems (from the point of view of Kolmogorov complexity) are amenable groups of exponential growth.

The idea to use a version of Kolmogorov complexity to the word problem has appeared in a few recent articles [KS05, Nab96].

Also the group G started to be popular among specialist in cryptography (the first time it was observed that G could be useful in cryptography was in [GZ91]). This is based on new developments in commercial cryptography which requires a groups with easy word problems but difficult conjugacy problems. The conjugacy problem for G will be considered in detail in the next section and this problem is certainly much harder than the word problem in G.

The group G is well-behaved not only with respect to the WP, but also to other algorithmic problems. One of them is the membership problem (MP) (sometimes called the generalized word problem). The membership problem asks for an algorithm which, given elements $g, h_1, \ldots, h_m$, decides if g belongs to the subgroup generated by the elements $h_1, \ldots, h_m$. This problem is also decidable in G. This follows from the subgroup separability property (every finitely generated subgroup is closed in the profinite topology i.e. can be presented as an intersection of subgroups of finite index) [GW03a] and the solvability of WP.

Problem 4.9. *Is it correct that every finitely generated branch group is subgroup separable?*

Problem 4.10. *Find a condition that insure that a self-similar group is subgroup separable (and hence has decidable membership problem).*

An example of self-similar group with undecidable MP is a direct product $F_7 \times F_7$ of two copies of a free group of rank 7 which follows from possibility represent $F_7 \times F_7$ as a group generated by states of finite automaton. Namely, first do this for F_7 [GM03], then for the direct product use a standard construction in the theory of automaton groups and use the Mikhailova construction [LS01], which is discussed below. It would be interesting to find a reasonable decision problem which is undecidable for G.

An interesting circle of problems initiated by study of the group G and other self-similar groups is the question about presentability of G and other self-similar groups of branch type by finite L-presentations.

By a finite L-presentation we mean a presentation of the form

$$M = \langle a_1, \ldots, a_m \mid R, \tau^n(Q), n \geq 0 \rangle \tag{4.6}$$

where R and Q are finite sets of elements of a free group F_m with basis $a_1, \ldots, a_m$ and τ an endomorphism of this group (i.e. a substitution $\tau\colon a_i \to w_i(a_\mu)$, $i = 1, \ldots, m$ determined by some set of elements $w_i \in F_m$). An L-presentation (4.6) is called pure if R is empty set and is called ascending if it is pure and τ induces an injective endomorphism $\tau\colon M \to M$.

The first example of a self-similar group with L-presentation was found by I. Lysionok (Lysenok) [Lys85] who showed that G has the following presentation

$$G \simeq \langle a, b, c, d \mid 1 = a^2 = b^2 = c^2 = d^2 = bcd = \sigma^n((ad)^4) = \sigma^n((adacac)^4), n \geq 0 \rangle \tag{4.7}$$

where the substitution σ is defined as

$$\sigma\colon \begin{cases} a \to aca \\ b \to d \\ c \to b \\ d \to c \end{cases} \tag{4.8}$$

The presentation (4.7) is not pure but can be modified to the pure presentation

$$G \simeq \langle a, b, c, d \mid 1 = \sigma^n(r), n \geq 0, r \in \{a^2, b^2, c^2, d^2, bcd, (ad)^4, (adacac)^4\} \rangle.$$

This presentation is also ascending as σ induces the injective endomorphism of G.

The group G is not finitely presented and moreover the presentation (4.7) is minimal (the deletion of any relator changes the group). This fact was used in [Gri99] to show that the Schur multiplicator (i.e. the second homology group of G with coefficient in trivial G-module $\mathbb{Z}$) is infinite dimensional and is isomorphic to the direct sum of infinitely many copies of $\mathbb{Z}/2\mathbb{Z}$.

Infinitely presented groups given by finite L-presentation can be viewed as "finitely presented in generalized sense". Ascending L-presentations are useful because allow to embed a group "in a nice way" in a finitely presented group. Namely the ascending HNN-extension

$$\widetilde{M} = \langle M, t \mid t^{-1}mt = \tau(m), m \in M \rangle$$

of a group M given by ascending L-presentation

$$M = \langle a_1, \ldots, a_m \mid \tau^n(Q), n \geq 0 \rangle$$

is a finitely presented group with the presentation

$$\widetilde{M} = \langle a_1, \ldots, a_m, t \mid Q, t^{-1}a_i t = \tau(a_i), i = 1, \ldots, m \rangle.$$

The embedding $M \hookrightarrow \widetilde{M}$ preserves some of the properties of M (for instance to be amenable). In case of G

$$\widetilde{G} = \langle a, b, c, d, t \mid 1 = a^2 = b^2 = c^2 = d^2 = bcd, t^{-1}at = aca, t^{-1}bt = d, t^{-1}ct = b,$$
$$t^{-1}dt = c \rangle$$

$$\simeq \langle a, t \mid a^2 = TaTatataTatataTataT = (Tata)^8 = (T^2ataTat^2aTata)^4 = 1 \rangle$$

where T stands for t^{-1} (the last presentation was found by L. Bartholdi [CSGdlH99]). $\widetilde{G}$ shares with G the property to be amenable but not elementary amenable but is a group of exponential growth in contrast to G which has intermediate growth between polynomial and exponential (Section 8). A more general point of view on L-presentations is presented in [Bar03].

Being amenable the group $\widetilde{G}$ does not contain a free subgroup on two generators. The theorem of R. Bieri and R. Strebel [Bau93] states that a finitely presented indicable group either is an ascending HNN extension with a finitely generated base group or contains a free subgroup of rank two (a group is indicable if it can be mapped onto $\mathbb{Z}$).

Problem 4.11. *Is it correct that a finitely presented indicable group not containing a free subgroup of rank 2 is an ascending HNN extension of a group with finite L-presentation?*

Problem 4.12. a) *Which self-similar groups have finite L-presentations?*
b) *Which self-similar groups have ascending finite L-presentations?*

We are now going to briefly describe the main steps in the existing method for finding L-presentations developed in [Lys85, Sid87b, Gri98, Gri99, Bar03]. Let M be a self-similar group.

Step 1. Find a suitable branch algorithm for solving the WP in M. This requires finding a suitable finitely presented group with solvable WP that covers M. In case of G one such group is the group Γ from (3.3). However, for our purposes it is better to replace Γ by another group, by adding one more relator $(ad)^4$. The important property that should be achieved by the choice of analogue of covering group is the contraction of the length after the projections.

Step 2. Let Γ be the chosen covering group and let $M \simeq \Gamma/\Omega$. The second step consists of presenting Ω as a union $\bigcup_{n-1}^{\infty} \Omega_n$ where $\Omega_n \triangleleft \Gamma$ is the set of elements for which the branch algorithm stops its work (and "accepts" the element) on or before level n of the tree, and then analyzing the structure of the groups Ω_n, $n = 1, 2, \ldots$. Assume that the cardinality of the alphabet (used to determine the self-similar structure on M) is d, that

$$\psi: \begin{cases} M \longrightarrow M \wr_{\times d} S_d \\ st_M(1) \longrightarrow \underbrace{M \times \cdots \times M}_{d} \end{cases}$$

is the embedding of type (3.10). Let $\nu\colon \Gamma \to M$ be the covering used in the branch algorithm, denote $\Xi = \gamma^{-1}(st_M(1))$ and let

$$\theta\colon \Xi \longrightarrow \underbrace{\Gamma \times \cdots \times \Gamma}_{d}$$

be the lift of ψ to Ξ. Denote further $\Omega_1 = \mathrm{Ker}\,\theta = \theta^{-1}(1)$ and $\Omega_n = \theta^{-n}(1)$ where 1 represent here the identity element in the direct product of copies of Γ.

One way of finding a generating set for Ω_1 (as normal subgroup) is through finding a presentations for Ξ and $\theta(\Xi)$ and comparing them. In this approach a question arises about presentations of finitely generated subgroups in direct products (which we will discuss a bit later). If $\theta(\Xi)$ has finite index in Γ^d then it is a finitely presented group whose presentation can be found by the standard Reidemeister–Schreier method. For instance, in case of G and covering group Γ given by the presentation

$$\langle a, b, c, d \mid a^2 = b^2 = c^2 = d^2 = bcd = (ad)^4 = 1 \rangle$$

$\theta(\Xi)$ has finite index in $\Gamma \times \Gamma$ (this is precisely why the relation $(ad)^4 = 1$ was added) and hence is a finitely presented group. In this case $\Omega_1 = \langle (ac)^8, (adacac)^4 \rangle^\Gamma$.

Step 3. Finding a generating set for Ω_n requires finding a substitution σ which well "cooperates" with the embedding ψ and projections. In case of G the substitution σ given by (4.8) has the properties that $p_1\psi\sigma = id$ and $p_0\psi\sigma(r) = 1$ for any relator r in G (p_0, p_1 are the projections). Hence

$$\sigma^n((ad)^4) = (1, \sigma^{n-1}((ad)^4))$$
$$\sigma^n((adacac)^4) = (1, \sigma^{n-1}((adacac)^4))$$

for $n \geq 1$ and this leads to the relations

$$\Omega_n = \langle \sigma^i((ad)^4), 1 \leq i \leq n, \sigma^j((adacac)^4), 0 \leq j \leq n - 1 \rangle,$$
$$\theta(\Omega_n) = \Omega_{n-1} \times \Omega_{n-1}$$

and eventually to the presentation (4.7).

While for the group G the choice of the substitution σ is rather easy, for most self-similar groups it is quite difficult to find suitable Γ, σ and other tools which can be used in finding a L-presentation.

The above discussion have been touched the question about finite presentability of subgroups in direct products. As the groups Γ used in all known examples of computations of L-presentations are not far from virtually free groups we naturally come to the following question.

Problem 4.13. *Let Γ be a (nonelementary) hyperbolic group and $d \geq 2$. Which finitely generated subgroups of Γ^d have finite presentation?*

Recall that hyperbolic groups were defined by M.Gromov [Gro87]. In the above question we can distinguish as a special case the case when a subgroup is a subdirect product (i.e. its projection on each factor is Γ).

For the case when Γ is a free group and $d = 2$ we have the complete answer given by the theorem of G. Baumslag and J. Roseblade [BR84], [BW99]. Generalizations of this result to product of copies of surface groups are done in [BW99].

An interesting class of groups can be defined while studying the word problem in self-similar groups. Let me call them companions of self-similar groups. Namely, given a self-similar group M and a finitely presented group Γ which covers M let the companion group be $\mathcal{C}(M, \Gamma) = \Gamma/\Omega$ where $\Omega = \bigcup_{n=1}^{\infty} \Omega_n$, $\Omega_n = \theta^{-n}(1)$ and

$$\theta \colon \Xi \longrightarrow \underbrace{\Gamma \times \cdots \times \Gamma}_{d}$$

is the homomorphism determined by the automaton. In a contracting situation $\mathcal{C}(M, \Gamma) = M$ but in general $\mathcal{C}(M, \Gamma)$ and M may be different. Companions of self-similar groups do not have the (T)-property of Kazhdan (private communication of A. Erschler).

Computation of presentations of companion groups starts with computation of $\Omega_1 = \mathrm{Ker}\,\theta$ and, as mentioned above, is related to finding a presentation of a subgroup in direct products. In case of $\Gamma = F_m$ and $d = 2$ in many examples the arising subgroups in $F_m \times F_m$ are of Mikhailova type [LS01] (i.e. subgroups generated by diagonal elements $(a_1, a_1), \ldots, (a_m, a_m)$, where $a_1, \ldots, a_m$ form a basis of F_m and elements of the form $(1, u_i)$, $u_i \in F_m$, $i = 1, \ldots, k$).

For instance let us consider the automaton

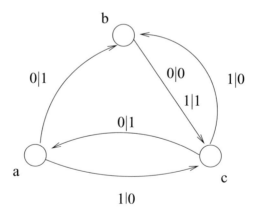

FIGURE 4.2

that is $a = (b, c)\varepsilon$, $b = (c, c)$, $c = (a, b)\varepsilon$ where ε is nonidentity element of S_2. Let $M = \langle a, b, c \rangle$. Then

$$st_M(1) = \langle a^2, b, ca^{-1}, ac, aba^{-1} \rangle$$

and for the canonical embedding $\psi \colon M \to M \wr S_2$

$$\psi \colon \begin{cases} s_1 = a^2 \longrightarrow (bc, cb) \\ s_2 = b \longrightarrow (c, c) \\ s_3 = ca^{-1} \longrightarrow (ab^{-1}, bc^{-1}) \\ s_4 = ac \longrightarrow (b^2, ca) \\ s_5 = aba^{-1} \longrightarrow (bcb^{-1}, c). \end{cases}$$

Performing a Nielsen transformation on the set s_1, s_2, s_3, s_4, s_5 one can get a new set of generators u_1, u_2, u_3, u_4, u_5 of $st_M(1)$ for which

$$\psi \colon \begin{cases} u_1 \longrightarrow (a, a) \\ u_2 \longrightarrow (b, b) \\ u_3 \longrightarrow (c, c) \\ u_4 \longrightarrow (1, acb^{-2}) \\ u_5 \longrightarrow (1, cbc^{-1}b^{-1}) \end{cases} \tag{4.9}$$

so the image of θ in $F_3 \times F_3$ is a subgroup generated by pairs (4.9).

Problem 4.14. *What kind of presentations can be obtained for the Mikhailova subgroups in $F_m \times F_m$ determined by finite automata?*

Mikhailova subgroup $M < F_m \times F_m$ has the form of the semidirect product $(N \times N) \rtimes D$, where $N = \langle u_1, \ldots, u_k \rangle^{F_m}$ and $D = \langle (a_1, a_1), \ldots, (a_m, a_m) \rangle$ is the diagonal subgroup. Therefore its presentation can be described by finding a generating set for the free subgroup $N < F_m$, then for the product $N \times N$ and calculation of the action of the generators of D by conjugation on the generators of $N \times N$.

The generators for N can be calculated in terms of a spanning subtree of the Cayley graph of the group

$$\langle a_1, \ldots, a_m \mid u_1 = \cdots = u_k = 1 \rangle. \tag{4.10}$$

It may happen however that this group has unsolvable WP (which is equivalent to the fact that corresponding subgroup in $F_m \times F_m$ is not a recursive subset, which was the original idea of Mikhailova to show that the membership problem is undecidable in $F_m \times F_m$ [LS01]).

Problem 4.15. *Is there a finite invertible automaton over $\{0, 1\}$ for which the corresponding group (4.10) has undecidable WP?*

5. The conjugacy problem and the isomorphism problem

While the word problem was solved for G immediately after the group was discovered [Gri80] the conjugacy problem (CP) was solved only in the end of the 90-ies [Leo98, Roz98]. For p-groups G_ω with p odd and ω a periodic sequence, and for Gupta-Sidki p-groups the problem was solved in [WZ97] where it was shown that these groups are conjugacy separable.

The solution given in [Leo98, Roz98] is based on a different idea and in certain sense is more direct. As a byproduct it also leads to the conclusion that the group G has the conjugacy separability property. The result of [Leo98] is more general than the one given in [Roz98] as it deals with the whole class of groups $G_\omega, \omega \in \Omega_2$.

Theorem 5.1. *The conjugacy problem is solvable in the group G_ω if and only if the sequence ω is recursive.*

We will describe the algorithm for the case of the group G modifying the exposition from [BGŠ03].

Let $x, g, h \in G$ and

$$g^x = h \tag{5.1}$$

In this case either both g, h are in $\mathrm{St}_G(1)$ or both g, h are not in $\mathrm{St}_G(1)$. We will view (5.1) as an equation in G with variable x, and g and h are considered to be the coefficients of the equation. The next lemma shows how (5.1) can be rewritten in terms of projections.

Lemma 5.1. 1. *Let $g, h \in \mathrm{St}_G(1), g = (g_0, g_1), h = (h_0, h_1)$.*
(α) If $x \in \mathrm{St}_G(1), x = (x_0, x_1)$ then equation (5.1) is equivalent to the system of equations

$$\begin{cases} g_0{}^{x_0} = h_0 \\ g_1{}^{x_1} = h_1. \end{cases} \tag{5.2}$$

(β) If $x \notin st_G(1), x = (x_0, x_1)a$ then (5.1) is equivalent to

$$\begin{cases} g_0{}^{x_0} = h_1 \\ g_1{}^{x_1} = h_0. \end{cases} \tag{5.3}$$

2. *Let $g, h \notin \mathrm{St}_G(1), g = (g_0, g_1)a, h = (h_0, h_1)a$.*
(α) If $x \in \mathrm{St}_G(1), x = (x_0, x_1)$ then (5.1) is equivalent to

$$\begin{cases} (g_0 g_1)^{x_0} = h_0 h_1 \\ x_1 = g_1 x_0 (h_1)^{-1}. \end{cases} \tag{5.4}$$

(β) If $x \notin st_G(1), x = (x_0, x_1)a$ then (5.1) is equivalent to

$$\begin{cases} (g_0 g_1)^{x_0} = h_1 h_0 \\ x_1 = (g_0)^{-1} x_0 h_1. \end{cases} \tag{5.5}$$

The solutions of the systems (5.2), (5.3), (5.4), (5.5) are considered under the condition that (x_0, x_1) is an element of G.

The cases listed in the lemma are demonstrated by the diagrams in Figure 5.1.

We call the diagrams representing Case 1 independent, while the diagrams representing Case 2 are called dependent (the right vertex on the bottom is dependent on the left vertex). Also we call vertices labelled by a conjugacy equation "c-type" and other vertices ''d-type". This diagram will be used later to draw solution trees for CP.

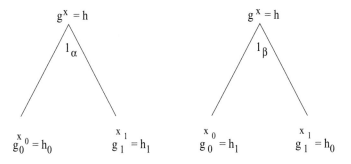

Case 1

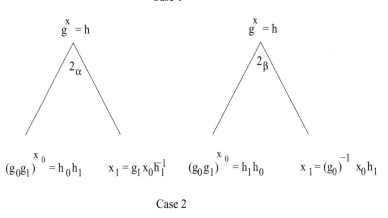

Case 2

FIGURE 5.1

Consider the groups $K = \langle (ab)^2 \rangle^G$, $K > K_1$, $K_1 = K \times K$ where each factor acts on the corresponding subtree, and recall that K_1 has finite index in G (see Section 6). Thus there are only finitely many cosets Kx.

Let

$$Q(g,h) = \{Kx : g^x = h\}. \tag{5.6}$$

The next lemma contains the main idea of the algorithm: finding the set $Q(g,h)$ from the equations given in the bottom of the diagrams in Figure 5.1.

Lemma 5.2. 1. *Let* $g = (g_0, g_1), h = (h_0, h_1) \in G$ *and let for some* $u_i, v_j, w_s \in G, i \in I, j \in J, s \in S$, *where* I, J *and* S *are some indexing sets,*

$$Q(g,h) = \{Ku_i : i \in I\},$$
$$Q(g_0, h_0) = \{Kv_j : j \in J\},$$
$$Q(g_1, h_1) = \{Kw_s : s \in S\}.$$

Then for every $i \in I$ *there is* $j \in J$ *and* $s \in S$ *such that the element* $z = (v_j, w_s)$ *is in* G *and* $Ku_i = Kz$.

2. *Let $g = (g_0, g_1)a, h = (h_0, h_1)a \in G$ and let*

$$Q(g, h) = \{Ku_i : i \in I\}$$
$$Q(g_0g_1, h_0h_1) = \{Kv_j : j \in J\}.$$

Then for every $i \in I$ there exist $j, s \in J$ such that the element $z = (v_j, g_1 v_s h_0^{-1})$ is in G and $Ku_i = Kz$.

This lemma corresponds to the case (α) of the Lemma 5.1. The statement corresponding to the case (β) is similar and we omit it. Later when we quote Lemma 5.2 we will assume that it covers all the cases and hope that this will not lead to any confusion. Let us call the sets defined by (5.6) the Q-sets.

While the diagrams drawn in Figure 5.1 show how to move from the equation on the top to the equations on the bottom, the last lemma provides the way of moving in the opposite direction (by which we mean the computing of the Q-set on the top of the diagram using the Q-sets on the bottom).

In case the equations on the bottom are simpler than the equation on the top this suggests the following construction. Given the equation (5.1) construct for each $t \geq 1$ the set of trees (we call them solution trees) of depth $\leq t$ by application to the root vertex any of the rules described by Figure 5.1 and then iterating this process independently at every new arising c-type vertex in such a way that the tree will not exceed the t-th level. This procedure leads to a set of decorated trees whose vertices are labelled by the equations of conjugacy type or by the relations corresponding to vertices of d-type. If for some t there is a possibility to compute for any tree of described type the Q-sets for c-type vertices which are leaves of the tree, then the Q-sets of the other vertices can be computed as well. This leads to the computation of Q-sets of all vertices including the root vertex.

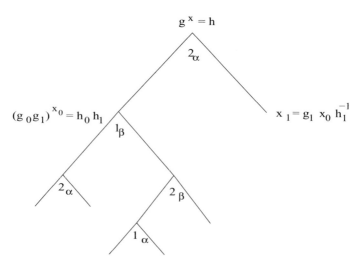

FIGURE 5.2

Observe that g and h are conjugate if and only if $Q(g,h)$ is nonempty. This leads to a solution of CP in our situation when we have the reduction of the total length $\tau = |g| + |h|$ of coefficients of (5.1) (in case $\tau > 2$).

Indeed, in the light of Lemma 3.1, the length reduction is obvious for Case 1 of Lemma 5.1. For Case 2 we have only the estimate (3.7) but if we assume that g and h have reduced form $a*a*\cdots a*$ (and we can always assume this, if not we just conjugate one or both coefficients by a suitable element from the set $\{a,b,c,d\}$), then $|g_0g_1| + |h_0h_1| \le |g| + |h|$ (which follows from Lemma 3.1). If d occurs at least in one of coefficients, then we again get a reduction of the length when moving in the tree one level down, as projecting d produces the empty symbol in one of two directions. If this is not the case then moving a level down may not decrease τ but at least the parameter τ will not increase. Further, projecting c produces d and projecting b produces c, so after at most three moves down we get a reduction of τ in all equations on the third level, for any decorated tree.

Thus after at most $t = 3\tau$ steps down in the tree we will get equations with $\tau \le 2$. If $\tau = 0$ then the equation is just the relation $1 = 1$. In the other cases the coefficients belong to the set $\{a,b,c,d\}$. As there is only finitely many such equations, we may assume when proving the decidability of CP that we know Q-sets for all such equations. Indeed, we can compute these sets explicitly. Namely, it is known that a,b,c,d belong to different conjugacy classes. Simple arguments show that

$$Q(a,a) = \{1, a, dad, (ad)^2\}$$
$$Q(b,b) = \{1, b, c, d\}$$
$$Q(c,c) = \{1, b, c, d\}$$
$$Q(d,d) = \{1, b, c, d, ada, (ad)^2, bada, badad\}.$$

Also we have the following table of lifts of pairs of cosets (we only show the representatives for each coset; the symbol $-$ means that the corresponding pair does not belong to G))

	1	a	b	c	d	ada	$(ad)^2$	$bada$	$badad$
1	1	—	d	—	—	—	—	—	—
a	—	—	—	b	c	—	—	—	—
b	ada	—	$(ad)^2$	—	—	—	—	—	—
c	—	b	—	—	—	—	—	—	—
d	—	$bada$	—	—	—	—	—	—	—
ada	—	—	—	—	—	—	—	—	—
$(ad)^2$	—	—	—	—	—	—	—	1	—
$bada$	—	—	—	—	—	—	—	—	—
$badad$	—	—	—	—	—	d	—	—	$(ad)^2$

So to solve CP in G for the pairs of elements g, h we do as follows. Compute τ; construct all solution trees of depth not greater 3τ which have leafs of c-type labelled by equations with coefficients belonging to the set $1, a, b, c, d$ (if one of coefficients is 1 then the second is also 1 and the equation becomes $1 = 1$, in which case the Q-set consists of all cosets Kx); for each such a tree find Q-sets of the leafs and the Q-sets of all other vertices finishing the computation in the root vertex. This completes the description of the algorithm. The algorithm also allows to prove the following

Theorem 5.2. *The group G is conjugacy separable.*

Recall that being conjugacy separable means that two elements are conjugate if and only if they are conjugate in each finite quotient. In other words, to show that if two elements g, h are not conjugate in G then they are not conjugate in some finite quotient one has to consider the images of any g, h in $G/\operatorname{St}_G(3\tau + 1)$. They will not be conjugate either, as for the quotient group a similar algorithm of verification of conjugacy works.

Although the algorithm is quite nice, it looks as if it has large complexity both in space and in time. Namely one has to construct all trees of the sort we consider of depth $\le 3\tau$ and a rough estimate for the number of such trees is two iterations of the exponential function. Then for each tree one has to solve the problem of lifting of Q-sets. So a rough upper bound for complexity both in space and in time is e^{e^τ}.

This makes the group G a good candidate for use in cryptography where groups with easy WP but difficult CP play a special role . But maybe the CP is indeed not so difficult in G? This is our next question.

Problem 5.1. *What is the complexity of CP for G? Is it subexponential, exponential or superexponential?*

Problem 5.2.   a) *Is conjugacy problem solvable for a group generated by a finite set of finite invertible automata?*
  b) *In particular is it solvable for a self-similar group?*

Problem 5.3. *Are all self-similar groups with solvable CP conjugacy separable?*

It is well known that the classical Dehn isomorphism problem is undecidable and there is no algorithm which, given a finite presentation, determines if a group is trivial or not. Nevertheless the isomorphosm problem can be considered for special classes of groups and special type of presentations and then it may happen to be decidable.

Problem 5.4. *Is the isomorphism problem decidable for the class of self-similar groups?*

Problem 5.5. *Is the isomorphism problem decidable for recursively presented branch groups?*

I would expect that the answer to both questions is negative. But there are some results showing that for smaller subclasses an algorithm exists.

One can consider the isomorphism problem for a class of groups containing non recursively presented groups for instance for the class G_ω. In 1984 I observed that for each $\omega \in \Omega$ there are at most countably many sequences $\eta \in \Omega$ such that $G_\omega \simeq G_\eta$. The proof of this fact was based on the statement that the bijection between canonical set of generators of G_ω and G_η extends to a homomorphism $G_\omega \rightarrow G_\eta$ if and only if $\omega \simeq \eta$. The problem of isomorphism for the class G_ω, $\omega \in \Xi$, was recently solved by V. Nekrashevych [Nek05] who showed that G_ω is isomorphic to G_η if and only if ω can be obtained from η by some permutation of symbols 0,1,2 applied to all entries. The proof is based on the technique of [LN02] or on the result from [GW03b], stating that under certain conditions an action of a branch group on a rooted tree is essentially unique (unique up to the procedure of deletion of levels) .

A new idea to the solution of isomorphism problem is advanced by V. Nekrashevych in [Nek05]. Namely, to certain self-similar groups of branch type he associates a topological dynamical system and proves that the groups are isomorphic if and only if the corresponding systems are conjugate. In some cases he is able to show that the groups are not isomorphic by showing that the corresponding dynamical systems are not conjugate.

This is the first case when non-isomorphism of groups is established by means of dynamical systems.

One can also consider a classification problem for the self-similar groups. For instance one can try to classify them when fixing a complexity.

By a complexity we mean the parameter (m, n) where m is the cardinality of the states of the automaton generating the corresponding group while n is the cardinality of the alphabet. There are six groups of complexity (2,2). There is a hope that soon we will have the classification of self-similar groups of complexity (3,2). The next steps would be the consideration of groups of complexity (2,3) and (3,3).

6. Subgroup structure and branch groups

The group G has very nice and interesting subgroup structure. The main property is that its lattice of subnormal subgroups has branching structure, following the structure of the tree on which the group acts. Such groups are called branch groups (see relevant definitions below). Moreover, the group G is regular branch over the subgroup $K = \langle (ab)^2 \rangle^G$. The latter means that K has finite index in G and, if we denote by K_n the subgroup of $\mathcal{G} = \mathrm{Aut}\, T$ equal to the direct product of 2^n copies of K (each factor acts on a subtree of T with root at the n-th level), then K_n is a subgroup of K of finite index.

This automatically implies that all K_n, $n = 1, 2, \ldots$ are subgroups of G and that $\{K_n\}_{n=1}^\infty$ is a descending sequence of normal subgroups of finite index in G with trivial intersection. The existence of such a sequence plays a crucial role in many considerations, in particular in proving that G is a just infinite group, i.e., is infinite but every proper quotient is finite [Gri00b]. Thus all normal subgroups in G have finite index.

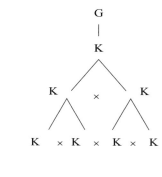

FIGURE 6.1

The question of existence of a residually finite just infinite group different from classical examples like $SL(n, \mathbb{Z})$/centre, $n \geq 3$ [Men65], [BMS67] was raised in [CM82] and answered in [GS83, Gri84a] by producing examples of just infinite groups of branch type. Indeed branch groups constitute one of three classes into which the class of just infinite groups naturally splits [Gri00b]. The other two classes are related to hereditary just infinite groups and to simple groups [Gri00b].

Let L be a group acting on a rooted tree T. Consider the following subgroups called respectively stabilizer of level n, rigid stabilizer of a vertex u and rigid stabilizer of level n:

$$st_L(n) = \{g \in L \colon u^g = u, \forall u \in V_n\},$$

$$\mathrm{rist}_L(u) = \{g \in L \colon u^g = u \quad \forall u \in T \backslash T_u\},$$

$$\mathrm{rist}_L(n) = \langle \mathrm{rist}_L(u) \colon u \in V_n \rangle = \prod_{u \in V_n} \mathrm{rist}_L(u),$$

where V_n denotes the set of vertices on level n, and T_u denotes the subtree with root at vertex u)

Although a stabilizer always has finite index, a rigid stabilizer can have infinite index and can even be trivial. We get the following classification of actions:

1) $\mathrm{rist}_L(n)$ has finite index in L for all $n = 1, 2, \ldots$
2) $\mathrm{rist}_L(n)$ has infinite index starting from some n, but is an infinite subgroup for all $n = 1, 2, \ldots$
3) $\mathrm{rist}_L(n)$ is finite group starting at some level n (and hence is trivial starting at some level $m \geq n$).

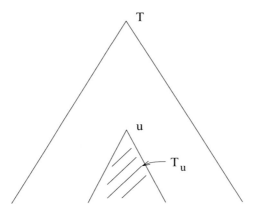

FIGURE 6.2. $\mathrm{rist}_G(u)$ acts nontrivially only on a subtree T_u

A group L is called a branch group if for some sequence $\overline{m} = \{m_n\}_{n=1}^{\infty}$ of integers $m_n \geq 2$ (called branch index) it has a faithful, spherically transitive (i.e., transitive on levels) action of type 1) on spherically homogeneous rooted tree $T_{\overline{m}}$ defined by the sequence $\overline{m}$ [Gri00a]. An action of type 1) is called a branch action, an action of type 2) is called weakly branch and action of type 3) is called diagonal.

A group L acting faithfully on a d-regular rooted tree $T = T_d$ (that is, with $m_n = d$, $n = 1, 2, \ldots$) is called regular branch over a subgroup $M < L$ if $[L : M] < \infty$ and M contains as a subgroup of finite index the group

$$M_1 = \underbrace{M \times \cdots \times M}_{d} \tag{6.1}$$

i-th factor in (6.1) acts on the corresponding subtree T_i of T with root u_i at first level and we use the natural identification of T_i and T to induce the action of M from T to T_i.

A regular branch group L is also a branch group because $\mathrm{rist}_L(n) \geq M_n$ where

$$M_n = \underbrace{M \times \cdot \times M}_{d^n}$$

(here each factor acts on the corresponding subtree with root at n-th level), and clearly $[G : M_n] < \infty$, $n = 1, 2, \ldots$.

A group L is called weakly regular branch over a subgroup M if $M \neq \{1\}$ and M contains M_1 as a subgroup (but not necessary of finite index).

We do not introduce the notion of a diagonal group since every countable residually finite group acts faithfully on some spherically homogeneous tree. Just take a descending sequence of normal subgroups of finite index intersecting trivially, form a tree in which the vertices are the left cosets and the incidence is induced by inclusion (the root is the whole group), and let the group act on the coset tree by left multiplication.

The group G is regular branch group for its action on the binary rooted tree as defined in Section 1. Indeed, $\mathrm{rist}_G(n) \geq K_n$ for any $n = 1, 2, \ldots$ where $K = \langle (ab)^2 \rangle^G$.

Other examples of regular branch groups include Gupta–Sidki p-groups and $IMG(z^2+i)$ [BGN03, Gri00a, BGŠ03]. The Basilica group (which is $IMG(z^2-1)$)) is an example of a weakly branch group [GŻ02a].

Let us now give more information about the structure of stabilizers and rigid stabilizers. The abelization G_{ab} is isomorphic to $\mathbb{Z}/2\mathbb{Z} \times \mathbb{Z}/2\mathbb{Z} \times \mathbb{Z}/2\mathbb{Z}$ and the group G has 7 subgroups of index 2:

$$\langle b, ca \rangle, \quad \langle c, ad \rangle, \quad \langle d, ab \rangle$$
$$\langle b, a, a^c \rangle, \quad \langle c, a, a^d \rangle, \quad \langle d, a, a^b \rangle$$
$$st_G(1) = H = \langle b, d, b^a, d^a \rangle,$$

listed in accordance with the number of generators 2, 3 or 4. Normal closures of generators and factors by them are

$$A = \langle a \rangle^G = \langle a, a^b, a^c, a^d \rangle, \quad G/A \simeq \mathbb{Z}/2\mathbb{Z} \times \mathbb{Z}/2\mathbb{Z}$$
$$B = \langle b \rangle^G = \langle b, b^a, b^{ad}, b^{ada} \rangle, \quad G/B \simeq D_8$$
$$C = \langle c \rangle^G = \langle c, c^a, c^{ad}, c^{ada} \rangle, \quad G/C \simeq D_8$$
$$D = \langle d \rangle^G = \langle d, d^a, d^{ac}, d^{aca} \rangle, \quad G/D \simeq D_{16}$$

(here D_{2n} is the dihedral group of order $2n$). To formulate the next statement and give more information on the top part of the lattice of normal subgroups consider the following subgroups of G

$$K = \langle (ab)^2 \rangle^G, \quad L = \langle (ac)^2 \rangle^G, \quad M = \langle (ad)^2 \rangle^G$$
$$\overline{B} = \langle B, L \rangle, \quad \overline{C} = \langle C, K \rangle, \quad \overline{D} = \langle D, K \rangle$$
$$T = K^2 = \langle (ab)^4 \rangle^G$$
$$T_m = \underbrace{T \times \cdots \times T}_{2^n}$$
$$K_m = \underbrace{K \times \cdots \times K}_{2^m}$$
$$M_m = T_{m-1}K_m.$$

In T_m and K_m each factor acts on a corresponding subtree with roots at the m-th level.

Theorem 6.1 ([Roz93], [BG02a]). *The following hold in G.*

(i)

$$st_G(n) = \begin{cases} H & \text{if } n = 1 \\ \langle D, T \rangle & \text{if } n = 2 \\ \langle M_2, (ab)^4 (adacac)^2 \rangle & \text{if } n = 3 \\ \underbrace{st_G(3) \times \cdots \times st_G(3)}_{2^{n-3}} & \text{if } n \geq 4 \end{cases}.$$

(ii)

$$rist_G(n) = \begin{cases} D & \text{if } n = 1 \\ K_n & \text{if } n \geq 2 \end{cases}.$$

(iii) *in the lower central series* $\{\gamma_n(G)\}_{n=1}^{\infty}$

$$\gamma_{2^m+1}(G) = M_m, \qquad m = 1, 2, \ldots,$$

(iv) *the derived series* $\{G^{(n)}\}_{n=1}^{\infty}, \{K^{(n)}\}_{n=1}^{\infty}$ *of groups* G *and* K *respectively*

$$G^{(n)} = rist_G(2n - 3), \qquad n \geq 3$$

$$K^{(2)} = rist_G(2n), \qquad n \geq 2.$$

It is important for the study of various properties of the group G to get as full information as possible about the lattice of subgroups of finite index, normal subgroups, subgroups closed in the profinite topology. For the moment the lattice of normal subgroups is tolerably well understood, and its upper part below the subgroup $H = st_G(1)$ is given in Figure 6.3 (see [BG02a]).

Detailed study of normal subgroups of index $\leq 2^{11}$ is carried out in [CSST01] where it is shown in particular that for every normal subgroup $N \lhd G$, $N \neq \{1\}$, there exists n such that $M_{n+1} < N < st_G(n)$. The problem is solved in [Bar05] via using Lie algebras constructed from the lower central series of G. It follows from his work that the normal subgroup growth of G is equal to $n^{\log_2 3}$, and that every normal subgroup of G is generated as a normal subgroup by one or two elements.

The following problem remains open:

Problem 6.1. *What is the subgroup growth of* G?

On subgroup growth and related topics see [LS03].

Maximal subgroups play a special role in group theory. The question whether all maximal subgroup of G have finite index (let us call such a property (P)) was open for a long time before it was answered positively by E. Pervova [Per00]. She proved that only subgroups of index 2 (listed above) are maximal subgroups in G and hence the Frattini subgroup $\phi(G)$ is equal to the commutator subgroup G'. This gives a very simple and efficient criterion to determine when a set of elements $S \subset G$ generates G, namely if and only if its image $\overline{S}$ in $G/\Phi(G) \simeq \mathbb{Z}/2\mathbb{Z} \times \mathbb{Z}/2\mathbb{Z} \times \mathbb{Z}/2\mathbb{Z}$ generates the quotient. It is obvious that this is a necessary condition. To show sufficiency let us assume that $\overline{S}$ generates $G/\Phi(G)$ but $\langle S \rangle$ is a proper subgroup L in G. Then L is a subgroup of some maximal subgroup $\tilde{L}$ in G which has index 2 so the image of L in $G/\Phi(G)$ will be of index 2. Contradiction.

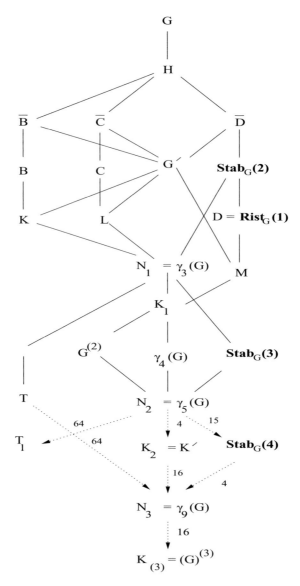

FIGURE 6.3. The top of the lattice of normal subgroups of G below H.

Pervova extended her result to Gupta–Sidki p-groups and some other groups [Per05]. Her proof is quite tricky and doesn't seem to work for all branch groups.

Problem 6.2. *Is there a finitely generated branch group with a maximal subgroup of infinite index?*

In [GW03a] Pervova's result is improved by showing that not only G has property (P) but also any group L abstractly commensurable with G has property (P) (recall that two groups are abstractly commensurable if they have isomorphic subgroups of finite index). In particular, every subgroup of G of finite index possesses (P). By the same arguments as in [GW03b] one can show that if L is a branch group and has (P) then any group abstractly commensurable with L has (P).

Weakly maximal subgroups constitute another important class of subgroups. (These are subgroups of infinite index maximal with respect to this property, i.e., adding to the group a single element from its complement extends it to a subgroup of finite index). The group G has many such subgroups. For instance, denote by ∂T the boundary of the tree, that is the set of all geodesic rays connecting the root with infinity. The action of G on T extends naturally to an action on ∂T. Then for any $\xi \in \partial T$, the stabilizer

$$P_\xi = st_G(\xi) = \{g \in G \colon \xi^g = \xi\}$$

is a weakly maximal subgroup [BG02a]. We call P_ξ a parabolic subgroup.

It is also easy to produce an example of a weakly maximal subgroup in G not of parabolic type. For instance, let $L = (Q \cdot U) \rtimes \langle a \rangle$ where Q is the diagonal subgroup of G consisting of elements (g, g), $g \in B$ (g has to belong to the subgroup $B \cdot \langle (ad)^2 \rangle$ in order for the pair (g, g) belong to G) and $U = \langle c, c^a \rangle$. Then L is a finitely generated weakly maximal subgroup of G. Indeed let us show that if $x \notin L$ then $\langle L, x \rangle$ has finite index in G. As $a \in L$ we can assume $x \in st_G(1)$, $x = (x_0, x_1)$. Let $x_0 = yz$ where $y \in B$ and $z \in \langle a, d \rangle$. The element (y, y) belongs to Q so we can replace x by element $(z, y^{-1} x_1)$. Taking an element in U of the form (z, w) we can replace x by $(1, w^{-1} y^{-1} x_1)$, with $w^{-1} y^{-1} x_1 \neq 1$. Let $N = \langle w^{-1} y^{-1} x_1 \rangle^G$. Then N is a normal subgroup of finite index and the group $\langle L, x \rangle$ contains the group $\psi^{-1}(N \times N)$ and therefore has finite index in G.

Problem 6.3. *Describe all weakly maximal subgroups in G.*

The following description of the parabolic subgroup $P = P_G(1^\infty)$ for the rightmost path $\xi = 1 \cdots 1 \cdots = 1^\infty$ is given in [BG02a]. Denote $B = \langle b \rangle^G$, $Q = B \cap P$, $R = K \cap P$ where $K = \langle (ab)^2 \rangle^G$. Then the following decompositions hold

$$P = (B \times Q) \rtimes \langle c, (ac)^4 \rangle$$
$$Q = (K \times R) \rtimes \langle b, (ac)^4 \rangle \qquad (6.2)$$
$$R = (K \times R) \rtimes \langle (ac)^4 \rangle$$

where the factors $B \times Q$, $K \times R$ act on subtrees with roots at the first level. (6.2) should be viewed as recursive formulas describing P, Q, R in terms of action of the same groups on corresponding subtrees. Here we use again self-similarity of involved groups. By iterating formulae (6.2) we get the following decomposition for P

$$P = (B \times ((K \times ((K \times \cdots) \rtimes \langle (ac)^4 \rangle)) \rtimes \langle b, (ac)^4 \rangle)) \rtimes \langle c, (ac)^4 \rangle$$

where each factor from the direct or semi-direct product acts on the corresponding subtree as indicated in Figure 6.4. The product $B \times K \times K \times \cdots \times K \times \cdots$, where

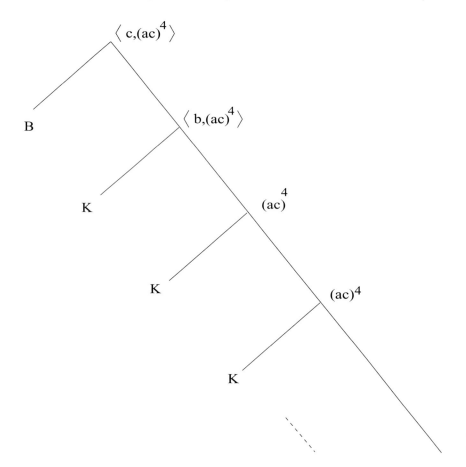

FIGURE 6.4. Decomposition of parabolic subgroup.

each factor acts on the corresponding subtree according to 6.4, is a subgroup of the group P on part of which the factors $\langle c, (ac)^4 \rangle$, $\langle b, (ac)^4 \rangle$, $\langle (ac)^4 \rangle \ldots$ from the decomposition above act. These factors are not, formally speaking, elements of G (because they act not on the whole tree but only on certain subtrees). In order to get a corresponding element of G we need to lift the projection by using the i-th power of the endomorphism σ (see (4.8)).

Thus active factors $\langle b, (ac)^4 \rangle$, $\langle (ac)^4 \rangle$, $\langle ac \rangle^4 \ldots$ which correspond to the right-most ray on the diagram 6.4 will be represented by subgroups $\sigma(\langle b, (ac)^4 \rangle)$, $\sigma^2(\langle (ac)^4 \rangle)$, $\sigma^3(\langle ac \rangle^4) \ldots$. This is explained by the fact that for an arbitrary element $g \in \langle b, (ac)^4 \rangle$ the following holds: $\psi \sigma(g) = (1, g)$.

A parabolic subgroup P_ξ can be presented as an intersection of subgroups of finite index

$$P_\xi = \bigcap_{n=1}^{\infty} P_{n,\xi}$$

where $P_{n,\xi}$ is the stabilizer of the vertex of level n belonging to the path ξ. This implies that P_ξ is closed in profinite topology. More generally, any weakly maximal subgroup $L < G$ can also be presented as an intersection of subgroups of finite index. Indeed, there is a maximal subgroup Q_1 of index 2 which contains L, and, as it has property (P), there is a subgroup Q_2 of index 2 in Q_1 containing L, and so on. We get a descending sequence $\{Q_n\}_{n=1}^{\infty}$ of subgroups of finite index which contain L.

$$\bar{L} = \bigcap_{n=1}^{\infty} Q_n$$

If we suppose that their intersection is different from L we get a contradiction with the weak maximality of L. Hence $L = \bigcap_{n=1}^{\infty} Q_n$ and L is closed in the profinite topology.

Problem 6.4. *Describe in algebraic terms the subgroups of G which are closed in profinite topology.*

As was already mentioned in Section 4, finitely generated subgroups of G are closed in profinite topology (see [GW03a]). Groups with this property are called subgroup separable (SS). Free groups, polycyclic groups, surface groups, some 3-manifold groups and free products of groups with property (SS) are known to be (SS). One of the consequences of the (SS) property is the solvability of the membership problem.

The following problem is inspired by an analogous result of L. Ribes and P. Zaleski [RZ93] in the case of a free group.

Problem 6.5. *Is the product $A_1 A_2 \ldots A_n$ of finitely generated subgroups of G closed in the profinite topology?*

The next property that we are going to discuss is the congruence subgroup property (CSP). We say that a group L acting on rooted tree $T_{\overline{m}}$ has the CSP if every subgroup $M < L$ of finite index contains $st_L(n)$ for some $n \geq 1$. Thus we consider $\{st_L(n)\}_{n=1}^{\infty}$ as the principal congruence subgroups sequence. The choice of the descending sequence of stabilizers looks natural but is not the only possible. For instance, for branch groups, one can also study the congruence subgroup problem with respect to the sequence of rigid stabilizers.

The group G has the CSP. Indeed every subgroup L of finite index in G contains a normal subgroup of finite index and hence by Theorem 4 [Gri00b] contains the commutator subgroup $(\text{rist}_G(n))'$ of a rigid stabilizer for sufficiently large n. Therefore it contains the group $K_n' \cong \underbrace{K' \times \cdots \times K'}_{2^n}$. But K' contains

$st_G(5)$, hence $L > \underbrace{st_G(5) \times \cdots \times st_G(5)}_{2^n} \geq st_G(n+5)$ (each factor in the above products acts on the corresponding subtree with root at the n-th level).

An example of a branch p-group acting on the p-regular rooted tree (p odd prime) and without CSP is constructed in [Per02].

The CSP for groups acting on trees is important first of all because it allows to identify the closure of the group in Aut T with its profinite completion. For groups without CSP the problem of description of profinite completion is harder. In some cases it is solved in [Per04].

Let L be a residually finite group and $\{M_n\}_{n=1}^{\infty}$ be a descending sequence of subgroups of finite index with intersection that has a trivial core (i.e., $\cap_{g \in G} P^g = \{1\}$, where P denotes $\cap_{n=1}^{\infty} M_n$) . Call such a sequence a T-sequence. Then a tree of type $T_{\overline{m}}$ and a faithful action of L on T can be constructed in canonical way. Namely as the set of vertices of T one takes the set of cosets gM_n, $g \in L$, $n = 0, 1, 2, \ldots$ ($M_0 = L$) (the root vertex is represented by the coset $1 \cdot L$), and two vertices gM_n and hM_k, $n \leq k$ are joined by an edge if and only if $k = n+1$ and $hM_k \subset gM_n$. The action of L on the set of vertices is by left multiplication.

Conversely, given a tree $T = T_{\overline{m}}$ and faithful action of L on T one gets a sequence of subgroups of finite index of the form $\{st_L(u_n)\}_{n=1}^{\infty}$ where $\{u_n\}_{n=1}^{\infty}$ is the sequence of vertices of a geodesic ray $\xi \in \partial T$ joining the root vertex with infinity.

If $\{M_n\}_{n=1}^{\infty}$ is a sequence of normal subgroups, then $M_n = st_L(n)$, $n = 1, 2, \ldots$ Thus the study of actions on rooted trees is closely related to the study of descending sequences of subgroups of finite index whose intersection has trivial core. Call two sequences $\overline{M} = \{M_n\}_{n=1}^{\infty}$, $\overline{N} = \{N_n\}_{n=1}^{\infty}$ equivalent and write $\{M_n\}_{n=1}^{\infty} \sim \{N_n\}_{n=1}^{\infty}$ if for any n there is k s.t. $M_n < N_k$ and vice versa, for any k there is n s.t. $M_n > N_k$.

Problem 6.6.   (i) *Is there a branch group with infinitely many equivalence classes of normal T-sequences?*

(ii) *If the answer to the above question is positive how complicated can be the lattice of classes?*

Let us say that two actions of L on trees $T_{\overline{M}}$ and $T_{\overline{N}}$ given by a normal T-sequences $\overline{M} = \{M_n\}_{n=1}^{\infty}$ and $\overline{N} = \{N_n\}_{n=1}^{\infty}$ are equivalent if $\overline{M} \sim \overline{N}$. The study of actions of a group on a rooted tree up to this equivalence is an interesting topic. For a group of branch type there should not be too many different actions. One result in this direction is obtained in [GW03b] where it is showm that under certain conditions a branch group has only one action up to deletion of levels.

An important role in the study of lattice of subgroups of a group L belongs to the group of automorphisms of L. For many groups L of branch type Aut L coincides with the normalizer $N_{\text{Aut }T}(L)$ of L in the group of automorphisms of the tree. There are at least three approaches to proving this. One was given by Sidki [Sid87a] in the case of the 3-group of Gupta and Sidki, another by Lavreniuk and Nekrashevich [LN02] for branch groups which have a so-called saturated

sequence of characteristic subgroups, and a third one [GW03b] via the property
of uniqueness of actions mentioned above.

Problem 6.7. *Is it correct that for any branch group L acting on a tree T*

$$Aut\ L = N_{Aut\ T}(L)?$$

If true, this helps calculating Aut L, as was used in [Sid87a] for Gupta–Sidki
3-groups and in [GS04] for G, where it was shown that Out $G \cong \mathbb{Z}/2\mathbb{Z} \oplus \mathbb{Z}/2\mathbb{Z} \oplus$
$\mathbb{Z}/2\mathbb{Z}\dots$ is an elementary 2-group of infinite rank (the generators of this group are
explicitly calculated). As a consequence of the result in [GS04] we get that every
normal subgroup in G is characteristic.

It is shown in [LN02] if L is a weakly branch group then

$$Aut\ L = N_{Homeo\ \partial T}(L),$$

where Homeo $\partial T(L)$ is the group of homeomorphisms of the boundary ∂T.

An interesting approach to the study of normal and characteristic subgroups
of groups acting on rooted trees is due to Kaloujnin [Kal45, Kal48] and is based
on the use of so-called tables and their characteristic called height. The modern
exposition of Kaloujnin method leading to some new results is done in [CSLST04]
and [LNS05]. It would be interesting to apply Kaloujnin method for studying
characteristic and normal subgroups of finitely generated self-similar groups of
branch type.

7. Finite quotients and groups of finite type

Any group L acting faithfully on a rooted tree T is residually finite: the approximating sequence of finite quotient groups is $L_n = L/st_L(n)$, $n = 1, 2, \dots$. The
group L_n acts on the subtree $T_{[n]}$ (part of T up to n-th level) where the action is
induced by "forgetting" the action of L below level n. The study of the sequence
$\{L_n\}$ can be useful for the study of L and its profinite completion.

Let us consider the sequence $G_n = G/st_G(n)$ approximating the group G.
The first three groups are $G_1 = C_2$, $G_2 = C_2 \wr C_2$, $G_3 = C_2 \wr C_2 \wr C_2$ (so they are
the full groups of automorphisms of the trees $T_{[1]}, T_{[2]}, T_{[3]}$). Starting from the
fourth level the character of the sequence $\{G_n\}$ completely changes, in particular
the number of generators remains to be 3. If we keep the notation a, b, c, d for the
generating elements of G_n then the following statement holds [Gri00b, dlH00].

Proposition 7.1. *When $n \geq 3$ the map* (3.4) *induces an embedding*

$$\psi^{(n)}\colon\ G_{n+1} \longrightarrow G_n \wr C_2 = (G_n \times G_n) \rtimes \langle a \rangle \qquad (7.1)$$

the image of which has index 8 and has the form

$$((B_n \times B_n) \rtimes \langle (a, c), (c, a) \rangle) \rtimes \langle a \rangle$$

where B_n is the image of B in G_n.

Corollary 7.1.
$$|G_n| = 2^{5 \cdot 2^{n-3}+2}.$$

Indeed $|G_3| = 2^7$ and (7.1) imply

$$|G_{n+1}| = \frac{1}{4}|G_n|^2,$$

for $n \geq 3$.

The above proposition and corollary can be used in two different directions. First, it can be used for computation of the Hausdorff dimension of the closure $\overline{G}$ of G in Aut T (recall that $\overline{G}$ is isomorphic to the profinite completion $\hat{G}$ because of the congruence subgroup property).

By definition the Hausdorff dimension of a closed subgroup L in a profinite group M with a descending chain $\{M_n\}$ of open normal subgroups is a number from $[0,1]$ defined as

$$\text{Dim } L = \liminf_{n \to \infty} \frac{\log |LM_n/M_n|}{\log |M/M_n|}.$$

In our case $M = \text{Aut } T, M_n = st(n)$,

$$|M/M_n| = 2^{1+2+\cdots+2^{n-1}} = 2^{2^n - 1}$$

and hence

$$\text{Dim } \overline{G} = \lim_{n \to \infty} \frac{5 \cdot 2^{n-3} + 2}{2^n - 1} = \frac{5}{8}.$$

It was mentioned in the introduction that profinite groups of the form $\overline{L}$ where L is a self-similar group play an important role in Galois Theory and in the theory of profinite groups.

Problem 7.1.   (i) *What is the set of values of Hausdorff dimensions of groups of the form $\overline{L}$ where L is a self-similar group?*
  (ii) *In particular, are the numbers* Dim $\overline{L}$ *always rational?*
 (iii) *Is there an algorithm that, given an automaton A, computes* Dim $\overline{L}$?

We note that M. Abert and B. Virag have shown that, with probability 1, the Hausdorff dimension of the closure of three randomly chosen tree automorphisms is 1 [AV05]. Z. Sunik has constructed concrete examples of regular branch self-similar groups of Hausdorff dimension arbitrary close to 1 [Sun05].

The second application of Proposition 7.1 leads to the idea of groups of finite type which we are going to explore now. But before doing this let us consider a finitary version.

If $g_0, g_1 \in G_n$ and $x \in S_2$ then the element $(g_0, g_1)x \in \text{Aut } T_{[n+1]}$ belongs to G_{n+1} if and only if it belongs to the image of the map (7.1). In the language of portraits it means that the portrait (the labelling of vertices of $T_{[n+1]}$ up to level n by elements of S_2) of the element $(g_0, g_1)x$, has label x at the root vertex, has the portrait of the element g_0 in left subtree and the portrait of the element g_1 in right subtree (the left and the right subtrees have their roots on the first level of $T_{[n+1]}$).

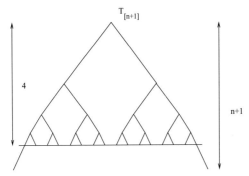

Moreover the labelling inside the subtree $T_{[4]}$ must coincide with a portrait of some element of the group G_4. Vice versa, if this condition is satisfied then $(g_0, g_1)x$ belongs to G_{n+1}. Indeed, since the index of the image $\overline{B}$ of B in the group G_3 coincides with the index $|G : B| = 8$ it follows that $B \geq st_G(3)$. If $(h_0, h_1)y \in G_{n+1}$ has the same portrait as $(g_0, g_1)x$ on the subtree $T_{[4]}$ then $x = y$ and $(g_0, g_1)x((h_0, h_1)x)^{-1} = (g_0 h_0^{-1}, g_1 h_1^{-1}) \in st_{G_{n+1}}(4)$. Therefore $g_0 h_0^{-1}$, $g_1 h_1^{-1} \in st_{G_n}(3)$ and so $g_0 h_0^{-1}$, $g_1 h_1^{-1} \in B_n$. But this implies that $(g_0 h_0^{-1}, g_1 h_1^{-1})$ belongs to the image of $\psi^{(n+1)}$, and therefore $(g_0, g_1)x((h_0, h_1)x)^{-1} \in G_{n+1}$ and $(g_0, g_1)x \in G_{n+1}$.

From these considerations we get

Proposition 7.2. *A labelling of $T_{[n+1]}$ is a portrait of some element in G_{n+1} if and only if for each subtree $T' \subset T_{[n+1]}$ which is isomorphic to $T_{[4]}$ the pattern of the labelling inside T' determines an element of the group G_4 (after identification of T' with $T_{[4]}$).*

This statement is illustrated in Figure 7.1.

Let us call a configuration on $T_{[4]}$ which correspond to elements in Aut $T_{[4]} \backslash G_4$ forbidden patterns. Then what was written in the previous paragraphs can be reformulated as follows.

Proposition 7.3. *A labelling of $T_{[n+1]}$ is a portrait of some element of G_{n+1} if and only if it doesn't contain a forbidden pattern.*

In a way analogous to the case of the group G one can define a core portrait of an element of G_n and the depth $d(g)$, $g \in G_n$. Again as in 3.16 the inequality $d_n(g) \leq \log|g| + 1$ holds. The next problem is analogous to the Problem 3.4 (iii).

Problem 7.2. (i) *Find a description of the portraits of elements of fixed length in the groups G_n, $n = 1, 2, \dots$?*

(ii) *What is the maximal length of the elements in G_n and how do the the portraits of the elements of maximal length look like?*

In all cases we consider the length of the elements in G_n with respect to the systems of generators that is the image of the canonical one.

Similar arguments which were used to prove the Proposition 7.1 lead to the following

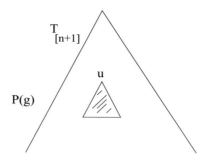

FIGURE 7.1. $\forall T' \subseteq T_{[n+1]}$ the corresponding pattern is not forbidden.

Proposition 7.4. *A core portrait $P_{core}(g)$, $g \in \mathrm{Aut}\ T$ determines an element of G if and only if the corresponding portrait $P(g)$ of g does not contain a forbidden pattern.*

Corollary 7.2. *A portrait $P(g)$, $g \in \mathrm{Aut}\ T$ determines an element of the closure $\overline{G}$ of G in $\mathrm{Aut}\ T$ if and only if it does not contain a forbidden pattern.*

Remark. The verification if $P_{\mathrm{core}}(g)$ satisfies the condition of Proposition 7.2 requires only verification of the condition on the subtree $T_{[d+4]}$ where d is the depth of $P_{\mathrm{core}}(g)$.

In Ergodic Theory there is a notion of a subshift of finite type [Kit98]. Given an alphabet X, and a finite set $F \subset X^*$ of words (called forbidden words) one considers the subspace $\Omega_F \subset \Omega$ of the space $\Omega = X^{\mathbb{N}}$ of infinite sequences consisting of those sequences $x = x_0 x_1 \ldots x_n \ldots$ that do not contain forbidden subwords. This set is closed with respect to the shift τ, where

$$(\tau x)_n = x_{n+1}.$$

The dynamical system (τ, Ω_F) is called a subshift of finite type.

In our situation the "ray" $\mathbb{N}$ is replaced by a tree and the set F is replaced by a set of forbidden patterns.

Remark. In the definition of a subshift of finite type the finite set F can be replaced by a finite set of forbidden words of the same length. This corresponds to our case because all forbidden patterns have the same size, namely the size of the tree $T_{[4]}$.

The example of G suggests the following definition. Let $T = T_m$ be an m-regular rooted tree, A be a subgroup of $\mathrm{Aut}\ T_{[d]}$ for some $d \geq 1$, $B = \mathrm{Aut}\ T_{[d]} \backslash A$ and let F be the set of portraits of the elements in B. The elements of the set F will be called forbidden patterns and d will be called the depth of the forbidden patterns.

Definition 7.1. *A closed subgroup $M < \mathrm{Aut}\ T_m$ is called a subgroup of finite type if there is a set F of the type described in the previous paragraphs such that an* *element* $g \in$ Aut T_m *belongs to* M *if and only if its portrait* $P(g)$ *does not contain a forbidden pattern.*

The examples of groups of finite type are the full group of automorphisms of the tree and the closure $\overline{G}$ of G in Aut T_2. The group $\overline{IMG}(z^2 + i)$ is of finite type, as are the completions of Gupta–Sidki p-groups.

Proposition 7.5. *Let* $M <$ Aut T_m *be a closed subgroup of finite type acting transitively on levels. Then* M *is a regular branch weakly self-similar group.*

Proof. It is obvious that M is a weakly self-similar group. Let $L = st_M(d)$ where d is the depth of the forbidden patterns determining the group and let $L_1 = \underbrace{L \times \cdots \times L}_{m}$ where each factor acts on the corresponding subtree with root at the first level.

We claim that L_1 is a subgroup of L. Indeed if $k \in L_1, k = (k_1, \ldots, k_m), k \in L$ and $P(k)$ is a portrait of k then $P(k)$ has no forbidden patterns with the root at level ≥ 1. But all the labels in a pattern with root at the zero level are equal to the identity element. As the identity element belongs to A it does not belong to B so such a pattern is not forbidden. Therefore L_1 is a subgroup of M and even a subgroup of L. L is an open (normal) subgroup in M (and so has finite index). As $L = st_M(d)$, $L_1 \geq st_M(d+1)$ so L_1 also is an open subgroup in Aut T therefore $[M : L_1] < \infty$ and M is a regular branch over L. $\square$

Problem 7.3. *Let* M *be a profinite group of finite type.*

(i) *What conditions on the set of forbidden patterns* F *imply that* M *is finitely generated as a profinite group?*

(ii) *What conditions on* F *imply that* M *contains a dense abstract branch subgroup?*

Remark. A necessary condition for M to be finitely generated is that the abelianization M_{ab} is finitely generated. Let $C_2 = \{1, -1\}$ be the multiplicative group of order 2. There is a canonical homomorphism $M \xrightarrow{\xi} \prod_{n=0}^{\infty} C_2$ (Cartesian product) which in component n sends an element $g \in M$ to the element of C_2 equal to the product of the signatures of the labels of the vertices on level n. Of course, to have M finitely generated the image has to be a finitely generated.

If M is given by a set of forbidden patterns then the image of ξ is not necessarily determined by a finite set of forbidden patterns as shown in the example below.

In the case of G, after changing $\{1, -1\}$ to $\{0, 1\}$, we have

$$\alpha = \xi(a) = 100\ldots$$
$$\beta = \xi(b) = 0\ 110\ 110\ldots$$
$$\gamma = \xi(c) = 0\ 101\ 101\ldots$$
$$\delta = \xi(d) = 0\ 011\ 011\ldots,$$

and these elements generate the group $C_2 \times C_2 \times C_2$.

The description of the elements of a group of finite type by a set of forbidden patterns allows us to "visualize" them and hence to get a better understanding of the structure of the group. Subshifts of finite type are particular case of sofic systems [LM95] for which the set of forbidden words constitutes a regular language (i.e. can be described by a finite automaton).

Problem 7.4. (i) *What is the analogue of sofic systems for self-similar profinite groups or for regular branch profinite groups?*

(ii) *What is the class of "languages" of forbidden patterns that serves to describe any self-similar profinite group or any regular branch profinite group?*

For instance this question is interesting even for the Basilica group given by the automaton in Figure 9.1.

One of the simplest examples of a group of a finite type which is different from Aut T and acts spherically transitively is the following. Take a binary tree and consider the subgroup $A < $ Aut $T_{[2]}$ which acts transitively on $T_{[2]}$, has index 2 in Aut $T_{[2]}$ and whose elements are given by the configurations

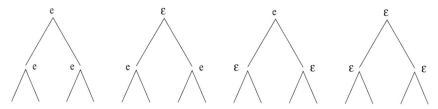

Then the forbidden patterns are

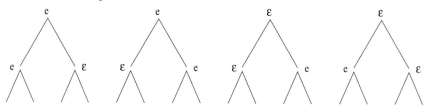

For such a set of forbidden patterns the image of ξ in $\bigoplus_{n=0}^{\infty} \mathbb{Z}/2\mathbb{Z}$ has only two elements, namely $0000\ldots$ and $1000\ldots$.

8. Around Milnor's question

Let L be a finitely generated group with a (finite) system of generators S and let $|g|$ be the length of the element $g \in L$ with respect to S i.e. the smallest n such that

$$g = s_{i_1} \ldots s_{i_n},$$

for some $s_{i_j} \in S \cup S^{-1}, j = 1, \ldots, n$.

The growth function $\gamma(n) = \gamma_L^S(n)$ is the function

$$\gamma(n) = \#\{g \in L: \ |g| \le n\}$$

defined already in Section 3 by (3.15) in a more general situation and let

$$\gamma^{sph}(n) = \#\{g \in L: \ |g| = n\}$$

be the spherical growth function.

The group L has polynomial growth if there are constants c and d such that $\gamma(n) \le cn^d$. A theorem of Gromov [Gro81] states that the class of finitely generated virtually nilpotent groups is exactly the class of groups of polynomial growth.

The free group $\mathbb{F}_2$ of rank 2 has exponential growth, as does any group containing $\mathbb{F}_2$ as a subgroup (or even a free submonoid on 2 generators). For instance, finitely generated solvable groups which are not virtually nilpotent contain a free submonoid on 2 generators and hence have exponential growth. For more on growth see [Gri90, GdlH97, dlH00]. The growth cannot be superexponential because of the trivial estimate $\gamma(n) \le (2|S|)^{n+1}$.

Two functions $\gamma_1(n)$ and $\gamma_2(n)$ are called equivalent if there is a constant c such that $\gamma_1(n) \le c\gamma_2(cn)$, $\gamma_2(n) \le c\gamma_1(cn)$, $n = 1, 2, \ldots$. This equivalence is denoted by $\sim$ while the corresponding preorder is denoted by $\precsim$. The following problem was raised by Milnor.

Problem 8.1. [Mil68] *Is it correct that the growth function of a finitely generated group is either equivalent to a polynomial function n^d or to an exponential function 2^n?*

This problem was solved in [Gri83, Gri84a] where the first examples of groups of intermediate between polynomial and exponential growth where constructed. Indeed the group G happened to be the first known example of a group of intermediate growth.

The idea of the proof of the fact that G has intermediate growth with estimates

$$e^{n^\alpha} \le \gamma_G(n) \le e^{n^\beta} \tag{8.1}$$

for some constants $0 < \alpha, \beta < 1$ is the following.

That G has a super polynomial growth follows from the fact that G is an infinite finitely generated torsion group and from Gromov's description of groups of polynomial growth (indeed finitely generated nilpotent torsion group is finite; this is a classical result of group theory). But such an argument doesn't gives a lower bound in (8.1). To get such a bound one can observe that G is (abstractly) commensurable with $G \times G$ (i.e. G and $G \times G$ contain subgroups of finite index which are isomorphic). To see this let us use the embedding (3.4). As the image Im ψ contains $B \times B$ where $B = \langle b \rangle^G$ and as $\psi^{-1}(B \times B) = D = \langle d \rangle^G$ and both B and D have finite index in G the commensurability of G and $G \times G$ follows.

As $\gamma_{G \times G}(n) \sim \gamma_G^2(n)$ and a group and a subgroup of finite index have equivalent growth functions we get the equivalence $\gamma_G(n) \sim \gamma_G^2(n)$ from which the lower bound in (8.1) easily follows.

More precise but still easy arguments show that α can be taken to be 0.5. This was done in [Gri84a] by showing that $\psi(B_1(4n))$ "almost" contains $B_1(n) \times B_1(n)$ where $B_1(n)$ is the ball of radius n centered at the identity element in the Cayley graph of G i.e. for any pair of elements $g_0, g_1 \in G$, $|g_i| \leq n$, $i = 0, 1$ such that $(g_0, g_1) \in G$ there is an element $g \in G$ of length $|g| \leq 4n$ such that $\psi(g) = (g_0, g_1)$. Indeed this is a consequence of Lemma 3.2. Much more delicate arguments show that α can be taken as 0.504 [Leo01], [Bar01]. For the upper bound we again use the embedding ψ but now iterating it three times.

Namely let $L = st_G(3)$ be the stabilizer of the third level in the binary tree $T = T_2$. The group L is a subgroup of finite index in G and let

$$\psi_3 \colon \; L \to \underbrace{G \times \cdots \times G}_{8}$$

be a monomorphism corresponding to the threefold iteration of ψ. The crucial lemma is the

Lemma 8.1. [Gri84a] *Let* $g \in L$ *and*

$$\psi(g) = (g_1, \dots, g_8).$$

Then

$$\sum_{i=1}^{\infty} |g_i| \leq \lambda |g| + 10 \tag{8.2}$$

where $\lambda = 3/4$.

Actually ψ_3 can be extended to a monomorphism

$$\psi_3 \colon \; G \to G \wr_Y (C_2 \wr C_2 \wr C_2)$$

where $Q = C_2 \wr C_2 \wr C_2$ is identified with the group Aut $T_{[3]}$, $T_{[3]}$ is the part of T up to the third level, and Y is the set of leaves of $T_{[3]}$ (i.e. the set of vertices of the third level) on which Q acts.

If

$$\psi(g) = (g_1, \dots, g_8) \cdot \sigma \tag{8.3}$$

where $\sigma \in Q$ then the inequality (8.2) still holds.

A hint for getting the upper bound of type (8.1) will be given below. Let us now show how to deduce from (8.2) that G has subexponential growth.

Because $\gamma(n)$ is a semimultiplicative function (i.e. $\gamma(m + n) \leq \gamma(m) \cdot \gamma(n)$) the limit

$$\delta = \lim_{n \to \infty} \sqrt[n]{\gamma(n)}$$

exists and $\delta = 1$ if and only if the group has subexponential growth.

The inequality (8.2) implies the existence of a constant C such that

$$\gamma(n) \leq C \sum \gamma(n_1) \cdots \gamma(n_8) \tag{8.4}$$

where the sum is taken over 8-tuples $(n_1, \dots, n_8)$ with $n_1 + \cdots + n_8 \leq \frac{3}{4}n + 10$. As the number of such tuples can be estimated by a polynomial $\mathcal{P}(n)$ (of degree 8) and as $\gamma(n_i) \sim \delta^{n_i}$ when n_i is large the inequality (8.4) leads to the inequality $\delta \leq \delta^{3/4}$ which implies $\delta = 1$, so G has subexponential growth. Additional considerations show that (8.4) is enough to conclude the existence of $\beta < 1$ for which the upper bound in (8.1) holds. For details see [Gri84a, Gri98]) where the bound is given by $\beta = \log_{32} 31$. A better technique was used in [Bar98], [MP01], [BŠ01] to improve the value of the bound $\beta = 0.767\ldots$.

The smaller λ and the smaller the level ℓ of the tree for which an inequality of type (8.2) holds, the better estimate for β can be obtained (in our case $\lambda = \frac{3}{4}$, $\ell = 3$). As was pointed out to me by Z. Sunik the constant $3/4$ can be replaced by $2/3$. Let us consider briefly Sunik's arguments.

If g is an element stabilizing the third level and g is represented by a reduced word w of length equal to $|g|$, then we can draw a tree of projections for w:

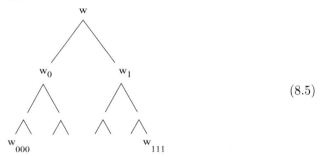

(8.5)

where $w_i = r(\varphi_i(w))$, $i = 0, 1$, are the results of the rewriting processes obtained through the reduction described in Section 4. The words w_{ij} are obtained in a similar way from w_i, and w_{ijk} are obtained from w_{ij}. Thus w_{ijk} are reduced words (perhaps not geodesic in G) representing the elements $g_1, \ldots, g_8$ from (8.3). We are going to show that

$$\sum_{ijk} |w_{ijk}| \leq \frac{2}{3}|w| + C$$

for some constant C (the value of constant C is not important for the estimate of the growth). Let us split all the occurrences of letters from the set $\{b, c, d\}$ in $w_{000}, \ldots, w_{111}$ into two sets E and F. In the first set we include those occurrences of letters that came from w (under the rewriting process) as a result of a collision (reduction) of the form (ii) i.e. $xy = z$. Such a letter is called a "killer".

The letters (i.e. the corresponding occurrences) that are not "killers" are called "neutral" elements. They constitute the set F. Let $\xi = |E|$ and $\eta = |F|$. Then

$$\sum_{i,j,k} |w_{ijk}|_* \approx (|E| + |E|) = \xi + \eta,$$

where $|w|_*$ denotes the *-length (that is a number of occurrences of the symbols b, c, d and therefore the *-length is approximately half of the actual length) and $\approx$ means "equality mod some additive constant". Denote $t = \sum_{i,j,k} |w_{ijk}|_*$ and assume that $|w| = 2n$, so that $|w|_* \approx n$.

As each "killing" reduces one letter we have $t \leq n - \xi$. Let δ_0 be the number of occurrences of the letter d in w, δ_1 the total number of occurrences of the letter d in w_0 and w_1, and δ_2 the total number of occurrences of the letter d in w_{00}, w_{01}, w_{10} and w_{11} (so δ_i is the total number of d-symbols in the words at level i in the tree in (8.5), $i = 0, 1, 2$).

As each neutral element in the decomposition from the root vertex (alma mater) to the second level has once to be equal to d (this follows from the rewriting rules) $\delta_0 + \delta_1 + \delta_2 \geq \eta$. As d produces a symbol 1 for one of its two projections this gives a reduction of the number of a's in the next level which also implies reduction of the $*$-length (this is correct up to ± 1 for each word in the decomposition. Therefore

$$t \precsim n - (\delta_0 + \delta_1 + \delta_2) \leq n - \eta.$$

We have a system

$$\begin{cases} t \precsim n - \xi \\ t \precsim n - \eta \\ t \approx \xi + \eta \end{cases}$$

Adding these relations we get

$$2t \precsim \frac{2}{3}(2n).$$

But $2t$ represents the total length of the eight projections and $2n$ was the length of w, hence we are done.

Improvements of the growth estimate for the group G are provided in [Bar98], [MP01] and [BŠ01]. Let us focus on the estimate given in [Bar98]. The idea is to replace the word length $|g|$ by the weight length $|g|_\omega$ defined on the generating elements $s \in S \cup S^{-1}$ by positive numbers $\omega(s)$ with extension to arbitrary elements $g \in G$ by

$$|g|_\omega = \min[\omega(s_1) + \ldots + \omega(s_n)]$$

where the minimum is taken over all the presentations of g as a product $g = s_1 \ldots s_n$ of generators $s_i \in S \cup S^{-1}$ and then consider the corresponding growth function. It is easy to see the rate of growth of a group does not depend on the choice of the weight so one can try to get a better estimate using an appropriate weight. In order to have a good contracting property for projections, the weight has to satisfy special properties. In case of G this is realized as follows.

Let $\eta \approx 0.811$ be the real root of the polynomial $x^3 + x^2 + x - 2$ and let ω be a weight on G defined by

$$\omega(a) = 1 - \eta^3 = \eta^2 + \eta - 1$$
$$\omega(b) = \eta^3 = 2 - \eta - \eta^2$$
$$\omega(c) = 1 - \eta^2$$
$$\omega(d) = 1 - \eta.$$

Then the following statement holds [Bar98].

Lemma 8.2. *Let* $g \in st_G(1)$ *with* $\psi(g) = (g_0, g_1)$. *Then*

$$\eta(|g|_\omega + |a|_\omega) \geq |g_0|_\omega + |g_1|_\omega. \tag{8.6}$$

To see this observe that

$$|g|_\omega + |c|_\omega > |d|_\omega$$
$$|b|_\omega + |d|_\omega > |c|_\omega$$
$$|c|_\omega + |d|_\omega = |b|_\omega$$

and therefore the minimal (with respect to ω) presentation of g as a product of generators has the form (3.2). Let the word u be a ω-minimal presentation of g. Construct the words u_0, u_1 using ψ just as in the case of the standard word length. Note that

$$\eta(|a|_\omega + |b|_\omega) = |a|_\omega + |c|_\omega$$
$$\eta(|a|_\omega + |c|_\omega) = |a|_\omega + |d|_\omega \tag{8.7}$$
$$\eta(|a|_\omega + |d|_\omega) = |b|_\omega.$$

As $\psi(b) = (a, c)$ and $\psi(aba) = (c, a)$, each b in u contributes $|a|_\omega + |c|_\omega$ to the total weight of u_0 and u_1; similar arguments apply to c and d.

Grouping together pairs of generators in (3.2) such as ba, ca, da we see that $\eta|g|_\omega$ is a sum of left-hand terms (with a possible difference in a single term $\eta|a|_\omega$). At the same time $|g_0|_\omega + |g_1|_\omega$ is bounded by the total weight of the letters in u_0 and u_1, which is the sum of the corresponding right-hand terms in (8.7) and this leads to (8.2).

The difference between Lemma 8.1 and Lemma 8.2 is that in (8.2) we get essential reduction of the word length only after we project on the third level, while for the length $|g|_\omega$ the reduction is present already at the first level. This leads to the better upper estimate $\beta \leq 0.767$. (See Prop. 4.3 in [Bar98], Th. 5.9 in [MP01] and Th. 6.1 in [BŠ01].)

We formulate now some open questions on the growth of G.

Problem 8.2. (i) *Does the limit*

$$\lim_{n \to \infty} \frac{\gamma_G(n+1)}{\gamma_G(n)} \tag{8.8}$$

exist, where $\gamma_G(n)$ *is the growth function with respect to the canonical generating set* $\{a, b, c, d\}$ *of* G?

(ii) *Is it correct that the limit* (8.8) *exists for every finite system of generators of* G?

(iii) *Answer the corresponding questions for the spherical growth function* $\gamma_G^{sph}(n)$.

Remark. Of course, the subexponential growth of G implies that if the limit (8.8) exists, it is equal to 1.

Problem 8.3. (i) *Is there an* α *such that*

$$\gamma_G(n) \sim e^{n^\alpha}?$$

(ii) *If the answer to (i) is NO what are the limits*

$$\varlimsup_{n\to\infty} \frac{\log\log\gamma_G(n)}{\log n}$$

$$\lim_{n\to\infty} \frac{\log\log\gamma_G(n)}{\log n}?$$

Problem 8.4. (i) *Is there a constant $\alpha > 0$ such that*

$$\gamma_G^{sph}(n) \succeq e^{n^\alpha}?$$

(ii) *Is it correct that $\gamma_G^{sph}(n) \sim \gamma_G(n)$?*

Problem 8.5. (i) *Is it correct that the generating series $\Gamma_G(z) = \sum\limits_{n=0}^{\infty} \gamma_G(n)z^n$*
satisfies a some differential—functional equation?
(ii) *What can be said about the singular point 1 of $\Gamma_G(z)$? For instance, is it*
correct that $\Gamma(z)$ behaves as $e^{c/(z-1)^\alpha}$ when $z \to 1$ in the interval $(1-\varepsilon, 1)$,
$\varepsilon > 0$, where α and c are some constants?
(iii) *What is the set of singular points of $\Gamma_G(z)$ on the unit circle? Is the unit disc*
$\{z : |z| < 1\}$ the natural domain of convergence for $\Gamma_G(z)$?

Remark. (i) In case

$$\lim_{z\nearrow 1} \frac{\Gamma(z)}{e^{c(z-1)^\alpha}} = B$$

for some constant B the asymptotic of $\gamma(n)$ has the form

$$\gamma(n) \approx e^{Dn^\delta}$$

with some constants D and δ so we get a positive answer to some of the above
problems.
(ii) As the coefficients of a McLauren series of an algebraic function grow either
polynomially or exponentially $\Gamma_G(z)$ is a transcendental function.

The next discussion based on [Gri88] relates the problem of calculation of
the growth of G to a similar problem for the automaton given by Figure 1.4. Let
A be a finite automaton of the type defined in Section 1. Let $A^{(n)} = A \circ \cdots \circ A$
be the n-fold composition of A and $\tilde{A}^{(n)} = \min(A^{(n)})$ be the minimization of A^n
(see [Eil74] for a minimization algorithm). Let $\gamma_A(n)$ be the number of states of
$\tilde{A}^{(n)}$. The function $\gamma_A(n)$ is called the growth function of the automaton A.

Remark. Let $S = S(A) = \langle A_q : q \in Q \rangle$ be the semigroup generated by states of
A and let
$$\gamma_S^{middle}(n) = \#\{s \in S \mid s = A_{q_1} \ldots A_{q_n}, q_i \in Q\}$$
be the function counting the number of elements which can be presented as a
product of exactly n generators. Clearly
$$\gamma_S^{sph}(n) \le \gamma_S^{middle}(n) \le \gamma_S(n).$$
The following proposition is obvious.

Proposition 8.1.

$$\gamma_{S(A)}^{middle}(n) = \gamma_A(n).$$

Indeed, the states of the automaton $A^{(n)}$ are in bijection with products $\{A_{q_1} \ldots A_{q_n}\}$ of length n, while the states of the automaton $\min(A^{(n)})$ are in bijection with elements of the semigroup $S(A)$ which can be represented by products of length n.

In case A is the automaton given by Figure 1.4 the semigroup $S(G)$ coincides with the group $G(A)$ generated by the automaton A and the middle growth function coincides with the ordinary growth function (because the identity element is represented by one of the states). Therefore the study of the growth of the automaton reduces to the study of the growth function of G and we get the example of an invertible automaton of intermediate growth as observed in [Gri88]. It would be interesting to develop methods to study the growth of automata and apply them to study the growth of groups. Interesting examples of noninvertible automata of intermediate growth are constructed in [RS02a, RS02b].

Let us conclude this section by an example which on one side is very similar to G but at the same time has different asymptotics of growth as shown by A. Erschler [Ers04a].

Let $\mathcal{E} = G(E)$ be a group generated by the automaton E given by Figure 8.1.

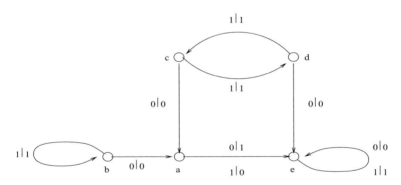

FIGURE 8.1. Automaton E

Then the generators E_a, E_b, E_c, E_d have order two, $E_b E_c E_d = 1$ and the group $\mathcal{E} = G(E)$ is a group of intermediate growth with growth estimates

$$e^{\frac{n}{\log^{2+\varepsilon} n}} \preceq \gamma_{\mathcal{E}}(n) \preceq e^{\frac{n}{\log^{1-\varepsilon} n}} \tag{8.9}$$

for any $\varepsilon > 0$. Therefore the automaton E also has intermediate growth with the same estimates as in (8.9). The group $\mathcal{E}$ is isomorphic to the group G_ω, where $\omega = (01)^\infty$. As it was already mentioned, the groups $G_\omega, \omega \in \Xi$) have intermediate growth.

The methods used in [Ers04b] are based on the study of asymptotic properties of random walks on groups and the corresponding Poisson boundary.

An important role in the study of formal languages is played by formal power series [SS78]. In the context of group theory analogous role is played by the complete growth functions studied in [GN97], where their rationality is shown in the case of hyperbolic groups, and also a relation to the growth function of the language of all geodesics and the problem of the spectrum of the discrete Laplace operator (Section 13) is indicated. Formal power series and their connections to regular and context-free languages are well studied [SS78]. On the other hand, recent results of K. Roever, D. Holt, and M. Elder, M. Gutierrez, Z. Sunik indicate that indexed languages are useful in the study of various properties of G. It would be good to develop formal power series methods for such languages as well.

9. Amenability and elementary classes

Amenable groups were introduced by J. von Neumann in 1929 [vN29] in his study of the algebraic roots of the phenomenon known as Banach–Tarskii Paradox [Gre69, Wag93]. Recall that a group M is called amenable if there is a measure μ defined on all subsets of M, satisfying the properties

1. $0 \leq \mu(E) \leq 1, \mu(M) = 1$ (positivity and normalization),
2. $\mu(gE) = \mu(E)$ (left invariance),
3. $\mu(E \cup F) = \mu(E) + \mu(F)$ if $E \cap F = \emptyset$ (additivity).

Here $E, F \subset M$ are arbitrary subsets and $g \in M$ is an arbitrary element.

Finite groups and commutative groups are amenable (the first fact is obvious, the second is nontrivial and based on the axiom of choice [Gre69]). We emphasize that the measure μ is assumed to be only finitely additive, so it should not be confused with Haar measure, which for discrete groups is a measure that has positive mass at each point and hence cannot satisfy the conditions imposed on μ in the case of an infinite group. Von Neumann also observed that the free group F_k of rank $k \geq 2$ is not amenable and that the class of amenable groups is closed with respect to the operations

(i) taking a subgroup,
(ii) taking a factor group,
(iii) extension,
(iv) direct limit.

Let AG be the class of amenable groups, NF be the class of groups without free subgroups of rank ≥ 2 and EG be the class of elementary amenable groups, i.e. the smallest class of groups containing finite groups, commutative groups and closed with respect to the operations (i)–(iv). Then the inclusions

$$EG \subseteq AG \subseteq NF$$

hold. The questions whether the equalities $EG = AG$ or $AG = NF$ hold were raised in [Day57]. The second question in the form of the conjecture "the group is non-amenable if and only if it contains a free subgroup with two generators" is sometimes called von Neumann conjecture [Ver82] (although there is no written confirmation that von Neumann indeed asked this question). This question was answered negatively in [Ol'80], [Adi82] (see more on this problem and the history of it solution in [GK93, GS02]).

Surprisingly even through the class EG was introduced (not explicitly) already by von Neumann in 1929 and the theory of invariant means and their applications was intensively developed [Gre69, Pat88] the first examples of amenable groups not in the class EG appeared only in 1983 when the first examples of groups of intermediate growth were discovered. For instance, G belongs to $AG \backslash EG$. Indeed it was already known in 1957 [AVŠ57] and later rediscovered a few times that groups of subexponential growth are amenable. At the same time it was shown in [Cho80] that groups from EG cannot have intermediate growth. They also cannot be finitely generated infinite torsion groups (i.e. Burnside groups). So the fact that $G \notin EG$ follows from several properties of G.

One more property that prevents the group G from being elementary amenable is its branch property, discussed in Section 6.

Theorem 9.1. *Let M be a finitely generated branch group. Then $M \notin EG$.*

The attempts initiated by J. von Neumann and M. Day to give an algebraic description of the class of amenable groups failed but a new idea may come and a new investigations in this direction are needed. An important step in understanding of the algebraic features of amenable groups is to understand amenable just infinite groups, since any finitely generated amenable group has a just infinite quotient, which also must be amenable. As the class of just infinite groups splits into three subclasses: branch, almost hereditarily just infinite and almost simple [Gri00b], a natural problem is to describe in algebraic terms the amenable groups inside each subclass.

Problem 9.1.   a) *Is it correct that a branch group is non-amenable if and only if it contains a free subgroup with two generators?*
  b) *Is there a hereditarily just infinite amenable group which does not belong to the class EG?*
  c) *Is there an infinite finitely generated amenable simple group?*
  d) *Is there a finitely generated infinite torsion amenable group of bounded exponent?*

It seems unlikely to have a positive answer on questions b) and c). Most known examples of finitely generated branch groups are amenable but S. Sidki and J.W. Wilson recently constructed an examples of finitely generated branch groups with free subgroups on two generators [SW03].

A rich source of amenable but not elementary amenable groups is the class of self-similar groups. As was already mentioned the self-similar groups of intermediate growth (such as G) are of this type. Remarkably, there are also amenable but not elementary amenable automaton groups that "have nothing to do" with the class of groups of intermediate growth, as shown by L. Bartholdi and B. Virag in [BV03]. The group they have been studying is the so-called Basilica group B, defined by the automaton in Figure 9.1, introduced for the first time in [GŻ02a, GŻ02b] as the group generated by the automaton given by Figure 9.1. The

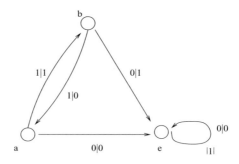

FIGURE 9.1. The automaton generating Basilica group

group B is a torsion free group of exponential growth in which a and b generate a free monoid. It was shown in [GŻ02a] that B does not contain a free subgroup on two generators and does not belong to the class SG of subexponentially amenable groups. The latter is the class of groups containing all groups of subexponential growth and closed with respect to operations (i)–(iv). As the groups of subexponential growth are amenable the inclusion $SG \subseteq AG$ holds and in [Gri98] (see also [CSGdlH99, GZ03]) the question about the equality $SG = AG$ was raised.

Negative answer to this question was provided in [BV03] by showing that B is amenable. The proof has probabilistic flavor and uses Kesten's criterion of amenability [Kes59]. In a paper by V. Kaimanovich the class of amenable self-similar groups is extended by new examples and the methods of entropy for random walks are applied in the proof of amenability [Kai].

A large class of self-similar groups without a free subgroup of rank 2 was found by S. Sidki in [Sid04]. This is the class of groups generated by automata determining a language of polynomial growth. In [BKNV] the amenability is shown for the subclass consisting of groups generated by bounded automata.

Problem 9.2. a) *Is there an algorithm which allows to determine which self-similar groups are amenable?*

b) *Is there a non-amenable self-similar group without free subgroups on two generators?*

We see that branch groups and groups generated by finite automata give fundamental contributions to the study of the phenomenon of amenability.

One of the important topics is the study of growth of Følner sequences in amenable groups. Recall that a sequence $\{E_n\}_{n=1}^{\infty}$ of finite subsets of a group M is Følner sequence if for any $g \in M$

$$\frac{|gE_n \,\triangle\, E_n|}{|E_n|} \xrightarrow[n\to\infty]{} 0$$

($\triangle$ denotes symmetric difference). A group M is amenable if and only if there is such a sequence [Føl57]. The next notion was introduced by A. Vershik [Ver82].

Let M be a finitely generated amenable group with a system of generators S and let $f_M(n)$, $n = 1, 2, \ldots$ be the infimum of the cardinalities of the finite subsets $E \subset M$ with the property

$$\frac{|sE \triangle E|}{|E|} < \frac{1}{n}$$

for $s \in S$. The function $f_M(n)$ is called the Følner function for G. The rate of growth of $f_M(n)$ when $n \to \infty$ does not depend on the choice of the finite generating set, is an important asymptotic invariant and, until recently, very little was known about it. Recent results in this direction were obtained in [Ers03] where the invariant is computed for wreath products in terms of its values for the factors. New progress is achieved in [Ers] were, based on a generalization of automaton groups (the author call her groups "piecewise automatic groups"), A. Erschler showed that given any function $f \colon \mathbb{N} \to \mathbb{N}$ there exists a finitely generated group M of intermediate growth for which the Følner function satisfies $f_M(n) \geq f(n)$ for all sufficiently large n. Erschler's construction uses some features of the construction of groups G_ω and, in particular, it uses the topology in the space of finitely generated groups discussed in Section 2.

The book in progress by C. Pittet and L. Saloff-Coste [PSC01] also contributes to the subject.

Problem 9.3.   a) *What are the possible types of growth of Følner functions in the case of finitely generated amenable branch groups?*
   b) *The same question for self-similar groups.*

Although we know that all groups G_ω, $\omega \in \Omega$ are amenable (they have intermediate growth if $\omega \in \Xi$ and are virtually metabilian if $\omega \in \Omega \backslash \Xi$ there is a number of interesting open questions about these groups related to the amenability phenomenon.

Problem 9.4.   a) *Find the growth of the Følner function $f_\omega(n)$ for the group G_ω?*
   b) *What is the shape of optimal Følner sets in G_ω?*
   c) *Answer the above questions for the group G.*

We even do not know if the sequence of balls $\{B_1^G(n)\}_{n=1}^\infty$ in the group G is a Følner sequence (although we know that some subsequence $B_1^G(n_i)$ must be a Følner sequence). The corresponding question is equivalent to the question stated in Problem 8.2(i).

To formulate the next question, let us present the groups G_ω in the form of quotients F_4/N_ω where F_4 is a free group of rank four with generators a, b, c, d and N_ω consisting of all words $w(a^{\pm 1}, b^{\pm 1}, c^{\pm 1}, c^{\pm 1})$ representing the identity in G_ω. Let

$$N = \bigcap_{\omega \in \Omega} N_\omega.$$

Problem 9.5.   a) *Is the group $\mathcal{N} = F_4/N$ finitely presented?*
   b) *Is $\mathcal{N} = F_4/N$ amenable?*

The next observation is due to V. Nekrashevych

Theorem 9.1. *The group $\mathcal{N}$ is a contracting self-similar group generated by the automaton given in Figure 9.2.*

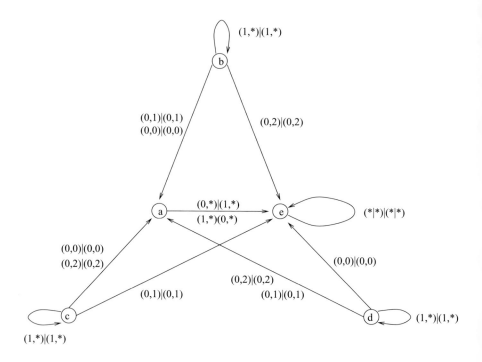

$$\text{Figure 9.2}$$

The automaton described by Figure 9.2 has five states representing the generators a, b, c, d of $\mathcal{N}$ and the identity element e and it is an automaton over the alphabet $\mathcal{A} = \{0, 1\} \times \{0, 1, 2\}$ consisting of six elements. The star $*$ represents any element of $\{0, 1, 2\}$. The proof of Theorem 9.1 is straightforward if one recalls the definition of the groups G_ω and the rule of action of initial automata on sequences described in Section 1.

The action of $\mathcal{N}$ on the 6-regular rooted tree T_6 given by the above automaton presentation of G is not level transitive. For each $\omega \in \Omega$ there is a binary $\mathcal{N}$-invariant subtree T_ω in T_6 consisting of vertices $(u, v) \in \mathcal{A}^*$ whose second coordinate follows the path ω and the restriction of the action of $\mathcal{N}$ on T_ω factorized by the kernel of the action is isomorphic to the action of G_ω on the binary tree. Thus the above presentation of $\mathcal{N}$ models a simultaneous action of the groups G_ω for all $\omega \in \Omega$. The group $\mathcal{N}$ has exponential growth, it is residually finite and has uncountably many homomorphic images. In [Gri84b] a residually finite group of intermediate growth with uncountably many homomorphic images was constructed.

10. Representations

In the case a group is not virtually abelian the description of all unitary representations is a "wild" problem. But even if there are no examples of classification of all irreducible unitary representations of a group that is not of type (I) [Tho68] a lot of work has been done in the study and use of particular representations or classes of representations.

One of the most important representations is the left regular representation λ_G and many problems in mathematics can be solved by answering a particular question about such representations. An important role belongs to the finite dimensional representations, especially in the case when the group in question is a MAP-group ("maximally almost periodic"), that is, the finite dimensional representations separate the elements of the group. For a finitely generated group this is the case if and only if the group is residually finite.

Groups acting faithfully on rooted trees and, in particular, self-similar groups belong to this class. A natural sequence of finite dimensional representations that can be associated with a group M acting on a tree $T = T_{\overline{m}}$ is the sequence of permutational representations $\{\pi_n\}_{n=1}^\infty$ given by the action of the group on the levels of the tree.

Another important class of representations are the quasiregular representations $\lambda_{M/H}$ in the space $\ell^2(M/H)$ given by the left action of M on the set M/H of left cosets gH. For a group M acting on a rooted tree T a natural first choice for a subgroup is the parabolic subgroup $P = P_\xi$ defined as the stabilizer $st_M(\xi)$ of some point ξ on the boundary ∂T of the tree. In case M acts spherically transitively the particular choice of ξ usually doesn't play an important role and the representations λ_{M/P_ξ} share many common properties. On the other hand, when M is a branch group λ_{M/P_ξ} are irreducible representations and no two of them are isomorphic [BG00a]. The family of representations $\{\lambda_{M/P_\xi}, \xi \in \partial T\}$ separates the elements of M since

$$\bigcap_{\xi \in \partial T} P_\xi = \{1\}.$$

The groups P_ξ already appeared in Section 6 as examples of weakly maximal subgroups. These subgroups are closed in the profinite topology as

$$P_\xi = \bigcap_{n=1}^\infty P_{n,\xi}$$

where $P_{n,\xi} = st_M(u_n)$ and u_n is the unique vertex on level n that belongs to the path ξ. Thus the coefficients of the representation λ_{M/P_ξ} are approximated by the coefficients of the finite dimensional representations $\lambda_{M/P_{n,\xi}}$ (which are isomorphic to the permutational representations π_n). The weak maximality of P_ξ and the fact the P_ξ is closed in the profinite topology make the representations λ_{M/P_ξ} especially important from many points of view.

An interesting new development is the appearance of Gelfand pairs associated to self-similar groups. Gelfand pairs play important role in many topics, in particular in the study of random walks [Dia88]. It is shown in [BG02a] that (M_n, P_n), $n = 2, 3, \ldots$ are Gelfand pairs where $M_n = M/st_M(n)$, $P_n = st_{M_n}(u_n)$ (u_n any vertex on level n) and the group M is either G or Gupta–Sidki p-group. In [BG02a] a direct proof is provided showing that the corresponding Hecke algebra $\mathcal{L}(M, P)$ is commutative, while in [BdlH03] it is observed that Gelfand pairs can be constructed for those branch groups whose action on the tree is 2-point transitive on levels (transitive on the sets of pairs of vertices of the same level with fixed distance between them.) The action of G is 2-point transitive while the action of Gupta–Sidki p-groups is not. Perhaps at the moment there is no example of a branch group M such that (M, P_n) are not Gelfand pairs.

Problem 10.1. *Is there a branch group M such that for some n the pair (M_n, P_n) is not a Gelfand pair?*

Branch groups allow not only to produce examples of Gelfand pairs consisting of finite groups but also to produce examples of Gelfand pairs consisting of profinite groups. Namely for G the pair $(\overline{G}, \overline{P}_n)$ where the bar denotes the closure in Aut T is such an example.

The idea to use group actions on rooted trees to study Gelfand pairs is very fruitful. It allows not only construction of new examples of such pairs, but also clarifies the classical results by presenting them in a new angle, as it was done in [CSSFa, CSSFb].

A part of the process of studying the quasiregular representations (finite dimensional or infinite dimensional), looking for Gelfand pairs and understanding the dynamics of a group acting on rooted trees is the investigation of the action on level n of the stabilizer $st_M(u_n)$, where u_n is a vertex at level n. For instance, in case of G the action of $st_G(u_n)$ has $n + 1$ orbits consisting of $1, 1, 2, 2^2, \ldots, 2^{n-1}$ points and these orbits are depicted in Figure 10.1.

An important class of representations arises from actions of groups by measure preserving transformations. For any group M acting on a rooted tree T one can consider the dynamical system $\mathcal{D} = (M, \partial T, \nu)$ where ν is uniform measure on the boundary ∂T of the tree (i.e. ν is the product $\bigotimes_{n=1}^{\infty} \nu_n$ where ν_n is the uniform measure $\{\frac{1}{m_n}, \ldots, \frac{1}{m_n}\}$ supported on a set of cardinality m_n).

The corresponding unitary representation ρ in the Hilbert space $\mathcal{H} = L^2(\partial T, \nu)$ given by

$$(\rho(g)f)(x) = f(g^{-1}x)$$

is a sum of finite dimensional representations. Indeed, let $\mathcal{H}_n$ be the subspace spanned by the characteristic functions of cylinder sets of rank n (each such set

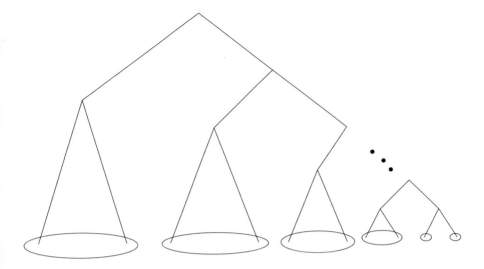

FIGURE 10.1. Orbits of the action of $st_G(1^n)$ on n-th level.

consists of those $\xi \in \partial T$ that pass through some fixed vertex at level n). Then $\mathcal{H}_n$, $n = 0, 1, 2, \ldots$ are $\rho(M)$ invariant subspaces and $\mathcal{H}_n$ naturally embeds into $\mathcal{H}_{n+1}$.

Let $\mathcal{H}_n^\perp$ be the corresponding orthogonal complement to H_n in H_{n+1}, $n = 0, 1, \ldots$. Then

$$\mathcal{H} = \mathbb{C} \oplus \bigoplus_{n=0}^{\infty} \mathcal{H}_n^\perp$$

and

$$\rho = 1_\mathbb{C} \oplus \bigoplus_{n=0}^{\infty} \rho_n^\perp \tag{10.1}$$

where $\rho_n^\perp$ is the restriction of ρ on $\mathcal{H}_n^\perp$. The finite dimensional representation $\rho_n = \rho|_{\mathcal{H}_n}$ allows a decomposition

$$\rho_n = 1_C \oplus \bigoplus_{i=0}^{n-1} \rho_i^\perp$$

and is isomorphic to the permutational representation π_n, which in turn is isomorphic to the quasiregular representation λ_{M/P_n}. There is a big difference between the representations λ_{M/P_ξ}, $\xi \in \partial T$ and ρ, but they are both approximated by the same sequence of finite dimensional representations ρ_n.

Problem 10.2. *Is there an algorithm which, given a finite automaton, finds a decomposition of the representation ρ_n into irreducible components?*

The following question may be useful as steps in solving the above problem.

Problem 10.3. a) *Is there an algorithm which, given an automaton A, describes the orbits of the action of G(A) on the levels of the corresponding tree?*

b) *Is there an algorithm which, given an automaton A, describes the orbits of the action of the stabilizer $P_n = st_{G(A)}(u_n)$ (where u_n the rightmost vertex at level n) on the n-th level, $n = 1, 2, \ldots$?*

Problem 10.4. a) *Is there an algorithm which, given a finite automaton A, finds all irreducible representations of the finite quotients $M_n = M/st_M(n)$, $n = 1, 2, \ldots$, where $M = G(A)$?*

b) *Find all irreducible representations of the finite quotients of G.*

The solution of Problem 10.4 b) would give a complete description of the irreducible representations of the profinite completion $\widehat{G}$ since all such representations are in natural bijection with the irreducible representations of the groups G_n, $n \geq 1$. This follows from the following three facts about G and $\widehat{G}$:

(i) congruence subgroup property for G,
(ii) just infiniteness of $\widehat{G}$,
(iii) absence of faithful finite dimensional representations for G.

The last fact follows from the following observation obtained jointly with T. Delzant.

Theorem 10.1. *Let M be a branch group. Then M is not linear (i.e. it does not have a faithful finite dimensional representation over any field).*

The study of representations over fields different from $\mathbb{C}$ (in particular modular representations) of the group G (and other self-similar groups and groups acting on trees) is important for many topics in algebra. The first results in this direction are obtained in [Sid97] where an infinite dimensional irreducible representation of the Gupta–Sidki 3-group over a field of characteristic 3 is constructed. It is not known if such a representation exists for G (replace the characteristic 3 by 2).

Problem 10.5. *Does the group G have an irreducible representation over a field of characteristic 2?*

This question as well as the next one is related to Kaplansky Conjecture on Jacobson radical [Kap70] (see also [Pas98]). Recall that the Jacobson radical is the intersection of all maximal left ideals of the group algebra over the given field.

Problem 10.6. *Is it correct that the fundamental ideal $w(\mathbb{F}_2[G])$ of the group algebra $\mathbb{F}_2[G]$ of G over the field $\mathbb{F}_2$ of two elements coincides with the Jacobson radical $J(\mathbb{F}_2[G])$ of this algebra?*

Positive answer to this question would give a counterexample to Kaplansky Conjecture. A theorem of Lichtman [Lih63] states:

Let M be a finitely generated group and let $\mathbb{K}$ be a filed of positive characteristic p. If $T(\mathbb{K}[G]) = w(\mathbb{K}[G])$ then

(i) M is a p-group.

(ii) If $M \neq 1$, then $M \neq M'$.

(iii) If M is an infinite group, then M has an infinite, residually finite homomorphic image.

(iv) Any maximal subgroup of M is normal of index p.

(v) If H is a subgroup of M of infinite index, then there exists a chain of subgroups $M = M_0 > M_1 > M_2 > \cdots > H$ with $M_{i+1} \triangleright M_i$ and $|M_{i+1}/M_i| = p$.

The group G satisfies these condition with $p = 2$. The properties listed in (i), (ii), (iii) hold because G is an infinite residually finite 2-group. The property (iv) is established by E. Pervova [Per00] and (v) follows from a generalization of the result of Pervova on subgroups of finite index given in [GW03a]. Indeed, let H be a subgroup of infinite index. Then H is contained in a maximal subgroup G_1 of G which has index 2 by (iv). Now apply the same argument to the pair (G_1, H) etc. The above discussion shows that the group G is a candidate for a counterexample to the Kaplansky Conjecture.

11. C^*-algebras

One can naturally associate at least three C^*-algebras $C^*_\rho(M)$, $C^*_{M/P}(M)$, $C^*_r(M)$ to a group M acting on a rooted tree. These algebras are defined as (norm) closure of the corresponding unitary representations ρ, $\lambda_{M/P}$ or λ_M of the group algebra $\mathbb{C}[M]$. The algebra $C^*_r(M)$ is called reduced C^*-algebra of the group M.

It is easy to check that, for the corresponding norms $\| \ \|_\rho$, $\| \ \|_{M/P}$, the inequality

$$\|x\|_\rho \geq \|x\|_{M/P}$$

holds for any element $x \in \mathbb{C}[M]$. Therefore there is a canonical $*$-homomorphism $\varphi \colon C^*_\rho \to C^*_{M/P}$. In case the pair $(M, M/P)$ is amenable φ is an isomorphism [BG00a, Nek05]. In case M is amenable group $C^*_\rho(M)$ and $C^*_{M/P}(M)$ are quotients of $C^*_r(M)$ (because any unitary representation of amenable group is weakly contained in the regular representation [Gre69]).

Reduced C^*-algebras play a fundamental role in the theory of C^*-algebras and in the study of unitary representations of groups. Unfortunately very little is known about them. One of rare results is the simplicity of C^*_r in case of a free group of finite or countable rank and in case of groups having the so-called Powers property [Pow75], [BCdlH94].

The study of the algebras $C^*_{M/P}(M)$ is interesting not only because of the relations to quasiregular representations but also because of their use in the study of spectral properties of regular graphs which will be discussed in the next section. We concentrate our attention here on $C^*_\rho(M)$.

Proposition 11.1. $C^*_\rho(M)$ is residually finite dimensional algebra (i.e. embeds into $\bigoplus_{i=0}^{\infty} M_{n_i}(\mathbb{C})$ for some sequence n_i of integers [BL00]).

This follows from the decomposition (10.1).

Just infinite groups play important role in group theory [Gri00b, BGŠ03]. Perhaps a similar role should belong to just infinite-dimensional C^*-algebras.

Problem 11.1. *Let M be a self-similar just-infinite group.*

(a) *Is it correct that $C_\rho^*(M)$ has only finite-dimensional proper images?*

(b) *Answer the above question for G.*

In case of the group G the representations $\rho_n^\perp$ (of dimension 2^n) are irreducible [BG00a] and the embedding

$$C_\rho^*(G) \underset{\theta}{\hookrightarrow} \mathbb{C} \oplus \bigoplus_{n=0}^{\infty} M_{2^i}(\mathbb{C}) \tag{11.1}$$

holds with the image being a subdirect sum (its projection on each summand is onto). We are going to describe the embedding more precisely and to show that the algebra $C_\rho^*(G)$ is self-similar in the sense that the embedding

$$\psi: \; G \hookrightarrow G \wr S_2$$

from (3.4) has a natural extension to the embedding

$$\widetilde{\psi}: \; C_\rho^*(G) \hookrightarrow M_2(C_\rho^*(G)).$$

To see this let us mention that the partition $[0,1] = [0,\frac{1}{2}] \cup [\frac{1}{2}, 1]$ (or $\partial T = \partial T_0 \sqcup \partial T_1$ where T_0, T_1 are subtrees with a root at first level) leads to the decomposition $H = H_0 \oplus H_1$ where

$$H = L^2([0,1), m)$$

$$H_0 = L^2\left(\left[0, \frac{1}{2}\right], m_0\right)$$

$$H_1 = L^2\left(\left[\frac{1}{2}, 1\right], m_1\right),$$

and m, m_0, m_1 are restrictions of the Lebesgue measure on corresponding intervals. The natural isomorphisms $H \simeq H_i$, $i = 0, 1$ allow to represent an operator in H by a 2×2 operator matrix. This leads to the following recursive relations for the generators of G:

$$\rho(a) = \begin{pmatrix} 0 & 1 \\ 1 & 0 \end{pmatrix}, \quad \rho(b) = \begin{pmatrix} \rho(a) & 0 \\ 0 & \rho(c) \end{pmatrix},$$

$$\rho(c) = \begin{pmatrix} \rho(a) & 0 \\ 0 & \rho(d) \end{pmatrix}, \quad \rho(d) = \begin{pmatrix} 1 & 0 \\ 0 & \rho(b) \end{pmatrix}. \tag{11.2}$$

Similar recursions hold for the matrices $\rho_n(a)$, $\rho_n(b)$, $\rho_n(c)$, $\rho_n(d)$, $n = 1, 2, \ldots$. For $n = 0$

$$\rho_0(a) = \rho_0(b) = \rho(c) = \rho_0(d) = 1$$

where 1 is the scalar unity and for $n \geq 1$

$$\rho_n(a) = \begin{pmatrix} 0_{n-1} & 1_{n-1} \\ 1_{n-1} & 0_{n-1} \end{pmatrix}, \quad \rho_n(b) = \begin{pmatrix} \rho_{n-1}(a) & 0_{n-1} \\ 0_{n-1} & \rho_{n-1}(c) \end{pmatrix}$$

$$\rho_n(c) = \begin{pmatrix} \rho_{n-1}(a) & 0_{n-1} \\ 0_{n-1} & \rho_{n-1}(d) \end{pmatrix}, \quad \rho_n(d) = \begin{pmatrix} 1_{n-1} & 0_{n-1} \\ 0_{n-1} & \rho_{n-1}(b) \end{pmatrix}$$

(11.3)

where 0_n and 1_n are respectively a zero matrix and identity matrix of order 2^n.

The representation ρ_n is a subrepresentation of ρ_{n+1} because H_n naturally embeds in H_{n+1} as it was mentioned above. To a function f on level n one can associate a function $\tilde{f}$ on level $(n+1)$ given by

$$\tilde{f}(u0) = \tilde{f}(u1) = f(u),$$

where u is any vertex of level n. The complement of H_n in H_{n+1} consists of a function f which satisfies

$$f(u0) = -f(u1),$$

$|u| = n$. This shows that the restrictions of ρ_{n+1} on $\mathcal{H}_n^{\perp}$ (which we have denoted $\rho_n^{\perp}$) satisfy, in an appropriate basis, relations similar to those in (11.3)

$$\rho_0(a) = -1, \quad \rho_0(b) = \rho_0(c) = \rho_0(d) = 1$$

and for $n \geq 1$

$$\rho_n^{\perp}(a) = \begin{pmatrix} 0_{n-1} & 1_{n-1} \\ 1_{n-1} & 0_{n-1} \end{pmatrix}, \qquad \rho_n^{\perp}(b) = \begin{pmatrix} \rho_{n-1}^{\perp}(a) & 0_{n-1} \\ 0_{n-1} & \rho_{n-1}^{\perp}(c) \end{pmatrix}$$

$$\rho^{\perp}(c) = \begin{pmatrix} \rho_{n-1}^{\perp}(a) & 0_{n-1} \\ 0_{n-1} & \rho_{n-1}^{\perp}(d) \end{pmatrix}, \qquad \rho_n^{\perp}(d) = \begin{pmatrix} 1_{n-1} & 0_{n-1} \\ 0_{n-1} & \rho_{n-1}^{\perp}(b) \end{pmatrix}.$$

(11.4)

The relations (11.2) can be interpreted as ψ-images of operators $\rho(a)$, $\rho(b)$, $\rho(c)$, $\rho(d)$. They extend to any operator $\rho(x)$, $x \in \mathbb{C}[G]$ and hence to any element of C_{ρ}.

The relations (11.4) can be used for construction of matrices corresponding to the generators a, b, c, d in the representations $\rho_n^{\perp}$, $n = 1, 2, \dots$. Namely for appropriate choices of bases in $H_n^{\perp}$, $n \geq 1$, the elements $\theta(\rho(s))$, $s \in \{a, b, c, d\}$, have the form:

$$\theta(\rho(a)) = \bigoplus_{n=-1}^{\infty} a_n, \quad a_{-1} = 1, a_1 = -1$$

$$a_n = \begin{pmatrix} 0 & 1_{n-1} \\ 1_{n-1} & 0 \end{pmatrix}, \qquad 1_{n-1} - \text{identity matrix of size } 2^{n-1}$$

$$\theta(\rho(b)) = \bigoplus_{n=1}^{\infty} b_n, \quad b_{-1} = b_0 = 1, \qquad b_n = \begin{pmatrix} a_{n-1} & 0 \\ 0 & c_{n-1} \end{pmatrix},$$

$$\theta(\rho(c)) = \bigoplus_{n=-1}^{\infty} c_n, \quad c_{-1} = c_0 = 1, \qquad c_n = \begin{pmatrix} a_{n-1} & 0 \\ 0 & d_{n-1} \end{pmatrix},$$

$$\psi(\rho(d)) = \bigoplus_{n=-1}^{\infty} d_n, \quad d_{-1} = d_0 = 1, \qquad d_n = \begin{pmatrix} 1_{n-1} & 0 \\ 0 & b_{n-1} \end{pmatrix}.$$

For instance

$$C_4 = \left(\begin{array}{cccc:cccc}
0 & 0 & 0 & 0 & 1 & 0 & 0 & 0 \\
0 & 0 & 0 & 0 & 0 & 1 & 0 & 0 \\
0 & 0 & 0 & 0 & 0 & 0 & 1 & 0 \\
0 & 0 & 0 & 0 & 0 & 0 & 0 & 1 \\
\hdashline
1 & 0 & 0 & 0 & 0 & 0 & 0 & 0 \\
0 & 1 & 0 & 0 & 0 & 0 & 0 & 0 \\
0 & 0 & 1 & 0 & 0 & 0 & 0 & 0 \\
0 & 0 & 0 & 1 & 0 & 0 & 0 & 0
\end{array}
\quad \begin{array}{c} \bigcirc \end{array}\right.$$

The elements $\theta(\rho(s))$, $s \in \{a, b, c, d\}$ generate $C_\rho^*(G)$. Even though the generators of $C_\rho^*(G)$ are described explicitly, it is not easy to study the properties of the algebra and at the moment very little is known about $C_\rho^*(G)$.

In a similar way an embedding

$$\psi: \ L \hookrightarrow L \wr S_d$$

of self-similar group L given by relations of the type (3.10) can be extended to an embedding

$$\tilde{\psi}: \ C_\rho^*(L) \hookrightarrow M_d(C_\rho^*(L))$$

and a matrix and operator recursions of the type similar to (11.2), (11.3) hold. The above construction leads to a large family of self-similar algebras.

Any residually finite dimensional algebra embedded in $\bigoplus\limits_{i=1}^{\infty} M_{n_i}(\mathbb{C})$ has a faithful trace of the form

$$\mathrm{tr}\ x = \sum_{i=1}^{\infty} \frac{1}{2^{n_i}} \mathrm{tr}\ x_i.$$

Unfortunately, in our situation this trace does not agree with the embedding $\tilde{\psi}$. Having a self-similar algebra C^* and an embedding

$$\xi: \ C^* \to M_d(C^*), \quad \xi(c) = (c_{ij})_{i,j=1}^{d}.$$

it is natural to work with a trace τ such that

$$\tau(\xi(c)) = \frac{1}{d} \sum_{i=1}^{d} \tau(c_{ii})$$

for any $c \in C^*$.

If M is a group generated by a finite automaton over an alphabet on d letters then the space $\mathcal{H}_n$ has dimension d^n and the limit

$$\lim_{n \to \infty} \frac{1}{d^n} Tr_n \rho_n(g) \tag{11.5}$$

exists for any element $g \in M$, where Tr_n is the standard (not normalized) trace on matrix algebra $M_{d^n}(\mathbb{C})$. The value $\tau(g)$ of this limit has the properties of a trace and can be extended to $\mathbb{C}[M]$ and to $C_\rho^*(M)$. The obtained trace agrees with the embedding ψ.

The trace τ is group-like if $\tau(g) = 1$ for $g = 1$ and $\tau(g) = 0$, for $g \neq 1$.

Problem 11.2. *Let τ be a trace on $C_\rho^*(M)$ defined by (11.5) where M is a group generated by finite automaton.*

(a) *Is τ unique normalized trace that agrees with the embedding $\psi: C_\rho^* \to M_d(C_\rho^*)$ given by the automaton structure'*

(b) *Under which conditions τ is faithful?*

(c) *Under which conditions τ is group-like trace?*

The values $\tau(g)$, $g \in M$ satisfy the system of equations

$$\tau(g) = \sum_{i=1}^{d} \tau(g_{ii}) \tag{11.6}$$

where

$$\psi(g) = (g_{ij})_{i,j=1}^{d}$$

and the matrix elements g_{ij} belong to the set $0 \cup M$. If g_{ij} is nonzero, then it is an element of the group M of length not greater then the length of g. This shows that the system S of equations (11.6), where g runs over M, splits as union $\bigcup_{n=1}^{\infty} S_n$ of finite systems of equations where S_n consist only of the equations with $|g| \leq n$ (the length is considered with respect to the generating set of M given by the set of the states of the automaton). The question 11.2 (a) becomes a question on uniqueness of the solution of a system of equation. In case M is a contracting group everything reduces to the system (11.6) with g running through the core. For instance for the group G we get the system

$$\begin{cases} \tau(a) = 0 \\ \tau(b) = \dfrac{1}{2}(\tau(a) + \tau(c)) \\ \tau(c) = \dfrac{1}{2}(\tau(a) + \tau(d)) \\ \tau(d) = \dfrac{1}{2}(\tau(1) + \tau(b)) \end{cases}$$

with unique solution $\tau(b) = \frac{6}{7}$, $\tau(c) = \frac{5}{7}$, $\tau(d) = \frac{3}{7}$ and $\tau(g)$ can be effectively computed for any element $g \in G$ (by use of core portraits).

Orthogonal to the contracting case is the class of automatons determining group-like trace. A condition which implies such a situation is a local nontriviality of the action of M on rooted tree T_d by which we mean the triviality of the kernel of any homomorphism

$$st_M(u) \longrightarrow st_M(u)|_{T_u}$$

given by restriction of $st_M(u)$ on the subtree with a root at u. In other words if an element fixes a vertex u and acts trivially on the subtree T_u then it acts trivially on the whole tree.

Example of such actions are given by automata from Figures 1.5 and 1.6 [GŻ01].

In case τ is group-like trace on C_ρ the factor algebra $C_\rho/\ker \tau$, where

$$\ker \tau = \{x \colon \tau(xx^*) = 0\}$$

is isomorphic to $C_r^*(G)$.

Problem 11.3. (a) *Is there an algorithm which, given an automaton, checks if the action of corresponding group is locally nontrivial?*
(b) *Give an example of automaton for which the trace τ is group-like and faithful.*

Perhaps the lamplighter group given by automaton from Figure 1.6 answers positively the question 11.3(b). At the same time the trace determined by the automaton given by Figure 1.5 is not faithful. This follows from the fact that reduced C^*-algebra $C_r^*(F_m)$ of a free group of rank $m \geq 2$ is simple.

12. Schreier graphs

Schreier graphs associated to a group acting on a rooted tree are very interesting combinatorial objects.

Given a group L, a subgroup $H < L$ and a finite system of generators S of M the Schreier graph $\Gamma(M, H, S)$ consists of vertices represented by cosets gH, $g \in M$ and of edges of the form (gH, sgH) where $s \in S \cup S^{-1}$. To complete the definition one should consider them as oriented graphs with edge labelling by the elements of $S \cup S^{-1}$, but this is often ignored. Schreier graphs are k-regular graphs (viewed as non-oriented graphs) where the degree k of each vertex is equal to $2|S|$ (loops contribute 2 to the degree). Every k-regular graph with even degree $k = 2m$ can be realized as a Schreier graph of the form $\Gamma(F_m, H, S)$ where F_m is a free group of rank m and S is a basis of F_m [dlH00].

Let L be a group generated by a finite automaton over an alphabet on d letters. Then M naturally acts on d-regular rooted tree $T = T_d$ as was described in Section 1 and for any point of the boundary $\xi \in \partial T$ (represented by geodesic ray joining the root with infinity) one can consider the Schreier graphs Γ, and Γ_n, $n = 1, \ldots$, where

$$\Gamma = \Gamma(L, P, S),$$
$$\Gamma_n = \Gamma(L, P_n, S),$$

$P = st_L(\xi)$, $P_n = st_L(u_n)$ and u_n is the unique vertex at level n belonging to the path ξ_n. The graphs Γ_n are finite because P_n has finite index in M and Γ is infinite graph if the M-orbit of ξ is infinite (this holds for instance when the action is level transitive). As $P = \bigcap_{n=1}^{\infty} P_n$ and as $\{P_n\}$ is descending sequence, the graph Γ is the limit of the sequence $\{\Gamma_n\}$ in the sense of topology on the space of graphs considered in [GŻ99] (which is analogous to the topology discussed in Section 2). More precisely

$$(\Gamma, v) = \lim_{n \to \infty} (\Gamma_n, v_n) \tag{12.1}$$

where $v = P$, $v_n = P_n$ are reference points and convergence in the space of marked graphs (i.e. graphs with reference point) means that the neighborhoods of v_n of radius R in Γ_n stabilize to the corresponding neighborhood in Γ when $n \to \infty$.

The graph Γ and the approximating sequence $\{\Gamma_n\}$ keep important information about the group. Recent investigations show that the study of the asymptotic properties of Γ and $\{\Gamma_n\}$ is related to many topics in computer science, combinatorics, geometry, dynamics and other areas of mathematics.

To a k-regular graph Γ one can associate a Markov operator M acting on the Hilbert space $\ell^2(\Gamma)$ of square summable functions defined on the vertices of Γ by

$$(Mf)(x) = \frac{1}{k} \sum_{y \sim x} f(y)$$

where $y \sim x$ is the adjacency relation. M is a self-adjoint contraction and the spectrum $sp\ M \subseteq [-1, 1]$. If Γ is finite connected graph then 1 is simple eigenvalue of M. If Γ is infinite connected graph then $1 \in sp\ M$ if and only if Γ is amenable graph [CSGdlH99].

Using the spectral decomposition

$$M = \int\limits_{-1}^{1} \lambda dE(\lambda)$$

(where $E(\lambda)$ is a family of orthoprojectors) one can define for each vertex u the spectral function $\sigma_u(\lambda) = \langle E(\lambda)\delta_u, \delta_u \rangle$ where δ_u is delta function and $\langle\ ,\ \rangle$ denotes the standard scalar product in the Hilbert space $\ell^2(\Gamma)$. Let μ_n be the corresponding measure $(\mu([a,b]) = \sigma_n(a+) - \sigma_u(b))$. The observation made in [GŻ99] shows that μ_u^Γ is a (vector-valued) continuous function on the space of marked k-regular graphs.

In particular, the relation

$$\mu_v^\Gamma = \lim_{n \to \infty} \mu_{v_n}^{\Gamma_n}$$

is a consequence of (12.1). The next three questions are the most important in spectral theory of graphs:

(a) What is the norm (i.e. the spectral radius) of operator M?
(b) What is the spectrum $sp\ M$ (as a subset of $[-1, 1]$)?
(c) What can be said about the spectral measures μ_v? In particular, how μ_v decomposes in absolutely continuous, singular and discrete parts?

There are very few examples of computation of spectra of infinite Schreier graphs given in [BG00a], [GŻ01], [Sun05].

In case Γ is transitive graph (i.e. Aut Γ acts transitively on the set of vertices) the measures μ does not depend on v. This holds, for instance, when P is trivial subgroup and hence Γ is the Cayley graph in this case.

The spectral measure μ_v has moments

$$P_{v,v}^{(n)} = \int\limits_{-1}^{1} \lambda^n d\mu_v$$

equal to probabilities of return to v for a simple random walk on Γ with beginning in v. It is clear that in a non-transitive case this probabilities depends on v, therefore the measure μ_v depends on v in general.

In our situation one can attach to Γ a measure μ_* which is defined as

$$\mu_* = \lim_{n \to \infty} \mu_n \qquad (12.2)$$

where μ_n is a counting (or cumulative) measure on Γ_n which counts the ratio of the number of eigenvalues (including multiplicities) of the Markov operator on M_n which belong to a given interval and the total number of eigenvalues (i.e. the size of Γ_n).

The limit (12.2) exist for any covering system of graphs (in our case Γ_{n+1} covers Γ_n because P_{n+1} is a subgroup of P_n). This follows from one theorem of Serre [Ser97] (see [BG00a, GŻ04]). The measure μ_* was named KNS-spectral measure (Kesten–von Neumann–Serre) in [GŻ04]. The same measure was named empiric measure in [BG00a] where the first calculations of μ_* appeared. An open question is the question on the relation between the supports of the measures μ_* and μ_v, $v \in V(\Gamma)$. The union of supp μ_v gives the spectrum of M as a set. Another question that can be asked is to find conditions under which the support of μ_* coincides with $sp\, M$? Some results in this direction are provided in [GŻ99].

The graphs Γ and $\{\Gamma_n\}$ associated with the group G, or with the Gupta–Sidki 3-group and in some other cases are substitutional graphs, i.e. there is a substitutional rule which allows us to construct Γ_{n+1} from Γ_n by application of substitutions. A formal definition can be found in [BG00a], along with other references. Without getting into the details let us demonstrate this on the example of G.

The graph Γ_1 in this case has the form

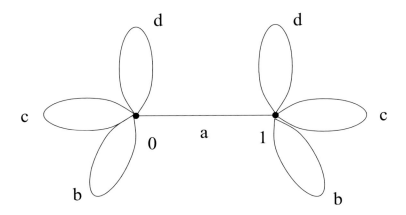

and the substitutional rule is

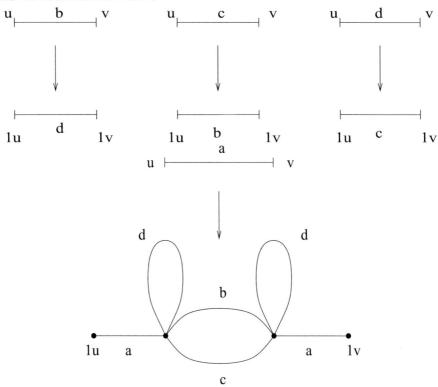

For instance Γ_2 and Γ_3 look like

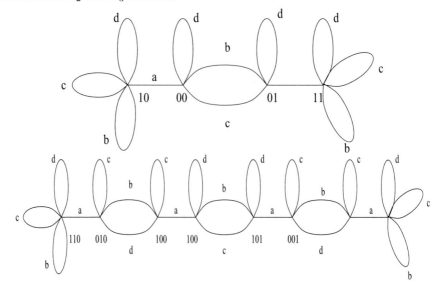

The infinite graph Γ looks as

if the sequence $\xi \in \partial T$ (determining the subgroup $P = st_G(\xi)$) is different from $1\,1\cdots 1\cdots$ and looks as

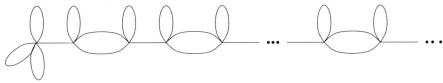

in the opposite case (we draw Γ without labelling of edges).

Problem 12.1. *For which automata the sequence $\{\Gamma_n\}$ of associated Schreier graphs is substitutional (i.e. can be described by a substitutional rule)?*

Important characteristics of a graph are its diameter and the girth (i.e. the size of the shortest nontrivial cycle).

Problem 12.2. *Describe all possible asymptotic behaviors of diameters and girths in graphs of the form $\Gamma_n = \Gamma(L, P_n, S)$, $n = 1, 2, \ldots$ where L is a self-similar group.*

Problem.   a) *Describe all possible rates of growth of graphs $\Gamma = \Gamma(L, P, S)$ where L is a self-similar group.*
  b) *Describe the set of all degrees of polynomial growth of graphs of type $\Gamma = \Gamma(L, P, S)$, where L is a self-similar group.*

In case of the group G the diameters of Γ_n grow exponentially while the growth of limit graph Γ is linear. In case of Gupta–Sidki 3-group or Fabrikowskii–Gupta group the sequence $\{\Gamma_n\}$ also is substitutional, the diameters grow exponentially but the degree of polynomial growth is $\log_2 3$ [BG00a]. Indeed the graph $\Gamma = \Gamma(L, P, S)$ always has polynomial growth in case the group L is contracting [BG02a], [BGŠ03]. An approach to calculation of the degree of polynomial growth of graphs γ of the above type and of diameters of the corresponding sequences $\{\Gamma_n\}$ is recently found by Bondarenko and Nekrashevych.

An interesting example of an automaton (and a group) determining a graph Γ of intermediate growth is constructed in [BCSN]. For the Basilica group the graph Γ is shown in Figure 12.1, which particularly explains the name of the group (more on this in [Kai]).

There are many examples when the graph Γ grows exponentially. For instance, the two-state automaton given by Figure 1.6 determines the Cayley graph of the lamplighter group for almost every point $\xi \in \partial T$ (with respect to uniform

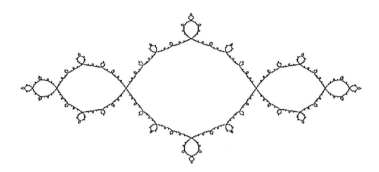

FIGURE 12.1. Graph Γ for Basilica group

measure) and hence has the exponential growth. Indeed Γ has exponential growth for any point $\xi \in \partial T$ since P is either trivial or cyclic [GŻ01].

Let X_n be a sequence of finite connected k-regular graphs and $|X_n| \to \infty$. Let λ_n be the second largest eigenvalue (after 1) of the Markov operator on X_n. Then the sequence $\{X_n\}_{n-1}^{\infty}$ is called a sequence of expanders if there exists $\delta > 0$ such that $\lambda_n \leq 1 - \delta$, $n = 1, 2, \ldots$. The bigger δ is the better expanding properties the family of graphs $\{X_n\}$ has . But δ can not be larger than $1 - \frac{2\sqrt{k-1}}{k}$ (this follows from Alon–Boppana Theorem [Lub94, GŻ99]).

When the absolute values of the second largest and the second smallest eigenvalues of the Markov operator on a connected k-regular graph are not greater than $\frac{2\sqrt{k-1}}{k}$ the graph is called Ramanujan graph. Expanders and Ramanujan play important roles in many topics including rather concrete applications (for more on this see in [Lub94, LPS88]). However, it is difficult to give explicit constructions and currently there are very few conceptually different constructions.

There are indications that sequences of the form $\{\Gamma_n\}$ constructed with the help of some particular finite automata are sequences of expanders (and perhaps in some cases even of Ramanujan graphs). For instance this holds for the automaton given by Figure 1.5 (computer experiments) with a rigorous proof of a weaker fact (namely that they are the so called asymptotic Ramanujan graphs).

By the Chung inequality [Ter99, Chu97] the diameter D of a finite connected graph X can be estimated using the second largest eigenvalue μ:

$$D \leq \frac{\log |X| - 1}{-\log \mu} + 1.$$

Thus if the group $L = G(A)$ generated by a finite automaton A acts transitively on levels and determines a sequence $\{\Gamma_n\}_{n=1}^{\infty}$ of expanders then the corresponding sequence of diameters grows linearly, since $|\Gamma_n| = d^n$ in this case (d is the cardinality of the alphabet).

Problem 12.3.   a) *Find an automaton producing a sequence $\{\Gamma_n\}_{n=1}^{\infty}$ of expand-ers.*

b) *Find an automaton producing a sequence $\{\Gamma_n\}_{n=1}^{\infty}$ of Ramanujan graphs.*

Problem 12.4. *Is there an algorithm which, given an automaton, determines the type of growth (polynomial, intermediate or exponential) of the diameters of Γ_n?*

13. Spectra and rational maps

Given a finite automaton A one can associate with it at least three spectra: the spectrum of the Cayley graph $\Gamma(G(A), Q)$ of the group $G(A)$ generated by the automaton A with respect to the system of generators given by the set of states, the spectrum of the Schreier graph $\Gamma(G(A), P, Q)$ where P is a stabilizer of a point on the boundary of the tree T_d (d-cardinality of the alphabet) and finally the spectrum of the Hecke type operator $\mathcal{M}$ in $L^2(\partial T, \nu)$ defined by

$$(\mathcal{M}f)(x) = \frac{1}{2|Q|} \sum_{q \in Q} (\rho(q) + \rho(q^{-1})) f(x)$$

where ρ is the representation described in Section 10.

As ρ is a direct sum of finite dimensional representations $\mathcal{M}$ has pure point spectrum. At the same time for most known examples the spectrum of the Markov operator on an infinite regular graph is continuous or at least is very far from discrete. However, there are exceptions, for instance the lamplighter group represented by automaton given by Figure 1.6 for which the spectrum is discrete.

The three defined spectra correspond to representations λ_G, $\lambda_{G/P}$ and ρ. If $\Gamma = \Gamma(G(A), P, Q)$ is amenable then its spectrum coincides with the spectrum of $\mathcal{M}$ [BG00a]. In general only the inclusion $sp \, M_\Gamma \supset sp \, \mathcal{M}$ holds. This follows from the existence of a surjective $*$-homomorphism $C_\rho^* \to C_{\lambda_{G/P}}^*$ discussed in Section 11. The computation of the spectrum of $\mathcal{M}$ is based on the relation

$$sp \, \mathcal{M} = \overline{\bigcup_{n=1}^{\infty} sp \, \mathcal{M}_n} \tag{13.1}$$

where $\mathcal{M}_n = \mathcal{M}|_{\mathcal{H}_n}$ and $\mathcal{H}_n$ is the space of dimension d^n defined in Section 10. The operator $\mathcal{M}_n$ is similar to the Markov operator on the finite Schreier graph $\Gamma_n = \Gamma(G(A), P_n, Q)$. The union (13.1) is increasing union since Γ_{n+1} covers Γ_n.

The computation of $sp \, \mathcal{M}$ reduces to computation of spectra of finite matrices defined by a recursion given by the diagram of A, as shown by example in Section 11. There is no general method which allows us to solve this problem efficiently. However, in some cases explicit computation is possible. Let us demonstrate this with a few examples.

For the group G we have the recursion (11.3) which can be rewritten in the form

$$a_0 = b_0 = c_0 = d_0 = 1$$

$$a_n = \begin{pmatrix} 0 & 1 \\ 1 & 0 \end{pmatrix}, \quad b_n = \begin{pmatrix} a_{n-1} & 0 \\ 0 & c_{n-1} \end{pmatrix}$$

$$c_n = \begin{pmatrix} a_{n-1} & 0 \\ 0 & d_{n-1} \end{pmatrix}, \quad d_n = \begin{pmatrix} 1 & 0 \\ 0 & b_{n-1} \end{pmatrix},$$

where we suppress ρ as well as the dimensions of the identity and zero matrices. Then

$$\mathcal{M}_n = a_n + b_n + c_n + d_n = \begin{pmatrix} 2a_{n-1} + 1 & 1 \\ 1 & \mathcal{M}_{n-1} - a_{n-1} \end{pmatrix}$$

and it is not clear how this relation could be used to get a recursion for the spectrum. But, surprisingly, the problem can be solved if instead of considering the standard spectral problem of inversion of the matrix $\mathcal{M}_n - \lambda I_n$ one considers the problem of inversion of the pencil of matrices

$$R_n^{(1)}(\lambda, \mu) = -\lambda a_n + b_n + c_n + d_n - (\mu + 1)I_n$$
$$= \mathcal{M}_n - (\lambda + 1)a_n - (\mu + 1)I_n$$

(the coefficients $-(\lambda + 1)$ and $\mu + 1$ are chosen instead of λ and μ because of simplification in the resulting formulas). The next recurrence for the determinant $|R_n^{(1)}(\lambda, \mu)|$ is the first step in the calculation of the spectrum.

Lemma 13.1 ([BG00a]). *For $n \geq 2$ we have*

$$|R_n^{(1)}(\lambda, \mu)| = (4 - \mu^2)^{2^{n-2}} \left| R_{n-1}^{(1)} \left(\frac{2\lambda^2}{4 - \mu^2}, \mu + \frac{\mu\lambda^2}{4 - \mu^2} \right) \right|. \qquad (13.2)$$

Proof. Observe that as $a_n^2 = 1$ (1 stands for the identity matrix)

$$(2a_{n-1} - \mu)(2a_{n-1} + \mu) = 4 - \mu^2$$

and

$$|2a_{n-1} - \mu| = \begin{vmatrix} -\mu & 2 \\ 2 & -\mu \end{vmatrix}_{2^{n-1}} = (\mu^2 - 4)^{2^{n-2}}.$$

Therefore

$$|R_n^{(1)}(\lambda, \mu)| = \begin{vmatrix} 2a_{n-1} - \mu & -\lambda \\ -\lambda & \mathcal{M}_{n-1} - a_{n-1} - (\mu + 1) \end{vmatrix}$$

$$= \begin{vmatrix} 2a_{n-1} - \mu & -\lambda \\ 0 & \mathcal{M}_{n-1} - a_{n-1} - (\mu + 1) - \frac{\lambda^2}{4 - mu^2}(2a_{n-1} + \mu) \end{vmatrix}$$

$$= |2a_{n-1} - \mu| \cdot \left| \mathcal{M}_{n-1} - \left(1 + \frac{2\lambda^2}{4 - \mu^2}\right)a_{n-1} - \left(1 + \mu + \frac{\mu\lambda^2}{4 - \mu^2}\right) \right|,$$

which proves the claim. $\qquad \square$

Using the relation (13.2) $(n-1)$ times we come to the relation

$$|R_n^{(1)}(\lambda, \mu)| = (4 - \mu^2)^{2^{n-2}+2^{n-3}+\cdots+2+1^0} R_1^{(1)}(F_1^{(n-1)}(\lambda, \mu))$$

where $F_1: \mathbb{R}^2 \to \mathbb{R}^2$ is the rational map

$$F_1: \begin{cases} \lambda \longrightarrow \dfrac{2\lambda^2}{4-\mu^2} \\[3mm] \mu \longrightarrow \mu + \dfrac{\mu\lambda^2}{4-\mu^2} \end{cases}$$

and $F^{(n)} = \underbrace{F \circ \cdots \circ F}_{n}$ is the n-fold composition of F. Thus the spectrum Σ_n of the pencil $R_n^{(1)}(\lambda, \mu)$ (i.e. the set of (λ, μ) for which $R_n(\lambda, \mu)$ is not invertible) heavily depends on the dynamics of the map F_1.

If Σ is the closure of $\overset{\infty}{\underset{n=1}{\bigcup}} \Sigma_n$ then the spectrum of $\mathcal{M}$ is the projection on the μ-axis of the intersection of Σ and the line $\lambda = -1$. On the other hand Lemma 13.1 shows that the spectrum Σ is F_1-invariant, i.e $F_1^{-1}(\Sigma) = \Sigma$. We see that at this point the spectral problem meets a problem from dynamics requiring a description of invariant subsets. In some examples arising from self-similar groups these invariant sets have quite complicated form, as we will see later.

Fortunately in case of G the problem of finding invariant sets and the study of the dynamics of F_1 has an easy solution because of the following observation.

Lemma 13.2. *The map F_1 is semi-conjugate to the map $\alpha: x \to 2x^2 - 1$ and the diagram*

$$\begin{array}{ccc} \mathbb{R}^2 & \overset{F_1}{\longrightarrow} & \mathbb{R}^2 \\ \psi \downarrow & & \downarrow \psi \\ \mathbb{R} & \underset{\alpha}{\longrightarrow} & \mathbb{R} \end{array}$$

where $\psi(\lambda, \mu) = \frac{4-\mu^2+\lambda^2}{4\lambda}$ is commutative.

The map $x \to 2x^2 - 1$ is known as Chebyshev–von Neumann–Ulam map. It is conjugate to the map $x \to x^2 - 2$. Lemma 13.2 shows that the map is integrable in certain sense and has invariant family of hyperbolas

$$H_\theta(\lambda, \mu) = 4 - \mu^2 + \lambda^2 + 4\lambda\theta$$

showed in Figure 13.2 (right). The F_1-preimage of H_θ is the union of H_{θ_1}, and H_{θ_2} where θ_1, θ_2 are α-preimages of θ. In other words the following relation holds:

$$H_\theta(F_1(\lambda, \mu)) = H_{\theta_1}(\lambda, \mu) H_{\theta_2}(\lambda, \mu).$$

Moreover, induction allows us to show that the factorization

$$|R_n^{(1)}(\lambda, \mu)| = (2 - \lambda - \mu)(2 + \lambda - \mu) \prod_\theta H_\theta(\lambda, \mu)$$

holds, where the product is taken over

$$\theta \in \bigcup_{i=0}^{n-2} \alpha^{-i}(0).$$

This leads to computation of counting measures μ_n on the graphs Γ_n and KNS spectral measure $\mu = \lim_{n\to\infty} \mu_n$ which in this case is supported on $[-\frac{1}{2}, 0] \cup [\frac{1}{2}, 1]$ and is absolutely continuous with respect to Lebesgue measure:

$$d\mu(x) = \frac{|1 - 4x| \; dx}{2\pi \sqrt{x(2x-1)(2x+1)(1-x)}}$$
$$= \{(\lambda, \mu): \; 4 - \lambda^2 + \mu^2 - \mu\xi = 0\}.$$

Its density is shown in Figure 13.1.

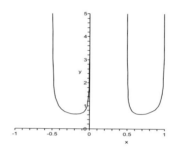

FIGURE 13.1. KNS spectral measure related to G.

There is another invariant family

$$U_\xi = 4 - \lambda^2 + \mu^2 - \mu\xi$$

of hyperbolas transversal to H_θ and shown in Figure 13.2 (left). Indeed not only the family $\{U_\xi\}$ is F_1-invariant but each hyperbola U_ξ is F_1-invariant so in this case the diagram

$$
\begin{array}{ccc}
\mathbb{R}^2 & \xrightarrow{\;F\;} & \mathbb{R}^2 \\
{\scriptstyle\varphi}\big\downarrow & & \big\downarrow{\scriptstyle\varphi} \\
\mathbb{R} & \xrightarrow{\;id\;} & \mathbb{R}
\end{array}
$$

is commutative, where

$$\varphi(\lambda, \mu) = \frac{4 - \lambda^2 + \mu^2}{\mu}.$$

Another example is provided by the lamplighter group L realized as a group generated by the automaton from Figure 1.6. Let u, v be the generators of L given by the states of the automaton and $w = v^{-1}u$. Then w acts on sequences by

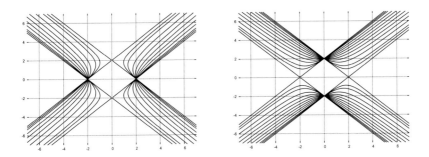

FIGURE 13.2

changing the first symbol and hence w coincides with the transformation a from the previous example. The operator recursion in this case is

$$w = \begin{pmatrix} 0 & 1 \\ 1 & 0 \end{pmatrix}, \quad u = \begin{pmatrix} 0 & u \\ v & 0 \end{pmatrix}, \quad v = \begin{pmatrix} u & 0 \\ 0 & v \end{pmatrix}.$$

Let

$$R^{(2)}(\lambda, \mu) = u + u^{-1} + v + v^{-1} - \lambda - \mu w$$

be the pencil of operators (for the representation ρ in $L^2(\partial T, \nu)$) and let $R_n^{(2)}$ be a finite dimensional approximation given by a restriction of the action on level n. Then [GŻ01]

$$|R_n^{(2)}(\lambda, \mu)| = (\mu - \lambda)^{2^{n-1}} |R_{n-1}^{(2)}(F_2(\lambda, \mu))|$$

if $n \geq 2$, where

$$F_2 : \begin{cases} \lambda \longrightarrow \dfrac{2 - \lambda^2 + \mu^2}{\mu - \lambda} \\ \mu \longrightarrow \dfrac{2}{\lambda - \mu}. \end{cases}$$

The map F_2 is also "integrable", namely it is semi-conjugate to the identity map via the map $\psi(\lambda, \mu) = \lambda + \mu$. The last fact allows to understand the dynamics of F_2 (more information can be found in [GŻ02b]) but the factorization of $|R_n(\lambda, \mu)|$ requires some extra work and relies on use of continuous fractions and (implicitly) Chebyshev polynomials.

The main result from [GŻ01] states that the Markov operator on the Cayley graph $\Gamma(L, \{u, v\})$ has a pure point spectrum concentrated on the points $\cos \frac{p}{q}\pi$, $q = 2, 3, \ldots, p \in \mathbb{Z}$. The spectral measure μ is therefore discrete and is concentrated on the above points with values

$$\mu\left(\cos \frac{p}{q}\pi\right) = \frac{1}{2^q - 1}, \quad (p, q) = 1.$$

(It is presented by Figure 13.3.)

As a consequence of this result a question of M. Atiyah [Ati76] was answered in [GLSŻ00] by providing a counterexample to the strong Atiyah conjecture [Lüc02].

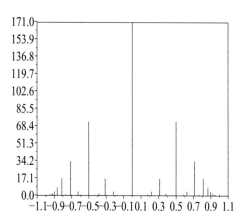

FIGURE 13.3

More sophisticated examples of integrable rational maps of $\mathbb{R}^2$ leading to computation of spectra of Schreier graphs of Gupta–Sidki 3-group and Fabrikovsky–Gupta groups are given in [BG00a].

The Basilica group B given by the automaton from Figure 9.1 is an example of intermediate situation when the recursion for the determinant exists but the dynamics of the corresponding map is unknown and we do not know what the spectrum in this case is (although we know that it has 1 is an accumulating point).

Let a, b be the two generators of B given by the (nonidentity) states of the automaton and let

$$R^{(3)}(\lambda, \mu) = a + a^{-1} + \lambda(b + b^{-1}) - \mu.$$

Then

$$|R_n^{(3)}(\lambda, \mu)| = \lambda^{2^n} |R_{n-1}^{(2)}(F_3(\lambda, \mu))|$$

where

$$F_3 : \begin{cases} \lambda \longrightarrow -2 + \dfrac{\lambda(\lambda - 2)}{\mu^2} \\ \\ \mu \longrightarrow \dfrac{\lambda - 2}{\mu^2}. \end{cases}$$

The spectrum of $R^{(3)}(\lambda, \mu)$ is F_3-invariant and diagrams made on a computer give an impression that here we deal with a strange attractor (see Figure 13.4).

Unfortunately, for most of the examples we are unable even to get a recursion for the determinant of the type exhibited in the above examples.

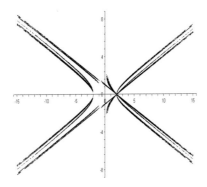

FIGURE 13.4

Problem 13.1.   a) *Is there an algorithm that, given an automaton A, finds a suitable pencil $R(\lambda_1, \ldots, \lambda_k)$ of matrices representing the set of generators for $G(A)$ and such that there is a rational map $F \colon \mathbb{R}^k \to \mathbb{R}^k$ (k is the number of parameters involved in the pencil) for which the recursion of type*

$$|R_n(\lambda_1, \ldots, \lambda_k)| = U \cdot V^{d^n} |R_{n-1}(F(\lambda_1, \ldots, \lambda_k))| \qquad (13.3)$$

holds, where U and V are some functions (and d is the cardinality of the alphabet)?
  b) *Give an example when such a recursion does not exist.*

Problem 13.2. *For the examples when the recursion of type (13.3) exists and, in particular, for Basilica group describe the topological structure of the invariant subsets.*

References

[Adi79]   S. I. Adian. *The Burnside problem and identities in groups*, volume 95 of *Ergebnisse der Mathematik und ihrer Grenzgebiete [Results in Mathematics and Related Areas]*. Springer-Verlag, Berlin, 1979.

[Adi82]   S. I. Adian. Random walks on free periodic groups. *Izv. Akad. Nauk SSSR Ser. Mat.*, 46(6):1139–1149, 1343, 1982.

[AHM]   W. Aitken, F. Hajir, and Chr. Maire. Finitely ramified iterated extensions.

[Ale72]   S. V. Alešin. Finite automata and the Burnside problem for periodic groups. *Mat. Zametki*, 11:319–328, 1972.

[Ati76]   M. F. Atiyah. Elliptic operators, discrete groups and von Neumann algebras. In *Colloque "Analyse et Topologie" en l'Honneur de Henri Cartan (Orsay, 1974)*, pages 43–72. Astérisque, No. 32–33. Soc. Math. France, Paris, 1976.

[AV05]   M. Abert and B. Virag. Dimension and randomness in groups acting on rooted trees. *J. Amer. Math. Soc.*, 18(1):157–192, 2005. (available at *http://www.arxiv.org/abs/math.GR/0212191*).

[AVŠ57]   G. M. Adel′son-Vel′skiĭ and Yu. A. Šreĭder. The Banach mean on groups. *Uspehi Mat. Nauk (N.S.)*, 12(6(78)):131–136, 1957.

[Bar98]   L. Bartholdi. The growth of Grigorchuk's torsion group. *Internat. Math. Res. Notices*, (20):1049–1054, 1998.

[Bar01]   L. Bartholdi. Lower bounds on the growth of a group acting on the binary rooted tree. *Internat. J. Algebra Comput.*, 11(1):73–88, 2001.

[Bar03]   L. Bartholdi. Endomorphic presentations of branch groups. *J. Algebra*, 268(2):419–443, 2003.

[Bar05]   L. Bartholdi. Lie algebras and growth in branch groups. *Pacific J. Math.*, 218(2):241–282, 2005.

[Bau93]   G. Baumslag. *Topics in combinatorial group theory*. Lectures in Mathematics ETH Zürich. Birkhäuser Verlag, Basel, 1993.

[BCdlH94]   M. Bekka, M. Cowling, and P. de la Harpe. Some groups whose reduced C^*-algebra is simple. *Inst. Hautes Études Sci. Publ. Math.*, (80):117–134 (1995), 1994.

[BCSN]   I. Bondarenko, T. Ceccherini-Silberstein, and V. Nekrashevych. Amenable graphs with dense non-cocompact pseudogroup of holonomy. In preparation.

[BdlH03]   M. Bachir Bekka and P. de la Harpe. Irreducibility of unitary group representations and reproducing kernels Hilbert spaces. *Expo. Math.*, 21(2):115–149, 2003. Appendix by the authors in collaboration with Rostislav Grigorchuk.

[BG00a]   L. Bartholdi and R. I. Grigorchuk. On the spectrum of Hecke type operators related to some fractal groups. *Tr. Mat. Inst. Steklova*, 231(Din. Sist., Avtom. i Beskon. Gruppy):5–45, 2000.

[BG00b]   L. Bartholdi and R. I. Grigorchuk. Spectra of non-commutative dynamical systems and graphs related to fractal groups. *C. R. Acad. Sci. Paris Sér. I Math.*, 331(6):429–434, 2000.

[BG02a]   L. Bartholdi and R. I. Grigorchuk. On parabolic subgroups and Hecke algebras of some fractal groups. *Serdica Math. J.*, 28(1):47–90, 2002.

[BG02b]   M. R. Bridson and R. H. Gilman. Context-free languages of sub-exponential growth. *J. Comput. System Sci.*, 64(2):308–310, 2002.

[BGN03]   L. Bartholdi, R. Grigorchuk, and V. Nekrashevych. From fractal groups to fractal sets. In *Fractals in Graz 2001*, Trends Math., pages 25–118. Birkhäuser, Basel, 2003.

[BGŠ03]   L. Bartholdi, R. I. Grigorchuk, and Z. Šunik. Branch groups. In *Handbook of algebra, Vol. 3*, pages 989–1112. North-Holland, Amsterdam, 2003.

[BKNV]   L. Bartholdi, V. Kaimanovich, V. Nekrashevych, and B Virag. Amenability of groups generated by bounded automata. In preparation.

[BL00]   M. B. Bekka and N. Louvet. Some properties of C^*-algebras associated to discrete linear groups. In *C^*-algebras (Münster, 1999)*, pages 1–22. Springer, Berlin, 2000.

[BMS67]   H. Bass, J. Milnor, and J.-P. Serre. Solution of the congruence subgroup problem for SL_n ($n \geq 3$) and Sp_{2n} ($n \geq 2$). *Inst. Hautes Études Sci. Publ. Math.*, (33):59–137, 1967.

[BOERT96] H. Bass, M. V. Otero-Espinar, D. Rockmore, and C. Tresser. *Cyclic renormalization and automorphism groups of rooted trees*, volume 1621 of *Lecture Notes in Mathematics*. Springer-Verlag, Berlin, 1996.

[Bos00]   N. Boston. p-adic Galois representations and pro-p Galois groups. In *New horizons in pro-p groups*, volume 184 of *Progr. Math.*, pages 329–348. Birkhäuser Boston, Boston, MA, 2000.

[BR84]   G. Baumslag and J. E. Roseblade. Subgroups of direct products of free groups. *J. London Math. Soc. (2)*, 30(1):44–52, 1984.

[BŠ01]   L. Bartholdi and Z. Šuník. On the word and period growth of some groups of tree automorphisms. *Comm. Algebra*, 29(11):4923–4964, 2001.

[BV03]   L. Bartholdi and B. Virag. Amenability via random walks. (available at *http://arxiv.org/abs/math.GR/0305262*), 2003.

[BW99]   M. R. Bridson and D. T. Wise. VH complexes, towers and subgroups of $F \times F$. *Math. Proc. Cambridge Philos. Soc.*, 126(3):481–497, 1999.

[CG00]   C. Champetier and V. Guirardel. Monoïdes libres dans les groupes hyperboliques. In *Séminaire de Théorie Spectrale et Géométrie, Vol. 18, Année 1999–2000*, volume 18 of *Sémin. Théor. Spectr. Géom.*, pages 157–170. Univ. Grenoble I, Saint, 2000.

[Cha91]   C. Champetier. *Properiétés génériques des groupes de type fini*. Thèse de doctorat, Université de Lyon I, 1991.

[Cha00]   C. Champetier. L'espace des groupes de type fini. *Topology*, 39(4):657–680, 2000.

[Cho80]   Ching Chou. Elementary amenable groups. *Illinois J. Math.*, 24(3):396–407, 1980.

[Chu97]   Fan R. K. Chung. *Spectral graph theory*, volume 92 of *CBMS Regional Conference Series in Mathematics*. Published for the Conference Board of the Mathematical Sciences, Washington, DC, 1997.

[CM82]   B. Chandler and W. Magnus. *The history of combinatorial group theory*, volume 9 of *Studies in the History of Mathematics and Physical Sciences*. Springer-Verlag, New York, 1982.

[CSGdlH99] T. Ceccherini-Silberstein, R. I. Grigorchuk, and P. de la Harpe. Amenability and paradoxical decompositions for pseudogroups and discrete metric spaces. *Tr. Mat. Inst. Steklova*, 224(Algebra. Topol. Differ. Uravn. i ikh Prilozh.):68–111, 1999.

[CSLST04] T. Ceccherini-Silberstein, Yu. Leonov, F. Scarabotti, and F. Tolli. Generalized Kaloujnine groups, universality and height of automorphisms. *Internat. J. Algebra Comput.*, 15(3):503–527, 2004.

[CSMS01]   T. Ceccherini-Silberstein, A. Machì, and F. Scarabotti. The Grigorchuk group of intermediate growth. *Rend. Circ. Mat. Palermo (2)*, 50(1):67–102, 2001.

[CSSFa] T. Ceccherini-Silberstein, F. Scarabotti, and Tolli F. Finite Gelfand pairs and their applications to probability and statistics. To appear.

[CSSFb] T. Ceccherini-Silberstein, F. Scarabotti, and Tolli F. Trees, wreath products and finite Gelfand pairs. To appear.

[Day57] M. M. Day. Amenable semigroups. *Illinois J. Math.*, 1:509–544, 1957.

[Dia88] P. Diaconis. *Group representations in probability and statistics.* Institute of Mathematical Statistics Lecture Notes—Monograph Series, 11. Institute of Mathematical Statistics, Hayward, CA, 1988.

[dlH00] P. de la Harpe. *Topics in geometric group theory.* Chicago Lectures in Mathematics. University of Chicago Press, Chicago, IL, 2000.

[dlH04] P. de la Harpe. Mesures finiment additives et paradoxes. *Panoramas et Synthesis*, 18:39–61, 2004.

[Eil74] S. Eilenberg. *Automata, languages, and machines. Vol. A.* Academic Press [A subsidiary of Harcourt Brace Jovanovich, Publishers], New York, 1974.

[Ers] A. Erschler. Piecewise automatic groups. to appear.

[Ers03] A. Erschler. On isoperimetric profiles of finitely generated groups. *Geom. Dedicata*, 100:157–171, 2003.

[Ers04a] A. Erschler. Boundary behavior for groups of subexponential growth. *Annals of Math.*, 160(3):1183–1210, 2004.

[Ers04b] A. Erschler. Not residually finite groups of intermediate growth, commensurability and non-geometricity. *J. Algebra*, 272(1):154–172, 2004.

[Føl57] E. Følner. Note on groups with and without full Banach mean value. *Math. Scand.*, 5:5–11, 1957.

[GdlH97] R. Grigorchuk and P. de la Harpe. On problems related to growth, entropy, and spectrum in group theory. *J. Dynam. Control Systems*, 3(1):51–89, 1997.

[GK93] R. I. Grigorchuk and P. F. Kurchanov. Some questions of group theory related to geometry. In *Algebra, VII*, volume 58 of *Encyclopaedia Math. Sci.*, pages 167–232, 233–240. Springer, Berlin, 1993.

[GL02] R. I. Grigorchuk and I. Lysenok. Burnside problem. In *The Consice handbook of algebra, A. V. Mikhalev and Gunter F. Pilz (eds.)*, pages 111–115. Kluwer Academic Publishers, 2002.

[GLSŻ00] R. I. Grigorchuk, P. Linnell, T. Schick, and A. Żuk. On a question of Atiyah. *C. R. Acad. Sci. Paris Sér. I Math.*, 331(9):663–668, 2000.

[GM99] R. I. Grigorchuk and A. Machì. An example of an indexed language of intermediate growth. *Theoret. Comput. Sci.*, 215(1-2):325–327, 1999.

[GM03] Y. Glasner and S. Mozes. Automata and square complexes. (available at *http://arxiv.org/abs/math.GR/0306259*), 2003.

[GN97] R. I. Grigorchuk and T. Nagnibeda. Complete growth functions of hyperbolic groups. *Invent. Math.*, 130(1):159–188, 1997.

[GNS00] R. I. Grigorchuk, V. V. Nekrashevich, and V. I. Sushchanskiĭ. Automata, dynamical systems, and groups. *Tr. Mat. Inst. Steklova*, 231(Din. Sist., Avtom. i Beskon. Gruppy):134–214, 2000.

[GP72] F. Gecseg and I. Peák. *Algebraic theory of automata.* Akadémiai Kiadó, Budapest, 1972.

[Gre69]   F. P. Greenleaf. *Invariant means on topological groups and their applications.* Van Nostrand Mathematical Studies, No. 16. Van Nostrand Reinhold Co., New York, 1969.

[Gri80]   R. I. Grigorchuk. On Burnside's problem on periodic groups. *Funktsional. Anal. i Prilozhen.*, 14(1):53–54, 1980.

[Gri83]   R. I. Grigorchuk. On the Milnor problem of group growth. *Dokl. Akad. Nauk SSSR*, 271(1):30–33, 1983.

[Gri84a]   R. I. Grigorchuk. Degrees of growth of finitely generated groups and the theory of invariant means. *Izv. Akad. Nauk SSSR Ser. Mat.*, 48(5):939–985, 1984.

[Gri84b]   R.I. Grigorchuk. The construction of p-groups of intermediate growth having continuum factor groups. *Algebra i Logika*, 23(4):383–394, 1984.

[Gri85]   R. I. Grigorchuk. A relationship between algorithmic problems and entropy characteristics of groups. *Dokl. Akad. Nauk SSSR*, 284(1):24–29, 1985.

[Gri88]   R. I. Grigorchuk. Semigroups with cancellations of degree growth. *Mat. Zametki*, 43(3):305–319, 428, 1988.

[Gri90]   R. I. Grigorchuk. Growth and amenability of a semigroup and its group of quotients. In *Proceedings of the International Symposium on the Semigroup Theory and its Related Fields (Kyoto, 1990)*, pages 103–108, Matsue, 1990. Shimane Univ.

[Gri96]   R. I. Grigorchuk. Weight functions on groups, and criteria for the amenability of Beurling algebras. *Mat. Zametki*, 60(3):370–382, 479, 1996.

[Gri98]   R. I. Grigorchuk. An example of a finitely presented amenable group that does not belong to the class EG. *Mat. Sb.*, 189(1):79–100, 1998.

[Gri99]   R. I. Grigorchuk. On the system of defining relations and the Schur multiplier of periodic groups generated by finite automata. In *Groups St. Andrews 1997 in Bath, I*, volume 260 of *London Math. Soc. Lecture Note Ser.*, pages 290–317. Cambridge Univ. Press, Cambridge, 1999.

[Gri00a]   R. I. Grigorchuk. Branch groups. *Mat. Zametki*, 67(6):852–858, 2000.

[Gri00b]   R. I. Grigorchuk. Just infinite branch groups. In *New horizons in pro-p groups*, volume 184 of *Progr. Math.*, pages 121–179. Birkhäuser Boston, Boston, MA, 2000.

[Gro81]   M. Gromov. Groups of polynomial growth and expanding maps. *Publ. Math. IHES*, 53:53–78, 1981.

[Gro87]   M. Gromov. Hyperbolic groups. In *Essays in group theory*, volume 8 of *Math. Sci. Res. Inst. Publ.*, pages 75–263. Springer, New York, 1987.

[GS83]   N. Gupta and S. Sidki. Some infinite p-groups. *Algebra i Logika*, 22(5):584–589, 1983.

[GS02]   R. Grigorchuk and Z. Sunik. Featured review on the paper "Non-amenable finitely presented torsion-by-cyclic groups" by Olshanskii, A. and Sapir, M. Publ. Math. Inst. Hautes E'tudes Sci. no. 96 (2002), 43–169 (2003)., 2002.

[GS04]   R. I. Grigorchuk and S. N. Sidki. The group of automorphisms of a 3-generated 2-group of intermediate growth. *Internat. J. Algebra Comput.*, 14(5-6):667–676, 2004.

[Gup89]   N. Gupta. On groups in which every element has finite order. *Amer. Math. Monthly*, 96(4):297–308, 1989.

[GW03a]   R. I. Grigorchuk and J. S. Wilson. A structural property concerning abstract commensurability of subgroups. *J. London Math. Soc. (2)*, 68(3):671–682, 2003.

[GW03b]   R. I. Grigorchuk and J. S. Wilson. The uniqueness of the actions of certain branch groups on rooted trees. *Geom. Dedicata*, 100:103–116, 2003.

[GZ91]   Max Garzon and Yechezkel Zalcstein. The complexity of Grigorchuk groups with application to cryptography. *Theoret. Comput. Sci.*, 88(1):83–98, 1991.

[GŻ99]   R. I. Grigorchuk and A. Żuk. On the asymptotic spectrum of random walks on infinite families of graphs. In *Random walks and discrete potential theory (Cortona, 1997)*, Sympos. Math., XXXIX, pages 188–204. Cambridge Univ. Press, Cambridge, 1999.

[GŻ01]   R. I. Grigorchuk and A. Żuk. The lamplighter group as a group generated by a 2-state automaton, and its spectrum. *Geom. Dedicata*, 87(1-3):209–244, 2001.

[GŻ02a]   R. I. Grigorchuk and A. Żuk. On a torsion-free weakly branch group defined by a three state automaton. *Internat. J. Algebra Comput.*, 12(1-2):223–246, 2002.

[GŻ02b]   R. I. Grigorchuk and A. Żuk. Spectral properties of a torsion-free weakly branch group defined by a three state automaton. In *Computational and statistical group theory (Las Vegas, NV/Hoboken, NJ, 2001)*, volume 298 of *Contemp. Math.*, pages 57–82. Amer. Math. Soc., Providence, RI, 2002.

[GZ03]   R. I. Grigorchuk and A. Zuk. Elementary classes and amenability. In *Proceedings of The Congress of Ukrainian Mathematicians, Functional Analysis*, pages 59–84, 2003.

[GŻ04]   R. I. Grigorchuk and A. Żuk. The Ihara zeta function of infinite graphs, the KNS spectral measure and integrable maps. In *Random walks and geometry*, pages 141–180. Walter de Gruyter GmbH & Co. KG, Berlin, 2004.

[Hoř63]   J. Hořeiš. Transformations defined by finite automata (in Russian). *Problemy Kibernetiki*, 9:23–26, 1963.

[HR03]   D. F. Holt and C. E. Röver. On real-time word problems. *J. London Math. Soc. (2)*, 67(2):289–301, 2003.

[HU79]   J. E. Hopcroft and J. D. Ullman. *Introduction to automata theory, languages, and computation*. Addison-Wesley Publishing Co., Reading, Mass., 1979.

[Inc01]   R. Incitti. The growth function of context-free languages. *Theoret. Comput. Sci.*, 255(1-2):601–605, 2001.

[Kai]   V. Kaimanovich. "Münchhausen trick" and amenability of self-similar groups. To appear in IJAC (2005), special issue "International Conference on Infinite Groups, Gaeta, 2003".

[Kal45]   L. Kaloujnine. Sur les p-groupes de Sylow du groupe symétrique du degré p^m. *C. R. Acad. Sci. Paris*, 221:222–224, 1945.

[Kal48]   L. Kaloujnine. La structure des p-groupes de Sylow des groupes symétriques finis. *Ann. Sci. École Norm. Sup. (3)*, 65:239–276, 1948.

[Kap70]   I. Kaplansky. "Problems in the theory of rings" revisited. *Amer. Math. Monthly*, 77:445–454, 1970.

[Kec95]   A. S. Kechris. *Classical descriptive set theory*, volume 156 of *Graduate Texts in Mathematics*. Springer-Verlag, New York, 1995.

[Kes59]   H. Kesten. Symmetric random walks on groups. *Trans. Amer. Math. Soc.*, 92:336–354, 1959.

[Kit98]   B. P. Kitchens. *Symbolic dynamics*. Universitext. Springer-Verlag, Berlin, 1998.

[KM98]   O. Kharlampovich and A. Myasnikov. Irreducible affine varieties over a free group. I. Irreducibility of quadratic equations and Nullstellensatz. *J. Algebra*, 200(2):472–516, 1998.

[kou02]   *The Kourovka notebook*. Rossiĭskaya Akademiya Nauk Sibirskoe Otdelenie Institut Matematiki, Novosibirsk, 2002.

[KS05]   I. Kapovich and P. Schupp. Delzant's *t*-invariant, one-relator groups and Kolmogorov complexity. (available at *http://arxiv.org/abs/math.GR/0305353*), 2005.

[Leo98]   Yu. G. Leonov. The conjugacy problem in a class of 2-groups. *Mat. Zametki*, 64(4):573–583, 1998.

[Leo01]   Yu. G. Leonov. On a lower bound for the growth of a 3-generator 2-group. *Mat. Sb.*, 192(11):77–92, 2001.

[Lih63]   A. I. Lihtman. On group rings of *p*-groups. *Izv. Akad. Nauk SSSR Ser. Mat.*, 27:795–800, 1963.

[LM95]   D. Lind and B. Marcus. *An introduction to symbolic dynamics and coding*. Cambridge University Press, Cambridge, 1995.

[LN02]   Ya. Lavreniuk and V. Nekrashevych. Rigidity of branch groups acting on rooted trees. *Geom. Dedicata*, 89:159–179, 2002.

[LNS05]   Yu. Leonov, V. Nekrashevych, and V. Sushchansky. Kaloujnine method, 2005. Preprint.

[LPS88]   A. Lubotzky, R. Phillips, and P. Sarnak. Ramanujan graphs. *Combinatorica*, 8(3):261–277, 1988.

[LS01]   R. C. Lyndon and P. E. Schupp. *Combinatorial group theory*. Classics in Mathematics. Springer-Verlag, Berlin, 2001.

[LS03]   A. Lubotzky and D. Segal. *Subgroup growth*, volume 212 of *Progress in Mathematics*. Birkhäuser Verlag, Basel, 2003.

[Lub94]   A. Lubotzky. *Discrete groups, expanding graphs and invariant measures*, volume 125 of *Progress in Mathematics*. Birkhäuser Verlag, Basel, 1994.

[Lüc02]   W. Lück. L^2-*invariants: theory and applications to geometry and K-theory*, volume 44 of *Ergebnisse der Mathematik und ihrer Grenzgebiete. 3. Folge. A Series of Modern Surveys in Mathematics [Results in Mathematics and Related Areas. 3rd Series. A Series of Modern Surveys in Mathematics]*. Springer-Verlag, Berlin, 2002.

[LV97]   Ming Li and P. Vitányi. *An introduction to Kolmogorov complexity and its applications*. Graduate Texts in Computer Science. Springer-Verlag, New York, second edition, 1997.

[Lys85]   I. G. Lysënok. A set of defining relations for the Grigorchuk group. *Mat. Zametki*, 38(4):503–516, 634, 1985.

[Men65]   J. L. Mennicke. Finite factor groups of the unimodular group. *Ann. of Math. (2)*, 81:31–37, 1965.

[Mil68]   J. Milnor. Problem 5603. *Amer. Math. Monthly*, 75:685–686, 1968.

[MP01]   Roman Muchnik and Igor Pak. On growth of Grigorchuk groups. *Internat. J. Algebra Comput.*, 11(1):1–17, 2001.

[Nab96]   A. Nabutovsky. Fundamental group and contractible closed geodesics. *Comm. Pure Appl. Math.*, 49(12):1257–1270, 1996.

[Nek03]   V. V. Nekrashevich. Iterated monodromy groups. *Dopov. Nats. Akad. Nauk Ukr. Mat. Prirodozn. Tekh. Nauki*, (4):18–20, 2003.

[Nek04]   V. V. Nekrashevych. Cuntz-Pimsner algebras of group actions. *J. Operator Theory*, 52(2):223–249, 2004.

[Nek05]   V. V. Nekrashevych. Self-similar groups. 2005.

[Ol'80]   A. Ju. Ol'šanskiĭ. On the question of the existence of an invariant mean on a group. *Uspekhi Mat. Nauk*, 35(4(214)):199–200, 1980.

[Pas98]   D. S. Passman. The semiprimitivity of group algebras. In *Methods in ring theory (Levico Terme, 1997)*, volume 198 of *Lecture Notes in Pure and Appl. Math.*, pages 199–211. Dekker, New York, 1998.

[Pat88]   A. L. T. Paterson. *Amenability*, volume 29 of *Mathematical Surveys and Monographs*. American Mathematical Society, Providence, RI, 1988.

[Per00]   E. L. Pervova. Everywhere dense subgroups of a group of tree automorphisms. *Tr. Mat. Inst. Steklova*, 231(Din. Sist., Avtom. i Beskon. Gruppy):356–367, 2000.

[Per02]   E. L. Pervova. The congruence property of AT-groups. *Algebra Logika*, 41(5):553–567, 634, 2002.

[Per04]   E. L. Pervova. Profinite completions of some groups acting on trees. (available at *http://arxiv.org/abs/math.GR/0411192*), 2004.

[Per05]   E. L. Pervova. Maximal subgroups of non locally finite fractal groups., 2005. To appear in *Inter. J. of Alg. and Comput.*

[Pil00]   K. M. Pilgrim. Dessins d'enfants and Hubbard trees. *Ann. Sci. École Norm. Sup. (4)*, 33(5):671–693, 2000.

[Pow75]   R. T. Powers. Simplicity of the C^*-algebra associated with the free group on two generators. *Duke Math. J.*, 42:151–156, 1975.

[PSC01]   Ch. Pittet and L. Saloff-Coste. A survey on the relationships between volume growth, isoperimetry, and the behavior of simple random walk on Cayley graphs, with examples. (available at *http://www.math.cornell.edu/ lsc/surv.ps.gz*), 2001.

[Rev91]   G. E. Revesz. *Introduction to Formal Languages*. Dover Publications, Inc., 1991.

[Roz93]   A. V. Rozhkov. Centralizers of elements in a group of tree automorphisms. *Izv. Ross. Akad. Nauk Ser. Mat.*, 57(6):82–105, 1993.

[Roz98]   A. V. Rozhkov. The conjugacy problem in an automorphism group of an infinite tree. *Mat. Zametki*, 64(4):592–597, 1998.

[RS02a]   I. I. Reznikov and V. I. Sushchanskiĭ. Growth functions of two-state au-
          tomata over a two-element alphabet. *Dopov. Nats. Akad. Nauk Ukr. Mat.
          Prirodozn. Tekh. Nauki*, (2):76–81, 2002.

[RS02b]   I. I. Reznikov and V. I. Sushchanskiĭ. Two-state Mealy automata of inter-
          mediate growth over a two-letter alphabet. *Mat. Zametki*, 72(1):102–117,
          2002.

[RZ93]   L. Ribes and P. A. Zalesskii. On the profinite topology on a free group. *Bull.
          London Math. Soc.*, 25(1):37–43, 1993.

[Sel01]   Z. Sela. Diophantine geometry over groups. I. Makanin-Razborov diagrams.
          Publ. Math. Inst. Hautes Études Sci., (93):31–105, 2001.

[Ser97]   J.-P. Serre. Répartition asymptotique des valeurs propres de l'opérateur de
          Hecke T_p. *J. Amer. Math. Soc.*, 10(1):75–102, 1997.

[Sha00]   Y. Shalom. Rigidity of commensurators and irreducible lattices. *Invent.
          Math.*, 141(1):1–54, 2000.

[Sid87a] S. Sidki. On a 2-generated infinite 3-group: subgroups and automorphisms.
          J. Algebra, 110(1):24–55, 1987.

[Sid87b] S. Sidki. On a 2-generated infinite 3-group: the presentation problem. *J.
          Algebra*, 110(1):13–23, 1987.

[Sid97]   S. Sidki. A primitive ring associated to a Burnside 3-group. *J. London Math.
          Soc. (2)*, 55(1):55–64, 1997.

[Sid98]   S. Sidki. *Regular trees and their automorphisms*, volume 56 of *Monografías
          de Matemática [Mathematical Monographs]*. Instituto de Matemática Pura
          e Aplicada (IMPA), Rio de Janeiro, 1998.

[Sid04]   S. Sidki. Finite automata of polynomial growth do not generate a free group.
          Geom. Dedicata, 108:193–204, 2004.

[SS78]   A. Salomaa and M. Soittola. *Automata-theoretic aspects of formal power
          series*. Springer-Verlag, New York, 1978.

[Sun05]   Z. Sunik. Hausdorff dimension of branched automaton groups, 2005.
          Preprint.

[SW03]   S. Sidki and J. S. Wilson. Free subgroups of branch groups. *Arch. Math.
          (Basel)*, 80(5):458–463, 2003.

[Ter99]   A. Terras. *Fourier analysis on finite groups and applications*, volume 43 of
          London Mathematical Society Student Texts. Cambridge University Press,
          Cambridge, 1999.

[Tho68]   E. Thoma. Eine Charakterisierung diskreter Gruppen vom Typ I. *Invent.
          Math.*, 6:190–196, 1968.

[Ver82]   A. Vershik. Amenability and approximation of infinite groups. *Selecta Math.
          Soviet.*, 2(4):311–330, 1982.

[vN29]   J. von Neumann. Zur allgemeinen Theorie des Masses. *Fund. Math.*, 13:73–
          116 and 333, 1929. = *Collected works*, vol. I, pages 599–643.

[Vov00]   T. Vovkivsky. Infinite torsion groups arising as generalizations of the second
          Grigorchuk group. In *Algebra (Moscow, 1998)*, pages 357–377. de Gruyter,
          Berlin, 2000.

[VV05]   M. Vorobets and Ya. Vorobets. On a free group of transformations defined by an automaton, 2005. preprint.

[Wag93]   S. Wagon. *The Banach-Tarski paradox*. Cambridge University Press, Cambridge, 1993.

[Zel90]   E.I. Zel'manov. Solution of the restricted burnside problem for groups of odd exponent. *Izv. Akad. Nauk SSSR, Ser. Mat.*, 54(1):42–59, 1990.

[Zel91]   E.I. Zel'manov. Solution of the restricted burnside problem for 2-groups. *Mat. Sb.*, 182(4):568–592, 1991.

Rostislav Grigorchuk
Mathematics Department
Texas A&M University
College Station, TX 77843-3368
USA
e-mail: `grigorch@math.tamu.edu`

Progress in Mathematics, Vol. 248, 219–267
© 2005 Birkhäuser Verlag Basel/Switzerland

Cubature Formulas, Geometrical Designs, Reproducing Kernels, and Markov Operators

Pierre de la Harpe and Claude Pache

Abstract. Cubature formulas and geometrical designs are described in terms of reproducing kernels for Hilbert spaces of functions on the one hand, and Markov operators associated to orthogonal group representations on the other hand. In this way, several known results for spheres in Euclidean spaces, involving cubature formulas for polynomial functions and spherical designs, are shown to generalize to large classes of finite measure spaces (Ω, σ) and appropriate spaces of functions inside $L^2(\Omega, \sigma)$. The last section points out how spherical designs are related to a class of reflection groups which are (in general dense) subgroups of orthogonal groups.

Mathematics Subject Classification (2000). Primary 05B30, 65D32; Secondary 46E22.

Keywords. Cubature formulas, spherical designs, reproducing kernels, Markov operators, group representations, reflection groups.

Let (Ω, σ) be a finite measure space and let $\mathcal{F}$ be a vector space of integrable real-valued functions on Ω. It is a natural question to ask when and how integrals $\int_\Omega \varphi(\omega) d\sigma(\omega)$ can be computed, or approximated, by sums $\sum_{x \in X} W(x)\varphi(x)$, where X is a subset of Ω and $W : X \longrightarrow \mathbb{R}^*_+$ a weight function, for all $\varphi \in \mathcal{F}$.

When Ω is an interval of the real line, this is a basic problem of numerical integration with a glorious list of contributors: Newton (1671), Cotes (1711), Simpson (1743), Gauss (1814), Chebyshev (1874), Christoffel (1877), and Bernstein (1937), to quote but a few; see [45] for some historical notes, [44], and [31]. The theory is inseparable of that of orthogonal polynomials [97].

When Ω is a space of larger dimension, the problems involved are often of geometrical and combinatorial interest. One important case is that of spheres in Euclidean spaces, with rotation-invariant probability measure, as in [33] and [47]. Work related to integrations domains with $\dim(\Omega) > 1$ goes back to [75] and [1]; see also the result of Voronoi (1908) recalled in Item 1.15 below. Examples which

The authors acknowledge support from the *Swiss National Science Foundation*.

have been considered include various domains in Euclidean spaces (hypercubes, simplices, ..., Item 1.20), surfaces (e.g. tori), and Euclidean spaces with Gaussian measures (Item 3.11).

There are also interesting cases where Ω itself is a finite set [32].

Section 1 collects the relevant definitions for the general case (Ω, σ). It reviews several known examples on intervals and spheres.

Our main point is to show that there are two notions which are convenient for the study of cubature formulas, even if they are rarely explicitly introduced in papers of this subject.

First, we introduce in Section 2 the formalism of *reproducing kernel Hilbert spaces* ([3], [67], [20]), which is appropriate for generalizing to other spaces various results which are standard for spheres. See for example the existence theorem for cubature formulas (X, W) with bound on $|X|$ (Items 2.6–2.8), Propositions 2.10 & 3.4 on tight cubature formulas being geometrical designs, Proposition 2.12 on the set of distances $D_X = \{c \in]0, \infty[\mid c = d(x, y) \text{ for some } x, y \in X, x \neq y\}$ if Ω is a metric space and if "Condition (M)" is satisfied, and Items 2.13–2.14 for a single equality which guarantees that a pair (X, W) is a cubature formula of strength k.

Section 3 concentrates on spheres, on consequences of the existence of the central symmetry, and in particular on spherical designs of odd strengths. An example of "lattice construction" (Item 3.10) provides a cubature formula in $\mathbb{S}^7$ of strength 11 and size 2400. Item 3.11 hints at some very rudimentary constructions of *Gaussian designs*.

Secondly, we consider in Section 4 *Markov operators* (called elsewhere "convolution operators", "difference operators", "Hecke operators", or "discrete Laplace operators"). On the one hand, they provide an alternative definition of cubature formulas on spheres [81], and at least part of this can be generalized to other Riemannian symmetric spaces of the compact type (work in progress); also, they show a natural connection with the work of Lubotzky, Phillips, and Sarnak (see Theorem 4.7). On the other hand, they suggest an interesting class of examples of infinite *reflection groups*, as shown in Section 5 where we state and prove an unpublished result of B. Venkov on remarkable sets of generators of the orthogonal group $O(8, \mathbb{Z}[1/2])$.

The subject of cubature formulas has natural connections with several other subjects. We give brief mentions to:

- representations of finite groups (see [16], [56], and Example 1.14),
- lattices in Euclidean spaces (see [101], [102], and Examples 1.16 & 3.10),
- Lehmer's conjecture (Example 1.16) and Waring's problem (Item 3.9) from number theory,
- Dvoretzky's theorem from Banach space theory (see [73] and Proposition 3.12).

It is a pleasure to thank Boris Venkov, as well as Béla Bajnok, Eiichi Bannai, Gaël Collinet, and Martin Gander for many conversations and indications which have been most helpful during this work.

1. Cubature formulas, designs, and polynomial spaces

Let (Ω, σ) be a finite measure space. By a **space of functions** in $L^2(\Omega, \sigma)$, we mean a linear subspace of genuine [1] real-valued functions on Ω given with an embedding in $L^2(\Omega, \sigma)$.

Definition 1.1. A **cubature formula** for a finite dimensional space $\mathcal{F}$ of functions in $L^2(\Omega, \sigma)$ is a pair (X, W), where X is a finite subset of Ω and $W : X \longrightarrow \mathbb{R}_+^*$ is a weight function, such that

$$\sum_{x \in X} W(x)\varphi(x) = \int_\Omega \varphi(\omega)d\sigma(\omega)$$

for all $\varphi \in \mathcal{F}$. If $W : x \longmapsto \sigma(\Omega)/|X|$ is the uniform weight, the set X is called a **design** for $\mathcal{F}$.

This definition would make sense for $\mathcal{F} \subset L^1(\Omega, \sigma)$. But the examples we have in mind are in $L^2(\Omega, \sigma)$, and we will use Hilbert space methods for which we have to assume that $\mathcal{F} \subset L^2(\Omega, \sigma)$; two functions $f_1, f_2 \in \mathcal{F}$ have a scalar product given by

$$\langle f_1 \mid f_2 \rangle = \int_\Omega f_1(\omega)f_2(\omega)d\sigma(\omega).$$

There are other denominations for these and related notions, including "Chebyshev quadrature rule" (for our "designs") with appropriate "algebraic degree of exactness"; see, e.g., [44].

We write $\langle \cdot \mid \cdot \rangle$ to denote both the scalar product of two vectors in a Hilbert space of functions and the scalar product in the standard Euclidean space $\mathbb{R}^n$.

Example 1.2 (Simpson's formula). In the case of an interval $\Omega = [a, b]$ of the real line $(a < b)$ and Lebesgue measure, *Simpson's formula*

$$\frac{b-a}{6}\varphi(a) + \frac{4(b-a)}{6}\varphi\left(\frac{a+b}{2}\right) + \frac{b-a}{6}\varphi(b) = \int_a^b \varphi(t)dt$$

provides a cubature formula for the space $\mathcal{F}^{(3)}(a, b)$ of polynomial functions of degree at most 3 on $[a, b]$. More generally, if φ is of class $\mathcal{C}^4$ on $[a, b]$, we have

$$\left| \frac{b-a}{6}\varphi(a) + \frac{4(b-a)}{6}\varphi\left(\frac{a+b}{2}\right) + \frac{b-a}{6}\varphi(b) - \int_a^b \varphi(t)dt \right|$$

$$\leq \frac{(b-a)^5}{2880} \sup_{a \leq t \leq b} \left| \varphi^{(4)}(t) \right|.$$

---

[1] Not functions modulo equality almost everywhere!

Simpson's formula is most often used in its *compound form*, namely on the n subintervals $[a + \frac{j-1}{n}(b-a), a + \frac{j}{n}(b-a)]$, $1 \le j \le n$, for n large enough; in this form, it provides a very efficient tool in numerical analysis.

The following notion borrows some of the ingredients of the notion of "polynomial space" from [46].

Definition 1.3. A **sequence of polynomial spaces** on a finite measure space (Ω, σ) is a nested sequence of finite dimensional spaces

$$\mathbb{R} = \mathcal{F}^{(0)} \subset \mathcal{F}^{(1)} \subset \cdots \subset \mathcal{F}^{(k)} \subset \cdots$$

of functions in $L^2(\Omega, \sigma)$ with the following property: $\mathcal{F}^{(k)}$ is linearly generated by products $\varphi_1 \varphi_2$ with $\varphi_1 \in \mathcal{F}^{(1)}$ and $\varphi_2 \in \mathcal{F}^{(k-1)}$, for all $k \ge 1$.

Remark 1.4. (i) For $(\mathcal{F}^{(k)})_{k \ge 0}$ as in the definition, there is a natural mapping from the kth symmetric power of $\mathcal{F}^{(1)}$ onto $\mathcal{F}^{(k)}$. If $n + 1 = \dim_{\mathbb{R}}(\mathcal{F}^{(1)})$, it follows that $\dim_{\mathbb{R}}(\mathcal{F}^{(k)}) \le \binom{n+k}{k}$.

(ii) Many examples of sequences of polynomial spaces have one more property: the union $\bigcup_{k=0}^{\infty} \mathcal{F}^{(k)}$ is dense in $L^2(\Omega, \sigma)$.

Definition 1.5. Let (Ω, σ) and $(\mathcal{F}^{(k)})_{k \ge 0}$ be as in Definition 1.3. A **cubature formula of strength k on Ω** is a pair (X, W), consisting of a finite subset $X \subset \Omega$ and a weight $W : X \longrightarrow \mathbb{R}_+^*$, which is a cubature formula for $\mathcal{F}^{(k)}$. In case $W(x) = \sigma(\Omega)/|X|$ for all $x \in X$, the set X is a **geometrical [2] k-design** on Ω.

1.6. Organising questions. (i) For standard examples of finite subsets X of Ω, compute the largest strength k for which X is a geometrical k-design.

(ii) For k given, find designs and cubature formulas (X, W) of strength k with $|X|$ minimal. (More on this in Item 1.17, on tight spherical designs.) In case Ω is moreover a metric space, describe the distance set

$$D_X = \{c \in]0, \infty[\mid c = d(x, y) \quad \text{for some} \quad x, y \in X, x \ne y\}.$$

(iii) Dually, for given N, find designs and cubature formulas (X, W) with $|X| = N$ of maximum strength.

(iv) Asymptotics and equidistribution. Find sequences $(X^{(k)}, W^{(k)})_{k \ge 1}$, where each $(X^{(k)}, W^{(k)})$ is a cubature formula of strength k, such that the sequence of measures $(\sigma^{(k)})_{k \ge 1}$ converges to σ, where $\sigma^{(k)} = \sum_{x \in X^{(k)}} W^{(k)}(x)\delta_x$ and where δ_x denotes the Dirac measure of support $\{x\}$. Optimize in some sense the speed of convergence.

When each $X^{(k)}$ is a geometrical k-design, this is the standard question of σ-equidistribution of a sequence of finite subsets of a measure space; see [91].

---

[2] In particular cases, we replace "geometrical" by more suggestive adjectives such as "**interval**" in the situation of Example 1.10 with Ω an interval in $\mathbb{R}$, "**spherical**" in the situation of $\Omega = \mathbb{S}^{n-1}$, or "**Gaussian**" if Ω is a real vector space and σ a Gaussian measure. However, it should be kept in mind that the "Euclidean designs" of [80] and [35] (among others) are of a different nature, since the reference measure σ on the Euclidean space $\Omega = \mathbb{R}^n$ depends in this case on the weighted set (X, W) through some radial factor.

In the following proposition, $\ell^2(X,W)$ stands for the Hilbert space of real-valued functions on X, with scalar product defined by

$$\langle \psi_1 \mid \psi_2 \rangle = \sum_{x \in X} W(x)\psi_1(x)\psi_2(x).$$

Proposition 1.7. *Let (Ω, σ) be a finite measure space. Let $\mathcal{H}$ be a finite dimensional space of functions in $L^2(\Omega, \sigma)$. Assume that there exists a finite dimensional subspace $\mathcal{H}'$ of $L^2(\Omega, \sigma)$ such that $\mathcal{H}$ is linearly generated by products $\varphi_1\varphi_2$ with $\varphi_1, \varphi_2 \in \mathcal{H}'$. Let X be a finite subset of Ω and let $W : X \longrightarrow \mathbb{R}_+^*$ be a weight.*
Then (X, W) is a cubature formula for $\mathcal{H}$ if and only if the restriction mapping

$$\rho : \mathcal{H}' \longrightarrow \ell^2(X,W)$$

is an isometry. In particular, if (X, W) is a cubature formula for $\mathcal{H}$, then

(∗) $$|X| \geq \dim_{\mathbb{R}}(\mathcal{H}').$$

Proof. The condition for ρ to be an isometry is

$$\sum_{x \in X} W(x)\varphi_1(x)\varphi_2(x) = \int_\Omega \varphi_1(\omega)\varphi_2(\omega)d\sigma(\omega)$$

for all $\varphi_1, \varphi_2 \in \mathcal{H}'$. By hypothesis on $\mathcal{H}$ and $\mathcal{H}'$, this condition is equivalent to

$$\sum_{x \in X} W(x)\varphi(x) = \int_\Omega \varphi(\omega)d\sigma(\omega)$$

for all $\varphi \in \mathcal{H}$.   □

Definition 1.8. Let (Ω, σ) and $(\mathcal{F}^{(k)})_{k \geq 0}$ be as in Definition 1.3. A cubature formula (X, W) of strength $2l$ is **tight** if equality holds in (∗) above, with $\mathcal{H}' = \mathcal{F}^{(l)}$ and $\mathcal{H} = \mathcal{F}^{(2l)}$.
For a spherical cubature formula of strength $2l + 1$, "tight" is defined in 1.11.

Tight cubature formulas are rare: see Discussion 1.17.

1.9. Quadrature formulas on intervals. Let Ω be a real interval (a, b), with $-\infty \leq a < b \leq \infty$ (the interval can be open or closed). The spaces $\mathcal{F}^{(k)}(a, b)$ of polynomial functions on Ω of degree at most k provide the canonical example for Definition 1.3.
Consider in particular the case $\Omega = [-1, 1]$, with Lebesgue measure, and an integer $l \geq 1$. Let first

$$X^{(l)} = \{x_{1,l}, x_{2,l}, \ldots, x_{l,l}\} \subset [-1, 1]$$

be an arbitrary subset of l distinct points in $[-1, 1]$. Let $L_j(t) = \prod \frac{t - x_{i,l}}{x_{j,l} - x_{i,l}}$ (product over $i \in \{1, \ldots, l\}$, $i \neq j$) be the corresponding *elementary Lagrange interpolation polynomials*. Then $(X^{(l)}, W^{(l)})$ is a "cubature formula" for $\mathcal{F}^{(l-1)}(-1, 1)$, with the "weight" $W^{(l)}$ defined on $X^{(l)}$ by

$$W^{(l)}(x_{j,l}) = \int_{-1}^{1} L_j(t)dt \qquad (1 \leq j \leq l).$$

But quotation marks are in order since the "weight" values need not be positive; for example, in case $x_{j,l} = -1 + 2\frac{j-1}{l-1}$, these **Newton-Cotes formulas** have all weights positive if and only if either $l \leq 8$ or $l = 10$ (see, e.g., Chapter 2 of [29]).

Let now $X^{(l)}$ be precisely the set of the roots of the Legendre polynomial of degree l, namely of $P_l(t) = (-1)^l \frac{l!}{2l!} \frac{d^l}{dt^l} \left((1 - t^2)^l\right)$. The pair $(X^{(l)}, W^{(l)})$ is a cubature formula for $\mathcal{F}^{(2l-1)}(-1, 1)$, known as a **Gauss quadrature** [3] (or sometimes *Gauss-Jacobi mechanical quadrature*); this is the unique cubature formula with $\leq l$ points for $\mathcal{F}^{(2l-1)}(-1, 1)$ and Lebesgue measure. The weights $W^{(l)}(x_{j,l})$ of the Gauss formula are strictly positive; indeed, since the polynomial L_j defined above have now their *squares* in $\mathcal{F}^{(2l-1)}(-1, 1)$, we have

$$W^{(l)}(x_{j,l}) = \sum_{k=1}^{l} W^{(l)}(x_{k,l}) L_j(x_{k,l})^2 = \int_{-1}^{1} L_j(t)^2 dt > 0$$

for $j \in \{1, \ldots, l\}$. (See, e.g., [45].)

In general, there are analogous formulas for other intervals of the real line and other measures with finite moments and with $\dim_{\mathbb{R}} \left(L^2((a, b), \sigma)\right) = \infty$. This is a part of the theory of orthogonal polynomials; see Section 3.4 in [97]. There are also related results in larger dimensions; see Section 3.7 in [37].

1.10. Interval designs. It is a particular case of a theorem of Seymour and Zaslavsky that interval designs (named "averaging sets" by these authors) exist for any finite-dimensional space of continuous functions on $\Omega =]0, 1[$, with various measures; see [92] and [2].

An interval design for the space $\mathcal{F}^{(2l-1)}(-1, 1)$ of polynomial functions of degree at most $2l - 1$, and Lebesgue measure, requires strictly more than $\frac{1}{4}l^2$ points and there exists a constant c for which it is known that cl^2 points suffice (this is a 1937 result of Bernstein for which we refer to Section 1.2 in [65]). For low values of k, as an answer to a question of Chebyshev (1874), Bernstein has shown that there exist interval designs for $\mathcal{F}^{(k)}(-1, 1)$ with $|X| = k$ if and only if $k \leq 7$ or $k = 9$, and there is an explicit construction for these values of k (see Section 10.3 in [68]).

1.11. Cubature formulas on spheres and tightness. Let $\Omega = \mathbb{S}^{n-1}$ be the unit sphere in the Euclidean space $\mathbb{R}^n$, $n \geq 2$, and let σ denote the probability measure on $\mathbb{S}^{n-1}$ which is invariant by the orthogonal group $O(n)$.

For spheres, the standard example for Definition 1.3 is given by

$$\mathbb{R} = \mathcal{F}^{(0)}(\mathbb{S}^{n-1}) \subset \mathcal{F}^{(1)}(\mathbb{S}^{n-1}) \subset \cdots \subset \mathcal{F}^{(k)}(\mathbb{S}^{n-1}) \subset \cdots$$

where $\mathcal{F}^{(k)}(\mathbb{S}^{n-1})$ denotes the space of restrictions to the sphere of polynomial functions on $\mathbb{R}^n$ of degree at most k. In this paper, when Definition 1.8 is particularized to spherical cubature formulas and designs, it is always with respect to

---

[3] If Ω is an interval of $\mathbb{R}$, the term "quadrature" is often used instead of "cubature" as in Definition 1.1.

this sequence of polynomial spaces; thus, a cubature formula (X, W) or strength $2l$ is **tight** if

$$|X| = \dim_{\mathbb{R}} \left(\mathcal{F}^{(l)}(\mathbb{S}^{n-1}) \right) = \binom{n+l-1}{n-1} + \binom{n+l-2}{n-1}.$$

Consider the space $\mathcal{P}^{(k)}(\mathbb{S}^{n-1})$ of restrictions to $\mathbb{S}^{n-1}$ of polynomial functions on $\mathbb{R}^n$ which are homogeneous of degree k, and the space $\mathcal{H}^{(k)}(\mathbb{S}^{n-1})$ of restrictions to $\mathbb{S}^{n-1}$ of polynomial functions on $\mathbb{R}^n$ which are homogeneous of degree k and harmonic [4]. We have

$$\mathcal{F}^{(k)}(\mathbb{S}^{n-1}) = \mathcal{P}^{(k)}(\mathbb{S}^{n-1}) \oplus \mathcal{P}^{(k-1)}(\mathbb{S}^{n-1})$$

$$\mathcal{P}^{(k)}(\mathbb{S}^{n-1}) = \bigoplus_{j=0}^{[k/2]} \mathcal{H}^{(k-2j)}(\mathbb{S}^{n-1})$$

for all $k \geq 0$ (see e.g. Section IV.2 in [95]). For reasons related to Proposition 3.2 and Remark 3.3 below, a cubature formula (X, W) or strength $2l + 1$ is **tight** if

$$|X| = 2 \dim_{\mathbb{R}} \left(\mathcal{P}^{(l)}(\mathbb{S}^{n-1}) \right) = 2 \binom{n+l-1}{n-1}.$$

Another example for Definition 1.3 is given by

$$\mathbb{R} = \mathcal{P}^{(0)}(\mathbb{S}^{n-1}) \subset \mathcal{P}^{(2)}(\mathbb{S}^{n-1}) \subset \cdots \subset \mathcal{P}^{(2l)}(\mathbb{S}^{n-1}) \subset \cdots$$

(observe that $\mathcal{P}^{(k)}(\mathbb{S}^{n-1}) \subset \mathcal{P}^{(k+2)}(\mathbb{S}^{n-1})$ for all $k \geq 0$, since the restriction of $x_1^2 + \cdots + x_n^2$ to $\mathbb{S}^{n-1}$ is the constant 1).

A subset $X \subset \mathbb{S}^{n-1}$ is called **antipodal** if $-X = X$; such a X can always be written (non-uniquely) as a disjoint union of some Y inside X and of $-Y$. A cubature formula (X, W) on $\mathbb{S}^{n-1}$ is **antipodal** if X is antipodal and if $W(-x) = W(x)$ for all $x \in X$. If (X, W) is antipodal, $\sum_{x \in X} W(x)\varphi(x) = \int_{\mathbb{S}^{n-1}} \varphi(\omega)d\omega = 0$ for any homogenous polynomial function φ of odd degree; it follows that (X, W) is a cubature formula for $\mathcal{F}^{(2l+1)}(\mathbb{S}^{n-1})$ if and only if it is a cubature formula for $\mathcal{P}^{(2l)}(\mathbb{S}^{n-1})$.

1.12. On strength values for spherical designs and cubature formulas. The following existence result goes back essentially to the solution of Waring's problem by Hurwitz and Hilbert (see [61], [60], page 722 in [36], [40], and [78]):

for any $l \geq 0$, there exists a cubature formula $(Y^{(l)}, W^{(l)})$ for $\mathcal{P}^{(2l)}(\mathbb{S}^{n-1})$ with

$$\binom{n+l-1}{n-1} = \dim_{\mathbb{R}} \left(\mathcal{P}^{(l)}(\mathbb{S}^{n-1}) \right) \leq \left| Y^{(l)} \right|$$

$$\leq \dim_{\mathbb{R}} \left(\mathcal{P}^{(2l)}(\mathbb{S}^{n-1}) \right) - 1 = \binom{n+2l-1}{n-1} - 1.$$

The lower bound is that of Proposition 1.7; the upper bound is a particular case of Theorem 2.8 below, and improves by 1 the bound of Theorem 2.8 in [73]. The

---

[4] A smooth function φ on $\mathbb{R}^n$ is **harmonic** if $\sum_{j=1}^{n} \frac{\partial^2 \varphi}{\partial x_j^2} = 0$.

relation with Waring's problem is alluded to in Item 3.9; for the application to Waring's problem, it is necessary for the weights of the cubature formula to be *rational* positive numbers; our proof of Theorem 2.8 does not show this, and we refer to Chapter 3 of [78] for a complete proof.

It follows that *there exists an antipodal cubature formula* $(X^{(l)}, W^{(l)})$ *of strength* $2l + 1$ *with*

$$2 \binom{n+l-1}{n-1} \leq |X^{(l)}| \leq 2 \binom{n+2l-1}{n-1} - 2.$$

However, the proof of the existence of $(Y^{(l)}, W^{(l)})$ does not provide nice constructions. For $n \geq 3$ and $k \geq 4$, we do not know many good explicit cubature formulas of strength k on $\mathbb{S}^{n-1}$.

Note the relevant asymptotics. For n fixed:

$$\binom{n+l-1}{n-1} \approx \frac{l^{n-1}}{n!} \qquad \text{if} \quad l \to \infty.$$

For l fixed:

$$\binom{n+l-1}{n-1} \approx \frac{n^l}{l!} \qquad \text{if} \quad n \to \infty.$$

The lower bound $|X^{(l)}| \geq 2\binom{n+l-1}{n-1}$ has been improved in some cases, in particular for n fixed and $l \to \infty$ [104].

The general result of Seymour and Zaslavsky already quoted in Item 1.10 implies:

for any $n \geq 2$ *and* $k \geq 0$, *there exists a spherical* k-*design* $X \subset \mathbb{S}^{n-1}$.

The proof of [92] is not constructive and does not give any bound on the size $|X|$.

Let X be a non-empty finite subset of $\mathbb{S}^{n-1}$ and let W be a weight on X such that $\sum_{x \in X} W(x) = 1$.

(i) The pair (X, W) is always a cubature formula of strength 0. It is a cubature formula of strength 1 if and only if the weighted barycentre $\sum_{x \in X} W(x)x$ of (X, W) coincides with the origin of $\mathbb{R}^n$.

(ii) If (X, W) is a cubature formula of strength 2, then X generates $\mathbb{R}^n$. Indeed, if there exists a non-zero vector α orthogonal to X, the function $\omega \longmapsto \langle \alpha \mid \omega \rangle^2$ has a non-zero integral on $\mathbb{S}^{n-1}$ but vanishes identically on X.

Exercise. Denote by Z the *Gram matrix* of X, defined by $Z_{x,y} = \langle x \mid y \rangle$ for $x, y \in X$, and by J the matrix with rows and columns indexed by X and with all entries 1. Show that X is a spherical 2-design if and only if the three following conditions hold: (ii_a) $Z_{x,x} = 1$ for all $x \in X$, (ii_b) $ZJ = 0$, (ii_c) $Z^2 = n^{-1}|X| Z$. (Solution in Lemma 13.6.1 of [46].)

(iii) If (X, W) is a cubature formula of strength 4, then X cannot be a disjoint union of two orthogonal sets; more generally, in case there exist two vectors $\alpha, \beta \in \mathbb{S}^{n-1}$ such that $X \subset \alpha^{\perp} \cup \beta^{\perp}$, the function $\omega \longmapsto \langle \alpha \mid \omega \rangle^2 \langle \beta \mid \omega \rangle^2$ has a non-zero integral on $\mathbb{S}^{n-1}$ but vanishes identically on X.

(iv) If (X, W) is a cubature formula on $\mathbb{S}^{n-1}$ of strength $2l$ for some $l \geq 1$, then the set $P^{(l)}(X) = \{\omega \longrightarrow \langle \omega \mid x \rangle^l\}_{x \in X}$ linearly generates $\mathcal{P}^{(l)}(\mathbb{S}^{n-1})$. Indeed, assume that some $\varphi \in \mathcal{P}^{(l)}(\mathbb{S}^{n-1})$ is orthogonal to $P^{(l)}(X)$. If $\hat{\varphi} \in \mathcal{P}^{(l)}(\mathbb{S}^{n-1})$ is defined by

$$\hat{\varphi}(u) = \int_{\mathbb{S}^{n-1}} \langle \omega \mid u \rangle^l \varphi(\omega) d\sigma(\omega) \quad \forall \, u \in \mathbb{R}^n,$$

the hypothesis means that the restriction of $\hat{\varphi}$ to X vanishes; this implies that $\hat{\varphi} = 0$ by Proposition 1.7. But $\hat{\varphi} = 0$ implies $\varphi = 0$ since $\{\omega \longrightarrow \langle \omega \mid u \rangle^l\}_{u \in \mathbb{S}^{n-1}}$ linearly generates $\mathcal{P}^{(l)}(\mathbb{S}^{n-1})$. With a terminology borrowed from the theory of lattices (see Reminder 1.15), the special case of Property (iv) for $l = 2$ states that cubature formulas of strength 4 on $\mathbb{S}^{n-1}$ provide perfect sets.

For cubature formulas on spheres, the following equivalences are useful. Since many natural examples are provided by intersections of lattices with spheres of various radii (see 1.16), we find it useful to consider for each $\rho > 0$ the sphere $\rho \mathbb{S}^{n-1}$ of radius ρ and centre the origin in $\mathbb{R}^n$; we denote again by σ the $O(n)$-invariant probability measure on $\rho \mathbb{S}^{n-1}$.

Proposition 1.13. *Consider integers $n \geq 2$ and $k \geq 0$, a positive number ρ, a non-empty finite subset X of $\rho \mathbb{S}^{n-1}$, and a weight $W : X \longrightarrow \mathbb{R}^*_+$ such that $\sum_{x \in X} W(x) = 1$. The following conditions are equivalent* [5]

(i) $\sum_{x \in X} W(x)\varphi(x) = \int_{\rho \mathbb{S}^{n-1}} \varphi(\omega) d\sigma(\omega)$ *for all* $\varphi \in \mathcal{P}^{(j)}(\mathbb{R}^n)$ *and* $j \in \{0, 1, \ldots, k\}$, *namely* (X, W) *is a cubature formula of strength k on $\rho \mathbb{S}^{n-1}$.*

(ii) $\sum_{x \in X} W(x)\varphi(x) = \sum_{x \in X} W(x)\varphi(gx)$ *for all* $\varphi \in \mathcal{P}^{(j)}(\mathbb{R}^n)$, $j \in \{0, 1, \ldots, k\}$, *and* $g \in O(n)$.

(iii) $\sum_{x \in X} W(x)\varphi(x) = 0$ *for all* $\varphi \in \mathcal{H}^{(j)}(\mathbb{R}^n)$ *and* $j \in \{1, \ldots, k\}$.

(iv) *If l is the largest integer such that $2l \leq k$, there exists a constant c_{2l} such that*

(iv)$_a$
$$\sum_{x \in X} W(x)\langle x \mid u \rangle^{2l} = c_{2l} \, \rho^{2l} \langle u \mid u \rangle^l \quad \forall \, u \in \mathbb{R}^n,$$

and, if l' is the largest integer such that $2l' + 1 \leq k$, then

(iv)$_b$
$$\sum_{x \in X} W(x)\langle x \mid u \rangle^{2l'+1} = 0 \quad \forall \, u \in \mathbb{R}^n.$$

(v) *Condition (iv)$_a$ holds for any even integer $2l \leq k$ and Condition (iv)$_b$ holds for any odd integer $2l' + 1 \leq k$.*

Moreover, the constant in (iv)$_a$ is

$$c_{2l} = \frac{1 \times 3 \times 5 \times \cdots \times (2l - 1)}{n \, (n + 2) \, (n + 4) \, \cdots \, (n + 2l - 2)}$$

if $l > 0$, and $c_0 = 1$ if $l = 0$.

---

[5] Even if there are canonical identifications between $\mathcal{P}^{(k)}(\mathbb{R}^n)$, $\mathcal{P}^{(k)}(\mathbb{S}^{n-1})$, and $\mathcal{P}^{(k)}(\rho \mathbb{S}^{n-1})$, see Example 1.11, we choose here the first notation.

Proof. The proof of Theorem 3.2 in [102], for spherical designs, readily carries over to cubature formulas. □

Remarks. If (X, W) is antipodal, then Equality $(iv)_b$ is automatically satisfied.

Besides the conditions of Proposition 1.13, there are many other equivalent ones; see [33] and [47], as well as Item 2.14.

1.14. Examples of spherical designs: orbits of finite groups, inductive constructions, distance-regular graphs, and contact points of John's ellipsoids.

Group orbits.

It is easy to show that any orbit in $\mathbb{S}^{n-1}$ of any irreducible finite subgroup of $O(n)$ is a 2-spherical design; see [16], but the result can already be found in a 1940 paper by R. Brauer and H.S.M. Coxeter (Theorem 3.6.6 in [74]). Moreover, a $(2l)$-spherical design which is antipodal is also a $(2l + 1)$-spherical design. In particular, if $(e_1, \ldots, e_n)$ is the canonical basis of $\mathbb{R}^n$, then the set $\{\pm e_1, \ldots, \pm e_n\}$ is a spherical 3-design on $\mathbb{S}^{n-1}$ of cardinality $2n$ (hence a tight 3-design); the set $\{\pm e_1 \pm e_2 \pm \cdots \pm e_n\}$ (all choices of signs) is another spherical 3-design, of cardinality 2^n, and therefore non tight when $n \geq 3$; neither of these is a spherical 4-design.

Orbits of groups generated by reflections provide interesting spherical designs. In particular irreducible root systems of type A_n, D_n, E_n are spherical 3-designs (see Chapter V, § 6, Number 2 in [23]). Moreover, root systems of type A_2, D_4, E_6, E_7 provide spherical 5-designs and the root system of type E_8 provides a spherical 7-design.

For $n \geq 3$, there is a maximal strength $k_{\max}(n)$ for spherical designs in $\mathbb{S}^{n-1}$ which are orbits of finite subgroups of $O(n)$. It is moreover conjectured that $\sup_{n \geq 3} k_{\max}(n) < \infty$; the maximal value of k we know is $k = 19$ and occurs in dimension $n = 4$ (see 3.8 below). For more on spherical designs which are orbits of finite groups, see [16] and [56].

Inductive constructions.

Spherical designs can be constructed inductively as follows. On $\mathbb{S}^1$, the vertices of a regular N-gon constitute a k-design if and only if $N \geq k + 1$. For $n \geq 3$ and some given k, assume that we have a spherical k-design Y on $\mathbb{S}^{n-2}$, as well as an interval design $Z \subset \,]-1, 1[$ for the space $\mathcal{F}^{(k)}(-1, 1)$ of polynomial functions of degrees at most k and for the measure $d\sigma(t) = (1 - t^2)^{(n-3)/2}dt$. Then

$$X = \left\{ \left(\sqrt{1 - \|z\|^2}\, y, \, z \right) \in \mathbb{S}^{n-1} \, \middle| \, y \in Y \quad \text{and} \quad z \in Z \right\}$$

is a spherical k-design in $\mathbb{S}^{n-1}$. Denote by $M'_n(k)$ the smallest integer such that, for any $N \geq M'_n(k)$, there exists a spherical k-design of size N on $\mathbb{S}^{n-1}$. With the construction above, it can be shown that

$$M'_n(k) \leq O\left(k^{\frac{n(n+1)}{2} - 2} \right).$$

It is conjectured that $\frac{n(n+1)}{2} - 2$ can be reduced to $\frac{n(n-1)}{2}$. For all this, see [11] and other papers by Bajnok.

In case $n = 3$, it is a result of [66] that, for all $k \geq 0$, there exists a spherical k-design of size $O(k^3)$.

Constructions of spherical designs appear also in [77], [12], and [13].

Distance regular graphs.

Consider a distance-regular graph Γ, with vertex set $V(\Gamma)$, of valency k. Let θ be an eigenvalue of Γ, $\theta \neq k$, of some multiplicity m; let p denote the orthogonal projection of the space of functions $V(\Gamma) \longrightarrow \mathbb{R}$ to the θ-eigenspace of the adjacency matrix of Γ. Then $p(V(\Gamma))$ is a spherical 2-design in $\rho\mathbb{S}^{m-1}$ for the appropriate radius ρ (Corollary 13.6.2 in [46]).

Contact points of John's ellipsoids.

Let K be a convex body in n-space such that the maximal volume ellipsoid in K is the unit n-ball B. It is a theorem of John that there exist a finite subset X of $K \cap \partial B$ and a positive weight function W on X such that (X, W) is a cubature formula for $\partial B = \mathbb{S}^{n-1}$ of strength 2, indeed of strength 3 if K is assumed to be symmetric; see Lecture 3 in [14]. It could be interesting to investigate systematically polyhedra K such that $K \cap \partial B$ is a spherical k-design, or such that there exists a weight W for which $(K \cap \partial B, W)$ is a cubature formula of strength k, for various values of k.

1.15. A reminder on lattices. Let V be a Euclidean space of dimension $n \geq 1$. A **lattice** is a discrete subgroup Λ of V which generates V as a vector space, and n is the **rank** of Λ. If $\Lambda \subset V$ is a lattice, so is its **dual** $\Lambda^* = \{x \in V \mid \langle x \mid \Lambda \rangle \subset \mathbb{Z}\}$. A lattice Λ is **integral** if $\langle \Lambda \mid \Lambda \rangle \subset \mathbb{Z}$, equivalently if $\Lambda \subset \Lambda^*$. A lattice Λ is **unimodular** if $\mathrm{Vol}(V/\Lambda) = 1$; an integral lattice is unimodular if and only if $\Lambda = \Lambda^*$. A lattice Λ is **even** if $\langle \lambda \mid \lambda \rangle \in 2\mathbb{Z}$ for every $\lambda \in \Lambda$; an **odd** lattice is an integral lattice which is not even. Two lattices $\Lambda \subset V$, $\Lambda' \subset V'$ are **equivalent** if there exists an isometry g from V onto V' such that $g(\Lambda) = \Lambda'$. There is a natural notion of orthogonal direct sum of lattices, and a lattice is **irreducible** if it is not equivalent to a direct sum $\Lambda \oplus \Lambda'$ with $\Lambda, \Lambda' \neq 0$.

Let $\Lambda \subset V$ be an integral lattice; for $m \geq 1$, we denote by

$$\Lambda_m = \{\lambda \in \Lambda \mid \langle \lambda \mid \lambda \rangle = m\}$$

the **shell** of radius m (namely of radius $\sqrt{m}$ in the usual sense of Euclidean geometry). For any $\lambda \in \Lambda_1$, there is an integral lattice Λ' in the orthogonal subspace $\lambda^\perp$ such that $\Lambda = \mathbb{Z}\lambda \oplus \Lambda'$; in particular, $\Lambda_1 = \emptyset$ for an irreducible lattice of rank at least 2.

A standard example of an odd unimodular lattice is the **cubical lattice** $\mathbb{Z}^n \subset \mathbb{R}^n$. If $n \leq 11$, any odd unimodular lattice is equivalent to one of these.

Let n be a multiple of 4. In the standard Euclidean space $\mathbb{R}^n$, define the lattice $D_n = \{\lambda \in \mathbb{Z}^n \mid \sum_{i=1}^n \lambda_i \equiv 0 \pmod 2)\}$ and the **Witt lattice** $\Gamma_n = D_n \cup ((\frac{1}{2}, \ldots, \frac{1}{2}) + D_n)$. Then Γ_n is integral and unimodular; moreover, it is even if and only if n is a multiple of 8. Even unimodular lattices exist only in dimensions $n \equiv 0 \pmod 8$. If $n = 8$, any even unimodular lattice is equivalent to Γ_8, also known as the **root lattice of type E_8** or the **Korkine-Zolotareff lattice**. If $n = 16$, any even unimodular lattice is equivalent to either Γ_{16} (which is irreducible) or to $\Gamma_8 \oplus \Gamma_8$. If $n = 24$, there are 24 equivalence classes of even unimodular lattices (Niemeier's classification, 1968); the most famous of them is the lattice discovered by **Leech** (1964), which is the only even unimodular lattice Λ in dimension $n \le 31$ such that $\Lambda_2 = \emptyset$, and which has a remarkably high density (see, e.g., Table 1.5 in Chapter 1 of [26]). If $n = 32$, the number of equivalence classes of even unimodular lattices is larger than 8×10^7.

There is a classification of integral unimodular lattices of small rank. For $n \le 16$, Kneser (1957) has shown that the only irreducible integral unimodular lattices are $\mathbb{Z}$, Γ_n for $n = 8, 12, 16$, and three odd lattices in dimensions 14, 15, and 16. In particular, if $9 \le n \le 11$, any odd unimodular lattice is equivalent to either $\mathbb{Z}^n$ or $\Gamma_8 \oplus \mathbb{Z}^{n-8}$. Integral unimodular lattices have been later classified for $n \le 23$ (Conway and Sloane, 1982) and $n \le 25$ (Borcherds, 1984). For $n \le 24$, see Chapters 16 and 17 of [26]; minor corrections to previous tables are given in [7].

Let $\Lambda \subset V$ be a lattice (possibly neither integral nor unimodular). For $r > 0$, let B_r denote the ball or radius r centred at the origin of V. The **sphere packing** associated to Λ is $\bigcup_{\lambda \in \Lambda}(\lambda + B_r)$, where r denotes the largest real number such that the balls $\lambda + B_r$ have disjoint interiors ($\lambda \in \Lambda$); the **density** of Λ is the number

$$\lim_{R \to \infty} \mathrm{Vol}\left(\left(\bigcup_{\lambda \in \Lambda}(\lambda + B_r)\right) \cap B_R\right) \Big/ \mathrm{Vol}(B_R) = \frac{\mathrm{Vol}(B_r)}{\mathrm{Vol}(V/\Lambda)}.$$

The lattice Λ is **extreme** if this density is a local maximum in the space of all lattices of dimension n; since density is homothety-invariant, a unimodular lattice is extreme if its density is a local maximum in the space of all unimodular lattices of dimension n, a space which can be identified with the double coset space $SL(n, \mathbb{Z})\backslash SL(n, \mathbb{R})/O(n)$. Denote by

$$\Lambda_{\mathrm{short}} = \{\mu \in \Lambda \, , \, \mu \neq 0 \mid \langle \mu \mid \mu \rangle \le \langle \lambda \mid \lambda \rangle \quad \text{for all} \quad \lambda \in \Lambda \, , \, \lambda \neq 0\}$$

the shell of **short vectors** in Λ; the lattice Λ is **eutactic** if there exists a weight $W : \Lambda_{\mathrm{short}} \longrightarrow \mathbb{R}_+^*$ such that $(\Lambda_{\mathrm{short}}, W)$ is a spherical cubature formula of strength 3, and **perfect** if the set $\{\omega \longmapsto \langle \omega \mid \lambda \rangle^2 \mid \lambda \in \Lambda_{\mathrm{short}}\}$ linearly generates the space of homogeneous polynomial of degree 2 on V. The first result involving both lattices and cubature formulas is the following theorem of **Voronoi** (1908):

   a lattice is extreme if and only if it is both eutactic and perfect.

For Λ to be extreme, it is sufficient that Λ is **strongly perfect**, which means that Λ_{short} is a spherical 5-design (Theorem 6.4 in [102]). Strongly perfect lattices have been classified in dimensions $n \leq 11$, where they occur in dimensions $n = 1, 2, 4, 6, 7, 8, 10$ only [102], [79]; other examples occur in [9].

Standard references on lattices include [26], [38], [74], [76], [88], and [102].

1.16. Examples of spherical designs: lattice designs. Let Λ be a lattice in $\mathbb{R}^n$. Any non-empty shell Λ_m provides a spherical design of some strength in the sphere $\sqrt{m}\mathbb{S}^{n-1}$. This connection goes back to [101]. See [82] for many examples; let us indicate here a sample.

If $\Lambda \subset \mathbb{R}^8$ is a root lattice of type E_8, then Λ_{2m} is a spherical 7-design for any $m \geq 1$. Moreover, it can be shown that a shell Λ_{2m} is a 8-design if and only if the Ramanujan coefficient $\tau(m)$ of the modular form

$$\Delta_{24}(z) = q^2 \prod_{m=1}^{\infty} (1 - q^{2m})^{24} = \sum_{m=1}^{\infty} \tau(m) q^{2m}$$
$$= q^2 - 24q^4 + 256q^6 - 1472q^8 + \cdots \qquad (q = e^{i\pi z})$$

is zero. (See Example 3.10 below for a second appearance of this modular form Δ_{24} of weight 12.) Now it is a famous conjecture of Lehmer [70] that $\tau(m) \neq 0$ for all $m \geq 1$. If $\Gamma \subset \mathbb{R}^{16}$ is a Witt lattice (an irreducible even unimodular lattice, uniquely defined up to isometry by these properties in dimension 16), then Γ_{2m} is a spherical 3-design for any $m \geq 1$, and the condition for one of these shells to be a 4-design happens to be again the vanishing of the corresponding Ramanujan coefficients. The same holds for the reducible even unimodular lattice $\Lambda \oplus \Lambda \in \mathbb{R}^{16}$. Consequently, the following four claims are simultaneously true or not true:

 (i) *Lehmer's conjecture holds, namely $\tau(m) \neq 0$ for all $m \geq 1$;*
 (ii) *no shell of the root lattice Λ of type E_8 is a spherical 8-design;*
 (iii) *no shell of the lattice $\Lambda \oplus \Lambda \subset \mathbb{R}^{16}$ is a spherical 4-design;*
 (iv) *no shell of the Witt lattice $\Gamma \subset \mathbb{R}^{16}$ is a spherical 4-design.*

Claim (i) has been checked for $m \leq 10^{15}$ [90].

It is easy to formulate other equivalences of the same kind. For example, any shell of the Leech lattice L (see 1.15) is a spherical 11-design, and the condition for L_{2m} ($m \geq 2$) to be a 12-design is that the mth coefficient $\mu(m) = \sum_{j=1}^{m-1} \tau(j)\tau(m-j)$ of $(\Delta_{24})^2$ vanishes. Thus

$\mu_m \neq 0$ *for any $m \geq 2$ if and only if L_{2m} is not a spherical 12-design*

and, moreover, we have checked that $\mu_m \neq 0$ for $m \leq 1200$. We do not know of any other lattice with a shell which is a spherical design of strength $k \geq 11$.

Similarly, let now Λ denote a Niemeier lattice, namely an even unimodular lattice in dimension 24 with $\Lambda_2 \neq \emptyset$ (up to isometry, there are 23 such lattices).

Denote by Q the modular form of weight 4 defined by

$$Q(q) = (\theta_3(q))^8 - \frac{1}{16}(\theta_2(q)\theta_4(q))^4$$

$$= \left(\sum_{m \in \mathbb{Z}} q^{m^2}\right)^8 - \left(\sum_{m \in 1/2 + \mathbb{Z}} q^{m^2}\right)^4 \left(\sum_{m \in \mathbb{Z}} (-q)^{m^2}\right)^4$$

$$= 1 + 240q^2 + 2160q^4 + \cdots$$

and set

$$Q(q)\Delta_{24}(q) = \sum_{m=1}^{\infty} \nu_m q^{2m}.$$

Then

$\nu_m \neq 0$ *if and only if* Λ_{2m} *is not a spherical 4-design* $(m \geq 1)$

and, moreover, we have checked that $\nu_m \neq 0$ for $m \leq 1200$. (Observe that the condition is the same for any of the 23 Niemeier lattices.)

For one more example in this class, consider the cubical lattice $\mathbb{Z}^n$. The criterion stated in 1.14 in terms of irreducible representations of finite subgroups of $O(n)$ shows that all non-empty shells are spherical 3-designs. Moreover, it can be shown that there are two classes of "special shells" which are spherical 5-designs, namely $(\mathbb{Z}^4)_m$ for $m = 2a$ and $(\mathbb{Z}^7)_m$ for $m = 4^b(8a + 3)$. Let us restrict now for brevity to $n \geq 8$, and denote by $\Theta^{[n]}$ the modular form of weight $4 + n/2$ defined by

$$\Theta^{[n]}(q) = \frac{1}{16}(\theta_2(q)\theta_4(q))^4 (\theta_3(q))^n = \sum_{m=1}^{\infty} \kappa_m^{[n]} q^m.$$

Then

$\kappa_m^{[n]} \neq 0$ *if and only if* $(\mathbb{Z}^n)_m$ *is not a spherical 4-design* $(m \geq 1)$

and, moreover, we have checked that these hold for all $n \geq 8$ and $m \leq 1200$. (See [82], which contains a discussion including $1 \leq n \leq 7$.)

1.17. Tight spherical designs on spheres of dimension $n - 1 \geq 2$. (See [15] and references there.)

Even strengths.

Tight spherical $(2l)$-designs do not exist when $2l \geq 6$.

Tight spherical 4-designs in $\mathbb{S}^{n-1}$ are of size $\dim_{\mathbb{R}}(\mathcal{F}^{(2)}(\mathbb{S}^{n-1})) = \frac{1}{2}n(n+3)$. They cannot exist unless n is of the form $(2m+1)^2 - 3$. If $m = 1$ or $m = 2$ examples are known, and known to be unique up to isometry; they are respectively of size 27 in $\mathbb{S}^5$ and size 275 in $\mathbb{S}^{21}$ (see below). If $m = 3$ and $m = 4$ (and infinitely many larger values), non-existence has been proved.

Tight spherical 2-designs exist in all dimensions, and are regular simplices.

Odd strengths.

For $l \geq 0$, a tight spherical $(2l + 1)$-design is necessarily antipodal, by Theorem 5.12 in [33].

Tight spherical $(2l + 1)$-designs do not exist when $2l + 1 \geq 9$, up to one exception (which is unique up to isometry): the $196\,560$ short vectors of a Leech lattice which provide (after dividing all vectors by 2) an 11-design in $\mathbb{S}^{23}$. Observe that

$$196560 = 2\binom{28}{5} = 2\dim_{\mathbb{R}}(\mathcal{P}^{(5)}(\mathbb{S}^{23})).$$

Tight spherical 7-designs in $\mathbb{S}^{n-1}$ are of size $\frac{1}{3}n(n+1)(n+2)$. They cannot exist unless n is of the form $3m^2 - 4$. If $m = 2$ or $m = 3$, examples are known, and known to be unique up to isometry. (Up to homothety, they are respectively a root system of type E_8 and the short vertices of the unimodular integral lattice denoted by O_{23} in [102] [6].) If $m = 4$ and $m = 5$ (and infinitely many larger values), non-existence has been proved.

Tight spherical 5-designs in $\mathbb{S}^{n-1}$ are of size $n(n+1)$. They cannot exist unless n is either 3 or of the form $(2m+1)^2 - 2$. If $n = 3$, or $m = 1$, or $m = 2$, examples are known, and known to be unique up to isometry. (They are respectively a regular icosahedron and, up to homothety, the short vectors of a lattice which is dual to a root lattice of type E_7 and the short vectors of a lattice of type M_{23}^*, with the notation of [102].) If $m = 3$ and $m = 4$ (and infinitely many larger values), non-existence has been proved.

Let $X \subset \mathbb{S}^{n-1}$ be a tight spherical 5-design. It is a consequence of Proposition 3.4 that $\langle x \mid y \rangle = \pm \alpha$ for any $x, y \in X$ with $x \neq \pm y$, where $\alpha = 1/\sqrt{n+2}$. Choose $e \in X$ and set $X_0 = \{x \in X \mid \langle x \mid e \rangle = \alpha\}$. Then X_0 is a tight spherical 4-design in the sphere $\sqrt{1 - \alpha^2}\mathbb{S}^{n-2}$ centred at αe in the affine hyperplane $\{x \in \mathbb{R}^n \mid \langle x \mid e \rangle = \alpha\}$, by Theorem 8.2 in [33].

Tight spherical 3-designs exist in all dimensions, and are of the form $\{\pm e_1, \ldots, \pm e_n\}$, where $\{e_1, \ldots, e_n\}$ is an orthonormal basis of $\mathbb{R}^n$. Tight spherical 1-designs are of the form $\{\pm e_1\}$.

For results on tight designs on other spaces (Ω, σ), see [4] and [5].

1.18. Designs with few points on $\mathbb{S}^2$.

Given an integer $N \geq 1$, it is a natural question to ask what is the largest integer $k_N \geq 0$ for which there exists a spherical k_N-design with N points on a sphere, say here on $\mathbb{S}^2$. We report now some answers to this question, from [55].

---

[6]Let L be a Leech lattice. Let first $e \in L$ be a short vector, with $\langle e \mid e \rangle = 4$. Denote by p the orthogonal projection of $\mathbb{R}^{24}$ onto $e^{\perp}$ and set $L'_e = \{x \in L \mid \langle e \mid x \rangle \equiv 0 \pmod{2}\}$. Then $O_{23} = p(L'_e)$ is a unimodular integral lattice with $\min\{\langle x \mid x \rangle \mid x \in O_{23}, \ x \neq 0\} = 3$ and $|\{x \in O_{23} \mid \langle x \mid x \rangle = 3\}| = 4600$. Let then $f \in L$ be a vector such that $\langle f \mid f \rangle = 6$. Then $M_{23} = L \cap f^{\perp}$ is an integral lattice with $\min\{\langle x \mid x \rangle \mid x \in M_{23}, \ x \neq 0\} = 4$; if M_{23}^* denote its dual lattice, then $[M_{23}^* : M_{23}] = 4$, and there are 552 short vectors in M_{23}^*.

$N = 1$	$k_1 = 0$	single point
$N = 2$	$k_2 = 1$	pair of antipodal points
$N = 3$	$k_3 = 1$	equatorial equilateral triangle
$N = 4$	$k_4 = 2$	regular tetrahedron (tight)
$N = 5$	$k_5 = 1$	
$N = 6$	$k_6 = 3$	regular octahedron (tight)
$N = 7$	$k_7 = 2$	
$N = 8$	$k_8 = 3$	cube (non-tight)
$N = 9$	$k_9 = 2$	
$N = 10$	$k_{10} = 3$	
$N = 11$	$k_{11} = 3$	
$N = 12$	$k_{12} = 5$	regular icosahedron (tight)
$13 \le N \le 23$	$k_N \in \{3, 4, 5\}$	
$N = 20$	$k_{20} = 5$	regular dodecahedron (non tight)
$N = 24$	$k_{24} = 7$	improved snub cube
$\cdots$	$\cdots$	
$N = 60$	$k_{60} = 10$	

Remarks. (i) The table shows that k_N is *not* monotonic as a function of N.

(ii) We know from Propositions 1.7 and 3.2 that, on $\mathbb{S}^2$, any spherical $(2l)$-design has size $N \ge (l+1)^2$ and any spherical $(2l+1)$-design has size $N \ge (l+2)(l+1)$. The table above shows that these bounds are not sharp unless $2l \in \{0, 2\}$ or $2l + 1 \in \{1, 3, 5\}$. For example, a 4-design has minimal size $12 (> 9)$ and a 7-design has minimal size $24 (> 20)$.

(iii) The table refers to isolated examples; here is a continuous family, from [53]. Distribute the 12 points of a regular icosahedron into two poles N and S, and two sets P, Q of 5 points each in planes orthogonal to the diameter joining N to S. Let X_θ denote the union of the poles, the set P, and the image of the set Q by a rotation of angle θ fixing N and S. If θ is not a multiple of $2\pi/5$, then X_θ is a spherical 4-design which is not a 5-design.

(iv) The Archimedes' regular snub cube [7] (24 vertices) is a spherical 3-design (not a 4-design), while the improved snub cube reported to in the table is indeed a spherical 7-design.

(v) The regular truncated icosahedron (= soccer ball, 60 vertices) is a spherical 5-design (not a 6-design). The "improved soccer ball" of [48], [49], which is almost indistinguishable from the regular one with the naked eye, is a 9-design. The spherical 10-design of size 60 which appears in [55] is quite different.

(vi) The same paper shows a 9-design with $N = 48$ points; indeed $k_{48} = 9$ and $k_N < 9$ whenever $N < 48$ or $N \in \{49, 51, 53\}$. On the other hand, there exists a cubature formula on $\mathbb{S}^2$ of strength 9 and size 32 (Section 5 in [48], and Item 3.7 below).

---

[7]See for example http://mathworld.wolfram.com/SnubCube.html

(vii) It is conceivable that many values of N are relevant for extra-mathematical reasons:

$N \leq 12$ pores on pollen-grains (botany);

$N \geq 60$ atoms in various large carbon molecules (chemistry);

$N \sim 20\,000$ detectors for a PET tomography of the brain (medical imaging);

$100 \leq N \leq 10^{20}$ charged particles on a conducting sphere (electrostatics).

In Section 3 below, there are other examples of cubature formulas and designs on spheres.

1.19. Several quality criteria for spherical configurations. What is the best way to arrange a given number N of points on a sphere, say here on $\mathbb{S}^2$? The answer depends of course of what is meant by "the best". Besides maximizing the strength of the configuration viewed as a spherical design, we mention here two other criteria.

The **Tammes' problem** asks what are the configurations $\{x_1, \ldots, x_n\} \subset \mathbb{S}^2$ which maximize $\min_{1 \leq i,j \leq n, i \neq j}(\text{distance}(x_i, x_j))$, or equivalently minimize $\max_{1 \leq i,j \leq n, i \neq j}(\langle x_i \mid x_j \rangle)$. Configurations of N points on $\mathbb{S}^2$ are discussed in [98] for $N \leq 12$. Pieter L.M. Tammes (1903-1980) is a Dutch botanist who was interested in the distribution of places of exit on pollen grains (the most frequent case seems to be $N = 3$); he should not be confused with his aunt Tine Tammes (1871–1947), who made important contributions to early genetics [94].

Here are some of the configurations which are the best from the point of view of Tammes' problem: those of Example 1.18 for $N = 4, 6, 12$ (regular polytopes with triangular faces), but other configurations for $N = 8$ (square antiprisms), $N = 20$ (unknown configurations which are not regular dodecahedras [99]), and $N = 24$ (snub cubes, see ([87] and [84]). More on Tammes' problem in mathematics in [27], in Section 35 of [42], in Section 2.3 of Chapter 1 of [26], and in Chapter 3 of [41].

Let X be a finite subset of $\mathbb{S}^{n-1}$. When the emphasis is on properties like those of Proposition 1.13, the set X is called a *spherical design*. However, when the emphasis is on the distance set D_X defined in 1.6, the standard name for X is that of **spherical code**. For many constructions of spherical codes, see [41].

Problem 7 of Smale's *Mathematical problems for the next century* [93] asks what are the configurations which maximize $\prod_{1 \leq i < j \leq N} \text{distance}(x_i, x_j)$. The problem is motivated by complexity theory, and the search of algorithms related to the fundamental theorem of algebra.

For a discussion of related criteria, see [85].

1.20. Cubature formulas on other spaces. There are documented cubature formulas of strength $3, 5, \ldots$ (with respect to the space of polynomials) on the hypercube $[-1,1]^n$ of $\mathbb{R}^n$. For compact subsets of the plane with positive area, there are cubature formulas of strength k and of size $\frac{1}{2}(k+1)(k+2)$. See [31], in particular Section 5.7.

Let $\Omega = \mathbb{T}^2$ be a 2-torus of revolution embedded in the Euclidean space $\mathbb{R}^3$, together with its standard probability measure σ (up to a normalization factor, σ is the area given by the first fundamental form of the surface $\mathbb{T}^2$ embedded in 3-space). Let $\mathcal{F}^{(k)}(\mathbb{T}^2)$ be the space of functions on the torus which are restrictions of polynomial functions of total degree at most k on $\mathbb{R}^3$. Kuijlaars [69] has shown that there exist constants $C_1, C_2 > 0$ with the following property: for any $k \geq 0$, there exists a geometrical design on $\mathcal{F}^{(k)}(\mathbb{T}^2)$ with a number N of points satisfying $C_1 k^2 \leq N \leq C_2 k^2$.

2. Cubature formulas and reproducing kernels

This section begins with a reminder of standard material which goes back at least to [3] and [67]; see also [96] and [20].

Let $\mathcal{H}$ be a real Hilbert space of functions on a set Ω. We denote by $\langle \varphi_1 \mid \varphi_2 \rangle$ the scalar product of φ_1 and φ_2 in $\mathcal{H}$. We assume [8] that, for each $\omega \in \Omega$, the evaluation $\mathcal{H} \longrightarrow \mathbb{R}$, $\varphi \longmapsto \varphi(\omega)$ is continuous; this implies that there exists a function $\varphi_\omega \in \mathcal{H}$ such that $\varphi(\omega) = \langle \varphi \mid \varphi_\omega \rangle$ for all $\varphi \in \mathcal{H}$.

Definition 2.1. With the notation above, the **reproducing kernel** of $\mathcal{H}$ is the function $\Phi : \Omega \times \Omega \longrightarrow \mathbb{R}$ defined by $\Phi(\omega', \omega) = \langle \varphi_\omega \mid \varphi_{\omega'} \rangle$ for all $\omega, \omega' \in \Omega$, or equivalently by $\Phi(\cdot, \omega) = \varphi_\omega(\cdot)$.

The terminology, "reproducing", refers to the equality $\varphi(\omega) = \langle \varphi(\cdot) \mid \Phi(\cdot, \omega) \rangle$ for all $\varphi \in \mathcal{H}$ and $\omega \in \Omega$.

Proposition 2.2. *Let Ω, $\mathcal{H}$, $(\varphi_\omega)_{\omega \in \Omega}$, and Φ be as above.*

(i) *The kernel Φ is of positive type. In particular, its diagonal values $\Phi(\omega, \omega)$ are positive, and $|\Phi(\omega, \omega')|^2 \leq \Phi(\omega, \omega)\Phi(\omega', \omega')$ for all $\omega, \omega' \in \Omega$.*
(ii) *The family $(\varphi_\omega)_{\omega \in \Omega}$ generates $\mathcal{H}$.*
(iii) *If $\mathcal{H} \neq 0$, then $\Phi(\omega, \omega) \neq 0$ for some $\omega \in \Omega$.*
(iv) *If $(e_j)_{j \in J}$ is any orthonormal basis of $\mathcal{H}$, then*

$$\Phi(\omega, \omega') = \sum_{j \in J} e_j(\omega)e_j(\omega')$$

for all $\omega, \omega' \in \Omega$.
(v) *In case there exists a finite positive measure σ on Ω such that $\mathcal{H}$ is a closed subspace of $L^2(\Omega, \sigma)$, with $\int_{\Omega \times \Omega} |\Phi(\omega, \omega')|^2 d\sigma(\omega)d\sigma(\omega') < \infty$, then*

$$\dim_{\mathbb{R}}(\mathcal{H}) = \int_\Omega \Phi(\omega, \omega)d\sigma(\omega) < \infty.$$

(vi) *Assume moreover that Ω is a topological space and that the dimension of $\mathcal{H}$ is finite. Then the following conditions are equivalent:*
(vi$_a$) *all functions in $\mathcal{H}$ are continuous;*

---

[8] In all examples appearing below, $\mathcal{H}$ is finite dimensional and this condition is therefore automatically fulfilled. Hence, the reader can assume to start with that $\dim_{\mathbb{R}}(\mathcal{H}) < \infty$.

(vi$_b$) *the kernel* $\Phi : \Omega \times \Omega \longrightarrow \mathbb{R}$ *is continuous;*

(vi$_c$) *the mapping* $\omega \longmapsto \varphi_\omega$ *from* Ω *to* $\mathcal{H}$ *is continuous.*

Proof. (i) By definition, the kernel Φ is of positive type if it is symmetric and if

(∗) $$\sum_{j,k=1}^{n} \lambda_j \lambda_k \Phi(\omega_j, \omega_k) \geq 0$$

for all integers n, real numbers $\lambda_1, \ldots, \lambda_n$ and points $\omega_1, \ldots, \omega_n$ in Ω. This is clear here, since the left-hand term $\sum_{j,k=1}^{n} \lambda_j \lambda_k \langle \varphi_{\omega_k} \mid \varphi_{\omega_j} \rangle$ of (∗) is equal to the square of the Hilbert-space norm of the sum $\sum_{k=1}^{n} \lambda_k \varphi_{\omega_k}$. The last claim of (i) follows by the Cauchy-Schwarz inequality.

(ii) Observe that, for $\varphi \in \mathcal{H}$, the condition $\langle \varphi \mid \varphi_\omega \rangle = 0$ for all $\omega \in \Omega$ implies that $\varphi(\omega) = 0$ for all $\omega \in \Omega$.

(iii) Assume that $\Phi(\omega, \omega) = 0$ for all $\omega \in \Omega$. Then, by (i), $\Phi(\omega, \omega') = 0$ for all $\omega, \omega' \in \Omega$, and it follows from (ii) that $\mathcal{H} = 0$.

(iv) Evaluate the Fourier expansion $\varphi_{\omega'} = \sum_{j \in J} \langle \varphi_{\omega'} \mid e_j \rangle e_j$ at the point ω, and obtain $\Phi(\omega, \omega') = \sum_{j \in J} \langle \varphi_{\omega'} \mid e_j \rangle \langle e_j \mid \varphi_\omega \rangle = \sum_{j \in J} e_j(\omega) e_j(\omega')$.

(v) Denote by $K_\Phi : L^2(\Omega, \sigma) \longrightarrow L^2(\Omega, \sigma)$ the linear operator defined by the kernel Φ, namely by $(K_\Phi \varphi)(\omega) = \int_\Omega \Phi(\omega, \omega') \varphi(\omega') d\sigma(\omega')$ for all $\varphi \in \mathcal{H}$ and $\omega \in \Omega$. On the one hand, K_Φ is a Hilbert-Schmidt operator, because of the L^2-condition on Φ; on the other hand, K_Φ is the identity, since the kernel Φ is reproducing. It follows that the dimension of $\mathcal{H}$ is finite. Moreover, we have

$$\int_\Omega \Phi(\omega, \omega) d\sigma(\omega) = \sum_{j \in J} \int_\Omega |e_j(\omega)|^2 d\sigma(\omega) = \sum_{j \in J} \|e_j\|^2 = \dim_\mathbb{R}(\mathcal{H})$$

by (iv).

(vi) Since $\dim_\mathbb{R}(\mathcal{H}) < \infty$, the mapping of (vi$_c$) is continuous if and only if the real-valued functions $\omega \longmapsto \langle \varphi \mid \varphi_\omega \rangle = \varphi(\omega)$ are continuous for all $\varphi \in \mathcal{H}$, so that (vi$_a$) and (vi$_c$) are equivalent.

If (vi$_c$) holds, then Φ is continuous since it is the composition of the continuous mapping $(\omega', \omega) \longmapsto (\varphi_\omega, \varphi_{\omega'})$ from $\Omega \times \Omega$ to $\mathcal{H} \times \mathcal{H}$ with the scalar product. If (vi$_b$) holds, then $\varphi_\omega = \Phi(\cdot, \omega)$ is continuous for all $\omega \in \Omega$, and (vi$_a$) follows by (ii). $\square$

Example 2.3 (Atomic measure). Let X be a set and $W : X \longrightarrow \mathbb{R}_+^*$ a positive-valued function. The Hilbert space $\ell^2(X, W)$, with scalar product given by

$$\langle \psi_1 \mid \psi_2 \rangle = \sum_{x \in X} W(x) \psi_1(x) \psi_2(x) \quad \text{for all} \quad \psi_1, \psi_2 \in \ell^2(X, W),$$

gives rise to the functions

$$\psi_x = \frac{1}{W(x)} \delta_x : y \longmapsto \begin{cases} W(x)^{-1} & \text{if } y = x \\ 0 & \text{otherwise} \end{cases}$$

and to the reproducing kernel Ψ with values

$$\Psi(x,y) = \langle \psi_y \mid \psi_x \rangle = \begin{cases} W(x)^{-1} & \text{if } y = x \\ 0 & \text{otherwise.} \end{cases}$$

In particular, if X is finite and $W(x) = |X|^{-1}$ for all $x \in X$, then $|X|^{-1}\Psi$ is the characteristic function of the diagonal in $X \times X$.

Example 2.4 (Standard reproducing kernels on spheres). Let the notation be as in Example 1.11 and let $k \geq 0$. Each of the spaces $\mathcal{F}^{(k)}(\mathbb{S}^{n-1})$, $\mathcal{P}^{(k)}(\mathbb{S}^{n-1})$, $\mathcal{H}^{(k)}(\mathbb{S}^{n-1})$ is a subspace of the Hilbert space $L^2(\mathbb{S}^{n-1}, \sigma)$.

There exists a unique polynomial $Q^{(k)}(T) \in \mathbb{R}[T]$ of degree k with the following properties: for any $\omega \in \mathbb{S}^{n-1}$, the polynomial function defined on $\mathbb{R}^n$ which is homogeneous of degree k and of which the restriction to $\mathbb{S}^{n-1}$ is given by [9]

$$\varphi_\omega : \omega' \longmapsto Q^{(k)}(\langle \omega \mid \omega' \rangle)$$

is harmonic, and $Q^{(k)}(1) = \dim_{\mathbb{R}}(\mathcal{H}^{(k)}(\mathbb{S}^{n-1}))$; the polynomial $Q^{(k)}$ is a form of a Gegenbauer polynomial (see, e.g., Theorems IV.2.12 and IV.2.14 in [95], or Section IX.3 in [103]). It is routine to check that $Q^{(0)}(T) = 1$, $Q^{(1)}(T) = nT$, and

$$\frac{k+1}{n+2k} Q^{(k+1)}(T) = T Q^{(k)}(T) - \frac{n+k-3}{n+2k-4} Q^{(k-1)}(T)$$

for $k \geq 2$. Observe that $Q^{(k)}(T)$ is of the form

$$Q^{(k)}(T) = \sum_{j=0}^{[k/2]} (-1)^j c_j^{(k)} T^{k-2j}$$

with $c_0^{(k)}, c_1^{(k)}, \ldots > 0$. The reproducing kernel of the space $\mathcal{H}^{(k)}(\mathbb{S}^{n-1})$ is given by

$$\Phi^{(k)}(\omega, \omega') = Q^{(k)}(\langle \omega \mid \omega' \rangle)$$

for all $\omega, \omega' \in \mathbb{S}^{n-1}$. It is a restriction of the kernel

$$\mathbb{R}^n \times \mathbb{R}^n \ni (\omega, \omega') \longmapsto \sum_{j=0}^{[k/2]} (-1)^j c_j^{(k)} \langle \omega \mid \omega' \rangle^{k-2j} \langle \omega \mid \omega \rangle^j \langle \omega' \mid \omega' \rangle^j \in \mathbb{R}$$

which is homogeneous of degree k in each variable ω, ω' separately.

The reproducing kernels of the spaces $\mathcal{P}^{(k)}(\mathbb{S}^{n-1})$ and $\mathcal{F}^{(k)}(\mathbb{S}^{n-1})$ are given similarly in terms of the polynomials

$$C^{(k)}(T) = \sum_{j=0}^{[k/2]} Q^{(k-2j)}(T) \quad \text{and} \quad R^{(k)}(T) = \sum_{j=0}^{k} Q^{(j)}(T).$$

---

[9] Where $\langle \omega \mid \omega' \rangle = \sum_{i=1}^{n} \omega_i \omega_i'$.

Observe that

$$\dim_{\mathbb{R}}(\mathcal{H}^{(k)}(\mathbb{S}^{n-1})) = Q^{(k)}(1) = \binom{n+k-1}{n-1} - \binom{n+k-3}{n-1}$$

$$\dim_{\mathbb{R}}(\mathcal{P}^{(k)}(\mathbb{S}^{n-1})) = C^{(k)}(1) = \binom{n+k-1}{n-1}$$

$$\dim_{\mathbb{R}}(\mathcal{F}^{(k)}(\mathbb{S}^{n-1})) = R^{(k)}(1) = \binom{n+k-1}{n-1} + \binom{n+k-2}{n-1}$$

by Proposition 2.2.v.

This material is standard: see, e.g., [95] and [33].

Theorem 2.8 is the first general existence result for cubature formulas of this exposition. (It is a minor strengthening of Theorems 2.8 and 3.17 in [73].)

Lemma 2.5. *Let V be a finite dimensional real vector space, let μ be a probability measure on V, and let $b = \int_V v d\mu(v)$ denote its barycentre. If C is a convex subset of V such that $\mu(C) = 1$, then $b \in C$.*

Remarks. (i) The lemma is probably well-known, but we haven't been able to trace it in print. The proof below was shown to us by Yves Benoist.

(ii) The point of the lemma is that b is not only in the closure of C, but in C itself.

Proof. Let U be the minimal affine subspace of V such that $\mu(C \cap U) = 1$; upon restricting C, we can assume that $C \subset U$. Upon translating μ and C, we can assume that $b = 0$.

We claim that $b = 0$ is in the interior of C inside U. If this was not true, there would exist by the Hahn-Banach theorem a non-zero linear form ξ on U such that $\xi(c) \geq 0$ for all $c \in C$. The equality

$$0 = \xi(b) = \int_U \xi(w)d\mu(w) = \int_C \xi(c)d\mu(c)$$

would imply $\xi(c) = 0$ for almost every $c \in C$; thus, we would have $\mu(C \cap \mathrm{Ker}(\xi)) = 1$, in contradiction with the definition of U.

Hence b is in the interior of C inside U, and in particular $b \in C$. $\qquad\square$

Proposition 2.6. *Let (Ω, σ) be a finite measure space. Let $\mathcal{H}$ be a finite dimensional Hilbert space of functions on Ω which is a subspace of $L^2(\Omega, \sigma)$ and which contains the constant functions.*

Then there exists a cubature formula (X, W) for $\mathcal{H}$ such that

$$|X| \leq \dim_{\mathbb{R}}(\mathcal{H}).$$

Proof. Assume first that σ is a probability measure. Let $\mathcal{H}^0$ be the orthogonal supplement of the constants in $\mathcal{H}$ and let Φ^0 denote its reproducing kernel. Recall that, for $w \in \Omega$, the function $\varphi_w^0 : w' \longrightarrow \Phi^0(w', w)$ is in $\mathcal{H}^0$. The set

$$\tilde{\Omega} = \{\varphi \in \mathcal{H}^0 \mid \varphi = \varphi_w^0 \text{ for some } w \in \Omega\}$$

linearly generates $\mathcal{H}^0$ by Proposition 2.2.ii. Observe that

$$\int_\Omega \varphi_\omega^0(\omega')d\sigma(\omega) = \int_\Omega \varphi_{\omega'}^0(\omega)d\sigma(\omega) = \langle 1 \mid \varphi_{\omega'}^0 \rangle_{\mathcal{H}^0} = 0$$

for all $\omega' \in \Omega$, by definition of $\mathcal{H}^0$. It follows from the previous lemma, applied to the image of σ on $\tilde{\Omega}$, that 0 is in the convex hull of $\tilde{\Omega}$.

By Carathéodory's theorem (see e.g. Theorem 11.1.8.6 in [21]), there exist a finite subset X of Ω of cardinality at most $\dim_{\mathbb{R}}(\mathcal{H}^0) + 1 = \dim_{\mathbb{R}}(\mathcal{H})$ and a weight function $W : X \longrightarrow \mathbb{R}_+^*$ such that

$$\sum_{x \in X} W(x) = 1 \qquad \text{and} \qquad \sum_{x \in X} W(x)\varphi_x^0 = 0 \in \mathcal{H}^0.$$

For all $x \in X$ and $\omega' \in \Omega$, we have $\varphi_x^0(\omega') = \Phi^0(\omega', x) = \Phi^0(x, \omega') = \varphi_{\omega'}^0(x)$; hence

$$\sum_{x \in X} W(x)\varphi_{\omega'}^0(x) = 0 = \int_\Omega \varphi_{\omega'}^0(\omega)d\sigma(\omega)$$

for all $\omega' \in \Omega$. Since $\left(\varphi_{\omega'}^0\right)_{\omega' \in \Omega}$ generates $\mathcal{H}^0$, we have

$$\sum_{x \in X} W(x)\varphi(x) = \int_\Omega \varphi(\omega)d\sigma(\omega)$$

for all $\varphi \in \mathcal{H}^0$. Since the same equality holds obviously for constant functions, it holds also for all $\varphi \in \mathcal{H}$.

In case σ is not a probability measure, we can multiply all values of W by $\sigma(\Omega)$. Then the previous equality holds again for constant functions on Ω, and a posteriori for all $\varphi \in \mathcal{H}$. $\qquad\square$

In the situation of Proposition 2.6, suppose moreover that we have a cubature formula (X', W') for $\mathcal{H}$. Then there exists a cubature formula (X, W) for $\mathcal{H}$ such that

$$X \subset X' \qquad \text{and} \qquad |X| \leq \dim_{\mathbb{R}}(\mathcal{H}).$$

(This follows from the proof in [21] of Carathéodory's theorem. Alternatively, we can apply Proposition 2.6 to the subspace of $\ell^2(X', W')$ of restrictions to X' of functions in $\mathcal{H}$.)

Proposition 2.7. *In the situation of the previous proposition, assume moreover that Ω is a connected topological space and that $\mathcal{H}$ is a space of continuous functions. Then the bound on the size of X can be improved to*

$$|X| \leq \dim_{\mathbb{R}}(\mathcal{H}) - 1.$$

Proof. With the notation of the proof of Proposition 2.6, the subspace $\tilde{\Omega}$ of $\mathcal{H}^0$ is a continuous image of Ω by Proposition 2.2.vi, so that $\tilde{\Omega}$ is connected. In this case, the Carathéodory bound on X can be lowered by 1, by a classical theorem of Fenchel. See, e.g., Theorem 18 in [39]. $\qquad\square$

The following result is an immediate consequence of Propositions 1.7 and 2.7.

Theorem 2.8 (Existence of cubature formulas, with bounds on sizes). *Let Ω be a connected topological space with a finite measure σ and let $\left(\mathcal{F}^{(k)}\right)_{k \geq 0}$ be a sequence of polynomial spaces on (Ω, σ). Assume that functions in $\bigcup_{k \geq 0} \mathcal{F}^{(k)}$ are continuous on Ω. Choose $l \geq 0$.*

Then there exists a finite subset X of Ω such that

$$\dim_{\mathbb{R}}(\mathcal{F}^{(l)}) \leq |X| \leq \dim_{\mathbb{R}}(\mathcal{F}^{(2l)}) - 1.$$

and a weight $W : X \longrightarrow \mathbb{R}_+^*$ such that (X, W) is a cubature formula of strength $2l$ on Ω.

Our next proposition (2.10) shows properties of cubature formulas (X, W) for which the previous lower bound is an equality, namely which are tight (Definition 1.8).

Lemma 2.9. *Consider data consisting of*

(a) *a finite measure space (Ω, σ), a Hilbert space of functions $\mathcal{H} \subset L^2(\Omega, \sigma)$, and the corresponding reproducing kernel Φ;*

(b) *a finite subset X of Ω, a weight $W : X \longrightarrow \mathbb{R}_+^*$, the Hilbert space $\ell^2(X, W)$, and the kernel Ψ as in Example 2.3;*

(c) *the restriction mapping $\rho : \mathcal{H} \longrightarrow \ell^2(X, W)$, $\varphi \longmapsto \varphi \mid_X$, and the adjoint mapping $\rho^* : \ell^2(X, W) \longrightarrow \mathcal{H}$.*

With the previous notation, namely with φ_ω as in 2.1 and ψ_x as in 2.3, we have

(i) *$\rho^*(\psi_x) = \varphi_x$ for all $x \in X$;*

(ii) *ρ^* is an isometry if and only if Ψ is the restriction of Φ to $X \times X$;*

(iii) *in case $\dim_{\mathbb{R}}(\mathcal{H}) = |X|$, the mapping ρ is an isometry if and only if Ψ is the restriction of Φ to $X \times X$.*

Proof. (i) By definition of ρ^*, we have for all $\varphi \in \mathcal{H}$ and for all $x \in X$

$$\langle \varphi \mid \rho^*(\psi_x) \rangle = \langle \rho(\varphi) \mid \psi_x \rangle = (\rho(\varphi))(x) = \varphi(x)$$

and therefore $\rho^*(\psi_x) = \varphi_x$.

(ii) Let $x, y \in X$. On the one hand, $\langle \rho^*(\psi_x) \mid \rho^*(\psi_y) \rangle = \Phi(x, y)$ by (i); on the other hand, $\langle \psi_x \mid \psi_y \rangle = \Psi(x, y)$. Claim (ii) follows.

(iii) A linear mapping between two finite-dimensional Hilbert spaces of the same dimension is an isometry if and only if its adjoint is an isometry ; thus (iii) follows from (ii). $\square$

Proposition 2.10. *Let (Ω, σ) be a finite measure space and let $(\mathcal{F}^{(k)})_{k \geq 0}$ be a sequence of polynomial spaces on (Ω, σ); assume that the corresponding reproducing kernels $\Phi^{(k)}$ are constant on the diagonal of Ω.*

Choose $l \geq 0$ and let (X, W) be a cubature formula of strength $2l$ on Ω which is tight.

Then W is uniform, namely X is a tight geometrical $(2l)$-design.

Proof. The restriction mapping

$$\rho : \mathcal{F}^{(l)} \longrightarrow \ell^2(X, W)$$

is an isometry by Proposition 1.7. If (X, W) is tight, the domain and the range of ρ have the same dimension, so that ρ is onto. Thus, if $\Psi : X \times X \longrightarrow \mathbb{R}$ denotes as in Example 2.3 the reproducing kernel of $\ell^2(X, W)$, then

$$\Psi(x, x') = \Phi^{(l)}(x, x') \qquad \text{for all} \quad x, x' \in X$$

by Lemma 2.9. It follows that Ψ is constant on the diagonal of $X \times X$, so that the weight W is constant by Example 2.3; otherwise said, X is a geometrical $(2l)$-design. □

Definition 2.11. Let Ω be a metric space, let σ be a finite measure on Ω, let $\mathcal{H}$ be a finite dimensional Hilbert space of functions on Ω which is a subspace of $L^2(\Omega, \sigma)$, and let Φ denote the corresponding reproducing kernel. We say that **Condition (M) holds for $\mathcal{H}$ and a function** S if the values of the kernel depend only on the distances:

$$\Phi(\omega, \omega') = S(d(\omega, \omega'))$$

for all $\omega, \omega' \in \Omega$, and for some function $S : \mathbb{R}_+ \longrightarrow \mathbb{R}$.

Observe that, by Proposition 2.2.v, this implies that

$$\Phi(\omega, \omega) = S(0) = \sigma(\Omega)^{-1} \dim_{\mathbb{R}}(\mathcal{H})$$

for all $\omega \in \Omega$.

On spheres, Condition (M) holds [10] for the spaces $\mathcal{H}^{(l)}(\mathbb{S}^{n-1})$ [respectively $\mathcal{P}^{(l)}(\mathbb{S}^{n-1})$, $\mathcal{F}^{(l)}(\mathbb{S}^{n-1})$] with $S = Q^{(l)}$ [respectively $S = C^{(l)}$, $S = R^{(l)}$]; the notation is that of Example 2.4.

Proposition 2.12. *Let* (Ω, σ), $(\mathcal{F}^{(k)})_{k \geq 0}$, *and* $(\Phi^{(k)})_{k \geq 0}$ *be as in Proposition 2.10. Choose* $l \geq 0$ *and assume that Condition (M) holds for* $\mathcal{F}^{(l)}$ *and a function* $S^{(l)}$.

If X *is a tight geometrical $2l$-design on* Ω, *the set of distances*

$$D_X = \{c \in]0, \infty[\mid c = d(x, y) \quad \text{for some} \quad x, y \in X, x \neq y\}$$

is contained in the set of zeros of the function $S^{(l)}$.

Proof. Notation being as in the proof of Proposition 2.10, we have

$$\Psi(x, x') = \Phi^{(l)}(x, x') = S^{(l)}(d(x, x'))$$

for all $x, x' \in X$. Since $\Psi(x, x') = 0$ when $x \neq x'$ (see Example 2.3), the proposition follows. □

The last proposition of this section shows how the setting of reproducing kernels makes it straightforward to generalize a characterization which is well-known for spherical designs (Theorem 5.5 in [33], or Theorems 3.2 and 4.3 in [47]).

---

[10] Or rather the analogous condition for scalar products $\langle \omega \mid \omega' \rangle$ rather than distances $d(\omega, \omega')$.

Proposition 2.13. *Let (Ω, σ) be a finite measure space, let $\mathcal{H}$ be a finite dimensional Hilbert space of functions on Ω which is a subspace of $L^2(\Omega, \sigma)$, and let Φ denote the corresponding reproducing kernel. Let X be a non-empty finite subset of Ω and let W be a weight function on X.*

(i) *We have $\sum_{x,y \in X} W(x)W(y)\Phi(x,y) \geq 0$.*
(ii) *Equality holds in (i) if and only if $\sum_{x \in X} W(x)\varphi(x) = 0$ for all $\varphi \in \mathcal{H}$.*

Proof. The inequality holds in (i) since Φ is a kernel of positive type.

Assume now equality. For any orthonormal basis $(e_j)_{j \in J}$ of $\mathcal{H}$, Proposition 2.2.iv implies that

$$\sum_{x,y \in X} W(x)W(y)\Phi(x,y) = \sum_{x,y \in X} W(x)W(y) \sum_{j \in J} e_j(x)e_j(y)$$

$$= \sum_{j \in J} \left(\sum_{x \in X} W(x)e_j(x) \right)^2 = 0.$$

Hence $\sum_{x \in X} W(x)e_j(x) = 0$ for all $j \in J$, and therefore $\sum_{x \in X} W(x)\varphi(x) = 0$ for all $\varphi \in \mathcal{H}$.

The converse implication in (ii) is straightforward. $\qquad\square$

2.14. Particular case. Consider two integers $n \geq 2$ and $k \geq 1$, a non-empty finite subset X of $\mathbb{S}^{n-1}$, a weight $W : X \longrightarrow \mathbb{R}^*_+$ such that $\sum_{x \in X} W(x) = 1$, and the polynomials $Q^{(j)}$, $R^{(j)}$ defined in Example 2.4. Then

$$\sum_{x,y \in X} W(x)W(y)Q^{(j)}(\langle x \mid y \rangle) \overset{(*)}{\geq} 0 \quad \text{and} \quad \sum_{x,y \in X} W(x)W(y)R^{(j)}(\langle x \mid y \rangle) \overset{(**)}{\geq} 1$$

for all $j \geq 0$. Moreover, the following properties are equivalent:

(i) (X, W) is a cubature formula of strength k on $\mathbb{S}^{n-1}$;
(ii) equality holds in $(*)$ for all $j \in \{1, \ldots, k\}$;
(iii) equality holds in $(**)$ for $j = k$.

3. The case of spheres
Antipodal cubature formulas and spherical designs

In this section, we assume that $\Omega = \mathbb{S}^{n-1}$ is the unit sphere in $\mathbb{R}^n$, $n \geq 2$, that σ is the rotation-invariant probability measure on $\mathbb{S}^{n-1}$, and that the notation is as in Examples 1.11 and 2.4.

For tight $(2l)$-spherical designs, the set D_X of Proposition 2.12 is precisely known, and $|D_X| = l$. Instead of distances and D_X, it is more convenient to use scalar products and the set

$$A_X = \left\{ c \in [-1,1[\mid c = \langle x \mid y \rangle \quad \text{for some} \quad x, y \in X, \; x \neq y \right\}.$$

The following proposition is Theorem 5.1 of [47].

Proposition 3.1. *If X is a tight spherical $(2l)$-design on $\mathbb{S}^{n-1}$, then A_X coincides with the set of roots of the polynomial $R^{(l)}$ defined in 2.4.*

Proof. Let $Z_R^{(l)}$ denote the set of roots of $R^{(l)}$, which is of order l. Since $A_X \subset Z_R^{(l)}$ by the proof of Proposition 2.12, it is enough to show that $a = |A_X|$ is not less than l. More generally, let us show that $a \geq l$ for any spherical $(2l)$-design.

Define for each $x \in X$ a polynomial function γ_x of degree a by

$$\gamma_x(\omega) = \prod_{c \in A_X} \frac{\langle x \mid \omega \rangle - c}{1 - c} \quad \text{for all} \quad \omega \in \mathbb{R}^n.$$

Then $\gamma_x(x') = \delta_{x,x'}$ for all $x, x' \in X$, so that the family $(\gamma_x)_{x \in X}$ is linearly independent. Thus

$$\dim_{\mathbb{R}}(\mathcal{F}^{(a)}(\mathbb{S}^{n-1})) \geq |X|.$$

Since $|X| \geq \dim_{\mathbb{R}}(\mathcal{F}^{(l)}(\mathbb{S}^{n-1}))$ by Proposition 1.7, we have $a \geq l$.

In particular, $A_X = Z_R^{(l)}$ for a tight spherical $(2l)$-design. $\square$

Propositions 1.7, 2.10, 2.12, 3.1, and Theorem 2.8 apply to cubature formulas and spherical designs of *even strength*. We expose now the analogous facts for *odd strength*.

Proposition 3.2. (Compare with 1.7.) *Let $n \geq 2$, $l \geq 0$, and $\mathcal{P}^{(l)}(\mathbb{S}^{n-1}) \subset L^2(\mathbb{S}^{n-1}, \sigma)$ be as in Example 1.11. Let moreover $X = Y \sqcup (-Y)$ be a non-empty antipodal finite subset of the sphere and $W : X \longrightarrow \mathbb{R}_+^*$ be a symmetric weight.*

Then (X, W) is a cubature formula of strength $2l + 1$ if and only if the restriction mapping

$$\rho : \mathcal{P}^{(l)} \longrightarrow \ell^2(Y, 2W)$$

is an isometry. In particular, if (X, W) is a cubature formula of strength $2l + 1$, then

$$(\sharp) \qquad |X| \geq 2 \dim_{\mathbb{R}}(\mathcal{P}^{(l)}(\mathbb{S}^{n-1})) = 2 \binom{n + l - 1}{n - 1}.$$

Proof. Observe that $\psi(-\omega) = (-1)^j \psi(\omega)$ for all $j \geq 0$, $\psi \in \mathcal{P}^{(j)}(\mathbb{S}^{n-1})$, and $\omega \in \mathbb{S}^{n-1}$, so that

$$\sum_{x \in X} W(x) \varphi_1(x) \varphi_2(x) = 2 \sum_{y \in Y} W(y) \varphi_1(y) \varphi_2(y)$$

for $\varphi_1, \varphi_2 \in \mathcal{P}^{(l)}(\mathbb{S}^{n-1})$.

The condition for ρ to be an isometry is

$$2 \sum_{y \in Y} W(y) \varphi_1(y) \varphi_2(y) = \int_{\mathbb{S}^{n-1}} \varphi_1(\omega) \varphi_2(\omega) d\sigma(\omega)$$

for all $\varphi_1, \varphi_2 \in \mathcal{P}^{(l)}(\mathbb{S}^{n-1})$. As in the proof of Proposition 1.7, this can be written

$$2 \sum_{y \in Y} W(y) \varphi(y) = \int_{\mathbb{S}^{n-1}} \varphi(\omega) d\sigma(\omega)$$

or indeed

$$(*) \qquad \sum_{x \in X} W(x)\varphi(x) = \int_{\mathbb{S}^{n-1}} \varphi(\omega)d\sigma(\omega)$$

for all $\varphi \in \mathcal{P}^{(2l)}(\mathbb{S}^{n-1})$. Now $(*)$ holds for all $\varphi \in \mathcal{P}^{(2l+1)}(\mathbb{S}^{n-1})$, since both terms are zero in this case by symmetry reasons. Hence ρ is an isometry if and only if $(*)$ holds for all $\varphi \in \mathcal{F}^{(2l+1)}(\mathbb{S}^{n-1})$. $\qquad \square$

3.3. Remarks and definition. (i) Let (X, W) be an antipodal cubature formula on $\mathbb{S}^{n-1}$, and write $X = Y \sqcup (-Y)$. Observe that, since a function on X can be canonically written as the sum of an even function and an odd function, we have a decomposition of $\ell^2(X, W)$ as the orthogonal direct sum of two copies of $\ell^2(Y, 2W)$. It follows from Propositions 1.7 and 3.2 that the restriction mapping

$$\mathcal{F}^{(l)}(\mathbb{S}^{n-1}) = \mathcal{P}^{(l)}(\mathbb{S}^{n-1}) \oplus \mathcal{P}^{(l-1)}(\mathbb{S}^{n-1}) \longrightarrow \ell^2(X, W) = \ell^2(Y, 2W) \oplus \ell^2(Y, 2W)$$

is an isometry if and only if the restriction mapping $\mathcal{P}^{(l)}(\mathbb{S}^{n-1}) \longrightarrow \ell^2(Y, 2W)$ is an isometry.

(ii) For a spherical $(2l + 1)$-design X in $\mathbb{S}^{n-1}$ (not necessarily antipodal), it is known that $|X| \geq 2\dim_\mathbb{R}(\mathcal{P}^{(l)}(\mathbb{S}^{n-1}))$; moreover, in case equality holds, then X is antipodal. See Theorem 5.12 in [33]; the proof uses the "linear programming method", which is more powerful than the "Fisher type method" used in our proof of Proposition 3.2.

(iii) A cubature formula (X, W) of strength $2l + 1$ on $\mathbb{S}^{n-1}$ is **tight** (see the definition in 1.11) if equality holds in Equation ($\sharp$) of Proposition 3.2.

The following is Theorem 5.11 of [47].

Proposition 3.4. (Compare with 2.10, 2.12, and 3.1.) *Let (X, W) be a cubature formula of strength $2l + 1$ on $\mathbb{S}^{n-1}$ which is tight.*
Then W is uniform, namely X is a tight spherical $(2l + 1)$-design. Moreover the set

$$B_X = \big\{ c \in \,]-1, 1[\; | \; c = \langle x \, | \, y \rangle \quad \text{for some} \quad x, y \in X, \; x \neq \pm y \big\}$$

coincides with the set of roots of the polynomial $C^{(l)}$ defined in 2.4.

Proof. Let $Z_C^{(l)}$ denote the set of roots of $C^{(l)}$, which is of order l. Since $B_X \subset Z_C^{(l)}$ by the argument used in the proof of Proposition 2.12, it is enough to show that $b = |B_X|$ is not less than l. More generally, let us show that $b \geq l$ for any antipodal $(2l + 1)$-design.

Define for each $x \in X$ a polynomial function $\tilde{\gamma}_x$ of degree b by

$$\tilde{\gamma}_x(\omega) = \prod_{c \in B_X} \frac{\langle x \, | \, \omega \rangle - c}{1 - c} \quad \text{for all} \quad \omega \in \mathbb{R}^n.$$

As $-B_X = B_X$, we have

$$\tilde{\gamma}_x(\omega) = \frac{\prod_{c \in B_X}(\langle x \, | \, \omega \rangle + c)}{\prod_{c \in B_X}(1 - c)} = (-1)^b \frac{\prod_{c \in B_X}(\langle x \, | -\omega \rangle - c)}{\prod_{c \in B_X}(1 - c)} = (-1)^b \tilde{\gamma}_x(-\omega)$$

for all $\omega \in \mathbb{R}^n$. Thus the restriction of $\tilde{\gamma}_x$ to $\mathbb{S}^{n-1}$ is in $\mathcal{P}^{(b)}(\mathbb{S}^{n-1})$. Decompose X as $Y \sqcup (-Y)$ as in Item 1.11; since $\tilde{\gamma}_y(y') = \delta_{y,y'}$ for all $y, y' \in Y$, we have

$$\dim_{\mathbb{R}} \left(\mathcal{P}^{(b)}(\mathbb{S}^{n-1}) \right) \geq |Y| \geq \dim_{\mathbb{R}} \left(\mathcal{P}^{(l)}(\mathbb{S}^{n-1}) \right)$$

and therefore $b \geq l$.

In particular $B_X = Z_C^{(l)}$ for a tight antipodal spherical $(2l+1)$-design. $\quad\square$

Remark 3.5. Given n and l, let (X, W) be a cubature formula of strength $2l+1$ on $\mathbb{S}^{n-1}$ such that $|X| \leq |X'|$ for any cubature formula of strength $2l+1$ on $\mathbb{S}^{n-1}$. The weight W need not be uniform; see Example 3.7 for such a cubature formula of strength 7 on $\mathbb{S}^2$.

Proposition 3.6. (Compare with 2.8.) *For each $l \geq 0$, there exists a finite antipodal subset $X = Y \sqcup (-Y)$ of $\mathbb{S}^{n-1}$ such that*

$$\dim_{\mathbb{R}} \left(P^{(l)}(\mathbb{S}^{n-1}) \right) \leq |Y| \leq \dim_{\mathbb{R}} \left(P^{(2l)}(\mathbb{S}^{n-1}) \right) - 1$$

and a symmetric weight $W : X \longrightarrow \mathbb{R}_+^*$ such that (X, W) is an antipodal cubature formula of strength $2l+1$ on $\mathbb{S}^{n-1}$.

Proof. See Propositions 2.6 and 2.7.

Similar estimates are known for other spaces, in particular for hypecubes in $\mathbb{R}^n$; see page 366 of [31]. $\quad\square$

Example 3.7 (Cubature formula of strength 5, 7, and 9 on $\mathbb{S}^2$). A number of explicit cubature formulas can be collected from the literature, either directly or indirectly. Many of them are reviewed in [36] (see in particular pages 717–724) and [83]. Let us first write down an identity of Lucas (1877):

$$8(u_1^4 + u_2^4 + u_3^4) + \sum_{\epsilon_2, \epsilon_3 \in \{1, -1\}} (u_1 + \epsilon_2 u_2 + \epsilon_3 u_3)^4 = 12(u_1^2 + u_2^2 + u_3^2)^2.$$

Let $Y \subset \mathbb{S}^2$ be the set of size 7 containing the three vectors e_1, e_2, e_3 of the canonical orthonormal basis of $\mathbb{R}^3$ and the four vectors $3^{-1/2}(e_1 + \epsilon_2 e_2 + \epsilon_3 e_3)$, for $\epsilon_2, \epsilon_3 = 1, -1$. Define a weight $W : Y \longrightarrow \mathbb{R}_+^*$ by $W(e_i) = \frac{8}{60}$ and $W(3^{-1/2}(e_1 + \epsilon_2 e_2 + \epsilon_3 e_3)) = \frac{9}{60}$. Then Lucas' identity can be rewritten as

$$\sum_{y \in Y} W(y)\langle y \mid u \rangle^4 = \frac{1}{5}\langle u \mid u \rangle^2 \quad \text{for all} \quad u \in \mathbb{R}^3$$

so that (Y, W) is a cubature formula of size 7 for $\mathcal{P}^{(4)}(\mathbb{S}^2)$ by the argument of Proposition 1.13. If $X = Y \sqcup (-Y)$ and if W is extended by symmetry, $(X, \frac{1}{2}W)$ is a cubature formula of size 14 and strength 5.

Size $|Y| = 7$ [respectively $|X| = 14$] is not optimal for cubature formulas for $\mathcal{P}^{(4)}(\mathbb{S}^2)$ [respectively $\mathcal{F}^{(5)}(\mathbb{S}^2)$]. Indeed, as already reported in Items 1.17 and 1.18, the 12 vertices of a regular icosahedron provide a tight spherical 5-design, and such a design is unique up to isometry [33, p. 375]. The 6 vertices of a corresponding Y provide a design for the space $\mathcal{P}^{(4)}(\mathbb{S}^2)$. (Note that this is not a spherical 4-design, because $\frac{1}{|Y|}\sum_{y\in Y}\varphi(y)$ is not always equal to $\int_{\mathbb{S}^2}\varphi(w)d\sigma(w)$ for functions φ in the second summand of the decomposition $\mathcal{F}^{(4)}(\mathbb{S}^2) = \mathcal{P}^{(4)}(\mathbb{S}^2) \oplus \mathcal{P}^{(3)}(\mathbb{S}^2)$.)

From [83] and [55] (see also Item 1.18), we collect the following facts.

o There exists a pair (Y, W) with $|Y| = 11$ which is a cubature formula for $\mathcal{P}^{(6)}(\mathbb{S}^2)$, and 11 is the minimal possible size (pages 133–135 of [83]). Hence there exists a cubature formula $(X, \frac{1}{2}W)$ for $\mathcal{F}^{(7)}(\mathbb{S}^2)$ of size $|X| = 22$. On the other hand, a spherical design $X \subset \mathbb{S}^2$ of size $|X| \leq 22$ is of strength at most 5, and a spherical 7-design in $\mathbb{S}^2$ has size at least 24 (compare with the lower bound of Proposition 3.3 for 7-designs in $\mathbb{S}^2$, which is 20).

o There exists a pair (Y, W) with $|Y| = 16$ which is a cubature formula for $\mathcal{P}^{(8)}(\mathbb{S}^2)$, and 16 is the minimal possible size (pages 111 and 136 of [83], referring to Finden, Sobolev, and McLaren). Hence there exists a cubature formula $(X, \frac{1}{2}W)$ for $\mathcal{F}^{(9)}(\mathbb{S}^2)$ of size $|X| = 32$. On the other hand, a spherical design $X \subset \mathbb{S}^2$ of size $|X| \leq 32$ is of strength at most 7, and a spherical 9-design in $\mathbb{S}^2$ has size at least 48 (compare with the lower bound of Proposition 3.2 for 9-designs in $\mathbb{S}^2$, which is 30).

Example 3.8 (Cubature formulas of strengths 5, 7, 9 and 11 on $\mathbb{S}^3$). The following identities go back respectively to Liouville (1859), Kempner (1912), Hurwitz (1908), and J. Schur (1909):

$$\sum_{i=1}^{4}(2u_i)^4 + \sum_{\epsilon_i\in\{1,-1\}}(u_1 + \epsilon_2 u_2 + \epsilon_3 u_3 + \epsilon_4 u_4)^4 = 24\left(\sum_{i=1}^{4}u_i^2\right)^2,$$

$$\sum_{i=1}^{4}(2u_i)^6 + 8\sum_{I}(u_i + \epsilon u_j)^6 + \sum_{\epsilon_i\in\{1,-1\}}(u_1 + \epsilon_2 u_2 + \epsilon_3 u_3 + \epsilon_4 u_4)^6 = 120\left(\sum_{i=1}^{4}u_i^2\right)^3,$$

$$6\sum_{i=1}^{4}(2u_i)^8 + 60\sum_{I}(u_i + \epsilon u_j)^8 + \sum_{II}(2u_i + \epsilon_j u_j + \epsilon_k u_k)^8$$

$$+ 6\sum_{\epsilon_i\in\{1,-1\}}(u_1 + \epsilon_2 u_2 + \epsilon_3 u_3 + \epsilon_4 u_4)^8 = 5040\left(\sum_{i=1}^{4}u_i^2\right)^4,$$

$$9\sum_{i=1}^{4}(2u_i)^{10} + 180\sum_{I}(u_i + \epsilon u_j)^{10} + \sum_{II}(2u_i + \epsilon_j u_j + \epsilon_k u_k)^{10}$$

$$+ 9\sum_{\epsilon_i\in\{1,-1\}}(u_1 + \epsilon_2 u_2 + \epsilon_3 u_3 + \epsilon_4 u_4)^{10} = 22680\left(\sum_{i=1}^{4}u_i^2\right)^5.$$

Summations $\sum_I$ contain 12 terms $u_i + \epsilon u_j$, with $1 \leq i < j \leq 4$ and $\epsilon = \pm 1$. Summations $\sum_{II}$ contain 48 terms of the form $2u_i + \epsilon_j u_j + \epsilon_k u_k$, with i, j, k pairwise distinct in $\{1, 2, 3, 4\}$, $j < k$, and $\epsilon_j, \epsilon_k = \pm 1$. These four identities provide cubature formulas on $\mathbb{S}^3$ of sizes and strengths

$$
\begin{array}{lll}
24 & \text{and} & 5 & \text{(Liouville)}, \\
48 & \text{and} & 7 & \text{(Kempner)}, \\
144 & \text{and} & 9 & \text{(Hurwitz)}, \\
144 & \text{and} & 11 & \text{(Schur)}.
\end{array}
$$

Other examples appear in [36] (pages 717–724).

Observe that Liouville's identity provides a design for $\mathcal{P}^{(4)}(\mathbb{S}^3)$ of size 12 and a spherical 5-design of size 24. The latter is (up to homothety) a root system of type D_4; in other coordinates, it can also be written as

$$(L) \qquad \sum_{1\le i<j\le 4,\epsilon\in\{-1,1\}} (u_i + \epsilon u_j)^4 = 6\left(\sum_{i=1}^{4} u_i^2\right)^2$$

(apparently written down first by Lucas in 1876). This is *not* a design for $\mathcal{P}^{(6)}(\mathbb{S}^3)$, otherwise there would exist a cubature formula of strength 7 on $\mathbb{S}^3$ of size 24, but such cubature formulas cannot have size less than 40 by Proposition 3.2.

Kempner's identity provides a spherical 7-design in $\mathbb{S}^3$ of size 48. There is in [54] a cubature formula of strength 7 in $\mathbb{S}^3$ which is of size 46, conjecturally the optimal size.

Schur's identity is not optimal for cubature formulas of strength 11 on $\mathbb{S}^3$. Indeed, the 120 vertices of a regular polytope of Schläfli symbol [11] $\{3,3,5\}$ provide an antipodal spherical 11-design (compare with the lower bound 112 of Proposition 3.2 for 11-designs in $\mathbb{S}^3$).

Any orbit in $\mathbb{S}^3$ for the natural action of the Coxeter group H_4 is a 11-design, and there exists a particular orbit of size 1440 which is indeed a 19-design; see the end of Section 5 in [48], and [56]. There is another construction of cubature formula of strength 19, due to Salihov (1975), of size 720; see page 112 of [83].

3.9. A digression on Waring problem. Liouville used his identity (see 3.8) and Lagrange's theorem on the representation of integers as sums of four squares to show the following claim:

any positive integers is a sum of at most 53 *biquadrates (= fourth powers).*
Here is a proof of the claim, using Lucas' form (L) of Liouville's identity. Since any positive number is (by Lagrange's theorem) of the form $6(N_1^2 + N_2^2 + N_3^2 + N_4^2) + r$ with $N_1,\ldots,N_4 \in \mathbb{N}$ and $r \in \{0,1,2,3,4,5\}$, it is enough to check that any number of the form $6N^2$ is a sum of 12 biquadrates. Using Lagrange's theorem again, N can be written as a sum $n_1^2 + n_2^2 + n_3^2 + n_4^2$ of four squares. If $n \in \mathbb{Z}^4$ denotes the vector of coordinates n_1,n_2,n_3,n_4, we have $N = \langle n \mid n \rangle$ and

$$6N^2 = 6\langle n \mid n \rangle^2 = \sum_{1\le i<j\le 4,\epsilon\in\{-1,1\}} (n_i + \epsilon n_j)^4 = \text{sum of 12 fourth powers}$$

---

[11]See [28], page 153.

by (L), and this ends the proof.

Today, we know how to make this bound sharp, reducing it from 53 to 19 [18], [19], or even to 16 if a *finite number* of exceptions is allowed [30]. More precisely, there are exactly 96 numbers which are not sums of 16 biquadrates, and the maximum of them is $13\,792$; their list is shown in [34].

3.10. Lattice cubature formulas. Here is a variation on the construction of lattice designs described in Example 1.16.

Consider a lattice $\Lambda \subset \mathbb{R}^n$ which is even and unimodular. For an integer $l \geq 0$ and a harmonic homogeneous polynomial function $P \in \mathcal{H}^{(2l)}(\mathbb{R}^n)$, the **theta series** is defined by

$$\Theta_{\Lambda,P}(z) = \sum_{\lambda \in \Lambda} P(\lambda) q^{\langle \lambda | \lambda \rangle} = P(0) + \sum_{x \in \Lambda_2} P(x) q^2 + \sum_{x \in \Lambda_4} P(x) q^4 + \cdots$$

where $q = e^{i\pi z}$, $z \in \mathbb{C}$ with $\mathrm{Im}(z) > 0$, and $\Lambda_{2j} = \{x \in \Lambda \mid \langle x \mid x \rangle = 2j\}$. It is a modular form for $PSL(2, \mathbb{Z})$ of weight $\frac{n}{2} + 2l$, and a parabolic modular form if moreover $l > 0$. In particular, we have $\Theta_{\Lambda,P} = 0$ whenever $l > 0$ and $\frac{n}{2} + 2l \leq 10$ or $\frac{n}{2} + 2l = 14$ (see Corollary 3.3 in [38]).

Suppose now that $n = 8$, so that Λ is a root lattice of type E_8. The previous considerations show that Λ_{2j} is a spherical 7-design and a design for $\mathcal{H}^{(10)}(\mathbb{R}^8)$ for any $j \geq 1$. Moreover, for $P \in \mathcal{H}^{(8)}(\mathbb{R}^8)$, the parabolic modular form $\Theta_{\Lambda,P}$ is necessarily a constant multiple of

$$\Delta_{24}(z) = q^2 \prod_{m=1}^{\infty} (1 - q^{2m})^{24} = q^2 - 24q^4 + 252q^6 - 1472q^8 + \cdots \qquad (q = e^{i\pi z}).$$

In particular, $24 \sum_{x \in \Lambda_2} P(x) + \sum_{x \in \Lambda_4} P(x) = 0$. Observe that $P(x) = 2^4 P(x/\sqrt{2})$ for $x \in \Lambda_2$ and $P(x) = 2^8 P(x/2)$ for $x \in \Lambda_4$. It follows that

$$\frac{3}{2} \sum_{x \in \Lambda_2} P\left(\frac{x}{\sqrt{2}}\right) + \sum_{x \in \Lambda_4} P\left(\frac{x}{2}\right) = 0$$

for all $P \in \mathcal{H}^{(k)}(\mathbb{R}^8)$ and $1 \leq k \leq 11$. In other words, the first two shells of Λ provide a cubature formula in $\mathbb{S}^7$ of strength 11, with underlying set $\frac{1}{\sqrt{2}}\Lambda_2 \sqcup \frac{1}{2}\Lambda_4$, of size

$$|\Lambda_2| + |\Lambda_4| = 240 + 2160 = 2400,$$

and with exactly two different weights, namely $1/1680$ on $\frac{1}{\sqrt{2}}\Lambda_2$ and $1/2520$ on $\frac{1}{2}\Lambda_4$. This size compares favourably with the bounds of Proposition 3.6, which read here

$$2 \dim_{\mathbb{R}} \mathcal{P}^{(5)}(\mathbb{R}^8) = 1584 \leq N \leq 2 \dim_{\mathbb{R}} \mathcal{P}^{(10)}(\mathbb{R}^8) - 2 = 38894.$$

A similar lower bound shows that $\frac{1}{\sqrt{2}}\Lambda_2 \sqcup \frac{1}{2}\Lambda_4$ cannot enter a cubature formula of strength 13 on $\mathbb{S}^7$.

The previous construction indicates clearly a general procedure.

3.11. On Gaussian designs. Let l be an integer, $l \geq 0$, and let $\alpha \in \mathbb{R}^n$. On the one hand, it is classical that

$$\int_{\mathbb{S}^{n-1}} \langle \alpha \mid x \rangle^{2l} d\sigma(x) = \frac{(2l-1)!!}{n(n+2)\cdots(n+2l-2)} \langle \alpha \mid \alpha \rangle^l$$

(notation: $(2l-1)!! = \prod_{j=1}^{l}(2l-2j+1)$). Indeed, the left hand side is clearly a homogeneous polynomial of degree $2l$ in α which is invariant by the orthogonal group $O(n)$, and therefore a constant multiple of $\langle \alpha \mid \alpha \rangle^l$. If we apply both sides the Laplacian with respect to α, we find a recurrence relation for the constants, and this provides the result. On the other hand, we have

$$\frac{1}{\pi^{n/2}} \int_{\mathbb{R}^n} \langle \alpha \mid x \rangle^{2l} e^{-\|x\|^2} dx = \frac{(2l-1)!!}{2^l} \langle \alpha \mid \alpha \rangle^l$$

(calculus, using integration by parts in case $n=1$). It follows that

$$(*) \qquad \int_{\mathbb{S}^{n-1}} \varphi(x) d\sigma(x) = \frac{2^l}{n(n+2)\cdots(n+2l-2)} \frac{1}{\pi^{n/2}} \int_{\mathbb{R}^n} \varphi(x) e^{-\|x\|^2} dx$$

for all $\varphi \in \mathcal{P}^{(2l)}(\mathbb{R}^n)$, $l \geq 0$. [See the footnote in Proposition 1.13.] Observe that, for $\varphi \in \mathcal{P}^{(2l+1)}(\mathbb{R}^n)$, both integrals in $(*)$ vanish for symmetry reasons.

Let now X be a non-empty finite subset of $\mathbb{S}^{n-1}$ of size N. If X is a spherical t-design with $t \in \{1,2,3\}$, the set

$$X_G = \sqrt{\frac{n}{2}} X$$

is a *Gaussian t-design*. Indeed, for $\varphi \in \mathcal{P}^{(2)}(\mathbb{R}^n)$, we have

$$\frac{1}{N} \sum_{x \in X_G} \varphi(x) = \frac{n}{2N} \sum_{x \in X} \varphi(x) = \frac{n}{2} \int_{\mathbb{S}^{n-1}} \varphi(x) d\sigma(x) = \frac{1}{\pi^{n/2}} \int_{\mathbb{R}^n} \varphi(x) e^{-\|x\|^2} dx$$

(we leave to the reader the verifications with $\varphi \in \mathcal{P}^{(1)}(\mathbb{R}^n)$ and $\varphi \in \mathcal{P}^{(3)}(\mathbb{R}^n)$).

If X is now a spherical 5-design, set

$$\rho_1 = \sqrt{\frac{1}{2}(n - \sqrt{2n})} \qquad \text{so that} \qquad \frac{1}{2}(\rho_1^2 + \rho_2^2) = \frac{n}{2}$$

$$\rho_2 = \sqrt{\frac{1}{2}(n + \sqrt{2n})} \qquad \qquad \frac{1}{2}(\rho_1^4 + \rho_2^4) = \frac{n(n+2)}{4}.$$

Then

$$X_{GG} = \rho_1 X \sqcup \rho_2 X \qquad \text{(disjoint union)}$$

is a *Gaussian 5-design* of size $2N$. Indeed, for $\varphi \in \mathcal{P}^{(4)}(\mathbb{R}^n)$, we have

$$
\frac{1}{2N} \sum_{x \in X_{GG}} \varphi(x) = \frac{\rho_1^4}{2N} \sum_{x \in X} \varphi(x) + \frac{\rho_2^4}{2N} \sum_{x \in X} \varphi(x) = \frac{1}{2}(\rho_1^4 + \rho_2^4) \int_{\mathbb{S}^{n-1}} \varphi(x) d\sigma(x)
$$

$$
= \frac{1}{2}(\rho_1^4 + \rho_2^4) \frac{4}{n(n+2)} \frac{1}{\pi^{n/2}} \int_{\mathbb{R}^n} \varphi(x) e^{-\|x\|^2} dx
$$

$$
= \frac{1}{\pi^{n/2}} \int_{\mathbb{R}^n} \varphi(x) e^{-\|x\|^2} dx
$$

(where the last equality is a consequence of the values chosen for ρ_1 and ρ_2). Similarly

$$
\frac{1}{2N} \sum_{x \in X_{GG}} \varphi(x) = \frac{1}{\pi^{n/2}} \int_{\mathbb{R}^n} \varphi(x) e^{-\|x\|^2} dx
$$

for all $\varphi \in \mathcal{P}^{(2)}(\mathbb{R}^n)$. We leave to the reader the verifications with φ a homogeneous polynomial of odd degree.

We end the present chapter with a proposition and a construction suggesting one more connection with another subject.

For an integer $N \geq 1$ and a real number $p \geq 1$, we denote by $\ell^p(N)$ the classical Banach space of dimension N, with underlying space $\mathbb{R}^N$ and norm $\|x\|_p$, where

$$
\left\| (x_j)_{1 \leq j \leq N} \right\|_p = \sqrt[p]{\sum_{1 \leq j \leq N} |x_j|^p}.
$$

A linear mapping from a real vector space to $\ell^p(N)$ is said to be **degenerate** if its image is in a hyperplane of equation $x_j = 0$ for some $j \in \{1, \ldots, N\}$. The next proposition is from [73], and c_{2l} is the constant of our Proposition 1.13.

Proposition 3.12. *Consider integers $n \geq 2$, $l \geq 2$, and $N \geq 1$.*

   (i) *To any cubature formula (X, W) of size N for $\mathcal{P}^{(2l)}(\mathbb{S}^{n-1})$ corresponds a non-degenerate isometric embedding $J_{X,W} : \ell^2(n) \longrightarrow \ell^{2l}(N)$.*
   (ii) *Conversely, to any non-degenerate isometric embedding $J : \ell^2(n) \longrightarrow \ell^{2l}(N)$ corresponds a cubature formula of size N in $\mathcal{P}^{(2l)}(\mathbb{S}^{n-1})$.*

Proof. Let (X, W) be as in (i) and let $(x^{(1)}, \ldots, x^{(N)})$ be an enumeration of the points in X. The linear mapping $J_{X,W} : \mathbb{R}^n \longrightarrow \mathbb{R}^N$ defined by its coordinates

$$
J_{X,W}(u)_k = \left(\frac{W(x^{(k)})}{c_{2l}} \right)^{1/2l} \left\langle x^{(k)} \mid u \right\rangle, \qquad 1 \leq k \leq N,
$$

is an isometry $\ell^2(n) \longrightarrow \ell^{2l}(N)$ by the argument of Proposition 1.13. It is non-degenerate since $X^\perp = \{0\}$.

Let $J : \ell^2(n) \longrightarrow \ell^{2l}(N)$ be a non-degenerate isometry as in (ii). For each $k \in \{1, \ldots, N\}$, let $y^{(k)}$ be the unique vector in $\ell^2(n)$ such that $J(u)_k = \langle y^{(k)} \mid u \rangle$ for all $u \in \ell^2(n)$; observe that $y^{(k)} \neq 0$, by the non-degeneracy condition on N. Set $x^{(k)} = y^{(k)} / \|y^{(k)}\|_2$ and $W^{(k)} = c_{2l} \|y^{(k)}\|_2^{2l}$. Then $X = \{x^{(1)}, \ldots, x^{(N)}\}$ is a

N-subset of $\mathbb{S}^{n-1}$, the $W^{(k)}$ define a weight $X \longrightarrow \mathbb{R}_+^*$, and (X, W) is a cubature formula for $\mathcal{P}^{(2l)}(\mathbb{S}^{n-1})$ by Proposition 1.13. $\square$

4. Markov operators

Given a complex Hilbert space $\mathcal{H}$, we denote by $\mathcal{L}(\mathcal{H})$ the C*-algebra of all bounded linear operators on $\mathcal{H}$ and by $\mathcal{U}(\mathcal{H})$ its *unitary group*, consisting of all unitary operators on $\mathcal{H}$. Given a group Γ, we denote by $\mathbb{C}[\Gamma]$ the *group algebra* of all functions $\Gamma \longrightarrow \mathbb{C}$ of finite support, with multiplication the convolution; we consider Γ as a subset of $\mathbb{C}[\Gamma]$, by identifying $\gamma \in \Gamma$ with the function of value 1 on γ and 0 elsewhere. A *unitary representation* of a group Γ in a Hilbert space $\mathcal{H}$ is a group homomorphism $\pi : \Gamma \longrightarrow \mathcal{U}(\mathcal{H})$; it extends to a morphism of algebra $\mathbb{C}[\Gamma] \longrightarrow \mathcal{L}(\mathcal{H})$, denoted by π again, and defined by $\pi(f) = \sum_{\gamma \in \Gamma} f(\gamma)\pi(\gamma)$. Recall that the norm of a unitary operator is 1, so that $\|\pi(f)\| \leq \sum_{\gamma \in \Gamma} |f(\gamma)|$. Recall also that, if $f(\gamma^{-1}) = \overline{f(\gamma)}$ for all $\gamma \in \Gamma$, the operator $\pi(f)$ is self-adjoint; in particular, the spectrum of $\pi(f)$ is then a closed subset of the real interval $[-\|\pi(f)\|, \|\pi(f)\|]$.

Up to minor terminological changes, this holds for a *real* Hilbert space $\mathcal{H}$, its *orthogonal group* $\mathcal{O}(\mathcal{H})$, *orthogonal representations* $\Gamma \longrightarrow \mathcal{O}(\mathcal{H})$, and corresponding morphisms $\mathbb{R}[\Gamma] \longrightarrow \mathcal{L}(\mathcal{H})$.

Definition 4.1. Let Γ be a group, S a finite subset of Γ, and $W : S \longrightarrow \mathbb{R}_+^*$ a weight function; set $M_{S,W} = \sum_{s \in S} W(s)s \in \mathbb{R}[\Gamma]$. The **Markov operator** associated to the pair (S, W) and an orthogonal (or unitary) representation $\pi : \Gamma \longrightarrow \mathcal{U}(\mathcal{H})$ is the operator

$$\pi(M_{S,W}) = \sum_{s \in S} W(s)\pi(s)$$

on $\mathcal{H}$.

Markov operators (often with constant weights) play an important role in the study of random walks on Cayley graphs of groups [62], and more generally in connection with unitary representations of groups. See for example [17], [22], [43], [50], [51], [52], [57], and [58].

Consider a sigma-finite measure space (Ω, σ) and a group G acting on Ω by measurable transformations preserving the measure class of σ. The corresponding unitary representation ρ of G on $L^2(\Omega, \sigma)$ is defined by

$$(\rho(g)\varphi)(\omega) = \sqrt{\frac{dg^{-1}\sigma}{d\sigma}} \; \varphi(g^{-1}\omega)$$

for $g \in G$, $\varphi \in L^2(\Omega, \sigma)$, and $\omega \in \Omega$, where $\frac{dg^{-1}\sigma}{d\sigma}$ denotes the appropriate Radon-Nikodym derivative. Given a mapping $s : \Omega \longrightarrow G$, $\omega \longmapsto s_\omega$, a finite subset X of Ω, and a weight function $W : X \longrightarrow \mathbb{R}_+^*$, we have an element $M_{X,W} = \sum_{x \in X} W(x)s_x$ in the real group algebra of the subgroup of G generated by $s(X)$.

Given moreover a representation π of G in a Hilbert space $\mathcal{H}_\pi$, we have a Markov operator

$$\pi(M_{X,W}) = \sum_{x \in X} W(x)\pi(s_x) \in \mathcal{L}(\mathcal{H}_\pi).$$

The particular case studied in [81] is that of the orthogonal group $G = O(n)$ acting on the unit sphere $\Omega = \mathbb{S}^{n-1}$, with $\mathcal{H}_\pi$ one of the finite-dimensional spaces introduced in Example 1.11 above. In this case, the image of a point $x \in \mathbb{S}^{n-1}$ by the mapping s is the orthogonal reflection of $\mathbb{R}^n$ which fixes the hyperplane orthogonal to x.

Proposition 4.2. *Let $n \geq 2$, $l \geq 0$ be integers and let $\pi^{(l)}$ denote the natural representation of the group $O(n)$ in the space $\mathcal{H}^{(l)}(\mathbb{S}^{n-1})$ of harmonic homogeneous polynomials of degree l. For each $x \in \mathbb{S}^{n-1}$, let $s_x \in O(n)$ denote the reflection of $\mathbb{R}^n$ that fixes the hyperplane $x^\perp$ of $\mathbb{R}^n$.*

If (X, W) is a cubature formula of strength $2l$ on $\mathbb{S}^{n-1}$, then

$$\pi^{(l)}(M_{X,W}) = \frac{n-2}{2l+n-2} \, \mathrm{id}^{(l)}$$

where $\mathrm{id}^{(l)}$ denotes the identity operator on $\mathcal{H}^{(l)}(\mathbb{S}^{n-1})$.

Proof, repeated from [81] (see also [59]). For $x \in \mathbb{R}^n$, we define the operator $\tilde{s}_x$ on $\mathbb{R}^n$ by $\tilde{s}_x(u) = \langle x \mid x \rangle u - 2\langle x \mid u \rangle x$. Note that $\tilde{s}_x$ is selfadjoint, that $\tilde{s}_{\lambda x} = \lambda^2 \tilde{s}_x$ for $\lambda \in \mathbb{R}$, and that $\tilde{s}_x = s_x$ if $x \in \mathbb{S}^{n-1}$.

For $k \geq 0$, define the selfadjoint operator

$$\overline{M}^{(k)} = \int_{\mathbb{S}^{n-1}} \pi^{(k)}(s_\omega) d\sigma(\omega)$$

on $\mathcal{H}^{(k)}(\mathbb{R}^n)$. For $g \in O(n)$, we have $s_{g(\omega)} = g s_\omega g^{-1}$. Since the measure σ is $O(n)$-invariant and the representation $\pi^{(k)}$ is irreducible, it follows from Schur's lemma that $\overline{M}^{(k)}$ is a homothety. By a simple computation, we obtain the trace of $\pi^{(k)}(s_\omega)$, which is independent of ω, and therefore the trace of $\overline{M}^{(k)}$. It follows that

(1) $$\overline{M}^{(k)} = \frac{n-2}{2k+n-2} \, \mathrm{id}^{(k)}$$

for all $k \geq 0$.

For $\varphi_1, \varphi_2 \in \mathcal{H}^{(k)}(\mathbb{R}^n)$ and $u \in \mathbb{R}^n$, set

$$\Psi^{(k)}(\varphi_1, \varphi_2)(u) = \langle \varphi_1 \mid \varphi_2 \circ \tilde{s}_u \rangle = \int_{\mathbb{S}^{n-1}} \varphi_1(\omega)\varphi_2(\tilde{s}_u(\omega)) d\sigma(\omega).$$

It is easy to check that the function $\Psi^{(k)}(\varphi_1, \varphi_2)$ is polynomial, homogeneous of degree $2k$, and depends symmetrically on φ_1, φ_2. Thus, we have a linear mapping

$$\Psi^{(k)} : \mathrm{Sym}^2(\mathcal{H}^{(k)}(\mathbb{R}^n)) \longrightarrow \mathcal{P}^{(2k)}(\mathbb{R}^n).$$

Moreover $\int_{\mathbb{S}^{n-1}} \Psi^{(k)}(\varphi_1, \varphi_2)(\omega) d\sigma(\omega) = \left\langle \varphi_1 \mid \overline{M}^{(k)}(\varphi_2) \right\rangle.$

Consider now a non-empty finite subset X of $\mathbb{S}^{n-1}$, a weight $W : X \longrightarrow \mathbb{R}_+^*$, and an integer $l \geq 0$. Then

$$(2) \qquad \sum_{x \in X} W(x)\varphi(x) = \int_{\mathbb{S}^{n-1}} \varphi(\omega)d\sigma(\omega) \qquad \text{for all } \varphi \text{ in the image of } \Psi^{(l)}$$

if and only if

$$\sum_{x \in X} W(x)\langle \varphi_1 \mid \varphi_2 \circ s_x \rangle = \left\langle \varphi_1 \mid \overline{M}^{(l)}(\varphi_2) \right\rangle \qquad \text{for all } \varphi_1, \varphi_2 \in \mathcal{H}^{(l)}(\mathbb{R}^n),$$

namely, by the equality $\pi^{(l)}(M_{X,W})\varphi_2 = \sum_{x \in X} W(x)\varphi_2 \circ s_x$ defining $\pi^{(l)}(M_{X,W})$ and by (1), if and only if

$$(3) \qquad \qquad \pi^{(l)}(M_{X,W}) = \overline{M}^{(l)} = \frac{n-2}{2l+n-2} \operatorname{id}^{(l)}.$$

If (X, W) is a cubature formula of strength $2l$, then (2) holds for all $\varphi \in \mathcal{P}^{(2l)}(\mathbb{R}^n)$, and a fortiori for all φ in the image of $\Psi^{(l)}$, so that the proposition is proved. □

It is remarkable that a converse holds for $n > 2$.

Proposition 4.3. *Let $n \geq 3$, $l \geq 0$ be integers and let $\pi^{(l)}$ be as in the previous proposition. Let X be a non-empty antipodal finite subset of $\mathbb{S}^{n-1}$ and let $W : X \longrightarrow \mathbb{R}_+^*$ be a symmetric weight.*

If $\pi^{(l)}(M_{X,W})$ is a homothety, then (X, W) is a cubature formula of strength $2l+1$.

Proof. See [81]. □

More generally, consider a Riemannian symmetric pair (G, H) where G is a compact Lie group acting on $\Omega = G/H$, and the G-invariant probability measure σ on Ω. Let $s = s^{\text{symm}} : \Omega \longrightarrow G$ be the mapping which associates to a point $x \in G/H$ the symmetry of Ω fixing x. (For spheres, observe that $s^{\text{symm}}(x)$ is *minus* the reflection s_x fixing $x^\perp$.) There is an orthogonal decomposition

$$L^2(\Omega, \sigma) = \bigoplus_{\lambda \in \Lambda} V^\lambda$$

in irreducible G-spaces. For a finite subset X of Ω and a weight function W on X, there is an analogue of Proposition 4.2 concerning spaces V^λ for which (X, W) is a cubature formula and for which the Markov operator $\pi^{(\lambda)}(M_{X,W}) \in \operatorname{End}(V^{(\lambda)})$ is a constant multiple of the identity. For designs in G/H a Grassmannian, see [8] and [6].

Consider a non-empty finite subset S of $O(n)$ which is symmetric ($S^{-1} = S$) and denote by Γ_S the subgroup of $O(n)$ it generates.

From now on, we assume that $W(s) = |S|^{-1}$ for all $s \in S$ and we set $M_S = |S|^{-1} \sum_{s \in S} s \in \mathbb{R}[\Gamma_S]$.

Propositions 4.2 and 4.3 relate the spectra of Markov operators $\pi^{(k)}(M_S)$ for some small values of k to the cubature properties of (X, W). It happens that the spectra for $k \to \infty$ of the operators $\pi^{(k)}(M_S)$ are also important, as shown by Lubotzky, Phillips, and Sarnak [71], [72] (see also [24]). We formulate the weak Observation 4.4 before the stronger Proposition 4.5.

Let π_0 denote the natural representation of $O(n)$ in the space $L_0^2(\mathbb{S}^{n-1}, \sigma)$ of L^2-functions of zero average. The orthogonal sum $L_0^2(\mathbb{S}^{n-1}, \sigma) = \bigoplus_{k=1}^{\infty} \mathcal{H}^{(k)}(\mathbb{S}^{n-1})$ provides an orthogonal decomposition $\pi_0 = \bigoplus_{k=1}^{\infty} \pi^{(k)}$ into irreducible subrepresentations (see, e.g., [95]), so that

$$\|\pi_0(M_S)\| = \sup_{k \geq 1} \left\|\pi^{(k)}(M_S)\right\|.$$

Observation 4.4. For any finite symmetric subset S of $O(n)$, we have

$$1 \geq \|\pi_0(M_S)\| \geq \frac{1}{\sqrt{|S|}}.$$

Proof. (Compare with Proposition 2.3.2 in [86].) For all $s \in S$, we have $\|\pi_0(s)\| = 1$, since the representation π_0 is orthogonal. The upper bound follows.

Choose $\omega \in \mathbb{S}^{n-1}$ such that $s(\omega) \neq t(\omega)$ for all $s, t \in S \cup \{id\}$, $s \neq t$. There exists a neighbourhood U of ω in $\mathbb{S}^{n-1}$ such that $s(U) \cap t(U) = \emptyset$ for all $s, t \in S \cup \{id\}$, $s \neq t$, and a function $\varphi \in L_0^2(\mathbb{S}^{n-1}, \sigma)$ of norm 1 supported in U. Since $\|\pi_0(M_S)\varphi\|^2 = |S|^{-1}$, we have $\|\pi_0(M_S)\| \geq |S|^{-1/2}$. $\square$

Proposition 4.5. *Let S be a symmetric finite subset of $O(3)$. Then*

$$\|\pi_0(M_S)\| \geq \limsup_{k \to \infty} \left\|\pi^{(k)}(M_S)\right\| \geq \frac{2\sqrt{|S| - 1}}{|S|}.$$

The proofs of [71] and [24] are written up for the case of a subset S of $SO(n)$. The present generalization to $O(n)$ is rather straightforward. We isolate part of the proof in the following lemma.

Lemma 4.6. (i) *Let $g \in SO(3)$ be a rotation of angle $\theta_g \in [0, \pi]$. Then*

$$\text{trace}\left(\pi^{(k)}(g)\right) = \begin{cases} \dfrac{\sin[(2k+1)\theta_g/2]}{\sin[\theta_g/2]} & \text{if } \theta_g \neq 0 \\ 2k+1 & \text{if } \theta_g = 0. \end{cases}$$

(ii) *Let $g \in O(3)$ have eigenvalues $\exp(\pm i\theta_g)$ and -1. Then*

$$\text{trace}\left(\pi^{(k)}(g)\right) = \begin{cases} \dfrac{\cos[(2k+1)\theta_g/2]}{\cos[\theta_g/2]} & \text{if } \theta_g \neq \pi \\ (-1)^k & \text{if } \theta_g = \pi. \end{cases}$$

Proof. (i) The eigenvalues of $g \in SO(3)$ are $\exp(\pm i\theta_g)$ and 1. The space $\mathcal{P}^{(k)}(\mathbb{S}^2)$ has a linear basis of eigenvectors of the transformation induced by g of the form

$$\{x^a y^b z^{k-a-b} \mid a, b \geq 0 \text{ and } a + b \leq k\}.$$

For $j \in \{0, \ldots, k\}$, the trace of the linear endomorphism defined by g on the linear span of $\{x^a y^b z^{k-a-b} \mid a, b \geq 0 \text{ and } a + b = j\}$ is

$$\sigma_j(g) = e^{ij\theta_g} + e^{i(j-2)\theta_g} + \cdots + e^{-ij\theta_g} = \frac{\sin[(j+1)\theta_g]}{\sin[\theta_g]}$$

so that the trace of the linear endomorphism defined by g on $\mathcal{P}^{(k)}(\mathbb{S}^2)$ is $\sum_{j=0}^{k} \sigma_j(\theta_g)$. Since $\mathcal{P}^{(k)}(\mathbb{S}^2) = \bigoplus_{j \geq 0, k-2j \geq 0} \mathcal{H}^{(k-2j)}(\mathbb{S}^2)$, we have trace $\left(\pi^{(k)}(g)\right) = \sigma_k(\theta_g) + \sigma_{k-1}(\theta_g)$ and the formula of (i) follows.

(ii) Similarly, for $g \in O(3)$, $g \notin SO(3)$, the trace of the linear endomorphism defined by g on $\mathcal{P}^{(k)}(\mathbb{S}^2)$ is $\sum_{j=0}^{k} (-1)^{k-j} \sigma_j(\theta_g)$ so that

$$\text{trace}\left(\pi^{(k)}(g)\right) = \sigma_k(\theta_g) - \sigma_{k-1}(\theta_g).$$

(The *only* part of the proof which is not explicitly in [71] is (ii).)   □

Proof of Proposition 4.5. Step one. Recall that Γ_S is the subgroup of $O(3)$ generated by S. For each integer $N \geq 0$, let W_N denote the number of words [12] in letters of S, of length N. To each word $w \in W_N$ corresponds naturally an orthogonal transformation $g = g(w) \in \Gamma_S \subset O(3)$. We have $(M_S)^N = |S|^{-N} \sum_{w \in W_N} g(w)$ and therefore

$$|S|^N \text{ trace}\left(\pi^{(k)}((M_S)^N)\right)$$
$$= \sum_{\substack{w \in W_N \\ g(w) \in SO(3)}} \frac{\sin[(2k+1)\theta_g/2]}{\sin[\theta_g/2]} + \sum_{\substack{w \in W_N \\ g(w) \notin SO(3)}} \frac{\cos[(2k+1)\theta_g/2]}{\cos[\theta_g/2]}$$

by the previous lemma.

Let m_N denote the quotient by $|S|^N$ of the number of words $w \in W_N$ such that $g(w) = 1 \in O(3)$. We have

$$\lim_{k \to \infty} \frac{1}{2k+1} \text{ trace}\left(\pi^{(k)}((M_S)^N)\right) = m_N$$

for each $N \geq 0$. (This is Theorem 1.1 in [71]; observe however the change in notation: m_N there is $|S|^N$ times what m_N is here and in [62].)

Step two. Consider the Cayley graph [13] Cay_S of Γ_S with respect to S, the left regular representation λ_S of Γ_S, the corresponding Markov operator $\lambda_S(M_S)$, and its spectral measure μ_S. Kesten has shown that
$m_N = \int_{\mathbb{R}} t^N d\mu_S(t)$ for all $N \geq 0$,
$\|\lambda_S(M_S)\| = \limsup_{N \to \infty} \sqrt[N]{m_N}$,
$\|\lambda_S(M_S)\| \geq 2\sqrt{|S|-1}/|S|$ with equality if and only if Cay_S is a tree.
See [62].

---

[12]Words which need not be reduced in any sense.
[13]Two elements $\gamma_1, \gamma_2 \in \Gamma_S$ viewed as vertices of Cay_S are joined by an edge if $\gamma_1^{-1}\gamma_2 \in S$. We assume that $1 \notin S$, so that the Cayley graph is simple (= without loops), of degree $|S|$.

Step three. For each $k \geq 0$, let $\mu_{S,k}$ denote the quotient by $\dim_{\mathbb{R}} \mathcal{H}^{(k)}(\mathbb{S}^2) = 2k+1$ of the spectral measure of $\pi^{(k)}(M_S)$; this can be written

$$\mu_{S,k} = \frac{1}{2k+1} \sum_{j=1}^{2k+1} \delta(\lambda_{k,j})$$

where $\delta(\lambda)$ denotes a Dirac measure of support λ and where $(\lambda_{k,j})_{1 \leq j \leq 2k+1}$ are the eigenvalues of the endomorphism $\pi^{(k)}(M_S)$ of $\mathcal{H}^{(k)}(\mathbb{S}^2)$. Let $N \geq 0$; from the definition of $\mu_{S,k}$, we have

$$\frac{1}{2k+1} \operatorname{trace}\left(\pi^{(k)}((M_S)^N)\right) = \int_{\mathbb{R}} t^N d\mu_{S,k};$$

from the two previous steps, we have

$$\lim_{k \to \infty} \frac{1}{2k+1} \operatorname{trace}\left(\pi^{(k)}((M_S)^N)\right) = m_N = \int_{\mathbb{R}} t^N d\mu_S;$$

it follows that the sequence of measures $(\mu_{S,k})_{k \geq 1}$ (all of mass at most 1 since $\|\pi^{(k)}(M_S)\| \leq 1$ and $\|\lambda_S(M_S)\| \leq 1$) converges vaguely to μ_S, and in particular that

$$\limsup_{k \to \infty} \left\|\pi^{(k)}(M_S)\right\| \geq \frac{2\sqrt{|S|-1}}{|S|}. \qquad \square$$

Theorem 4.7 (Lubotzky-Phillips-Sarnak). *For each prime p, there exists a subset S_0 of $SO(3)$ of $p+1$ elements such that, if $S = S_0 \cup (S_0)^{-1}$,*

$$\left\|\pi^{(k)}(M_S)\right\| \leq \frac{2\sqrt{|S|-1}}{|S|} = \frac{\sqrt{2p+1}}{p+1}$$

for all $k \geq 1$.

4.8. Open problem. It is a natural problem, for each $l \geq 0$, to look for a finite set $S \subset O(3)$ of reflections (depending on l) such that the following conditions are fulfilled:

$$(*) \qquad \pi^{(k)}(M_S) = \frac{1}{2k+1} \operatorname{id}^{(k)} \qquad \text{for} \quad k \in \{1,\ldots,l\}$$

(see Propositions 4.2 and 4.3),

$$(**) \qquad \left\|\pi^{(k)}(M_S)\right\| \leq \frac{2\sqrt{|S|-1}}{|S|} \qquad \text{for } k \text{ large enough}$$

(see Proposition 4.5 and Theorem 4.7), and $|S|$ as small as possible.

Observe that $(*)$ is a way to write that all eigenvalues of $\pi^{(k)}(M_S)$ are equal to $1/(2k+1)$, for $k \in \{1,\ldots,l\}$, and that $(**)$ is a bound on the eigenvalues of $\pi^{(k)}(M_S)$, for large values of k.

5. Reflection groups

Let V be a Euclidean space. The *reflection* $s_x \in O(V)$ associated to $x \in V$, $x \neq 0$, is defined as above by

$$s_x(y) = y - 2\frac{\langle x \mid y \rangle}{\langle x \mid x \rangle}x \quad \text{for all} \quad y \in V.$$

Observe that $s_{x'} = s_x$ if and only if $\mathbb{R}x' = \mathbb{R}x$. In this chapter, a **reflection group** is a subgroup of $O(V)$ generated by a finite set of reflections; it *need not* act properly on V, contrarily to what is assumed most often in [23]. Thus, any finite subset X of $V \setminus \{0\}$ defines a reflection group $\Gamma_X \subset O(V)$ generated by $(s_x)_{x \in X}$.

In particular, let Λ be a lattice in V. We assume that Λ is integral, so that Λ is the disjoint union of the origin and of its non-empty shells $\Lambda_m, m \geq 1$, as defined in Item 1.15. Let $\Gamma_{\Lambda,m}$ denote the reflection group generated by $\{s_x \mid x \in \Lambda_m\}$. For example, if Λ is a root lattice (namely an even lattice Λ generated by Λ_2), then $\Gamma_{\Lambda,2}$ is a finite group, and indeed a direct product of Coxeter groups of types A_n ($n \geq 1$), D_n ($n \geq 4$), and E_n ($n = 6, 7, 8$); see for example [38]. In most cases however, $\Gamma_{\Lambda,m}$ is an infinite group.

Let us specialize to m a power of 2. The scalar product on V defines a symmetric $\mathbb{Z}$-bilinear form on Λ which extends to a symmetric $\mathbb{Z}[1/2]$-bilinear form

$$\beta : (\Lambda \otimes \mathbb{Z}[1/2]) \times (\Lambda \otimes \mathbb{Z}[1/2]) \longrightarrow \mathbb{Z}[1/2].$$

We denote by $O(\Lambda, \mathbb{Z}[1/2])$ the orthogonal group of β. Observe that $\Gamma_{\Lambda,m} \subset O(\Lambda, \mathbb{Z}[1/2])$.

In case of the cubical lattice $\mathbb{Z}^n \subset \mathbb{R}^n$, we write $O(n, \mathbb{Z}[1/2])$ rather than $O(\mathbb{Z}^n, \mathbb{Z}[1/2])$. Several lattices Λ define the same group $O(n, \mathbb{Z}[1/2])$; a sufficient condition for the group $O(\Lambda, \mathbb{Z}[1/2])$ to be isomorphic to $O(n, \mathbb{Z}[1/2])$ is that $2^k\mathbb{Z}^n \subset \Lambda \subset 2^{-l}\mathbb{Z}^n$ for some $k, l \geq 0$. In particular, if $\Lambda \subset \mathbb{R}^8$ is a root lattice of type E_8, then $O(\Lambda, \mathbb{Z}[1/2]) = O(8, \mathbb{Z}[1/2])$, since Λ and $\mathbb{Z}^8$ have a common sublattice of index 2 (which is a root lattice of type D_8). Here are some properties of the group $O(n, \mathbb{Z}[1/2])$:

(o) it is generated by reflections for any n;
(i) it is infinite if and only if $n \geq 5$;
(ii) it is naturally [14] a discrete cocompact subgroups of the 2-adic group $O(n, \mathbb{Q}_2)$;
(iii) it is finitely presented;
(iv) it is virtually torsion free, and more precisely $\text{Ker}\big(O(n, \mathbb{Z}[1/2]) \longrightarrow O(n, \mathbb{Z}/3)\big)$ is torsion free;
(v) it has finite-type homology—more precisely, for any Noetherian ring A and any $i \geq 0$, the A-module $H_i(O(n, \mathbb{Z}[1/2]), A)$ is finitely generated;

---

[14] More generally, if $\mathbb{P}$ is a finite set of distinct primes and if N denotes their product, $O(n, \mathbb{Z}[1/N])$ is naturally a discrete cocompact subgroup of the product of $O(n)$ with $\prod_{p \in \mathbb{P}} O(n, \mathbb{Q}_p)$, and therefore also a discrete cocompact subgroup of $\prod_{p \in \mathbb{P}} O(n, \mathbb{Q}_p)$. It follows that $O(n, \mathbb{Z}[1/N])$ is finitely presented [89].

(vi) it is of virtual cohomological dimension $\frac{1}{2}(n-\inf_{m\in\mathbb{Z}}|n-8m|)$, and the group $H_i(O(n,\mathbb{Z}[1/2]),\mathbb{Q})$ is zero in dimensions other than 0 or maximal;

(vii) the natural inclusion of $O(n,\mathbb{Z}[1/2])$ into the compact orthogonal group $O(n)$ has dense image if and only if the group $O(n,\mathbb{Z}[1/2])$ is infinite.

Several of these properties are straightforward consequences of (ii). Property (vii) follows from a strong approximation theorem due to Kneser [64]. For much more on these groups, see [25].

The next proposition shows a remarkable generating set of $O(8,\mathbb{Z}[1/2])$; it is an unpublished result of B. Venkov.

Proposition 5.1 (B. Venkov). *Let L be a root lattice of type E_8. Then the group $O(8,\mathbb{Z}[1/2])$ is generated by $\Gamma_{L,4}$ and a reflection with respect to one root, and also by the finite Weyl group $\Gamma_{L,2}$ of type E_8 and a reflection with respect to one element of L_4.*

Proof. Let $\mathcal{L}$ be the set of even unimodular lattices M in $V \approx \mathbb{R}^8$ such that $M\cap L$ is of index a power of 2 in both L and M (the same power since L and M are both unimodular). Then $\mathcal{L}$ is a metric space for the distance δ defined by

$$\delta(M,N) = \log_2([M:M\cap N]) = \log_2([N:M\cap N]).$$

It is a theorem of Kneser [63, p. 242] that $\mathcal{L}$ is connected by steps of length one; more precisely, given $M,N \in \mathcal{L}$ with $d = \delta(M,N)$, there exists a sequence $M = M^{(0)}, M^{(1)},\ldots, M^{(d)} = N$ in $\mathcal{L}$ such that $\delta(M^{(j-1)}, M^{(j)}) = 1$ for all $j \in \{1,\ldots,d\}$.

Set $G = O(8,\mathbb{Z}[1/2])$; this group operates naturally on $\mathcal{L}$. Here is the synopsis of the proof: for $g \in G$, set $M = g(L) \in \mathcal{L}$ and $d = \delta(L,M)$; we show by induction on d that g is a product of appropriate reflections. More precisely, for $M^{(0)} = L, M^{(1)},\ldots, M^{(d)} = M$ as above, we check that $M^{(j)} = s_x(M^{(j-1)})$ for some $x \in M^{(j-1)}$ with $\langle x \mid x \rangle = 4$.

Step one: $\langle \Gamma_{L,2}, \Gamma_{L,4}\rangle = \langle s_r, \Gamma_{L,4}\rangle = \langle \Gamma_{L,2}, s_x\rangle$ for any $r \in L_2$ and $x \in L_4$. (For a subset $S \subset O(V)$, we denote by $\langle S \rangle$ the subgroup of $O(V)$ generated by S.)

It is known that the group of all automorphisms of L coincides with the group $\Gamma_{L,2}$ generated by the symmetries with respect to the roots. For $g \in \mathrm{Aut}(L)$ and $x \in L$, $x \neq 0$, we have

$$(*)\qquad\qquad\qquad s_{g(x)} = gs_xg^{-1}.$$

As $\mathrm{Aut}(L)$ acts transitively on L_4, we have $\langle \Gamma_{L,2}, \Gamma_{L,4}\rangle = \langle \Gamma_{L,2}, s_x\rangle$ for any $x \in L_4$. (It is known that, more generally, $\mathrm{Aut}(L)$ acts transitively on each of L_2, L_4, L_6, the complement of $2L_2$ in L_8, L_{10}, and L_{12}; see page 122 of [26].)

Relations $(*)$ show that $\Gamma_{L,2}\cap\Gamma_{L,4}$ is a normal subgroup of $\Gamma_{L,2}$. Let $e_1,\ldots,e_8$ be an orthonormal basis of V such that $r = e_1 + e_2$ and $r' = e_1 - e_2$ are in L_2 (see page 268 of [23]); then $x = 2e_1$ and $x' = 2e_2$ are in L_4. A straightforward computation shows that $\mathrm{id}_V \neq s_rs_{r'} = s_xs_{x'} \in \Gamma_{L,2}\cap\Gamma_{L,4}$. As $\Gamma_{L,2}$ is almost simple (by Exercise 2 of § VI.4 in [23]), it follows that $\Gamma_{L,2}\cap\Gamma_{L,4}$ is a subgroup of index [15] at most 2 in $\Gamma_{L,2}$, so that $\Gamma_{L,2}$ is generated by $\Gamma_{L,2} \cap \Gamma_{L,4}$ and s_r for any $r \in L_2$. A fortiori $\langle \Gamma_{L,2}, \Gamma_{L,4} \rangle = \langle s_r, \Gamma_{L,4} \rangle$ for any $r \in L_2$. $\qquad\square$

Step two: a reminder on neighbours.

For this step, V can be a Euclidean space of any dimension $n \equiv 0 \pmod 8$ and M any even unimodular lattice in V. We view $\overline{M} = M/2M$ as a vector space of dimension n over the prime field $\mathbb{F}_2$. There is a nondegenerate quadratic form $q : \overline{M} \longrightarrow \mathbb{F}_2$ defined by $q(\overline{z}) = \frac{1}{2}\langle z \mid z \rangle \pmod 2$ for $z \in M$ representing $\overline{z} \in \overline{M}$. For each $z \in M$ such that $z \notin 2M$ and $\langle z \mid z \rangle \equiv 0 \pmod 4$, set

$$M_z = \{ m \in M \mid \langle m \mid z \rangle \equiv 0 \pmod 2 \} \quad \text{and} \quad M^z = M_z \sqcup (\frac{1}{2}z + M_z).$$

Then M^z is an integral unimodular lattice in V, and $M \cap M^z$ is of index 2 in both M and M^z. Moreover:

(i) M^z is even if $\langle z \mid z \rangle \equiv 0 \pmod 8$ and odd if $\langle z \mid z \rangle \equiv 4 \pmod 8$;

(ii) for two z, z' in M, not in $2M$, with $\langle z \mid z \rangle \equiv 0 \pmod 4$ and $\langle z' \mid z' \rangle \equiv 0 \pmod 4$, $M^z = M^{z'}$ if and only if $z' - z \in 2M$ and $\langle z \mid z \rangle \equiv \langle z' \mid z' \rangle \pmod 8$;

(iii) any integral unimodular lattice M' in V such that $M \cap M'$ is of index 2 in both M and M' appears as one of the lattices M^z ($z \in M$, $z \notin 2M$, $\langle z \mid z \rangle \equiv 0 \pmod 4$).

(For the analogous facts concerning an odd lattice M, see [100]). $\qquad\square$

Step three: short representatives in M of non-zero isotropic classes in $\overline{M}$.

Let V be again of dimension 8, so that two even unimodular lattices in V are always isomorphic, and let M be such a lattice. The 2^8 elements of $\overline{M}$ splits as

(a) the origin,

(b) 120 elements represented by pairs $\pm r$ of roots, which are nonisotropic for q,

(c) 135 nonzero isotropic elements.

Let $\psi : M_4 \longrightarrow \overline{M}$ be the restriction to M_4 of the canonical projection. We claim that each fiber of ψ has at most 16 elements. Indeed, for $x_1, x_2 \in M_4$ such that $x_1 \neq x_2$ and $\overline{x_1} = \overline{x_2}$, there exists $m \in M$ such that $x_2 - x_1 = 2m \neq 0$ and, upon changing x_2 to $-x_2$ if necessary, $\langle x_1 \mid x_2 \rangle \geq 0$. Then

(i) $\langle x_2 - x_1 \mid x_2 - x_1 \rangle = 4 + 4 - 2\langle x_1 \mid x_2 \rangle \leq 8$,

(ii) $\langle x_2 - x_1 \mid x_2 - x_1 \rangle = 4\langle m \mid m \rangle \geq 8$,

so that $\langle x_2 - x_1 \mid x_1 - x_2 \rangle = 8$ and $x_1 \perp x_2$. The claim follows.

Since M_4 has exactly $2160 = 135 \times 16$ elements and since the image of ψ is *a priori* inside the set of 135 nonzero isotropic elements of $\overline{M}$, it follows that ψ is onto this set. $\qquad\square$

Step four: Let $M, M' \subset V \approx \mathbb{R}^8$ be two even unimodular lattices such that $M \cap M'$ is of index 2 in both M and M'. Then there exists $x \in M_4$ such that $M' = s_x(M)$.

---

[15] We do not know if $\Gamma_{L,2} \cap \Gamma_{L,4}$ is the whole of $\Gamma_{L,2}$ or if it is the subgroup of elements which are products of even numbers of reflections.

Choose a nonzero isotropic element $\bar{z} \in \overline{M}$ represented by an element $x \in M_4$ (see Step three). Since M is unimodular, there exists a root $r \in M_2$ such that $\langle r \mid x \rangle = 1$. The element $z = x - 2r$ is also a representative of $\bar{z}$, and

$$\langle z \mid z \rangle = \langle x \mid x \rangle - 4\langle r \mid x \rangle + 4\langle r \mid r \rangle = 4 - 4 + 8 = 8.$$

Thus M^z is the even neighbour M' of M (see Step two).

To conclude the proof of Step four, we have to check that $s_x(M) = M^z$. For $m \in M$, we have

$$s_x(m) = m - \frac{1}{2}\langle z + 2r \mid m \rangle (z + 2r)$$

$$= m - \frac{1}{2}\langle z \mid m \rangle z - \langle z \mid m \rangle r - \langle r \mid m \rangle z - 2\langle r \mid m \rangle r$$

and

$$\langle z \mid s_x(m) \rangle \equiv \langle z \mid m \rangle - \langle z \mid m \rangle\langle z \mid r \rangle \pmod{2}.$$

If $\langle z \mid m \rangle$ is even, namely if $m \in M_z$, these formulas show that $s_x(m) \in M_z$, and consequently that $s_x(m) \in M^z$. If $\langle z \mid m \rangle$ is odd, namely if $m \in M \setminus M_z$, they show that $s_x(m) - \frac{1}{2}z \in M_z$, and we have again $s_x(m) \in M^z$. Thus $s_x(M) \subset M^z$. Since $s_x(M)$ and M^z are both unimodular, this shows that $s_x(M) = M^z$.

[Conversely, for any $x \in M_4$, we have $\delta(M, s_x(M)) = 1$.] $\qquad\square$

End of proof of Proposition 5.1

Consider, as in the beginning of this proof, an element $g \in O(8, \mathbb{Z}[1/2])$, the lattice $M = g(L) \in \mathcal{L}$ at distance d from L, and a sequence $\left(M^{(j)}\right)_{0 \le j \le d}$ such that $M^{(0)} = L$, $M^{(d)} = M$, and $M^{(j-1)} \cap M^{(j)}$ of index 2 in both $M^{(j-1)}$ and $M^{(j)}$. The previous steps show that there exists for each $j \in \{1, \ldots, d\}$ an element $x_j \in \left(M^{(j-1)}\right)_4$ such that $s_{x_j}(M^{(j-1)}) = M^{(j)}$. Set $g_j = \prod_{i=1}^{j-1} s_{x_i}$. Since $g^{-1}g_d$ is in $\mathrm{Aut}(L)$, and in particular in $O(8, \mathbb{Z}[1/2])$, by Step one, it is enough to show that $g_d \in O(8, \mathbb{Z}[1/2])$. We claim that $g_j \in O(8, \mathbb{Z}[1/2])$ for all $j \in \{1, \ldots, d\}$, and we prove the claim by induction on j.

The claim is clear for $j = 1$. Assume it holds for some value of $j \in \{1, \ldots, d-1\}$, and let us check it for $j+1$. Let $y_j \in L$ be the element such that $g_j(y_j) = x_j \in M^{(j)}$. Then $g_{j+1} = s_{x_j}g_j = g_j s_{y_j} g_j^{-1} g_j$ is indeed in G since $g_j \in G$ and $s_{y_j} \in \Gamma_{L,4}$. This shows that the claim holds for $j+1$. $\qquad\square$

Corollary 5.2. *Let L be a root lattice of type E_8. Then the subgroup $SO(8, \mathbb{Z}[1/2])$ of elements of determinant $+1$ in $O(8, \mathbb{Z}[1/2])$ is generated by the products $s_x s_{x'}$, where x, x' are in the shell L_4 of L.*

5.3. A set of 8 generators for $O(8, \mathbb{Z}[1/2])$. Let $(\epsilon_1, \ldots, \epsilon_8)$ denote the canonical orthonormal basis in $\mathbb{R}^8$. As in [23], set

$$\alpha_1 = \frac{1}{2}(\epsilon_1 + \epsilon_8) - \frac{1}{2}(\epsilon_2 + \cdots + \epsilon_7)$$

$$\alpha_2 = \epsilon_1 + \epsilon_2$$

$$\alpha_j = \epsilon_{j-1} - \epsilon_{j-2} \quad (j = 3, \ldots, 8)$$

P. de la Harpe and C. Pache

so that $\{\alpha_1, \ldots, \alpha_8\}$ is a basis of a root system of type E_8, and therefore also a basis of a root lattice $L \subset \mathbb{R}^8$ of type E_8. Let $s_1, \ldots, s_8$ denote the reflections associated to $\alpha_1, \ldots, \alpha_8$. Then $\mathrm{Aut}(L)$ has a Coxeter presentation with generators $s_1, \ldots, s_8$ and relations of the familiar form $(s_i s_j)^{m_{i,j}} = 1$.

Consider the vector $2\epsilon_2 \in L_4$ and the corresponding reflection $\tilde{s}_2 : x \longmapsto x - \langle x \mid \epsilon_2 \rangle \epsilon_2$. A simple computation shows that the conjugation by $\tilde{s}_2$ exchanges s_2 with s_3 and leaves s_j invariant for $j = 1, 4, 5, 6, 7, 8$. It follows from Proposition 5.1 that $O(8, \mathbb{Z}[1/2])$ has a generating set obtained from that of $\mathrm{Aut}(L)$ by replacing s_2 by $\tilde{s}_2$. Moreover, the order $\tilde{m}_{2,j}$ of $\tilde{s}_2 s_j$ in $O(8, \mathbb{Z}[1/2])$ is equal to

$$
\begin{array}{lll}
\infty & \text{for} & j = 1 \\
4 & \text{for} & j = 3, 4 \\
2 & \text{for} & j = 5, 6, 7, 8.
\end{array}
$$

Thus, for the generators of $O(8, \mathbb{Z}[1/2])$ described here, the orders $\tilde{m}_{i,j}$ of the products of two generators coincide with the corresponding orders $m_{i,j}$ for the Coxeter generators of $\mathrm{Aut}(L)$, this for all but three pairs of indices, namely for all but $(2, 1)$, $(2, 3)$, and $(2, 4)$.

References

[1] P. Appell, Sur une classe de polynômes à deux variables et le calcul approché des intégrales doubles, *Ann. Fac. Sci. Univ. Toulouse* **4** (1890), H1–H20.

[2] J. Arias de Reyna, A generalized mean-value theorem, *Mh. Math.* **106** (1988), 95–97.

[3] N. Aronszajn, Theory of reproducing kernels, *Trans. Amer. Math. Soc.* **68** (1950), 337–404.

[4] E. Bannai and E. Bannai, On Euclidean tight 4-designs, Preprint.

[5] E. Bannai and E. Bannai, Tight Gaussian 4-designs, *J. Algebraic Combinatorics* **22** (2005), 39–63.

[6] C. Bachoc, E. Bannai, and R. Coulangeon, Codes and designs in Grassmannian spaces, *Discrete Math.* (2004), 15–28.

[7] R. Bacher, Tables de réseaux entiers unimodulaires construits comme k-voisins de $\mathbb{Z}^n$, *J. Théor. Nombres Bordeaux* **9** (1997), 479–497.

[8] C. Bachoc, R. Coulangeon, and G. Nebe, Designs in Grassmannian spaces and lattices, *Journal of Algebraic Combinatorics* **16** (2002), 5–19.

[9] C. Bachoc and B. Venkov, Modular forms, lattices and spherical designs, in: "Réseaux euclidiens, designs sphériques et formes modulaires", *Monographie de l'Enseignement mathématique* **37** (2001), 87–111.

[10] B. Bajnok, Construction of designs on the 2-sphere, *Europ. J. Combinatorics* **12** (1991), 377–382.

[11] B. Bajnok, Chebyshev-type quadrature formulas on the sphere, *Congr. Numer.* **85** (1991), 214–218.

[12] B. Bajnok, Construction of spherical t-designs, *Geom. Dedicata* **43** (1992), 167–179.

[13] B. Bajnok, Construction of spherical 3-designs, *Graphs and Combinatorics* **14** (1998), 97–107.

[14] K. Ball, An elementary introduction to modern convex geometry, in: "Flavors of geometry", MSRI Publications, Silvio Levy, Editor, Cambridge Univ. Press, 1997, 1–58.

[15] E. Bannai, A. Munemasa, and B. Venkov, The nonexistence of certain tight spherical designs, *Algebra i Analyz* **16** no. 4 (2004), 1–23.

[16] E. Bannai, On some spherical t-designs, *J. Combinatorial Theory, Series A* **26** (1979), 157–161.

[17] L. Bartholdi and R. Grigorchuk, On the spectrum of Hecke operators related to some fractal groups, *Proc. Steklov Inst. Math.* **231** (2000), 1–41.

[18] R. Balasubramanian, J.-M. Deshouillers, and F. Dress, Problème de Waring pour les bicarrés. I. Schéma de la solution, *C. R. Acad. Sci. Paris Sér. I Math* **303** (1986), 85–88.

[19] R. Balasubramanian, J.-M. Deshouillers, and F. Dress, Problème de Waring pour les bicarrés. II. Résultats auxiliaires pour le théorème asymptotique, *C. R. Acad. Sci. Paris Sér. I Math* **303** (1986), 161–163.

[20] M. Bekka and P. de la Harpe (Appendix with R. Grigorchuk), Irreducibility of unitary group representations and reproducing kernels Hilbert spaces, *Expo. Math.* **21** (2003), 115–149.

[21] M. Berger, Géométrie, volume 3, Cedic / Fernand Nathan, 1978.

[22] C. Béguin, A. Valette and A. Zuk, On the spectrum of a random walk on the discrete Heisenberg group and the norm of Harper's operator, *J. Geometry and Physics* **21** (1997), 337–356.

[23] N. Bourbaki, Groupes et algèbres de Lie, chapitres 4, 5 et 6, Hermann, 1968.

[24] Y. Colin de Verdière, Distribution de points sur une sphère, *Séminaire Bourbaki 703 (1988), Astérisque* **177–178** (1989), 83–93.

[25] G. Collinet, Quelques propriétés homologiques du groupe $O_n(\mathbb{Z}[\frac{1}{2}])$, Thèse, 2002.

[26] J.H. Conway and N.J.A. Sloane, Sphere packings, lattices and groups, Third Edition, Springer, Grundlehren der mathematischen Wissenschaften **290**, 1999.

[27] H.S.M. Coxeter, The problem of packing a number of equal nonoverlapping circles on a sphere, *Trans. N.Y. Acad. Sci.* **24** (1962), 320–331.

[28] H.S.M. Coxeter, Regular polytopes, third edition, Dover, 1973.

[29] M. Crouzeix and A.L. Mignot, Analyse numérique des équations différentielles, Masson, 1984.

[30] H. Davenport, On Waring's problem for fourth powers, *Annals of Math. (2)* **40** (1939), 731–747. [= Collected Works, Volume III, 946–962].

[31] P.J. Davis and P. Rabinowitz, Methods of numerical integration, second edition, Academic Press, 1984.

[32] P. Delsarte, Hahn polynomials, discrete harmonics, and t-designs, *SIAM J. Appl. Math.* **34** (1978), 157–166.

[33] P. Delsarte, J.M. Goethals and J.J. Seidel, Spherical codes and designs, *Geometriae Dedicata* **6** (1977), 363–388.

[34] J.-M. Deshouillers, F. Hennecart, and B. Landreau, Waring's problem for sixteen biquadrates – numerical results, *J. Théor. Nombres Bordeaux* **12** (2000), 411–422 [see also MathReviews 2002b:11133 by Koichi Kawada].

[35] P. Delsarte and J.J. Seidel, Fisher type inequalities for Euclidean t-designs, *Linear Algebra and its Appl.* **114/115** (1989), 213–230.

[36] L.E. Dickson, History of the theory of numbers, Vol. II, Carnegie Institution of Washington, 1919 [reprinted by Cehlsea, 1966].

[37] C.F. Dunkl and Y. Xu, Orthogonal polynomials of several variables, Cambridge Univ. Press, 2001.

[38] W. Ebeling, Lattices and codes. A course partially based on lectures by F. Hirzebruch, Vieweg, 1994.

[39] H.G. Eggleston, Convexity, Cambridge University Press, 1958.

[40] W.J. Ellison, Waring's problem, *American Monthly* **78**[1] (1971), 10–36.

[41] T. Ericson and V. Zinoviev, Codes on Euclidean spheres, Elsevier, 2001.

[42] I. Fejes Tóth, Regular figures, Pergamon Press, 1964.

[43] A. Gambard, D. Jakobson and P. Sarnak, Spectra of elements in the group ring of $SU(2)$, *J. Eur. Math. Soc.* **1** (1999), 51–85.

[44] W. Gautschi, Advances in Chebyshev quadrature, in: "Numerical analysis, Dundee 1975", *Lecture Notes in Math.* **506** (1976), 100–121.

[45] W. Gautschi, Numerical analysis, an introduction, Birkhäuser, 1997.

[46] C.D. Godsil, Algebraic combinatorics, Chapman & Hall, 1993.

[47] J.M. Goethals and J.J. Seidel, Spherical designs, *Proc. Symp. Pure Math. A.M.S.* **34** (1979), 255–272.

[48] J.M. Goethals and J.J. Seidel, Cubature formulae, polytopes, and spherical designs, in: "The geometric vein, the Coxeter Festschrift", Springer, 1981, 203–218.

[49] J.M. Goethals and J.J. Seidel, The football, *Nieuw. Arch. Wiskunde* **29** (1981), 50–58.

[50] R. Grigorchuk and A. Zuk, The lamplighter group as a group generated by a 2-state automaton, and its spectrum, *Geom. Dedicata* **87** (2001), 209–244.

[51] R. Grigorchuk and A. Zuk, Spectral properties of a torsion-free weakly branch group defined by a three state automaton, in: "Computational and statistical group theory", *Contemp. Math.* **98**, Amer. Math. Soc., 2002, 57–82.

[52] R. Grigorchuk and A. Zuk, The Ihara zeta function of infinite graphs, the KNS spectral measure and integral maps, in "Random walks and geometry (Vienna 2001)", V.A. Kaimanovich, Editor, de Gruyter, 2004, 141–180.

[53] R.H. Hardin and N.J.A. Sloane, New spherical 4-designs, *Discrete Math.* **106/107** (1992), 255–264.

[54] R.H. Hardin and N.J.A. Sloane, Expressing $(a^2+b^2+c^2+d^2)^3$ as a sum of 23 sixth powers, *J. Combinatorial Theory, Series A* **68** (1994), 481–485.

[55] R.H. Hardin and N.J.A. Sloane, McLaren's improved snub cube and other new spherical designs in three dimensions, *Discrete Comput. Geom.* **15** (1996), 429–441.

[56] P. de la Harpe and C. Pache, Spherical designs and finite group representations (some results of E. Bannai), *Europ. J. Combinatorics* **25** issue 2, in memory of Jaap Seidel (2004), 213–227.

[57] P. de la Harpe, G. Robertson and A. Valette, On the spectrum of the sum of generators for a finitely generated group, *Israel J. Math.* **81** (1993), 65–96.

[58] P. de la Harpe, G. Robertson and A. Valette, On the spectrum of the sum of generators for a finitely generated group, part II, *Colloquium Math.* **65** (1993), 87–102.

[59] P. de la Harpe and B. Venkov, Groupes engendrés par des réflexions, designs sphériques et réseau de Leech, *C.R. Acad. Sc. Paris* **333** (2001), 745–750.

[60] D. Hilbert, Beweis für die Darstellbarkeit der ganzen Zahlen durch eine feste Anzahl n^{ter} Potenzen (Waringsches Problem), *Math. Annalen* **67** (1909), 281–300.

[61] A. Hurwitz, Über die Darstellung der ganzen Zahlen als Summen von n^{ten} Potenzen ganzer Zahlen, *Math. Annalen* **65** (1908), 424–427.

[62] H. Kesten, Symmetric random walks on groups, *Trans. Amer. Math. Soc.* **92** (1959), 336–354.

[63] M. Kneser, Klassenzahlen definiter quadratischer Formen, *Archiv der Mathematik* **8** (1957), 241–250.

[64] M. Kneser, Strong approximation, in: "Algebraic groups and discontinuous subgroups", A. Borel and G.D. Mostow, Editors, *Proc. Symp. Pure Math.* **9**, Amer. Math. Soc., 1968, 187–196.

[65] J. Korevaar (Notes by A.B.J. Kuijlaars), Chebyshev-type quadratures: use of complex analysis and potential theory, in: "Complex Potential Theory", P.M. Gauthier and G. Sabidussi, Editors, Kluwer Academic Publ., NATO ASI Series **439** (1994), 325–364.

[66] J. Korevaar and J.L.H. Meyers, Spherical Faraday cage for the case of equal point charges and Chebyshev-type quadrature on the sphere, *Integral transforms and special Functions* **1** (1993), 105–117.

[67] M.G. Krein, Hermitian-positive kernels on homogenous spaces, I & II, *Amer. Math. Soc. Translations (2)* **34** (1963), 69–108 & 109–164 [Original Russian paper in *Ukrain Math. Zurnal* **1** no. 4 (1949), 64–98 and **2** (1950), 10–59].

[68] V.I. Krylov, Approximate calculation of integrals, Macmillan, 1962.

[69] A. Kuijlaars, Chebyshev-type quadrature and partial sums of the exponential series, *Math. of Computation* **209** (1995), 251–263.

[70] D.H. Lehmer, The vanishing of Ramanujan's function $\tau(n)$, *Duke Math. J.* **14** (1947), 429–433.

[71] A. Lubotzky, R. Phillips, and P. Sarnak, Hecke operators and distributing points on the sphere I, *Comm. Pure Appl. Math.* **39** (1986), S149–S186.

[72] A. Lubotzky, R. Phillips, and P. Sarnak, Hecke operators and distributing points on the sphere II, *Comm. Pure Appl. Math.* **40** (1987), 401–420.

[73] Y.I. Lyubich and L.N. Vaserstein, Isometric embeddings between classical Banach spaces, cubature formulas, and spherical designs, *Geometriae Dedicata* **47** (1993), 327–362.

[74] J. Martinet, Perfect lattices in Euclidean spaces, Springer, Grundlehren der mathematischen Wissenschaften **237**, 2003.

[75] J.C. Maxwell, On approximate multiple integration between limits of summation, *Proc. Cambridge Philos. Soc.* **3** (1877), 39–47.

[76] J. Milnor and D. Husemoller, Symmetric bilinear forms, Springer, Ergebnisse der Mathematik und ihrer Grenzgebiete **73**, 1973.

[77] Y. Mimura, A construction of spherical 2-designs, *Graphs and Combinatorics* **6** (1990), 369–372.

[78] M.B. Nathanson, Additive number theory, the classical bases, Springer, Graduate Texts in Mathematics **164**, 1996.

[79] G. Nebe and B. Venkov, The strongly perfect lattices of dimension 10, *J. Théor. Nombres Bordeaux* **12** (2000), 503–518.

[80] A. Neumaier and J.J. Seidel, Discrete measures for spherical designs, eutactic stars and lattices, *Nederl. Akad. Wetensch. Proc. Ser. A 91, Indag. Math.* **50** (1988), 321–334.

[81] C. Pache, Sur le spectre des opérateurs de Markov de designs sphériques, *Europ. J. Combinatorics* **25** (2004), 591–620.

[82] C. Pache, Shells of selfdual lattices viewed as spherical designs, to appear in: *Internat. J. Algebra Comput.*, 2004.

[83] B. Reznick, Sums of even powers of real linear forms, *Memoirs of the Amer. Math. Soc.* **96**, 1992, no. 463.

[84] R.M. Robinson, Arrangement of 24 points on a sphere, *Math. Annalen* **144** (1961), 17–48.

[85] E.B. Saff and A.B.J. Kuijlaars, Distributing many points on a sphere, *Mathematical Intelligencer* **19** no. 1 (1997), 5–11.

[86] P. Sarnak, Some applications of modular forms, Cambridge University Press, 1990.

[87] K. Schütte and B.L. van der Waerden, Auf welcher Kugel haben $5, 6, 7, 8$ oder 9 Punkte mit Mindestabstand Eins Platz? *Math. Annalen* **123** (1951), 96–124.

[88] J-P. Serre, Cours d'arithmétique, Presses univ. de France, 1970.

[89] J-P. Serre, Cohomologie des groupes discrets, in: "Prospects in mathematics", *Annals of Math. Studies* **70**, Princeton Univ. Press, 1971, 77–169 [= Oeuvres, Volume II, 593–685, see also 725–727].

[90] J-P. Serre, Sur la lacunarité des puissances de η, *Glasgow Math. J.* **27** (1985), 203–221 [= Oeuvres, Volume IV, 66–84, see also 640].

[91] J-P. Serre, Répartition asymptotique des valeurs propres de l'opérateur de Hecke T_p, *Journal Amer. Math. Soc.* **10** (1997), 75–102 [= Oeuvres, Volume IV, 543–570].

[92] P.D. Seymour and T. Zaslavsky, Averaging sets: a generalization of mean values and spherical designs, *Adv. in Math.* **152** (1984), 213–240.

[93] S. Smale, Mathematical problems for the next century, *Mathematical Intelligencer* **20** no. 2 (1998), 7–15.

[94] I.H. Samhuis, A female contribution to early genetics: Tine Tammes and Mendel's laws for continuous characters, *J. for the History of Biology* **28** (1995), 495–531.

[95] E.M. Stein and G. Weiss, Introduction to Fourier analysis on Euclidean spaces, Princeton Univ. Press, 1971.

[96] J. Stewart, Positive definite functions and generalizations, an historical survey, *Rocky Mountain J. Math.* **6** (1976), 409–434.

[97] G. Szegö, Orthogonal polynomials, Colloquium Publications **23**, Amer. Math. Soc., 1939.

[98] P.M.L. Tammes, On the origin of number and arrangement of the places of exist on the surface of pollengrains, *Recueil des travaux botaniques néerlandais* **27** (1930), 1–84.

[99] B.L. van der Waerden, Punkte auf der Kugel. Drei Zusätze, *Math. Ann.* **123** (1952), 213–222.

[100] B. Venkov, Odd unimodular lattices, *Zap. Naučn. Sem. Leningrad. Otdel. Mat. Inst. Steklov (LOMI)* **86** (1979), 40–48.

[101] B. Venkov, On even unimodular extremal lattices, *Proc. Steklov Inst. Math.* **165** (1985), 47–52.

[102] B. Venkov (Notes par J. Martinet), Réseaux et designs sphériques, in: "Réseaux euclidiens, designs sphériques et formes modulaires", Monographie de l'Enseignement mathématique **37**, 2001, 10–86.

[103] N.J. Vilenkin, Special functions and the theory of group representations, Translations of Math. Monographs **22**, Amer. Math. Soc., 1968.

[104] V.A. Yudin, Lower bounds for spherical designs, *Investiya Math.* **61** no. 3 (1997), 213–223.

Pierre de la Harpe and Claude Pache
Section de Mathématiques
Université de Genève
C.P. 64
CH-1211 Genève 4
Switzerland
e-mail: `Pierre.delaHarpe@math.unige.ch`
      `Claude.Pache@math.unige.ch`

Progress in Mathematics, Vol. 248, 269–322

Survey on Classifying Spaces for Families of Subgroups

Wolfgang Lück

Abstract. We define for a topological group G and a family of subgroups $\mathcal{F}$ two versions for the classifying space for the family $\mathcal{F}$, the G-CW-version $E_{\mathcal{F}}(G)$ and the numerable G-space version $J_{\mathcal{F}}(G)$. They agree if G is discrete, or if G is a Lie group and each element in $\mathcal{F}$ compact, or if $\mathcal{F}$ is the family of compact subgroups. We discuss special geometric models for these spaces for the family of compact open groups in special cases such as almost connected groups G and word hyperbolic groups G. We deal with the question whether there are finite models, models of finite type, finite dimensional models. We also discuss the relevance of these spaces for the Baum-Connes Conjecture about the topological K-theory of the reduced group C^*-algebra, for the Farrell-Jones Conjecture about the algebraic K- and L-theory of group rings, for Completion Theorems and for classifying spaces for equivariant vector bundles and for other situations.

Mathematics Subject Classification (2000). 55R35, 57S99, 20F65, 18G99.

Keywords. Family of subgroups, classifying spaces.

Contents

0. Introduction

We define for a topological group G and a family of subgroups $\mathcal{F}$ two versions for the classifying space for the family $\mathcal{F}$, the G-CW-version $E_{\mathcal{F}}(G)$ and the numerable G-space version $J_{\mathcal{F}}(G)$. They agree, if G is discrete, or if G is a Lie group and each element in $\mathcal{F}$ is compact, or if each element in $\mathcal{F}$ is open, or if $\mathcal{F}$ is the family of compact subgroups, but not in general.

One motivation for the study of these classifying spaces comes from the fact that they appear in the Baum-Connes Conjecture about the topological K-theory of the reduced group C^*-algebra and in the Farrell-Jones Conjecture about the algebraic K- and L-theory of group rings and that they play a role in the formulations and constructions concerning Completion Theorems and classifying spaces for equivariant vector bundles and other situations. Because of the Baum-Connes Conjecture and the Farrell-Jones Conjecture the computation of the relevant K- and L-groups can be reduced to the computation of certain equivariant homology groups applied to these classifying spaces for the family of finite subgroups or the family of virtually cyclic subgroups. Therefore it is important to have nice geometric models for these spaces $E_{\mathcal{F}}(G)$ and $J_{\mathcal{F}}(G)$ and in particular for the orbit space $G\backslash E_{\mathcal{FIN}}(G)$.

The space $E_{\mathcal{F}}(G)$ has for the family of compact open subgroups or of finite subgroups nice geometric models for instance in the cases, where G is an almost connected group G, where G is a discrete subgroup of a connected Lie group, where G is a word hyperbolic group, arithmetic group, mapping class group, one-relator group and so on. Models are given by symmetric spaces, Teichmüller spaces, outer space, Rips complexes, buildings, trees and so on. On the other hand one can construct for any CW-complex X a discrete group G such that X and $G\backslash E_{\mathcal{FIN}}(G)$ are homotopy equivalent.

We deal with the question whether there are finite models, models of finite type, finite dimensional models. In some sense the algebra of a discrete group G is reflected in the geometry of the spaces $E_{\mathcal{FIN}}(G)$. For torsionfree discrete groups $E_{\mathcal{FIN}}(G)$ is the same as EG. For discrete groups with torsion the space $E_{\mathcal{FIN}}(G)$ seems to carry relevant information which is not present in EG. For instance for a discrete group with torsion EG can never have a finite dimensional model, whereas this is possible for $E_{\mathcal{FIN}}(G)$ and the minimal dimension is related to the notion of virtual cohomological dimension.

The space $J_{\mathcal{COM}}(G)$ associated to the family of compact subgroups is sometimes also called the classifying space for proper group actions. We will abbreviate it as $\underline{J}G$. Analogously we often write $\underline{E}G$ instead of $E_{\mathcal{COM}}(G)$. Sometimes the abbreviation $\underline{E}G$ is used in the literature, especially in connection with the Baum-Connes Conjecture, for the G-space denoted in this article by $\underline{J}G = J_{\mathcal{COM}}(G)$. This does not really matter since we will show that the up to G-homotopy unique G-map $\underline{E}G \to \underline{J}G$ is a G-homotopy equivalence.

A reader, who is only interested in discrete groups, can skip Sections 2 and 3 completely.

Group means always locally compact Hausdorff topological group. Examples are discrete groups and Lie groups but we will also consider other groups. Space always means Hausdorff space. Subgroups are always assumed to be closed. Notice that isotropy groups of G-spaces are automatically closed. A map is always understood to be continuous.

The author is grateful to Britta Nucinkis, Ian Leary and Guido Mislin for useful comments.

1. G-CW-Complex-Version

In this section we explain the G-CW-complex version of the classifying space for a family $\mathcal{F}$ of subgroups of a group G.

1.1. Basics about G-CW-Complexes

Definition 1.1 (G-CW-complex). *A G-CW-complex X is a G-space together with a G-invariant filtration*

$$\emptyset = X_{-1} \subseteq X_0 \subset X_1 \subseteq \ldots \subseteq X_n \subseteq \ldots \subseteq \bigcup_{n \geq 0} X_n = X$$

such that X carries the colimit topology with respect to this filtration (i.e. a set $C \subseteq X$ is closed if and only if $C \cap X_n$ is closed in X_n for all $n \geq 0$) and X_n is obtained from X_{n-1} for each $n \geq 0$ by attaching equivariant n-dimensional cells, i.e. there exists a G-pushout

$$
\begin{array}{ccc}
\coprod_{i \in I_n} G/H_i \times S^{n-1} & \xrightarrow{\coprod_{i \in I_n} q_i^n} & X_{n-1} \\
\downarrow & & \downarrow \\
\coprod_{i \in I_n} G/H_i \times D^n & \xrightarrow[\coprod_{i \in I_n} Q_i^n]{} & X_n
\end{array}
$$

The space X_n is called the *n-skeleton* of X. Notice that only the filtration by skeletons belongs to the G-CW-structure but not the G-pushouts, only their existence is required. An *equivariant open n-dimensional cell* is a G-component of $X_n - X_{n-1}$, i.e. the preimage of a path component of $G\backslash(X_n - X_{n-1})$. The closure of an equivariant open n-dimensional cell is called an *equivariant closed n-dimensional cell*. If one has chosen the G-pushouts in Definition 1.1, then the equivariant open n-dimensional cells are the G-subspaces $Q_i(G/H_i \times (D^n - S^{n-1}))$ and the equivariant closed n-dimensional cells are the G-subspaces $Q_i(G/H_i \times D^n)$.

Remark 1.2 (Proper G-CW-complexes). A G-space X is called *proper* if for each pair of points x and y in X there are open neighborhoods V_x of x and W_y of y in X such that the closure of the subset $\{g \in G \mid gV_x \cap W_y \neq \emptyset\}$ of G is compact. A G-CW-complex X is proper if and only if all its isotropy groups are compact [48, Theorem 1.23]. In particular a free G-CW-complex is always proper. However, not every free G-space is proper.

Remark 1.3 (G-CW-complexes with open isotropy groups). Let X be a G-space with G-invariant filtration

$$\emptyset = X_{-1} \subseteq X_0 \subseteq X_1 \subseteq \ldots \subseteq X_n \subseteq \ldots \subseteq \bigcup_{n \geq 0} X_n = X.$$

Then the following assertions are equivalent. i.) Every isotropy group of X is open and the filtration above yields a G-CW-structure on X. ii.) The filtration above yields a (non-equivariant) CW-structure on X such that each open cell $e \subseteq X$ and each $g \in G$ with $ge \cap e \neq \emptyset$ left multiplication with g induces the identity on e.

In particular we conclude for a discrete group G that a G-CW-complex X is the same as a CW-complex X with G-action such that for each open cell $e \subseteq X$ and each $g \in G$ with $ge \cap e \neq \emptyset$ left multiplication with g induces the identity on e.

Example 1.4 (Lie groups acting properly and smoothly on manifolds). If G is a Lie group and M is a (smooth) proper G-manifold, then an equivariant smooth triangulation induces a G-CW-structure on M. For the proof and for *equivariant smooth triangulations* we refer to [36, Theorem I and II].

Example 1.5 (Simplicial actions). Let X be a simplicial complex on which the group G acts by simplicial automorphisms. Then all isotropy groups are closed and open. Moreover, G acts also on the barycentric subdivision X' by simplicial automorphisms. The filtration of the barycentric subdivision X' by the simplicial n-skeleton yields the structure of a G-CW-complex what is not necessarily true for X.

A G-space is called *cocompact* if $G \backslash X$ is compact. A G-CW-complex X is *finite* if X has only finitely many equivariant cells. A G-CW-complex is finite if and only if it is cocompact. A G-CW-complex X is *of finite type* if each n-skeleton is finite. It is called *of dimension* $\leq n$ if $X = X_n$ and *finite dimensional* if it is of dimension $\leq n$ for some integer n. A free G-CW-complex X is the same as a G-principal bundle $X \to Y$ over a CW-complex Y (see Remark 2.8).

Theorem 1.6 (Whitehead Theorem for Families). *Let $f: Y \to Z$ be a G-map of G-spaces. Let $\mathcal{F}$ be a set of (closed) subgroups of G which is closed under conjugation. Then the following assertions are equivalent:*

(i) *For any G-CW-complex X, whose isotropy groups belong to $\mathcal{F}$, the map induced by f*

$$f_*: [X, Y]^G \to [X, Z]^G, \qquad [g] \mapsto [g \circ f]$$

between the set of G-homotopy classes of G-maps is bijective.

(ii) *For any $H \in \mathcal{F}$ the map $f^H: Y^H \to Z^H$ is a weak homotopy equivalence i.e. the map $\pi_n(f^H, y): \pi_n(Y^H, y) \to \pi_n(Z^H, f^H(y))$ is bijective for any base point $y \in Y^H$ and $n \in \mathbb{Z}, n \geq 0$.*

Proof. (i) $\Rightarrow$ (ii) Evaluation at $1H$ induces for any CW-complex A (equipped with the trivial G-action) a bijection $[G/H \times A, Y]^G \xrightarrow{\cong} [A, Y^H]$. Hence for any CW-complex A the map f^H induces a bijection

$$(f^H)_*: [A, Y^H] \to [A, Z^H], \qquad [g] \to [g \circ f^H].$$

This is equivalent to f^H being a weak homotopy equivalence by the classical non-equivariant Whitehead Theorem [84, Theorem 7.17 in Chapter IV.7 on page 182].

(ii) $\Rightarrow$ (i) We only give the proof in the case, where Z is G/G since this is the main important case for us and the basic idea becomes already clear. The general case is treated for instance in [78, Proposition II.2.6 on page 107]. We have to show for any G-CW-complex X such that two G-maps $f_0, f_1 \colon X \to Y$ are G-homotopic provided that for any isotropy group H of X the H-fixed point set Y^H is *weakly contractible* i.e. $\pi_n(Y^H, y)$ consists of one element for all base points $y \in Y^H$. Since X is $\operatorname{colim}_{n \to \infty} X_n$ it suffices to construct inductively over n G-homotopies $h[n] \colon X_n \times [0,1] \to Z$ such that $h[n]_i = f_i$ holds for $i = 0, 1$ and $h[n]|_{X_{n-1} \times [0,1]} = h[n-1]$. The induction beginning $n = -1$ is trivial because of $X_{-1} = \emptyset$, the induction step from $n - 1$ to $n \geq 0$ done as follows. Fix a G-pushout

$$
\begin{array}{ccc}
\coprod_{i \in I_n} G/H_i \times S^{n-1} & \xrightarrow{\coprod_{i \in I_n} q_i^n} & X_{n-1} \\
\downarrow & & \downarrow \\
\coprod_{i \in I_n} G/H_i \times D^n & \xrightarrow[\coprod_{i \in I_n} Q_i^n]{} & X_n
\end{array}
$$

One easily checks that the desired G-homotopy $h[n]$ exists if and only if we can find for each $i \in I$ an extension of the G-map

$$
f_0 \circ Q_i^n \cup f_1 \circ Q_i^n \cup h[n-1] \circ (q_i^n \times \operatorname{id}_{[0,1]}) \colon
$$
$$
G/H_i \times D^n \times \{0\} \cup G/H_i \times D^n \times \{1\} \cup G/H_i \times S^{n-1} \times [0,1] \to Y
$$

to a G-map $G/H_i \times D^n \times [0,1] \to Y$. This is the same problem as extending the (non-equivariant) map $D^n \times \{0\} \cup D^n \times \{1\} \cup S^{n-1} \times [0,1] \to Y$, which is given by restricting the G-map above to $1H_i$, to a (non-equivariant) map $D^n \times [0,1] \to Y^{H_i}$. Such an extension exists since Y^{H_i} is weakly contractible. This finishes the proof of Theorem 1.6. $\qquad\qquad\qquad\qquad\qquad\qquad\qquad\qquad\qquad\qquad\qquad\qquad\qquad\square$

A G-map $f \colon X \to Y$ of G-CW-complexes is a G-homotopy equivalence if and only if for any subgroup $H \subseteq G$ which occurs as isotropy group of X or Y the induced map $f^H \colon X^H \to Y^H$ is a weak homotopy equivalence. This follows from the Whitehead Theorem for Families 1.6 above.

A G-map of G-CW-complexes $f \colon X \to Y$ is *cellular* if $f(X_n) \subseteq Y_n$ holds for all $n \geq 0$. There is an equivariant version of the *Cellular Approximation Theorem*, namely, every G-map of G-CW-complexes is G-homotopic to a cellular one and each G-homotopy between cellular G-maps can be replaced by a cellular G-homotopy [78, Theorem II.2.1 on page 104].

1.2. The G-CW-Version for the Classifying Space for a Family

Definition 1.7 (Family of subgroups). A *family $\mathcal{F}$ of subgroups* of G is a set of (closed) subgroups of G which is closed under conjugation and finite intersections.

Examples for $\mathcal{F}$ are

$$
\begin{array}{lll}
\mathcal{TR} & = & \{\text{trivial subgroup}\}; \\
\mathcal{FIN} & = & \{\text{finite subgroups}\}; \\
\mathcal{VCYC} & = & \{\text{virtually cyclic subgroups}\}; \\
\mathcal{COM} & = & \{\text{compact subgroups}\}; \\
\mathcal{COMOP} & = & \{\text{compact open subgroups}\}; \\
\mathcal{ALL} & = & \{\text{all subgroups}\}.
\end{array}
$$

Definition 1.8 (Classifying G-CW-complex for a family of subgroups). *Let $\mathcal{F}$ be a family of subgroups of G. A model $E_{\mathcal{F}}(G)$ for the classifying G-CW-complex for the family $\mathcal{F}$ of subgroups is a G-CW-complex $E_{\mathcal{F}}(G)$ which has the following properties:*

(i) *All isotropy groups of $E_{\mathcal{F}}(G)$ belong to $\mathcal{F}$.*
(ii) *For any G-CW-complex Y, whose isotropy groups belong to $\mathcal{F}$, there is up to G-homotopy precisely one G-map $Y \to X$.*

We abbreviate $\underline{E}G := E_{\mathcal{COM}}(G)$ and call it the universal G-CW-complex for proper G-actions.

In other words, $E_{\mathcal{F}}(G)$ is a terminal object in the G-homotopy category of G-CW-complexes, whose isotropy groups belong to $\mathcal{F}$. In particular two models for $E_{\mathcal{F}}(G)$ are G-homotopy equivalent and for two families $\mathcal{F}_0 \subseteq \mathcal{F}_1$ there is up to G-homotopy precisely one G-map $E_{\mathcal{F}_0}(G) \to E_{\mathcal{F}_1}(G)$.

Theorem 1.9 (Homotopy characterization of $E_{\mathcal{F}}(G)$). *Let $\mathcal{F}$ be a family of subgroups.*

(i) *There exists a model for $E_{\mathcal{F}}(G)$ for any family $\mathcal{F}$.*
(ii) *A G-CW-complex X is a model for $E_{\mathcal{F}}(G)$ if and only if all its isotropy groups belong to $\mathcal{F}$ and for each $H \in \mathcal{F}$ the H-fixed point set X^H is weakly contractible.*

Proof. (i) A model can be obtained by attaching equivariant cells $G/H \times D^n$ for all $H \in \mathcal{F}$ to make the H-fixed point sets weakly contractible. See for instance [48, Proposition 2.3 on page 35].

(ii) This follows from the Whitehead Theorem for Families 1.6 applied to $f \colon X \to G/G$. $\qquad\square$

A model for $E_{\mathcal{ALL}}(G)$ is G/G. In Section 4 we will give many interesting geometric models for classifying spaces $E_{\mathcal{F}}(G)$, in particular for the case, where G is discrete and $\mathcal{F} = \mathcal{FIN}$ or, more generally, where G is a (locally compact topological Hausdorff) group and $\mathcal{F} = \mathcal{COM}$. In some sense $\underline{E}G = E_{\mathcal{COM}}(G)$ is the most interesting case.

2. Numerable G-Space-Version

In this section we explain the numerable G-space version of the classifying space for a family $\mathcal{F}$ of subgroups of group G.

Definition 2.1 ($\mathcal{F}$-numerable G-space). A $\mathcal{F}$-numerable G-space is a G-space, for which there exists an open covering $\{U_i \mid i \in I\}$ by G-subspaces such that there is for each $i \in I$ a G-map $U_i \to G/G_i$ for some $G_i \in \mathcal{F}$ and there is a locally finite partition of unity $\{e_i \mid i \in I\}$ subordinate to $\{U_i \mid i \in I\}$ by G-invariant functions $e_i \colon X \to [0,1]$.

Notice that we do not demand that the isotropy groups of a $\mathcal{F}$-numerable G-space belong to $\mathcal{F}$. If $f \colon X \to Y$ is a G-map and Y is $\mathcal{F}$-numerable, then X is also $\mathcal{F}$-numerable.

Lemma 2.2. *Let $\mathcal{F}$ be a family. Then a G-CW-complex is $\mathcal{F}$-numerable if each isotropy group is a subgroup of an element in $\mathcal{F}$.*

Proof. This follows from the Slice Theorem for G-CW-complexes [48, Theorem 1.37] and the fact that $G\backslash X$ is a CW-complex and hence paracompact [64]. □

Definition 2.3 (Classifying numerable G-space for a family of subgroups). *Let $\mathcal{F}$ be a family of subgroups of G. A model $J_{\mathcal{F}}(G)$ for the classifying numerable G-space for the family $\mathcal{F}$ of subgroups is a G-space which has the following properties:*

  (i) $J_{\mathcal{F}}(G)$ is $\mathcal{F}$-numerable.
  (ii) *For any $\mathcal{F}$-numerable G-space X there is up to G-homotopy precisely one G-map $X \to J_{\mathcal{F}}(G)$.*

We abbreviate $\underline{J}G := J_{\mathcal{COM}}(G)$ and call it the universal numerable G-space for proper G-actions, or briefly the universal space for proper G-actions.

In other words, $J_{\mathcal{F}}(G)$ is a terminal object in the G-homotopy category of $\mathcal{F}$-numerable G-spaces. In particular two models for $J_{\mathcal{F}}(G)$ are G-homotopy equivalent, and for two families $\mathcal{F}_0 \subseteq \mathcal{F}_1$ there is up to G-homotopy precisely one G-map $J_{\mathcal{F}_0}(G) \to J_{\mathcal{F}_1}(G)$.

Remark 2.4 (Proper G-spaces). A $\mathcal{COM}$-numerable G-space X is proper. Not every proper G-space is $\mathcal{COM}$-numerable. But a G-CW-complex X is proper if and only if it is $\mathcal{COM}$-numerable (see Lemma 2.2).

Theorem 2.5 (Homotopy characterization of $J_{\mathcal{F}}(G)$). *Let $\mathcal{F}$ be a family of subgroups.*

  (i) *For any family $\mathcal{F}$ there exists a model for $J_{\mathcal{F}}(G)$ whose isotropy groups belong to $\mathcal{F}$.*
  (ii) *Let X be a $\mathcal{F}$-numerable G-space. Equip $X \times X$ with the diagonal action and let $\mathrm{pr}_i \colon X \times X \to X$ be the projection onto the i-th factor for $i = 1, 2$. Then X is a model for $J_{\mathcal{F}}(G)$ if and only if for each $H \in \mathcal{F}$ there is $x \in X$ with $H \subseteq G_x$ and pr_1 and pr_2 are G-homotopic.*
  (iii) *For $H \in \mathcal{F}$ the H-fixed point set $J_{\mathcal{F}}(G)^H$ is contractible.*

Proof. (i) A model for $J_{\mathcal{F}}(G)$ is constructed in [78, Theorem I.6.6. on page 47] and [7, Appendix 1], namely, as the infinite join $*_{n=1}^{\infty} Z$ for $Z = \coprod_{H \in \mathcal{F}} G/H$. There G is assumed to be compact but the proof goes through for locally compact topological Hausdorff groups. The isotropy groups are finite intersections of the isotropy groups appearing in Z and hence belong to $\mathcal{F}$.

(ii) Let X be a model for the classifying space $J_{\mathcal{F}}(G)$ for $\mathcal{F}$. Then $X \times X$ with the diagonal G-action is a $\mathcal{F}$-numerable G-space. Hence pr_1 and pr_2 are G-homotopic by the universal property. Since for any $H \in \mathcal{F}$ the G-space G/H is $\mathcal{F}$-numerable, there must exist a G-map $G/H \to X$ by the universal property of $J_{\mathcal{F}}(G)$. If x is the image under this map of $1H$, then $H \subseteq G_x$.

Suppose that X is a G-space such that for each $H \in \mathcal{F}$ there is $x \in X$ with $H \subseteq G_x$ and pr_1 and pr_2 are G-homotopic. We want to show that then X is a model for $J_{\mathcal{F}}(G)$. Let $f_0, f_1 \colon Y \to X$ be two G-maps. Since $\mathrm{pr}_i \circ (f_0 \times f_1) = f_i$ holds for $i = 0, 1$, f_0 and f_1 are G-homotopic. It remains to show for any $\mathcal{F}$-numerable G-space Y that there exists a G-map $Y \to X$. Because of the universal property of $J_{\mathcal{F}}(G)$ it suffices to do this in the case, where $Y = *_{n=1}^{\infty} L$ for $L = \coprod_{H \in \mathcal{F}} G/H$. By assumption there is a G-map $L \to X$. Analogous to the construction in [7, Appendix 2] one uses a G-homotopy from pr_1 to pr_2 to construct a G-map $*_{n=1}^{\infty} L \to X$.

(iii) Restricting to $1H$ yields a bijection

$$[G/H \times J_{\mathcal{F}}(G)^H, J_{\mathcal{F}}(G)]^G \xrightarrow{\cong} [J_{\mathcal{F}}(G)^H, J_{\mathcal{F}}(G)^H],$$

where we consider X^H as a G-space with trivial G action. Since $G/H \times X^H$ is a $\mathcal{F}$-numerable G-space, $[J_{\mathcal{F}}(G)^H, J_{\mathcal{F}}(G)^H]$ consists of one element. Hence $J_{\mathcal{F}}(G)^H$ is contractible. $\square$

Remark 2.6. We do not know whether the converse of Theorem 2.5 (iii) is true, i.e. whether a $\mathcal{F}$-numerable G-space X is a model for $J_{\mathcal{F}}(G)$ if X^H is contractible for each $H \in \mathcal{F}$.

Example 2.7 (Numerable G-principal bundles). A *numerable (locally trivial) G-principal bundle* $p \colon E \to B$ consists by definition of a $\mathcal{TR}$-numerable G-space E, a space B with trivial action and a surjective G-map $p \colon E \to B$ such that the induced map $G \backslash E \to B$ is a homeomorphism. A numerable G-principal bundle $p \colon EG \to BG$ is *universal* if and only if each numerable G-bundle admits a G-bundle map to p and two such G-bundle maps are G-bundle homotopic. A numerable G-principal bundle is universal if and only if E is contractible. This follows from [26, 7.5 and 7.7]. More information about numerable G-principal bundles can be found for instance in [35, Section 9 in Chapter 4] [78, Chapter I Section 8].

If $p \colon E \to B$ is a universal numerable G-principal bundle, then E is a model for $J_{\mathcal{TR}}(G)$. Conversely, $J_{\mathcal{TR}}(G) \to G \backslash J_{\mathcal{TR}}(G)$ is a model for the universal numerable G-principal bundle. We conclude that a $\mathcal{TR}$-numerable G-space X is a model for $J_{\mathcal{TR}}(G)$ if and only if X is contractible (compare Remark 2.6).

Remark 2.8 (G-Principal bundles over CW-complexes). Let $p\colon E \to B$ be a (locally trivial) G-principal bundle over a CW-complex. Since any CW-complex is paracompact [64], it is automatically a numerable G-principal bundle. The CW-complex structure on B pulls back to G-CW-structure on E [48, 1.25 on page 18]. Conversely, if E is a free G-CW-complex, then $E \to G\backslash E$ is a numerable G-principal bundle over a CW-complex by Lemma 2.2

The classifying bundle map from p above to $J_{T\mathcal{R}}(G) \to G\backslash J_{T\mathcal{R}}(G)$ lifts to a G-bundle map from p to $E_{T\mathcal{R}}(G) \to G\backslash E_{T\mathcal{R}}(G)$ and two such G-bundle maps from p to $E_{T\mathcal{R}}(G) \to G\backslash E_{T\mathcal{R}}(G)$ are G-bundle homotopic. Hence for G-principal bundles over CW-complexes one can use $E_{T\mathcal{R}}(G) \to G\backslash E_{T\mathcal{R}}(G)$ as the universal object.

We will compare the spaces $E_{\mathcal{F}}(G)$ and $J_{\mathcal{F}}(G)$ in Section 3. In Section 4 we will give many interesting geometric models for $E_{\mathcal{F}}(G)$ and $J_{\mathcal{F}}(G)$ in particular in the case $\mathcal{F} = \mathcal{COM}$. In some sense $\underline{J}G = J_{\mathcal{COM}}(G)$ is the most interesting case.

3. Comparison of the Two Versions

In this section we compare the two classifying spaces $E_{\mathcal{F}}(G)$ and $J_{\mathcal{F}}(G)$ and show that the two classifying spaces $\underline{E}G$ and $\underline{J}G$ agree up to G-homotopy equivalence.

Since $E_{\mathcal{F}}(G)$ is a $\mathcal{F}$-numerable space by Lemma 2.2, there is up to G-homotopy precisely one G-map

$$u\colon E_{\mathcal{F}}(G) \quad \to \quad J_{\mathcal{F}}(G). \tag{3.1}$$

Lemma 3.2. *The following assertions are equivalent for a family $\mathcal{F}$ of subgroups of G:*

(i) *The map $u\colon E_{\mathcal{F}}(G) \to J_{\mathcal{F}}(G)$ defined in 3.1 is a G-homotopy equivalence.*
(ii) *The G-spaces $E_{\mathcal{F}}(G)$ and $J_{\mathcal{F}}(G)$ are G-homotopy equivalent.*
(iii) *The G-space $J_{\mathcal{F}}(G)$ is G-homotopy equivalent to a G-CW-complex, whose isotropy groups belong to $\mathcal{F}$.*
(iv) *There exists a G-map $J_{\mathcal{F}}(G) \to Y$ to a G-CW-complex Y, whose isotropy groups belong to $\mathcal{F}$.*

Proof. This follows from the universal properties of $E_{\mathcal{F}}(G)$ and $J_{\mathcal{F}}(G)$. □

Lemma 3.3. *Suppose either that every element $H \in \mathcal{F}$ is an open (and closed) subgroup of G or that G is a Lie group and $\mathcal{F} \subseteq \mathcal{COM}$. Then the map $u\colon E_{\mathcal{F}}(G) \to J_{\mathcal{F}}(G)$ defined in 3.1 is a G-homotopy equivalence.*

Proof. We have to inspect the construction in [78, Lemma 6.13 in Chapter I on page 49] and will use the same notation as in that paper. Let Z be a $\mathcal{F}$-numerable G-space. Let $X = \coprod_{H \in \mathcal{F}} G/H$. Then $*_{n=1}^{\infty} X$ is a model for $J_{\mathcal{F}}(G)$ by [78, Lemma 6.6 in Chapter I on page 47]. We inspect the construction of a G-map $f\colon Z \to *_{n=1}^{\infty} X$. One constructs a countable covering $\{U_n \mid n = 1, 2, \ldots\}$ of Z by G-invariant open subsets of Z together with a locally finite subordinate partition of unity $\{v_n \mid n = 1, 2, \ldots\}$ by G-invariant functions $v_n \colon Z \to [0, 1]$ and G-maps $\phi_n \colon U_n \to X$. Then one obtains a G-map

$$f \colon Z \to *_{n=1}^{\infty} X, \qquad z \mapsto (v_1(z)\phi_1(z), v_2(z)\phi_2(z), \ldots),$$

where $v_n(z)\phi_n(z)$ means $0x$ for any $x \in X$ if $z \notin U_n$. Let $i_k \colon *_{n=1}^{k} X \to *_{n=1}^{\infty} X$ and $j_k \colon *_{n=1}^{k} X \to *_{n=1}^{k+1} X$ be the obvious inclusions. Denote by $\alpha_k \colon *_{n=1}^{k} X \to \operatorname{colim}_{k \to \infty} *_{n=1}^{k} X$ the structure map and by $i \colon \operatorname{colim}_{k \to \infty} *_{n=1}^{k} X \to *_{n=1}^{\infty} X$ the map induced by the system $\{i_k \mid k = 1, 2, \ldots\}$. This G-map is a (continuous) bijective G-map but not necessarily a G-homeomorphism. Since the partition $\{v_n \mid n = 1, 2, \ldots\}$ is locally finite, we can find for each $z \in Z$ an open G-invariant neighborhood W_z of z in Z and a positive integer k_z such that v_n vanishes on W_z for $n > k_z$. Define a map

$$f'_z \colon W_z \to *_{n=1}^{k_z} X, \qquad z \mapsto (v_1(z)\phi_1(z), v_2(z)\phi_2(z), \ldots, v_{k_z}(z)\phi_{k_z}(z)).$$

Then $\alpha_{k_z} \circ f'_z \colon W_z \to \operatorname{colim}_{k \to \infty} *_{n=1}^{k} X$ is a well-defined G-map whose composition with $i \colon \operatorname{colim}_{k \to \infty} *_{n=1}^{k} X \to *_{n=1}^{\infty} X$ is $f|_{W_z}$. Hence the system of the maps $\alpha_{k_z} \circ f'_z$ defines a G-map

$$f' \colon Z \to \operatorname{colim}_{k \to \infty} *_{n=1}^{k} X$$

such that $i \circ f' = f$ holds.

Let

$$\Delta_{n-1} = \left\{ (t_1, t_2 \ldots t_n) \mid t_i \in [0, 1], \sum_{i=1}^{n} t_i = 1 \right\} \subseteq \prod_{n=1}^{k} [0, 1]$$

be the standard $(n-1)$-simplex. Let

$$p \colon \left(\prod_{n=1}^{k} X \right) \times \Delta_n \to *_{n=1}^{k} X, \qquad (x_1, \ldots, x_n), (t_1, \ldots, t_n) \mapsto (t_1 x_1, \ldots, t_n x_n)$$

be the obvious projection. It is a surjective continuous map but in general not an identification. Let $\overline{*}_{n=1}^{k} X$ be the topological space whose underlying set is the same as for $*_{n=1}^{k} X$ but whose topology is the quotient topology with respect to p. The identity induces a (continuous) map $\overline{*}_{n=1}^{k} X \to *_{n=1}^{k} X$ which is not a homeomorphism in general. Choose for $n \geq 1$ a (continuous) function $\phi_n \colon [0, 1] \to [0, 1]$ which satisfies $\phi_n^{-1}(0) = [0, 4^{-n}]$. Define

$$u_k \colon *_{n=1}^{k} X \to \overline{*}_{n=1}^{k} X,$$

$$(t_n x_n \mid n = 1, \ldots, k) \mapsto \left(\frac{\phi_n(t_n)}{\sum_{n=1}^{k} \phi_n(t_n)} x_n \;\middle|\; n = 1, \ldots, k \right).$$

It is not hard to check that this G-map is continuous. If $\overline{j}_k \colon \overline{*}_{n=1}^{k} X \to \overline{*}_{n=1}^{k+1} X$ is the obvious inclusion, we have $u_{k+1} \circ j_k = \overline{j}_k \circ u_k$ for all $k \geq 1$. Hence the system of the maps u_k induces a G-map

$$u \colon \operatorname{colim}_{k \to \infty} *_{n=1}^{k} X \to \operatorname{colim}_{k \to \infty} \overline{*}_{n=1}^{k} X.$$

Next we want to show that each G-space $\overline{*}_{n=1}^{k} X$ has the G-homotopy type of a G-CW-complex, whose isotropy groups belong to $\mathcal{F}$. We first show that $\overline{*}_{n=1}^{k} X$ is a $\left(\prod_{n=1}^{k} G\right)$-$CW$-complex. It suffices to treat the case $k = 2$, the general case follows by induction over k. We can rewrite $X \overline{*} X$ as a $G \times G$-pushout

$$
\begin{array}{ccc}
X \times X & \xrightarrow{\ i_1\ } & CX \times X \\
{\scriptstyle i_2}\downarrow & & \downarrow \\
X \times CX & \longrightarrow & X \overline{*} X
\end{array}
$$

where CX is the cone over X and i_1 and i_2 are the obvious inclusions. The product of two finite dimensional G-CW-complexes is in a canonical way a finite dimensional $(G \times G)$-CW-complex, and, if (B, A) is a G-CW-pair, C a G-CW-complex and $f \colon B \to C$ is a cellular G-map, then $A \cup_f C$ inherits a G-CW-structure in a canonical way. Thus $X \overline{*} X$ inherits a $(G \times G)$-CW-complex structure.

The problem is now to decide whether the $\left(\prod_{n=1}^{k} G\right)$-$CW$-complex $\overline{*}_{n=1}^{k} X$ regarded as a G-space by the diagonal action has the G-homotopy type of a G-CW-complex. If each $H \in \mathcal{F}$ is open, then each isotropy group of the G-space $*_{n=1}^{k} X$ is open and we conclude from Remark 1.3 that $\overline{*}_{n=1}^{k} X$ with the diagonal G-action is a G-CW-complex Suppose that G is a Lie group and each $H \in \mathcal{F}$ is compact. Example 1.4 implies that for any compact subgroup $K \subseteq \prod_{n=1}^{k} G$ the space $\left(\prod_{n=1}^{k} G\right)/K$ regarded as G-space by the diagonal action has the G-homotopy type of a G-CW-complex. We conclude from [48, Lemma 7.4 on page 121] that $\overline{*}_{n=1}^{k} X$ with the diagonal G-action has the G-homotopy type of a G-CW-complex. The isotropy groups $\overline{*}_{n=1}^{k} X$ belong to $\mathcal{F}$ since $\mathcal{F}$ is closed under finite intersections and conjugation. It is not hard to check that each G-map $\overline{j}_k$ is a G-cofibration. Hence $\operatorname{colim}_{k \to \infty} \overline{*}_{n=1}^{k} X$ has the G-homotopy type of a G-CW-complex, whose isotropy groups belong to $\mathcal{F}$.

Thus we have shown for every $\mathcal{F}$-numerable G-space Z that it admits a G-map to a G-CW-complex whose isotropy groups belong to $\mathcal{F}$. Now Lemma 3.3 follows from Lemma 3.2. $\qquad\square$

Definition 3.4 (Totally disconnected group). *A (locally compact topological Hausdorff) group G is called* totally disconnected *if it satisfies one of the following equivalent conditions:*

(T) *G is totally disconnected as a topological space, i.e. each component consists of one point.*

(D) *The covering dimension of the topological space G is zero.*

(FS) *Any element of G has a fundamental system of compact open neighborhoods.*

We have to explain why these three conditions are equivalent. The implication (T) $\Rightarrow$ (D) $\Rightarrow$ (FS) is shown in [33, Theorem 7.7 on page 62]. It remains to prove (FS) $\Rightarrow$ (T). Let U be a subset of G containing two distinct points g and h. Let V be a compact open neighborhood of x which does not contain y. Then U

is the disjoint union of the open non-empty sets $V \cap U$ and $V^c \cap U$ and hence disconnected.

Lemma 3.5. *Let G be a totally disconnected group and $\mathcal{F}$ a family satisfying $\mathcal{COMOP} \subseteq \mathcal{F} \subseteq \mathcal{COM}$. Then the following square commutes up to G-homotopy and consists of G-homotopy equivalences*

$$
\begin{array}{ccc}
E_{\mathcal{COMOP}}(G) & \xrightarrow{\ u\ } & J_{\mathcal{COMOP}}(G) \\
\downarrow & & \downarrow \\
E_{\mathcal{F}}(G) & \xrightarrow[u]{} & J_{\mathcal{F}}(G)
\end{array}
$$

where all maps come from the universal properties.

Proof. We first show that any compact subgroup $H \subseteq G$ is contained in a compact open subgroup. From [33, Theorem 7.7 on page 62] we get a compact open subgroup $K \subseteq G$. Since H is compact, we can find finitely many elements $h_1, h_2, \ldots, h_s$ in H such that $H \subseteq \bigcup_{i=1}^{s} h_i K$. Put $L := \bigcap_{h \in H} hKh^{-1}$. Then $hLh^{-1} = L$ for all $h \in H$. Since $L = \bigcap_{i=1}^{s} h_i K h_i^{-1}$ holds, L is compact open. Hence LH is a compact open subgroup containing H.

This implies that $J_{\mathcal{F}}(G)$ is $\mathcal{COMOP}$-numerable. Obviously $J_{\mathcal{COMOP}}(G)$ is $\mathcal{F}$-numerable. We conclude from the universal properties that $J_{\mathcal{COMOP}}(G) \to J_{\mathcal{F}}(G)$ is a G-homotopy equivalence.

The map $u \colon E_{\mathcal{COMOP}}(G) \to J_{\mathcal{COMOP}}(G)$ is a G-homotopy equivalence because of Lemma 3.3.

This and Theorem 2.5 (iii) imply that $E_{\mathcal{COMOP}}(G)^H$ is contractible for all $H \in \mathcal{F}$. Hence $E_{\mathcal{COMOP}}(G) \to E_{\mathcal{F}}(G)$ is a G-homotopy equivalence by Theorem 1.9 (ii).   □

Definition 3.6 (Almost connected group). *Given a group G, let G^0 be the normal subgroup given by the component of the identity and $\overline{G} = G/G^0$ be the component group. We call G almost connected if its component group $\overline{G}$ is compact.*

A Lie group G is almost connected if and only if it has finitely many path components. In particular a discrete group is almost connected if it is finite.

Theorem 3.7 (Comparison of $E_{\mathcal{F}}(G)$ and $J_{\mathcal{F}}(G)$). *The map $u \colon E_{\mathcal{F}}(G) \to J_{\mathcal{F}}(G)$ defined in 3.1 is a G-homotopy equivalence if one of the following conditions is satisfied:*

(i) *Each element in $\mathcal{F}$ is an open subgroup of G.*
(ii) *The group G is discrete.*
(iii) *The group G is a Lie group and every element $H \in \mathcal{F}$ is compact.*
(iv) *The group G is totally disconnected and $\mathcal{F} = \mathcal{COM}$ or $\mathcal{F} = \mathcal{COMOP}$.*
(v) *The group G is almost connected and each element in $\mathcal{F}$ is compact.*

Proof. Assertions (i), (ii), (iii) and (iv) have already been proved in Lemma 3.3 and Lemma 3.5. Assertion (v) follows from Lemma 3.2 and Theorem 4.3.   □

The following example shows that the map $u\colon E_{\mathcal{F}}(G) \to J_{\mathcal{F}}(G)$ defined in 3.1 is in general not a G-homotopy equivalence.

Example 3.8 (Totally disconnected groups and $\mathcal{TR}$). Let G be totally disconnected. We claim that $u\colon E_{\mathcal{TR}}(G) \to J_{\mathcal{TR}}(G)$ defined in 3.1 is a G-homotopy equivalence if and only if G is discrete. In view of Theorem 2.5 (iii) and Lemma 3.3 this is equivalent to the statement that $E_{\mathcal{TR}}(G)$ is contractible if and only if G is discrete. If G is discrete, we already know that $E_{\mathcal{TR}}(G)$ is contractible. Suppose now that $E_{\mathcal{TR}}(G)$ is contractible. We obtain a numerable G-principal bundle $G \to E_{\mathcal{TR}}(G) \to G\backslash E_{\mathcal{TR}}(G)$ by Remark 2.8. This implies that it is a fibration by a result of Hurewicz [84, Theorem on p. 33]. Since $E_{\mathcal{TR}}(G)$ is contractible, G and the loop space $\Omega(G\backslash E_{\mathcal{TR}}(G))$ are homotopy equivalent [84, 6.9* on p. 137, 6.10* on p. 138, Corollary 7.27 on p. 40]. Since $G\backslash E_{\mathcal{TR}}(G)$ is a CW-complex, $\Omega(G\backslash E_{\mathcal{TR}}(G))$ has the homotopy type of a CW-complex [62]. Hence there exists a homotopy equivalence $f\colon G \to X$ be from G to a CW-complex X. Then the induced map $\pi_0(G) \to \pi_0(X)$ between the set of path components is bijective. Hence the preimage of each path component of X is a path component of G and therefore a point since G is totally disconnected. Since X is locally path-connected each path component of X is open in X. We conclude that G is the disjoint union of the preimages of the path components of X and each of these preimages is open in G and consists of one point. Hence G is discrete.

Remark 3.9 (Compactly generated spaces). In the following theorem we will work in the category of compactly generated spaces. This convenient category is explained in detail in [73] and [84, I.4]. A reader may ignore this technical point in the following theorem without harm, but we nevertheless give a short explanation.

A Hausdorff space X is called *compactly generated* if a subset $A \subseteq X$ is closed if and only if $A \cap K$ is closed for every compact subset $K \subseteq X$. Given a topological space X, let $k(X)$ be the compactly generated topological space with the same underlying set as X and the topology for which a subset $A \subseteq X$ is closed if and only if for every compact subset $K \subseteq X$ the intersection $A \cap K$ is closed in the given topology on X. The identity induces a continuous map $k(X) \to X$ which is a homeomorphism if and only if X is compactly generated. The spaces X and $k(X)$ have the same compact subsets. Locally compact Hausdorff spaces and every Hausdorff space which satisfies the first axiom of countability are compactly generated. In particular metrizable spaces are compactly generated.

Working in the category of compactly generated spaces means that one only considers compactly generated spaces and whenever a topological construction such as the cartesian product or the mapping space leads out of this category, one retopologizes the result as described above to get a compactly generated space. The advantage is for example that in the category of compactly generated spaces the exponential map $\mathrm{map}(X \times Y, Z) \to \mathrm{map}(X, \mathrm{map}(Y, Z))$ is always a homeomorphism, for an identification $p\colon X \to Y$ the map $p \times \mathrm{id}_Z\colon X \times Z \to Y \times Z$ is always an identification and for a filtration by closed subspaces $X_1 \subset X_2 \subseteq \ldots \subseteq X$ such that X is the colimit $\mathrm{colim}_{n\to\infty} X_n$, we always get $X \times Y = \mathrm{colim}_{n\to\infty}(X_n \times Y)$.

In particular the product of a G-CW-complex X with a H-CW-complex Y is in a canonical way a $G \times H$-CW-complex $X \times Y$. Since we are assuming that G is a locally compact Hausdorff group, any G-CW-complex X is compactly generated.

The following result has grown out of discussions with Ralf Meyer.

Theorem 3.10 (Equality of $\underline{E}G$ and $\underline{J}G$). *Let G be a locally compact second countable topological Hausdorff group. Then the canonical G-map $\underline{E}G \to \underline{J}G$ is a G-homotopy equivalence.*

Proof. In the sequel of the proof we work in the category of compactly generated spaces (see Remark 3.9). Notice that the model mentioned in Theorem 2.5 (i) is metrizable and hence compactly generated (see [7, Appendix 1]). Because of Lemma 3.2 it suffices to construct a G-CW-complex Z with compact isotropy groups together with a G-map $\underline{J}G \to Z$.

Let G^0 be the component of the identity which is a normal closed subgroup. Let $p\colon G \to G/G^0$ be the projection. The groups G^0 and G/G^0 are locally compact second countable Hausdorff groups and G/G^0 is totally disconnected. We conclude from Lemma 3.5 that there is a G-map $\underline{J}(G/G^0) \to E_{\mathcal{COMOP}}(G/G^0)$. Since $\underline{J}G$ is $\mathcal{COMOP}$-numerable, the G/G^0-space $G^0\backslash(\underline{J}G)$ is $\mathcal{COM}$-numerable and hence there exists a G/G^0-map $G^0\backslash(\underline{J}G) \to \underline{J}(G/G^0)$. Thus we get a G-map $u\colon \underline{J}G \to \operatorname{res}_p E_{\mathcal{COMOP}}(G/G^0)$, where the G-CW-complex $\operatorname{res}_p E_{\mathcal{COMOP}}(G/G^0)$ is obtained from the G/G^0-CW-complex $E_{\mathcal{COMOP}}(G/G^0)$ by letting $g \in G$ act by $p(g)$. We obtain a G-map $\operatorname{id} \times f\colon \underline{J}G \to \underline{J}G \times \operatorname{res}_p E_{\mathcal{COMOP}}(G/G^0)$. Hence it suffices to construct a G-CW-complex Z with compact isotropy groups together with a G-map $f\colon \underline{J}G \times \operatorname{res}_p E_{\mathcal{COMOP}}(G/G^0) \to Z$. For this purpose we construct a sequence of G-CW-complexes $Z_{-1} \subseteq Z_0 \subseteq Z_1 \subseteq \ldots$ such that Z_n is a G-CW-subcomplex of Z_{n+1} and each Z_n has compact isotropy groups, and G-homotopy equivalences $f_n\colon \operatorname{res}_p E_{\mathcal{COMOP}}(G/G^0)_n \times \underline{J}G \to Z_n$. with $f_{n+1}|_{\operatorname{res}_p E_{\mathcal{COMOP}}(G/G^0)_n} = f_n$, where $E_{\mathcal{COMOP}}(G/G^0)_n$ is the n-skeleton of $E_{\mathcal{COMOP}}(G/G^0)$. The canonical G-map

$$\operatorname{colim}_{n \to \infty} \left(\underline{J}G \times \operatorname{res}_p E_{\mathcal{COMOP}}(G/G^0)_n\right) \to \underline{J}G \times \operatorname{res}_p E_{\mathcal{COMOP}}(G/G^0)$$

is a G-homeomorphism. The G-space $Z = \operatorname{colim}_{n \to \infty} Z_n$ is a G-CW-complex with compact isotropy groups. Hence we can define the desired G-map by $f = \operatorname{colim}_{n \to \infty} f_n$ after we have constructed the G-maps f_n. This will be done by induction over n. The induction beginning $n = -1$ is given by $\operatorname{id}\colon \emptyset \to \emptyset$. The induction step from n to $(n+1)$ is done as follows. Choose a G/G^0-pushout

$$
\begin{array}{ccc}
\coprod_{i \in I}(G/G^0)/H_i \times S^n & \longrightarrow & E_{\mathcal{COMOP}}(G/G^0)_n \\
\downarrow & & \downarrow \\
\coprod_{i \in I}(G/G^0)/H_i \times D^{n+1} & \longrightarrow & E_{\mathcal{COMOP}}(G/G^0)_{n+1}
\end{array}
$$

where each H_i is a compact open subgroup of G/G^0. We obtain a G-pushout

$$\coprod_{i\in I} \operatorname{res}_p \left((G/G^0)/H_i \times S^n\right) \times \underline{J}G \longrightarrow \operatorname{res}_p E_{\mathcal{COMOP}}(G/G^0)_n \times \underline{J}G$$
$$\downarrow \qquad\qquad\qquad\qquad\qquad\qquad \downarrow$$
$$\coprod_{i\in I} \operatorname{res}_p \left((G/G^0)/H_i \times D^{n+1}\right) \times \underline{J}G \longrightarrow \operatorname{res}_p E_{\mathcal{COMOP}}(G/G^0)_{n+1} \times \underline{J}G$$

In the sequel let $K_i \subseteq G$ be the open almost connected subgroup $p^{-1}(H_i)$. The G-spaces $\operatorname{res}_p(G/G^0)/H_i$ and G/K_i agree. We have the G-homeomorphism

$$G \times_{K_i} \operatorname{res}_G^{K_i} \underline{J}G \xrightarrow{\cong} G/K_i \times \underline{J}G, \quad (g,x) \mapsto (gK_i, gx).$$

Thus we obtain a G-pushout

$$\left(\coprod_{i\in I} G \times_{K_i} (\operatorname{res}_G^{K_i} \underline{J}G)\right) \times S^n \xrightarrow{\ w\ } \operatorname{res}_p E_{\mathcal{COMOP}}(G/G^0)_n \times \underline{J}G$$
$$\operatorname{id}\times i \downarrow \qquad\qquad\qquad\qquad\qquad\qquad \downarrow \qquad (3.11)$$
$$\left(\coprod_{i\in I} G \times_{K_i} (\operatorname{res}_G^{K_i} \underline{J}G)\right) \times D^{n+1} \longrightarrow \operatorname{res}_p E_{\mathcal{COMOP}}(G/G^0)_{n+1} \times \underline{J}G$$

where $i\colon S^n \to D^{n+1}$ is the obvious inclusion.

Let X be a $\mathcal{COM}$-numerable K_i-space. Then the G-space $G\times_{K_i}\underline{J}K_i$ is a $\mathcal{COM}$-numerable and hence admits a G-map to $\underline{J}G$. Its restriction to $\underline{J}K_i = K_i \times_{K_i} \underline{J}K_i$ defines a K_i-map $f\colon X \to \operatorname{res}_G^{K_i} \underline{J}G$. If f_1 and f_2 are K_i-maps $X \to \operatorname{res}_G^{K_i} \underline{J}G$, we obtain G-maps $\overline{f_k}\colon G \times_{K_i} X \to \underline{J}G$ by sending $(g,x) \to gf_k(x)$ for $k = 0,1$. By the universal property of $\underline{J}G$ these two G-maps are G-homotopic. Hence f_0 and f_1 are K_i-homotopic. Since $K_i \subseteq G$ is open, $\operatorname{res}_G^{K_i} \underline{J}G$ is a $\mathcal{COM}$-numerable K_i-space. Hence the K_i-space $\operatorname{res}_G^{K_i} \underline{J}G$ is a model for $\underline{J}K_i$. Since K_i is almost connected, there is a K_i-homotopy equivalence $\underline{E}K_i \to \operatorname{res}_G^{K_i} \underline{J}G$ by Theorem 3.7 (v). Hence we obtain a G-homotopy equivalence

$$u_i\colon G \times_{K_i} \underline{E}K_i \to G \times_{K_i} (\operatorname{res}_G^{K_i} \underline{J}G)$$

with a K_i-CW-complex with compact isotropy groups as source.

In the sequel we abbreviate

$$X_n := \operatorname{res}_p E_{\mathcal{COMOP}}(G/G^0)_n \times \underline{J}G;$$
$$Y := \coprod_{i\in I} G \times_{K_i} (\operatorname{res}_G^{K_i} \underline{J}G);$$
$$Y' = \coprod_{i\in I} G \times_{K_i} \underline{E}K_i.$$

Choose a G-homotopy equivalence $v\colon Y' \to Y$. By the equivariant cellular Approximation Theorem we can find a G-homotopy $h\colon Y' \times S^n \times [0,1] \to Z_n$ such that $h_0 = f_n \circ w \circ (v\times \operatorname{id}_{S^n})$ and the G-map $h_1\colon Y \times S^n \to Z_n$ is cellular. Consider the following commutative diagram of G-spaces

$$
\begin{array}{ccccc}
Y \times D^{n+1} & \xleftarrow{\ \mathrm{id}_Y \times i\ } & Y \times S^n & \xrightarrow{\ w\ } & X_n \\
\ \downarrow{\scriptstyle \mathrm{id}} & & \ \downarrow{\scriptstyle \mathrm{id}} & & \ \downarrow{\scriptstyle f_n} \\
Y \times D^{n+1} & \xleftarrow{\ \mathrm{id}_Y \times i\ } & Y \times S^n & \xrightarrow{\ f_n \circ w\ } & Z_n \\
\ \uparrow{\scriptstyle v \times \mathrm{id}_{D^{n+1}}} & & \ \uparrow{\scriptstyle v \times \mathrm{id}_{S^n}} & & \ \uparrow{\scriptstyle \mathrm{id}} \\
Y' \times D^{n+1} & \xleftarrow{\ \mathrm{id}_{Y'} \times i\ } & Y' \times S^n & \xrightarrow{\ f_n \circ w \circ (v \times \mathrm{id}_{S^n})\ } & Z_n \\
\ \downarrow{\scriptstyle j_0} & & \ \downarrow{\scriptstyle j_0} & & \ \downarrow{\scriptstyle \mathrm{id}} \\
Y' \times D^{n+1} \times [0,1] & \xleftarrow{\ \mathrm{id}_{Y'} \times i \times \mathrm{id}_{[0,1]}\ } & Y' \times S^n \times [0,1] & \xrightarrow{\ h\ } & Z_n \\
\ \uparrow{\scriptstyle j_1} & & \ \uparrow{\scriptstyle j_1} & & \ \uparrow{\scriptstyle \mathrm{id}} \\
Y' \times D^{n+1} & \xleftarrow{\ \mathrm{id}_{Y'} \times i\ } & Y' \times S^n & \xrightarrow{\ h_1\ } & Z_n
\end{array}
$$

where j_0 and j_1 always denotes the obvious inclusions. The G-pushout of the top row is X_{n+1} by (3.11). Let Z_{n+1} be the G-pushout of the bottom row. This is a G-CW-complex with compact isotropy groups containing Z_n as G-CW-subcomplex. Let W_2 and W_3 and W_4 be the G-pushout of the second, third and fourth row. The diagram above induces a sequence of G-maps

$$
X_{n+1} \xrightarrow{u_1} W_2 \xleftarrow{u_2} W_3 \xrightarrow{u_3} W_4 \xleftarrow{u_4} Z_{n+1}
$$

The left horizontal arrow in each row is a G-cofibration as i is a cofibration. Each of the vertical arrows is a G-homotopy equivalence. This implies that each of the maps u_1, u_2, u_3 and u_4 are G-homotopy equivalences. Notice that we can consider Z_n as a subspace of W_2, W_3, W_4 such that the inclusion $Z_n \to W_k$ is a G-cofibration. Each of the maps u_2, u_3 and u_4 induces the identity on Z_n, whereas u_1 induces f_n on X_n. By a cofibration argument one can find G-homotopy inverses u_2^{-1} and u_4^{-1} of u_2 and u_4 which induce the identity on Z_n. Now define the desired G-homotopy equivalence $f_{n+1} \colon X_{n+1} \to Z_{n+1}$ to be the composition $u_4^{-1} \circ u_3 \circ u_2^{-1} \circ u_1$. This finishes the proof of Theorem 3.10.   □

4. Special Models

In this section we present some interesting geometric models for the space $E_{\mathcal{F}}(G)$ and $J_{\mathcal{F}}(G)$ focussing on $\underline{E}G$ and $\underline{J}G$. In particular we are interested in cases, where these models satisfy finiteness conditions such as being finite, finite dimensional or of finite type.

One extreme case is, where we take $\mathcal{F}$ to be the family $\mathcal{ALL}$ of all subgroups. Then a model for both $E_{\mathcal{ALL}}(G)$ and $J_{\mathcal{ALL}}(G)$ is G/G. The other extreme case is the family $\mathcal{TR}$ consisting of the trivial subgroup. This case has already been treated in Example 2.7, Remark 2.8 and Example 3.8.

286   Wolfgang Lück

4.1. Operator Theoretic Model

Let G be a locally compact Hausdorff topological group. Let $C_0(G)$ be the Banach space of complex valued functions of G vanishing at infinity with the supremum-norm. The group G acts isometrically on $C_0(G)$ by $(g \cdot f)(x) := f(g^{-1}x)$ for $f \in C_0(G)$ and $g, x \in G$. Let $PC_0(G)$ be the subspace of $C_0(G)$ consisting of functions f such that f is not identically zero and has non-negative real numbers as values.

The next theorem is due to Abels [1, Theorem 2.4].

Theorem 4.1 (Operator theoretic model). *The G-space $PC_0(G)$ is a model for $\underline{J}G$.*

Remark 4.2. Let G be discrete. Another model for $\underline{J}G$ is the space

$$X_G = \{f \colon G \to [0,1] \mid f \text{ has finite support, } \sum_{g \in G} f(g) = 1\}$$

with the topology coming from the supremum norm [7, page 248]. Let $P_\infty(G)$ be the geometric realization of the simplicial set whose k-simplices consist of $(k+1)$-tuples $(g_0, g_1, \ldots, g_k)$ of elements g_i in G. This also a model for $\underline{E}G$ [1, Example 2.6]. The spaces X_G and $P_\infty(G)$ have the same underlying sets but in general they have different topologies. The identity map induces a (continuous) G-map $P_\infty(G) \to X_G$ which is a G-homotopy equivalence, but in general not a G-homeomorphism (see also [80, A.2]).

4.2. Almost Connected Groups

The next result is due to Abels [1, Corollary 4.14].

Theorem 4.3 (Almost connected groups). *Let G be a (locally compact Hausdorff) topological group. Suppose that G is almost connected, i.e. the group G/G^0 is compact for G^0 the component of the identity element. Then G contains a maximal compact subgroup K which is unique up to conjugation. The G-space G/K is a model for $\underline{J}G$.*

The next result follows from Example 1.4, Theorem 3.7 (iii) and Theorem 4.3.

Theorem 4.4 (Discrete subgroups of almost connected Lie groups). *Let L be a Lie group with finitely many path components. Then L contains a maximal compact subgroup K which is unique up to conjugation. The L-space L/K is a model for $\underline{E}L$.*

If $G \subseteq L$ is a discrete subgroup of L, then L/K with the obvious left G-action is a finite dimensional G-CW-model for $\underline{E}G$.

4.3. Actions on Simply Connected Non-Positively Curved Manifolds

The next theorem is due to Abels [1, Theorem 4.15].

Theorem 4.5 (Actions on simply connected non-positively curved manifolds). *Let G be a (locally compact Hausdorff) topological group. Suppose that G acts properly and isometrically on the simply-connected complete Riemannian manifold M with non-positive sectional curvature. Then M is a model for $\underline{J}G$.*

4.4. Actions on CAT(0)-spaces

Theorem 4.6 (Actions on CAT(0)-spaces). *Let G be a (locally compact Hausdorff) topological group. Let X be a proper G-CW-complex. Suppose that X has the structure of a complete CAT(0)-space for which G acts by isometries. Then X is a model for $\underline{E}G$.*

Proof. By [13, Corollary II.2.8 on page 179] the K-fixed point set of X is a non-empty convex subset of X and hence contractible for any compact subgroup $K \subset G$.   □

This result contains as special case Theorem 4.5 and partially Theorem 4.7 since simply-connected complete Riemannian manifolds with non-positive sectional curvature and trees are CAT(0)-spaces.

4.5. Actions on Trees and Graphs of Groups

A *tree* is a 1-dimensional CW-complex which is contractible.

Theorem 4.7 (Actions on trees). *Suppose that G acts continuously on a tree T such that for each element $g \in G$ and each open cell e with $g \cdot e \cap e \neq \emptyset$ we have $gx = x$ for any $x \in e$. Assume that the isotropy group of each $x \in T$ is compact.*

Then G can be written as an extension $1 \to K \to G \to \overline{G} \to 1$ of a compact group containing G^0 and a totally disconnected group $\overline{G}$ such that K acts trivially and T is a 1-dimensional model for

$$E_{\mathcal{COM}}(G) = J_{\mathcal{COM}}(G) = E_{\mathcal{COMOP}}(G) = J_{\mathcal{COMOP}}(G).$$

Proof. We conclude from Remark 1.3 that T is a G-CW-complex and all isotropy groups are compact open. Let K be the intersection of all the isotropy groups of points of T. This is a normal compact subgroup of G which contains the component of the identity G^0. Put $\overline{G} = G/K$. This is a totally disconnected group. Let $H \subseteq G$ be compact. If e_0 is a zero-cell in T, then $H \cdot e_0$ is a compact discrete set and hence finite. Let T' be the union of all geodesics with extremities in $H \cdot e$. This is a H-invariant subtree of T of finite diameter. One shows now inductively over the diameter of T' that T' has a vertex which is fixed under the H-action (see [69, page 20] or [25, Proposition 4.7 on page 17]). Hence T^H is non-empty. If e and f are vertices in T^H, the geodesic in T from e to f must be H-invariant. Hence T^H is a connected CW-subcomplex of the tree T and hence is itself a tree. This shows that T^H is contractible. Hence T is a model for $E_{\mathcal{COM}}(G) = E_{\mathcal{COM}}(\overline{G})$. Now apply Lemma 3.5.   □

Let G be a locally compact Hausdorff group. Suppose that G acts continuously on a tree T such that for each element $g \in G$ and each open cell e with $g \cdot e \cap e \neq \emptyset$ we have $gx = x$ for any $x \in e$. If the G-action on a tree has possibly not compact isotropy groups, one can nevertheless get nice models for $E_{\mathcal{COMOP}}(G)$ as follows. Let V be the set of equivariant 0-cells and E be the set of equivariant 1-cells of T. Then we can choose a G-pushout

$$
\begin{array}{ccc}
\coprod_{e \in E} G/H_e \times \{-1, 1\} & \xrightarrow{\quad q \quad} & T_0 = \coprod_{v \in V} G/K_v \\
\downarrow & & \downarrow \\
\coprod_{e \in E} G/H_e \times [-1, 1] & \longrightarrow & T
\end{array}
\tag{4.8}
$$

where the left vertical arrow is the obvious inclusion. Fix $e \in E$ and $\sigma \in \{-1, 1\}$. Choose elements $v(e, \sigma) \in V$ and $g(e, \sigma) \in G$ such that q restricted to $G/H_e \times \{\sigma\}$ is the G-map $G/H_e \to G/K_{v(e,\sigma)}$ which sends $1H_e$ to $g(e, \sigma)K_{v(e,\sigma)}$. Then conjugation with $g(e, \sigma)$ induces a group homomorphism $c_{g(e,\sigma)} \colon H_e \to K_{v(e,\pm 1)}$ and there is an up to equivariant homotopy unique $c_{g(e,\sigma)}$-equivariant cellular map $f_{g(e,\sigma)} \colon E_{\mathcal{COMOP}}(H_e) \to E_{\mathcal{COMOP}}(K_{e(g,\sigma)})$. Define a G-map

$$
Q \colon \coprod_{e \in E} G \times_{H_e} E_{\mathcal{COMOP}}(H_e) \times \{-1, 1\} \ \to \ \coprod_{v \in V} G \times_{K_v} E_{\mathcal{COMOP}}(K_v)
$$

by requiring that the restriction of Q to $G \times_{H_e} E_{\mathcal{COMOP}}(H_e) \times \{\sigma\}$ is the G-map

$$
G \times_{H_e} E_{\mathcal{COMOP}}(H_e) \ \to G \times_{K_{v(e,\sigma)}} E_{\mathcal{COMOP}}(K_{g(e,\sigma)}), \qquad (g, x) \mapsto (g, f_{g(e,\sigma)}(x)).
$$

Let $T_{\mathcal{COMOP}}$ be the G-pushout

$$
\begin{array}{ccc}
\coprod_{e \in E} G \times_{H_e} E_{\mathcal{COMOP}}(H_e) \times \{-1, 1\} & \xrightarrow{\quad Q \quad} & \coprod_{v \in V} G \times_{K_v} E_{\mathcal{COMOP}}(K_v) \\
\downarrow & & \downarrow \\
\coprod_{e \in E} G \times_{H_e} E_{\mathcal{COMOP}}(H_e) \times [-1, 1] & \longrightarrow & T_{\mathcal{COMOP}}
\end{array}
$$

The G-space $T_{\mathcal{COMOP}}$ inherits a canonical G-CW-structure with compact open isotropy groups. Notice that for any open subgroup $L \subseteq G$ one can choose as model for $E_{\mathcal{COMOP}}(L)$ the restriction $\mathrm{res}^L_G E_{\mathcal{COMOP}}(G)$ of $E_{\mathcal{COMOP}}(G)$ to L and that there is a G-homeomorphism $G \times_L \mathrm{res}^L_G E_{\mathcal{COMOP}}(G) \xrightarrow{\cong} G/L \times E_{\mathcal{COMOP}}(G)$ which sends (g, x) to (gL, gx). This implies that $T_{\mathcal{COMOP}}$ is G-homotopy equivalent to $T \times E_{\mathcal{COMOP}}(G)$ with the diagonal G-action. If $H \subseteq G$ is compact open, then T^H is contractible. Hence $(T \times E_{\mathcal{COMOP}}(G))^H$ is contractible for compact open subgroup $H \subseteq G$. Theorem 1.9 (ii) shows

Theorem 4.9 (Models based on actions on trees). *The G-CW-complex $T_{\mathcal{COMOP}}$ is a model for $E_{\mathcal{COMOP}}(G)$.*

The point is that it may be possible to choose nice models for the various spaces $E_{\mathcal{COMOP}}(H_e)$ and $E_{\mathcal{COMOP}}(K_v)$ and thus get a nice model for $E_{\mathcal{COMOP}}(G)$. If all isotropy groups of the G-action on T are compact, we can choose all spaces $E_{\mathcal{COMOP}}(H_e)$ and $E_{\mathcal{COMOP}}(K_v)$ to be $\{\mathrm{pt.}\}$ and we rediscover Theorem 4.7.

Next we recall which discrete groups G act on trees. Recall that an oriented graph X is a 1-dimensional CW-complex together with an orientation for each 1-cell. This can be codified by specifying a triple $(V, E, s \colon E \times \{-1, 1\} \to V)$ consisting of two sets V and E and a map s. The associated oriented graph is the pushout

$$
\begin{array}{ccc}
E \times \{-1, 1\} & \xrightarrow{\;s\;} & V \\
\downarrow & & \downarrow \\
E \times [0, 1] & \longrightarrow & X
\end{array}
$$

So V is the set of vertices, E the set of edges, and for a edge $e \in E$ its initial vertex is $s(e, -1)$ and its terminal vertex is $s(e, 1)$. A *graph of groups* $\mathcal{G}$ on a connected oriented graph X consists of two sets of groups $\{K_v \mid v \in V\}$ and $\{H_e \mid e \in E\}$ with V and E as index sets together with injective group homomorphisms $\phi_{v,\sigma} \colon H_e \to K_{s(e,\sigma)}$ for each $e \in E$. Let $X_0 \subseteq X$ be some maximal tree. We can associate to these data *the fundamental group* $\pi = \pi(\mathcal{G}, X, X_0)$ as follows. Generators of π are the elements in K_v for each $v \in V$ and the set $\{t_e \mid e \in E\}$. The relations are the relations in each group K_v for each $v \in V$, the relation $t_e = 1$ for $e \in V$ if e belongs to X_0, and for each $e \in E$ and $h \in H_e$ we require $t_e^{-1}\phi_{e,-1}(h)t_e = \phi_{e,+1}(h)$. It turns out that the obvious map $K_v \to \pi$ is an injective group homomorphism for each $v \in V$ and we will identify in the sequel K_v with its image in π [25, Corollary 7.5 on page 33], [69, Corollary 1 in 5.2 on page 45]. We can assign to these data a tree $T = T(X, X_0, \mathcal{G})$ with π-action as follows. Define a π-map

$$
q \colon \coprod_{e \in E} \pi / \mathrm{im}(\phi_{e,-1}) \times \{-1, 1\} \;\to\; \coprod_{v \in V} \pi / K_v
$$

by requiring that its restriction to $\pi / \mathrm{im}(\phi_{e,-1}) \times \{-1\}$ is the π-map given by the projection $\pi / \mathrm{im}(\phi_{e,-1}) \to \pi / K_{s(e,-1)}$ and its restriction to $\pi / \mathrm{im}(\phi_{e,-1}) \times \{1\}$ is the π-map $\pi / \mathrm{im}(\phi_{e,-1}) \to \pi / K_{s(e,1)}$ which sends $g \, \mathrm{im}(\phi_{e,-1})$ to $gt_e \, \mathrm{im}(\phi_{e,1})$. Now define a 1-dimensional G-CW-complex $T = T(\mathcal{G}, X, X_0)$ using this π-map q and the π-pushout analogous to (4.8). It turns out that T is contractible [25, Theorem 7.6 on page 33], [69, Theorem 12 in 5.3 on page 52].

On the other hand, suppose that T is a 1-dimensional G-CW-complex. Choose a G-pushout (4.8). Let X be the connected oriented graph $G\backslash T$. It has a set of vertices V and as set of edges the set E. The required map $s \colon E \times \{-1, 1\} \to V$ sends $s(e, \sigma)$ to the vertex for which $q(G/H_e \times \{\sigma\})$ meets and hence is equal to $G/K_{s(e,\sigma)}$. Moreover, we get a graph of groups $\mathcal{G}$ on X as follows. Let $\{K_v \mid v \in V\}$ and $\{H_e \mid e \in E\}$ be the set of groups given by (4.8). Choose an element $g(e, \sigma) \in G$ such that the G-map induced by q from G/H_e to $G/K_{s(e,\sigma)}$ sends $1H_e$ to $g(e, \sigma)K_{s(e,\sigma)}$. Then conjugation with $g(e, \sigma)$ induces a group homomorphism $\phi_{e,\sigma} \colon H_e \to K_{s(e,\sigma)}$. After a choice of a maximal tree X_0 in X one obtains an isomorphism $G \cong \pi(\mathcal{G}, X, X_0)$. (Up to isomorphism) we get a bijective correspondence between pairs (G, T) consisting of a group G acting on an oriented tree T and a graph of groups on connected oriented graphs. For details we refer for instance to [25, I.4 and I.7] and [69, §5].

Example 4.10 (The graph associated to amalgamated products). Consider the graph D with one edge e and two vertices v_{-1} and v_1 and the map $s\colon \{e\} \times \{-1,1\} \to \{v_{-1},v_1\}$ which sends (e,σ) to v_σ. Of course this is just the graph consisting of a single segment which is homeomorphic to $[-1,1]$. Let $\mathcal{G}$ be a graph of groups on D. This is the same as specifying a group H_e and groups K_{-1} and K_1 together with injective group homomorphisms $\phi_\sigma\colon H_e \to K_\sigma$ for $\sigma \in \{-1,1\}$. There is only one choice of a maximal subtree in D, namely D itself. Then the fundamental group π of this graph of groups is the amalgamated product of K_{-1} and K_1 over H_e with respect to ϕ_{-1} and ϕ_1, i.e. the pushout of groups

$$
\begin{array}{ccc}
H_e & \xrightarrow{\ \phi_{-1}\ } & K_{-1} \\
{\scriptstyle \phi_1}\big\downarrow & & \big\downarrow \\
K_1 & \longrightarrow & \pi
\end{array}
$$

Choose ϕ_σ-equivariant maps $f_\sigma\colon \underline{E}H_e \to \underline{E}K_\sigma$. They induce π-maps

$$
F_\sigma\colon \pi \times_{H_e} \underline{E}H_e \to \pi \times_{K_\sigma} \underline{E}K_\sigma, \qquad (g,x) \mapsto (g,f_\sigma(x)).
$$

We get a model for $\underline{E}\pi$ as the π-pushout

$$
\begin{array}{ccc}
\pi \times_{H_e} \underline{E}H_e \times \{-1,1\} & \xrightarrow{\ F_{-1} \amalg F_1\ } & \pi \times_{K_{-1}} \underline{E}K_{-1} \amalg \pi \times_{K_1} \underline{E}K_1 \\
\big\downarrow & & \big\downarrow \\
\pi \times_{H_e} \underline{E}H_e \times [-1,1] & \longrightarrow & \underline{E}\pi
\end{array}
$$

Example 4.11 (The graph associated to an HNN-extension). Consider the graph S with one edge e and one vertex v. There is only one choice for the map $s\colon \{e\} \times \{-1,1\} \to \{v\}$. Of course this graph is homeomorphic to S^1. Let $\mathcal{G}$ be a graph of groups on S. It consists of two groups H_e and K_v and two injective group homomorphisms $\phi_\sigma\colon H_e \to K_v$ for $\sigma \in \{-1,1\}$. There is only one choice of a maximal subtree, namely $\{v\}$. The fundamental group π of $\mathcal{G}$ is the so called HNN-extension associated to the data $\phi_\sigma\colon H_e \to K_v$ for $\sigma \in \{-1,1\}$, i.e. the group generated by the elements of K_v and a letter t_v whose relations are those of K_v and the relations $t_v^{-1}\phi_{-1}(h)t_v = \phi_1(h)$ for all $h \in H_e$. Recall that the natural map $K_v \to \pi$ is injective and we will identify K_v with its image in π. Choose ϕ_σ-equivariant maps $f_\sigma\colon \underline{E}H_e \to \underline{E}K_v$. Let $F_\sigma\colon \pi \times_{\phi_{-1}} \underline{E}H_e \to \pi \times \underline{E}K_v$ be the π-map which sends (g,x) to $gf_{-1}(x)$ for $\sigma = -1$ and to $gt_ef_1(x)$ for $\sigma = 1$. Then a model for $\underline{E}\pi$ is given by the π-pushout

$$
\begin{array}{ccc}
\pi \times_{\phi_{-1}} \underline{E}H_e \times \{-1,1\} & \xrightarrow{\ F_{-1} \amalg F_1\ } & \pi \times_{K_v} \underline{E}K_v \\
\big\downarrow & & \big\downarrow \\
\pi \times_{\phi_{-1}} \underline{E}H_e \times [-1,1] & \longrightarrow & \underline{E}\pi
\end{array}
$$

Notice that this looks like a telescope construction which is infinite to both sides. Consider the special case, where $H_e = K_v$, $\phi_{-1} = $ id and ϕ_1 is an automorphism. Then π is the semidirect product $K_v \rtimes_{\phi_1} \mathbb{Z}$. Choose a ϕ_1-equivariant map $f_1 \colon \underline{E}K_v \to \underline{E}K_v$. Then a model for $\underline{E}\pi$ is given by the to both side infinite mapping telescope of f_1 with the $K_v \rtimes_{\phi_1} \mathbb{Z}$ action, for which $\mathbb{Z}$ acts by shifting to the right and $k \in K_v$ acts on the part belonging to $n \in \mathbb{Z}$ by multiplication with $\phi_1^n(k)$. If we additionally assume that $\phi_1 = $ id, then $\pi = K_v \times \mathbb{Z}$ and we get $\underline{E}K_v \times \mathbb{R}$ as model for $\underline{E}\pi$.

Remark 4.12. All these constructions yield also models for $EG = E_{\mathcal{TR}}(G)$ if one replaces everywhere the spaces $\underline{E}H_e$ and $\underline{E}K_v$ by the spaces EH_e and EK_v.

4.6. Affine Buildings

Let Σ be an *affine building*, sometimes also called *Euclidean building*. This is a simplicial complex together with a system of subcomplexes called *apartments* satisfying the following axioms:

(i) Each apartment is isomorphic to an affine Coxeter complex.
(ii) Any two simplices of Σ are contained in some common apartment.
(iii) If two apartments both contain two simplices A and B of Σ, then there is an isomorphism of one apartment onto the other which fixes the two simplices A and B pointwise.

The precise definition of an affine Coxeter complex, which is sometimes called also Euclidean Coxeter complex, can be found in [17, Section 2 in Chapter VI], where also more information about affine buildings is given. An affine building comes with metric $d \colon \Sigma \times \Sigma \to [0, \infty)$ which is non-positively curved and complete. The building with this metric is a CAT(0)-space. A simplicial automorphism of Σ is always an isometry with respect to d. For two points x, y in the affine building there is a unique line segment $[x, y]$ joining x and y. It is the set of points $\{z \in \Sigma \mid d(x, y) = d(x, z) + d(z, y)\}$. For $x, y \in \Sigma$ and $t \in [0, 1]$ let $tx + (1 - t)y$ be the point $z \in \Sigma$ uniquely determined by the property that $d(x, z) = td(x, y)$ and $d(z, y) = (1 - t)d(x, y)$. Then the map

$$r \colon \Sigma \times \Sigma \times [0, 1] \to \Sigma, \qquad (x, y, t) \mapsto tx + (1 - t)y$$

is continuous. This implies that Σ is contractible. All these facts are taken from [17, Section 3 in Chapter VI] and [13, Theorem 10A.4 on page 344].

Suppose that the group G acts on Σ by isometries. If G maps a non-empty bounded subset A of Σ to itself, then the G-action has a fixed point [17, Theorem 1 in Section 4 in Chapter VI on page 157]. Moreover the G-fixed point set must be contractible since for two points $x, y \in \Sigma^G$ also the segment $[x, y]$ must lie in Σ^G and hence the map r above induces a continuous map $\Sigma^G \times \Sigma^G \times [0, 1] \to \Sigma^G$. This implies together with Theorem 1.9 (ii), Example 1.5, Lemma 3.3 and Lemma 3.5

Theorem 4.13 (Affine buildings). *Let G be a topological (locally compact Hausdorff group). Suppose that G acts on the affine building by simplicial automorphisms such that each isotropy group is compact. Then each isotropy group is compact open, Σ is a model for $J_{\mathcal{COMOP}}(G)$ and the barycentric subdivision Σ' is a model for both $J_{\mathcal{COMOP}}(G)$ and $E_{\mathcal{COMOP}}(G)$. If we additionally assume that G is totally disconnected, then Σ is a model for both $\underline{J}G$ and $\underline{E}G$.*

Example 4.14 (Bruhat-Tits building). An important example is the case of a reductive p-adic algebraic group G and its associated affine Bruhat-Tits building $\beta(G)$ [75],[76]. Then $\beta(G)$ is a model for $\underline{J}G$ and $\beta(G)'$ is a model for $\underline{E}G$ by Theorem 4.13.

4.7. The Rips Complex of a Word-Hyperbolic Group

A metric space $X = (X, d)$ is called δ-*hyperbolic* for a given real number $\delta \geq 0$ if for any four points x, y, z, t the following inequality holds

$$d(x,y) + d(z,t) \leq \max\{d(x,z) + d(y,t), d(x,t) + d(y,z)\} + 2\delta. \qquad (4.15)$$

A group G with a finite symmetric set S of generators is called δ-*hyperbolic* if the metric space (G, d_S) given by G and the wordlength metric with respect to one (and hence all) finite symmetric set of generators is δ-hyperbolic.

The *Rips complex* $P_d(G, S)$ of a group G with a symmetric finite set S of generators for a natural number d is the geometric realization of the simplicial set whose set of k-simplices consists of $(k + 1)$-tuples $(g_0, g_1, \ldots g_k)$ of pairwise distinct elements $g_i \in G$ satisfying $d_S(g_i, g_j) \leq d$ for all $i, j \in \{0, 1, \ldots, k\}$. The obvious G-action by simplicial automorphisms on $P_d(G, S)$ induces a G-action by simplicial automorphisms on the barycentric subdivision $P_d(G, S)'$ (see Example 1.5). The following result is proved in [60], [61].

Theorem 4.16 (Rips complex). *Let G be a (discrete) group with a finite symmetric set of generators. Suppose that (G, S) is δ-hyperbolic for the real number $\delta \geq 0$. Let d be a natural number with $d \geq 16\delta + 8$. Then the barycentric subdivision of the Rips complex $P_d(G, S)'$ is a finite G-CW-model for $\underline{E}G$.*

A metric space is called *hyperbolic* if it is δ-hyperbolic for some real number $\delta \geq 0$. A finitely generated group G is called *hyperbolic* if for one (and hence all) finite symmetric set S of generators the metric space (G, d_S) is a hyperbolic metric space. Since for metric spaces the property hyperbolic is invariant under quasiisometry and for two symmetric finite sets S_1 and S_2 of generators of G the metric spaces (G, d_{S_1}) and (G, d_{S_2}) are quasiisometric, the choice of S does not matter. Theorem 4.16 implies that for a hyperbolic group there is a finite G-CW-model for $\underline{E}G$.

The notion of a hyperbolic group is due to Gromov and has intensively been studied (see for example [13], [29], [30]). The prototype is the fundamental group of a closed hyperbolic manifold.

4.8. Arithmetic Groups

Arithmetic groups in a semisimple connected linear $\mathbb{Q}$-algebraic group possess finite models for $\underline{E}G$. Namely, let $G(\mathbb{R})$ be the $\mathbb{R}$-points of a semisimple $\mathbb{Q}$-group $G(\mathbb{Q})$ and let $K \subseteq G(\mathbb{R})$ a maximal compact subgroup. If $A \subseteq G(\mathbb{Q})$ is an arithmetic group, then $G(\mathbb{R})/K$ with the left A-action is a model for $E_{\mathcal{FIN}}(A)$ as already explained in Theorem 4.4. The A-space $G(\mathbb{R})/K$ is not necessarily cocompact. The Borel-Serre completion of $G(\mathbb{R})/K$ (see [10], [68]) is a finite A-CW-model for $E_{\mathcal{FIN}}(A)$ as pointed out in [2, Remark 5.8], where a private communication with Borel and Prasad is mentioned.

4.9. Outer Automorphism Groups of Free groups

Let F_n be the free group of rank n. Denote by $\mathrm{Out}(F_n)$ the group of outer automorphisms of F_n, i.e. the quotient of the group of all automorphisms of F_n by the normal subgroup of inner automorphisms. Culler and Vogtmann [21], [81] have constructed a space X_n called *outer space* on which $\mathrm{Out}(F_n)$ acts with finite isotropy groups. It is analogous to the Teichmüller space of a surface with the action of the mapping class group of the surface. Fix a graph R_n with one vertex v and n-edges and identify F_n with $\pi_1(R_n, v)$. A *marked metric graph* (g, Γ) consists of a graph Γ with all vertices of valence at least three, a homotopy equivalence $g \colon R_n \to \Gamma$ called marking and to every edge of Γ there is assigned a positive length which makes Γ into a metric space by the path metric. We call two marked metric graphs (g, Γ) and (g', Γ') equivalent of there is a homothety $h \colon \Gamma \to \Gamma'$ such that $g \circ h$ and h' are homotopic. Homothety means that there is a constant $\lambda > 0$ with $d(h(x), h(y)) = \lambda \cdot d(x, y)$ for all x, y. Elements in outer space X_n are equivalence classes of marked graphs. The main result in [21] is that X is contractible. Actually, for each finite subgroup $H \subseteq \mathrm{Out}(F_n)$ the H-fixed point set X_n^H is contractible [44, Proposition 3.3 and Theorem 8.1], [83, Theorem 5.1].

The space X_n contains a *spine* K_n which is an $\mathrm{Out}(F_n)$-equivariant deformation retraction. This space K_n is a simplicial complex of dimension $(2n - 3)$ on which the $\mathrm{Out}(F_n)$-action is by simplicial automorphisms and cocompact. Actually the group of simplicial automorphisms of K_n is $\mathrm{Out}(F_n)$ [14]. Hence the barycentric subdivision K_n' is a finite $(2n - 3)$-dimensional model of $\underline{E}\,\mathrm{Out}(F_n)$.

4.10. Mapping Class groups

Let $\Gamma_{g,r}^s$ be the *mapping class group* of an orientable compact surface F of genus g with s punctures and r boundary components. This is the group of isotopy classes of orientation preserving selfdiffeomorphisms $F_g \to F_g$, which preserve the punctures individually and restrict to the identity on the boundary. We require that the isotopies leave the boundary pointwise fixed. We will always assume that $2g + s + r > 2$, or, equivalently, that the Euler characteristic of the punctured surface F is negative. It is well-known that the associated *Teichmüller space* $\mathcal{T}_{g,r}^s$ is a contractible space on which $\Gamma_{g,r}^s$ acts properly. Actually $\mathcal{T}_{g,r}^s$ is a model for $E_{\mathcal{FIN}}(\Gamma_{g,r}^s)$ by the results of Kerckhoff [42].

We could not find a clear reference in the literature for the to experts known statement that there exist a finite $\Gamma^s_{g,r}$-CW-model for $E_{\mathcal{FIN}}(\Gamma^s_{g,r})$. The work of Harer [32] on the existence of a spine and the construction of the spaces $T_S(\epsilon)^H$ due to Ivanov [37, Theorem 5.4.A] seem to lead to such models.

4.11. Groups with Appropriate Maximal Finite Subgroups

Let G be a discrete group. Let $\mathcal{MFIN}$ be the subset of $\mathcal{FIN}$ consisting of elements in $\mathcal{FIN}$ which are maximal in $\mathcal{FIN}$. Consider the following assertions concerning G:

(M) Every non-trivial finite subgroup of G is contained in a unique maximal finite subgroup.

(NM) $M \in \mathcal{MFIN}, M \neq \{1\} \Rightarrow N_G M = M$.

For such a group there is a nice model for $\underline{E}G$ with as few non-free cells as possible. Let $\{(M_i) \mid i \in I\}$ be the set of conjugacy classes of maximal finite subgroups of $M_i \subseteq Q$. By attaching free G-cells we get an inclusion of G-CW-complexes $j_1 \colon \coprod_{i \in I} G \times_{M_i} EM_i \to EG$, where EG is the same as $E_{\mathcal{TR}}(G)$, i.e. a contractible free G-CW-complex. Define $\underline{E}G$ as the G-pushout

$$\begin{array}{ccc}
\coprod_{i \in I} G \times_{M_i} EM_i & \xrightarrow{j_1} & EG \\
{\scriptstyle u_1}\downarrow & & \downarrow{\scriptstyle f_1} \quad EG \\
\coprod_{i \in I} G/M_i & \xrightarrow{k_1} & \underline{E}G
\end{array} \qquad (4.17)$$

where u_1 is the obvious G-map obtained by collapsing each EM_i to a point.

We have to explain why $\underline{E}G$ is a model for the classifying space for proper actions of G. Obviously it is a G-CW-complex. Its isotropy groups are all finite. We have to show for $H \subseteq G$ finite that $(\underline{E}G)^H$ contractible. We begin with the case $H \neq \{1\}$. Because of conditions (M) and (NM) there is precisely one index $i_0 \in I$ such that H is subconjugated to M_{i_0} and is not subconjugated to M_i for $i \neq i_0$ and we get

$$\left(\coprod_{i \in I} G/M_i\right)^H = (G/M_{i_0})^H = \{\text{pt.}\}.$$

Hence $\underline{E}G^H = \{\text{pt.}\}$. It remains to treat $H = \{1\}$. Since u_1 is a non-equivariant homotopy equivalence and j_1 is a cofibration, f_1 is a non-equivariant homotopy equivalence and hence $\underline{E}G$ is contractible (after forgetting the group action).

Here are some examples of groups Q which satisfy conditions (M) and (NM):

- Extensions $1 \to \mathbb{Z}^n \to G \to F \to 1$ for finite F such that the conjugation action of F on $\mathbb{Z}^n$ is free outside $0 \in \mathbb{Z}^n$.

  The conditions (M), (NM) are satisfied by [58, Lemma 6.3].
- Fuchsian groups F

  The conditions (M), (NM) are satisfied (see for instance [58, Lemma 4.5]).

In [58] the larger class of cocompact planar groups (sometimes also called cocompact NEC-groups) is treated.

- One-relator groups G

Let G be a one-relator group. Let $G = \langle (q_i)_{i \in I} \mid r \rangle$ be a presentation with one relation. We only have to consider the case, where G contains torsion. Let F be the free group with basis $\{q_i \mid i \in I\}$. Then r is an element in F. There exists an element $s \in F$ and an integer $m \geq 2$ such that $r = s^m$, the cyclic subgroup C generated by the class $\bar{s} \in G$ represented by s has order m, any finite subgroup of G is subconjugated to C and for any $g \in G$ the implication $g^{-1} C g \cap C \neq 1 \Rightarrow g \in C$ holds. These claims follows from [59, Propositions 5.17, 5.18 and 5.19 in II.5 on pages 107 and 108]. Hence G satisfies (M) and (NM).

4.12. One-Relator Groups

Let G be a one-relator group. Let $G = \langle (q_i)_{i \in I} \mid r \rangle$ be a presentation with one relation. There is up to conjugacy one maximal finite subgroup C which is cyclic. Let $p \colon *_{i \in I} \mathbb{Z} \to G$ be the epimorphism from the free group generated by the set I to G, which sends the generator $i \in I$ to q_i. Let $Y \to \bigvee_{i \in I} S^1$ be the G-covering associated to the epimorphism p. There is a 1-dimensional unitary C-representation V and a C-map $f \colon SV \to \mathrm{res}^C_G Y$ such that the following is true. The induced action on the unit sphere SV is free. If we equip SV and DV with the obvious C-CW-complex structures, the C-map f can be chosen to be cellular and we obtain a G-CW-model for $\underline{E}G$ by the G-pushout

$$
\begin{array}{ccc}
G \times_C SV & \xrightarrow{\bar{f}} & Y \\
\downarrow & & \downarrow \\
G \times_C DV & \longrightarrow & \underline{E}G
\end{array}
$$

where $\bar{f}$ sends (g, x) to $gf(x)$. Thus we get a 2-dimensional G-CW-model for $\underline{E}G$ such that $\underline{E}G$ is obtained from G/C for a maximal finite cyclic subgroup $C \subseteq G$ by attaching free cells of dimensions ≤ 2 and the CW-complex structure on the quotient $G \backslash \underline{E}G$ has precisely one 0-cell, precisely one 2-cell and as many 1-cells as there are elements in I. All these claims follow from [16, Exercise 2 (c) II. 5 on page 44].

If G is torsionfree, the 2-dimensional complex associated to a presentation with one relation is a model for BG (see also [59, Chapter III §§9 -11]).

4.13. Special Linear Groups of (2,2)-Matrices

In order to illustrate some of the general statements above we consider the special example $SL_2(\mathbb{R})$ and $SL_2(\mathbb{Z})$.

Let $\mathbb{H}^2$ be the 2-dimensional hyperbolic space. We will use either the upper half-plane model or the Poincaré disk model. The group $SL_2(\mathbb{R})$ acts by isometric diffeomorphisms on the upper half-plane by Moebius transformations, i.e. a matrix $\begin{pmatrix} a & b \\ c & d \end{pmatrix}$ acts by sending a complex number z with positive imaginary part to $\frac{az+b}{cz+d}$. This action is proper and transitive. The isotropy group of $z = i$ is $SO(2)$. Since $\mathbb{H}^2$ is a simply-connected Riemannian manifold, whose sectional curvature is constant -1, the $SL_2(\mathbb{R})$-space $\mathbb{H}^2$ is a model for $\underline{E}SL_2(\mathbb{R})$ by Theorem 4.5.

One easily checks that $SL_2(\mathbb{R})$ is a connected Lie group and $SO(2) \subseteq SL_2(\mathbb{R})$ is a maximal compact subgroup. Hence $SL_2(\mathbb{R})/SO(2)$ is a model for $\underline{E}SL_2(\mathbb{R})$ by Theorem 4.3. Since the $SL_2(\mathbb{R})$-action on $\mathbb{H}^2$ is transitive and $SO(2)$ is the isotropy group at $i \in \mathbb{H}^2$, we see that the $SL_2(\mathbb{R})$-manifolds $SL_2(\mathbb{R})/SO(2)$ and $\mathbb{H}^2$ are $SL_2(\mathbb{R})$-diffeomorphic.

Since $SL_2(\mathbb{Z})$ is a discrete subgroup of $SL_2(\mathbb{R})$, the space $\mathbb{H}^2$ with the obvious $SL_2(\mathbb{Z})$-action is a model for $\underline{E}SL_2(\mathbb{Z})$ (see Theorem 4.4).

The group $SL_2(\mathbb{Z})$ is isomorphic to the amalgamated product $\mathbb{Z}/4 *_{\mathbb{Z}/2} \mathbb{Z}/6$. From Example 4.10 we conclude that a model for $\underline{E}SL_2(\mathbb{Z})$ is given by the following $SL_2(\mathbb{Z})$-pushout

$$
\begin{array}{ccc}
SL_2(\mathbb{Z})/(\mathbb{Z}/2) \times \{-1,1\} & \xrightarrow{F_{-1} \amalg F_1} & SL_2(\mathbb{Z})/(\mathbb{Z}/4) \amalg SL_2(\mathbb{Z})/(\mathbb{Z}/6) \\
\downarrow & & \downarrow \\
SL_2(\mathbb{Z})/(\mathbb{Z}/2) \times [-1,1] & \longrightarrow & \underline{E}SL_2(\mathbb{Z})
\end{array}
$$

where F_{-1} and F_1 are the obvious projections. This model for $\underline{E}SL_2(\mathbb{Z})$ is a tree, which has alternately two and three edges emanating from each vertex. The other model $\mathbb{H}^2$ is a manifold. These two models must be $SL_2(\mathbb{Z})$-homotopy equivalent. They can explicitly be related by the following construction.

Divide the Poincaré disk into fundamental domains for the $SL_2(\mathbb{Z})$-action. Each fundamental domain is a geodesic triangle with one vertex at infinity, i.e. a vertex on the boundary sphere, and two vertices in the interior. Then the union of the edges, whose end points lie in the interior of the Poincaré disk, is a tree T with $SL_2(\mathbb{Z})$-action. This is the tree model above. The tree is a $SL_2(\mathbb{Z})$-equivariant deformation retraction of the Poincaré disk. A retraction is given by moving a point p in the Poincaré disk along a geodesic starting at the vertex at infinity, which belongs to the triangle containing p, through p to the first intersection point of this geodesic with T.

The tree T above can be identified with the Bruhat-Tits building of $SL_2(\widehat{\mathbb{Q}_p})$ and hence is a model for $\underline{E}SL_2(\widehat{\mathbb{Q}_p})$ (see [17, page 134]). Since $SL_2(\mathbb{Z})$ is a discrete subgroup of $SL_2(\widehat{\mathbb{Q}_p})$, we get another reason why this tree is a model for $SL_2(\mathbb{Z})$.

4.14. Manifold Models

It is an interesting question, whether one can find a model for $\underline{E}G$ which is a smooth G-manifold. One may also ask whether such a manifold model realizes the minimal dimension for $\underline{E}G$ or whether the action is cocompact. Theorem 4.5 gives some information about these questions for simply connected non-positively curved Riemannian manifolds and Theorem 5.24 for discrete subgroups of Lie groups with finitely many path components. On the other hand there exists a virtually torsionfree group G such that G acts properly and cocompactly on a contractible manifold (without boundary), but there is no finite G-CW-model for $\underline{E}G$ [23, Theorem 1.1].

5. Finiteness Conditions

In this section we investigate whether there are models for $E_{\mathcal{F}}(G)$ which satisfy certain finiteness conditions such as being finite, being of finite type or being of finite dimension as a G-CW-complex.

5.1. Review of Finiteness Conditions on BG

As an illustration we review the corresponding question for EG for a discrete group G. This is equivalent to the question whether for a given discrete group G there is a CW-complex model for BG which is finite, of finite type or finite dimensional.

We introduce the following notation. Let R be a commutative associative ring with unit. The trivial RG-module is R viewed as RG-module by the trivial G-action. A *projective resolution* or *free resolution* respectively for an RG-module M is an RG-chain complex P_* of projective or free respectively RG-modules with $P_i = 0$ for $i \leq -1$ such that $H_i(P_*) = 0$ for $i \geq 1$ and $H_0(P_*)$ is RG-isomorphic to M. If additionally each RG-module P_i is finitely generated and P_* is finite dimensional, we call P_* *finite*.

An RG-module M has *cohomological dimension* $\mathrm{cd}(M) \leq n$, if there exists a projective resolution of dimension $\leq n$ for M. This is equivalent to the condition that for any RG-module N we have $\mathrm{Ext}^i_{RG}(M, N) = 0$ for $i \geq n + 1$. A group G has *cohomological dimension* $\mathrm{cd}(G) \leq n$ over R if the trivial RG-module R has cohomological dimension $\leq n$. An RG-module M is of *type* FP_n, if it admits a projective RG-resolution P_* such that P_i is finitely generated for $i \leq n$ and of type FP_∞ if it admits a projective RG-resolution P_* such that P_i is finitely generated for all i. A group G is of type FP_n or FP_∞ respectively if the trivial $\mathbb{Z}G$-module $\mathbb{Z}$ is of type FP_n or FP_∞ respectively.

Here is a summary of well-known statements about finiteness conditions on BG. A key ingredient in the proof of the next result is the fact that the cellular RG-chain complex $C_*(EG)$ is a free and in particular a projective RG-resolution of the trivial RG-module R since EG is a free G-CW-complex and contractible, and that $C_*(EG)$ is n-dimensional or of type FP_n respectively if BG is n-dimensional or has finite n-skeleton respectively.

Theorem 5.1 (Finiteness conditions for BG). *Let G be a discrete group.*

(i) *If there exists a finite dimensional model for BG, then G is torsionfree.*

(ii) (a) *There exists a CW-model for BG with finite 1-skeleton if and only if G is finitely generated.*

(b) *There exists a CW-model for BG with finite 2-skeleton if and only if G is finitely presented.*

(c) *For $n \geq 3$ there exists a CW-model for BG with finite n-skeleton if and only if G is finitely presented and of type FP_n.*

(d) *There exists a CW-model for BG of finite type, i.e. all skeleta are finite, if and only if G is finitely presented and of type FP_∞.*

(e) *There exists groups G which are of type FP_2 and which are not finitely presented.*

(iii) *There is a finite CW-model for BG if and only if G is finitely presented and there is a finite free $\mathbb{Z}G$-resolution F_* for the trivial $\mathbb{Z}G$-module $\mathbb{Z}$.*

(iv) *The following assertions are equivalent:*

(a) *The cohomological dimension of G is ≤ 1.*

(b) *There is a model for BG of dimension ≤ 1;*

(c) *G is free.*

(v) *The following assertions are equivalent for $d \geq 3$:*

(a) *There exists a CW-model for BG of dimension $\leq d$.*

(b) *G has cohomological dimension $\leq d$ over $\mathbb{Z}$.*

(vi) *For Thompson's group F there is a CW-model of finite type for BG but no finite dimensional model for BG.*

Proof. (i) Suppose we can choose a finite dimensional model for BG. Let $C \subseteq G$ be a finite cyclic subgroup. Then $C\backslash \widetilde{BG} = C\backslash EG$ is a finite dimensional model for BC. Hence there is an integer d such that we have $H_i(BC) = 0$ for $i \geq d$. This implies that C is trivial [16, (2.1) in II.3 on page 35]. Hence G is torsionfree.

(ii) See [9] and [16, Theorem 7.1 in VIII.7 on page 205].

(iii) See [16, Theorem 7.1 in VIII.7 on page 205].

(iv) See [71] and [74].

(v) See [16, Theorem 7.1 in VIII.7 on page 205].

(vi) See [18]. □

5.2. Modules over the Orbit Category

Let G be a discrete group and let $\mathcal{F}$ be a family of subgroups. The *orbit category* $\mathrm{Or}(G)$ of G is the small category, whose objects are homogeneous G-spaces G/H and whose morphisms are G-maps. Let $\mathrm{Or}_\mathcal{F}(G)$ be the full subcategory of $\mathrm{Or}(G)$ consisting of those objects G/H for which H belongs to $\mathcal{F}$. A $\mathbb{Z}\mathrm{Or}_\mathcal{F}(G)$-*module* is a contravariant functor from $\mathrm{Or}_\mathcal{F}(G)$ to the category of $\mathbb{Z}$-modules. A morphism of such modules is a natural transformation. The category of $\mathbb{Z}\mathrm{Or}_\mathcal{F}(G)$-modules inherits the structure of an abelian category from the standard structure of an abelian category on the category of $\mathbb{Z}$-modules. In particular the notion of a projective $\mathbb{Z}\mathrm{Or}_\mathcal{F}(G)$-module is defined. The *free* $\mathbb{Z}\mathrm{Or}_\mathcal{F}(G)$-module $\mathbb{Z}\,\mathrm{map}(G/?, G/K)$ *based at the object* G/K is the $\mathbb{Z}\mathrm{Or}_\mathcal{F}(G)$-module that assigns to an object G/H the free $\mathbb{Z}$-module $\mathbb{Z}\operatorname{map}_G(G/H, G/K)$ generated by the set $\operatorname{map}_G(G/H, G/K)$. The key property of it is that for any $\mathbb{Z}\mathrm{Or}_{\mathcal{F}}(G)$-module N there is a natural bijection of $\mathbb{Z}$-modules

$$\operatorname{hom}_{\mathbb{Z}\mathrm{Or}_{\mathcal{F}}(G)}(\mathbb{Z}\operatorname{map}_G(G/?, G/K), N) \xrightarrow{\cong} N(G/K), \quad \phi \mapsto \phi(G/K)(\mathrm{id}_{G/K}).$$

This is a direct consequence of the Yoneda Lemma. A $\mathbb{Z}\mathrm{Or}_{\mathcal{F}}(G)$-module is *free* if it is isomorphic to a direct sum $\bigoplus_{i \in I} \mathbb{Z}\operatorname{map}(G/?, G/K_i)$ for appropriate choice of objects G/K_i and index set I. A $\mathbb{Z}\mathrm{Or}_{\mathcal{F}}(G)$-module is called *finitely generated* if it is a quotient of a $\mathbb{Z}\mathrm{Or}_{\mathcal{F}}(G)$-module of the shape $\bigoplus_{i \in I} \mathbb{Z}\operatorname{map}(G/?, G/K_i)$ with a finite index set I. Notice that a lot of standard facts for $\mathbb{Z}$-modules carry over to $\mathbb{Z}\mathrm{Or}_{\mathcal{F}}(G)$-modules. For instance, a $\mathbb{Z}\mathrm{Or}_{\mathcal{F}}(G)$-module is projective or finitely generated projective respectively if and only if it is a direct summand in a free $\mathbb{Z}\mathrm{Or}_{\mathcal{F}}(G)$-module or a finitely generated free $\mathbb{Z}\mathrm{Or}_{\mathcal{F}}(G)$-module respectively. The notion of a *projective resolution* P_* of a $\mathbb{Z}\mathrm{Or}_{\mathcal{F}}(G)$-module is obvious and notions like of cohomological dimension $\leq n$ or of type FP_∞ carry directly over. Each $\mathbb{Z}\mathrm{Or}_{\mathcal{F}}(G)$-module has a projective resolution. The trivial $\mathbb{Z}\mathrm{Or}_{\mathcal{F}}(G)$-module $\mathbb{Z}$ is the constant functor from $\mathrm{Or}_{\mathcal{F}}(G)$ to the category of $\mathbb{Z}$-modules, which sends any morphism to $\mathrm{id} \colon \mathbb{Z} \to \mathbb{Z}$. More information about modules over a category can be found for instance in [48, Section 9].

The next result is proved in [54, Theorem 0.1]. A key ingredient in the proof of the next result is the fact that the cellular $R\mathrm{Or}_{\mathcal{F}}(G)$-chain complex $C_*(E_{\mathcal{F}}(G))$ is a free and in particular a projective $R\mathrm{Or}_{\mathcal{F}}(G)$-resolution of the trivial $R\mathrm{Or}_{\mathcal{F}}(G)$-module R.

Theorem 5.2 (Algebraic and geometric finiteness conditions). *Let G be a discrete group and let $d \geq 3$. Then we have:*

  (i) *There is G-CW-model of dimension $\leq d$ for $E_{\mathcal{F}}(G)$ if and only if the trivial $\mathbb{Z}\mathrm{Or}_{\mathcal{F}}(G)$-module $\mathbb{Z}$ has cohomological dimension $\leq d$.*

 (ii) *There is a G-CW-model for $E_{\mathcal{F}}(G)$ of finite type if and only if $E_{\mathcal{F}}(G)$ has a G-CW-model with finite 2-skeleton and the trivial $\mathbb{Z}\mathrm{Or}_{\mathcal{F}}(G)$-module $\mathbb{Z}$ is of type FP_∞.*

(iii) *There is a finite G-CW-model for $E_{\mathcal{F}}(G)$ if and only if $E_{\mathcal{F}}(G)$ has a G-CW-model with finite 2-skeleton and the trivial $\mathbb{Z}\mathrm{Or}_{\mathcal{F}}(G)$-module $\mathbb{Z}$ has a finite free resolution over $\mathrm{Or}_{\mathcal{F}}(G)$.*

(iv) *There is a G-CW-model with finite 2-skeleton for $\underline{E}G = E_{\mathcal{FIN}}(G)$ if and only if there are only finitely many conjugacy classes of finite subgroups $H \subset G$ and for any finite subgroup $H \subset G$ its Weyl group $W_G H := N_G H/H$ is finitely presented.*

In the case, where we take $\mathcal{F}$ to be the trivial family, Theorem 5.2 (i) reduces to Theorem 5.1 (v), Theorem 5.2 (ii) to Theorem 5.1 (ii)d and Theorem 5.2 (iii) to Theorem 5.1 (iii), and one should compare Theorem 5.2 (iv) to Theorem 5.1 (ii)b.

Remark 5.3. Nucinkis [67] investigates the notion of $\mathcal{FIN}$-cohomological dimension and relates it to the question whether there are finite dimensional modules for $\underline{E}G$.

It gives another lower bound for the dimension of a model for $\underline{E}G$ but is not sharp in general [11].

5.3. Reduction from Topological Groups to Discrete Groups

The *discretization* G_d of a topological group G is the same group but now with the discrete topology. Given a family $\mathcal{F}$ of (closed) subgroups of G, denote by $\mathcal{F}_d$ the same set of subgroups, but now in connection with G_d. Notice that $\mathcal{F}_d$ is again a family. We will need the following condition

(S)   For any closed subgroup $H \subset G$ the projection $p: G \to G/H$ has a local cross section, i.e. there is a neighborhood U of eH together with a map $s: U \to G$ satisfying $p \circ s = \text{id}_U$.

Condition (S) is automatically satisfied if G is discrete, if G is a Lie group, or more generally, if G is locally compact and second countable and has finite covering dimension [65]. The metric needed in [65] follows under our assumptions, since a locally compact Hausdorff space is regular and regularity in a second countable space implies metrizability.

The following two results are proved in [54, Theorem 0.2 and Theorem 0.3].

Theorem 5.4 (Passage from totally disconnected groups to discrete groups). *Let G be a locally compact totally disconnected Hausdorff group and let $\mathcal{F}$ be a family of subgroups of G. Then there is a G-CW-model for $E_{\mathcal{F}}(G)$ that is d-dimensional or finite or of finite type respectively if and only if there is a G_d-CW-model for $E_{\mathcal{F}_d}(G_d)$ that is d-dimensional or finite or of finite type respectively.*

Theorem 5.5 (Passage from topological groups to totally disconnected groups). *Let G be a locally compact Hausdorff group satisfying condition (S). Put $\overline{G} := G/G^0$. Then there is a G-CW-model for $\underline{E}G$ that is d-dimensional or finite or of finite type respectively if and only if $\underline{E}\overline{G}$ has a $\overline{G}$-CW-model that is d-dimensional or finite or of finite type respectively.*

If we combine Theorem 5.2, Theorem 5.4 and Theorem 5.5 we get

Theorem 5.6 (Passage from topological groups to discrete groups). *Let G be a locally compact group satisfying (S). Denote by $\overline{\mathcal{COM}}$ the family of compact subgroups of its component group $\overline{G}$ and let $d \geq 3$. Then*

(i) *There is a d-dimensional G-CW-model for $\underline{E}G$ if and only if the trivial $\mathbb{Z}\text{Or}_{\overline{\mathcal{COM}}_d}(\overline{G}_d)$-module $\mathbb{Z}$ has cohomological dimension $\leq d$.*

(ii) *There is a G-CW-model for $\underline{G}$ of finite type if and only if $E_{\overline{\mathcal{COM}}_d}(\overline{G}_d)$ has a $\overline{G}_d$-CW-model with finite 2-skeleton and the trivial $\mathbb{Z}\text{Or}_{\overline{\mathcal{COM}}_d}(\overline{G}_d)$-module $\mathbb{Z}$ is of type FP_∞.*

(iii) *There is a finite G-CW-model for $\underline{E}G$ if and only if $E_{\overline{\mathcal{COM}}_d}(\overline{G}_d)$ has a $\overline{G}_d$-CW-model with finite 2-skeleton and the trivial $\mathbb{Z}\text{Or}_{\overline{\mathcal{COM}}_d}(\overline{G}_d)$-module $\mathbb{Z}$ has a finite free resolution.*

In particular we see from Theorem 5.5 that, for a Lie group G, type questions about $\underline{E}G$ are equivalent to the corresponding type questions of $\underline{E}\pi_0(G)$, since $\pi_0(G) = \overline{G}$ is discrete. In this case the family $\overline{\mathcal{COM}}_d$ appearing in Theorem 5.6. is just the family $\mathcal{FIN}$ of finite subgroups of $\pi_0(G)$.

5.4. Poset of Finite Subgroups

Throughout this Subsection 5.4 let G be a discrete group. Define the G-poset

$$\mathcal{P}(G) \ := \ \{K \mid K \subset G \text{ finite}, K \neq 1\}. \tag{5.7}$$

An element $g \in G$ sends K to gKg^{-1} and the poset-structure comes from inclusion of subgroups. Denote by $|\mathcal{P}(G)|$ the geometric realization of the category given by the poset $\mathcal{P}(G)$. This is a G-CW-complex but in general not proper, i.e. it can have points with infinite isotropy groups.

Let $N_G H$ be the *normalizer* and let $W_G H := N_G H / H$ be the *Weyl group* of $H \subset G$. Notice for a G-space X that X^H inherits a $W_G H$-action. Denote by CX the *cone* over X. Notice that $C\emptyset$ is the one-point-space.

If H and K are subgroups of G and H is finite, then G/K^H is a finite union of $W_G H$-orbits of the shape $W_G H/L$ for finite $L \subset W_G H$. Now one easily checks

Lemma 5.8. *The $W_G H$-space $\underline{E}G^H$ is a $W_G H$-CW-model for $\underline{E}W_G H$. In particular, if $\underline{E}G$ has a G-CW-model which is finite, of finite type or d-dimensional respectively, then there is a $W_G H$-model for $\underline{E}W_G H$ which is finite, of finite type or d-dimensional respectively.*

Notation 5.9 (The condition $b(d)$ and $B(d)$). *Let $d \geq 0$ be an integer. A group G satisfies the condition $b(d)$ or $b(<\infty)$ respectively if any $\mathbb{Z}G$-module M with the property that M restricted to $\mathbb{Z}K$ is projective for all finite subgroups $K \subset G$ has a projective $\mathbb{Z}G$-resolution of dimension d or of finite dimension respectively. A group G satisfies the condition $B(d)$ if $W_G H$ satisfies the condition $b(d)$ for any finite subgroup $H \subset G$.*

The length $l(H) \in \{0, 1, \ldots\}$ of a finite group H is the supremum over all p for which there is a nested sequence $H_0 \subset H_1 \subset \ldots \subset H_p$ of subgroups H_i of H with $H_i \neq H_{i+1}$.

Lemma 5.10. *Suppose that there is a d-dimensional G-CW-complex X with finite isotropy groups such that $H_p(X; \mathbb{Z}) = H_p(*, \mathbb{Z})$ for all $p \geq 0$ holds. This assumption is for instance satisfied if there is a d-dimensional G-CW-model for $\underline{E}G$. Then G satisfies condition $B(d)$.*

Proof. Let $H \subset G$ be finite. Then X/H satisfies $H_p(X/H; \mathbb{Z}) = H_p(*, \mathbb{Z})$ for all $p \geq 0$ [12, III.5.4 on page 131]. Let C_* be the cellular $\mathbb{Z}W_G H$-chain complex of X/H. This is a d-dimensional resolution of the trivial $\mathbb{Z}W_G H$-module $\mathbb{Z}$ and each chain module is a sum of $\mathbb{Z}W_G H$-modules of the shape $\mathbb{Z}[W_G H/K]$ for some finite subgroup $K \subset W_G H$. Let N be a $\mathbb{Z}W_G H$-module such that N is projective over $\mathbb{Z}K$ for any finite subgroup $K \subset W_G H$. Then $C_* \otimes_{\mathbb{Z}} N$ with the diagonal $W_G H$-operation is a d-dimensional projective $\mathbb{Z}W_G H$-resolution of N. □

Theorem 5.11 (An algebraic criterion for finite dimensionality). *Let G be a discrete group. Suppose that we have for any finite subgroup $H \subset G$ an integer $d(H) \geq 3$ such that $d(H) \geq d(K)$ for $H \subset K$ and $d(H) = d(K)$ if H and K are conjugate in G. Consider the following statements:*

(i) *There is a G-CW-model $\underline{E}G$ such that for any finite subgroup $H \subset G$*

$$\dim(\underline{E}G^H) = d(H).$$

(ii) *We have for any finite subgroup $H \subset G$ and for any $\mathbb{Z}W_G H$-module M*

$$H^{d(H)+1}_{\mathbb{Z}W_G H}(EW_G H \times (C|\mathcal{P}(W_G H)|, |\mathcal{P}(W_G H)|); M) = 0.$$

(iii) *We have for any finite subgroup $H \subset G$ that its Weyl group $W_G H$ satisfies $b(<\infty)$ and that there is a subgroup $\Delta(H) \subset W_G H$ of finite index such that for any $\mathbb{Z}\Delta(H)$-module M*

$$H^{d(H)+1}_{\mathbb{Z}\Delta(H)}(E\Delta(H) \times (C|\mathcal{P}(W_G H)|, |\mathcal{P}(W_G H)|); M) = 0.$$

Then (i) *implies both* (ii) *and* (iii). *If there is an upper bound on the length $l(H)$ of the finite subgroups H of G, then these statements* (i), (ii) *and* (iii) *are equivalent.*

The proof of Theorem 5.11 can be found in [49, Theorem 1.6]. In the case that G has finite virtual cohomological dimension a similar result is proved in [20, Theorem III].

Example 5.12. Suppose that G is torsionfree. Then Theorem 5.11 reduces to the well-known result [16, Theorem VIII.3.1 on page 190,Theorem VIII.7.1 on page 205] that the following assertions are equivalent for an integer $d \geq 3$:

(i) There is a d-dimensional CW-model for BG;
(ii) G has cohomological dimension $\leq d$;
(iii) G has virtual cohomological dimension $\leq d$.

Remark 5.13. If $W_G H$ contains a non-trivial normal finite subgroup L, then $|\mathcal{P}(W_G H)|$ is contractible and

$$H^{d(H)+1}_{\mathbb{Z}W_G H}(EW_G H \times (C|\mathcal{P}(W_G H)|, |\mathcal{P}(W_G H)|); M) = 0;$$
$$H^{d(H)+1}_{\mathbb{Z}\Delta(H)}(E\Delta(H) \times (C|\mathcal{P}(W_G H)|, |\mathcal{P}(W_G H)|); M) = 0.$$

The proof of this fact is given in [49, Example 1.8].

The next result is taken from [49, Theorem 1.10]. A weaker version of it for certain classes of groups and in l exponential dimension estimate can be found in [43, Theorem B] (see [49, Remark 1.12]).

Theorem 5.14 (An upper bound on the dimension). *Let G be a group and let $l \geq 0$ and $d \geq 0$ be integers such that the length $l(H)$ of any finite subgroup $H \subset G$ is bounded by l and G satisfies $B(d)$. Then there is a G-CW-model for $\underline{E}G$ such that for any finite subgroup $H \subset G$*

$$\dim(\underline{E}G^H) \leq \max\{3,d\} + (l - l(H))(d+1)$$

holds. In particular $\underline{E}G$ has dimension at most $\max\{3,d\} + l(d+1)$.

5.5. Extensions of Groups

In this subsection we consider an exact sequence of discrete groups $1 \to \Delta \to G \to \pi \to 1$. We want to investigate whether finiteness conditions about the type of a classifying space for $\mathcal{FIN}$ for Δ and π carry over to the one of G. The proof of the next Theorem 5.15 is taken from [49, Theorem 3.1]), the proof of Theorem 5.16 is an easy variation.

Theorem 5.15 (Dimension bounds and extensions). *Suppose that there exists a positive integer d which is an upper bound on the orders of finite subgroups of π. Suppose that $\underline{E}\Delta$ has a k-dimensional Δ-CW-model and $\underline{E}\pi$ has a m-dimensional π-CW-model. Then $\underline{E}G$ has a $(dk + m)$-dimensional G-CW-model.*

Theorem 5.16. *Suppose that Δ has the property that for any group Γ which contains Δ as subgroup of finite index, there is a k-dimensional Γ-CW-model for $\underline{E}\Gamma$. Suppose that $\underline{E}\pi$ has a m-dimensional π-CW-model. Then $\underline{E}G$ has a $(k + m)$-dimensional G-CW-model.*

We will see in Example 5.26 that the condition about Δ in Theorem 5.16 is automatically satisfied if Δ is virtually poly-cyclic.

The next two results are taken from [49, Theorem 3.2 and Theorem 3.3]).

Theorem 5.17. *Suppose for any finite subgroup $\pi' \subset \pi$ and any extension $1 \to \Delta \to \Delta' \to \pi' \to 1$ that $\underline{E}\Delta'$ has a finite Δ'-CW-model or a Δ'-CW-model of finite type respectively and suppose that $\underline{E}\pi$ has a finite π-CW-model or a π-CW-model of finite type respectively. Then $\underline{E}G$ has a finite G-CW-model or a G-CW-model of finite type respectively.*

Theorem 5.18. *Suppose that Δ is word-hyperbolic or virtually poly-cyclic. Suppose that $\underline{E}\pi$ has a finite π-CW-model or a π-CW-model of finite type respectively. Then $\underline{E}G$ has a finite G-CW-model or a G-CW-model of finite type respectively.*

5.6. One-Dimensional Models for $\underline{E}G$

The following result follows from Dunwoody [27, Theorem 1.1].

Theorem 5.19 (A criterion for 1-dimensional models). *Let G be a discrete group. Then there exists a 1-dimensional model for $\underline{E}G$ if and only the cohomological dimension of G over the rationals $\mathbb{Q}$ is less or equal to one.*

If G is finitely generated, then there is a 1-dimensional model for $\underline{E}G$ if and only if G contains a finitely generated free subgroup of finite index [41, Theorem 1]. If G is torsionfree, we rediscover the results due to Swan and Stallings stated in Theorem 5.1 (iv) from Theorem 5.19.

5.7. Groups of Finite Virtual Dimension

In this section we investigate the condition $b(d)$ and $B(d)$ of Notation 5.9 for a discrete group G and explain how our results specialize in the case of a group of finite virtual cohomological dimension.

Remark 5.20. There exists groups G with a finite dimensional model for $\underline{E}G$, which do not admit a torsionfree subgroup of finite index. For instance, let G be a countable locally finite group which is not finite. Then its cohomological dimension over the rationals is ≤ 1 and hence it possesses a 1-dimensional model for $\underline{E}G$ by Theorem 5.19. Obviously it contains no torsionfree subgroup of finite index. An example of a group G with a finite 2-dimensional model for $\underline{E}G$, which does not admit a torsionfree subgroup of finite index, is described in [11, page 493].

A discrete group G has *virtual cohomological dimension* $\leq d$ if and only if it contains a torsionfree subgroup Δ of finite index such that Δ has cohomological dimension $\leq d$. This is independent of the choice of $\Delta \subseteq G$ because for two torsionfree subgroups $\Delta, \Delta' \subseteq G$ we have that Δ has cohomological dimension $\leq d$ if and only if Δ' has cohomological dimension $\leq d$. The next two results are taken from [49, Lemma 6.1, Theorem 6.3, Theorem 6.4].

Lemma 5.21. *If G satisfies $b(d)$ or $B(d)$ respectively, then any subgroup Δ of G satisfies $b(d)$ or $B(d)$ respectively.*

Theorem 5.22 (Virtual cohomological dimension and the condition $B(d)$). *If G contains a torsionfree subgroup Δ of finite index, then the following assertions are equivalent:*

(i) *G satisfies $B(d)$.*
(ii) *G satisfies $b(d)$.*
(iii) *G has virtual cohomological dimension $\leq d$.*

Next we improve Theorem 5.14 in the case of groups with finite virtual cohomological dimension. Notice that for such a group there is an upper bound on the length $l(H)$ of finite subgroups $H \subset G$.

Theorem 5.23 (Virtual cohomological dimension and $\dim(\underline{E}G)$). *Let G be a discrete group which contains a torsionfree subgroup of finite index and has virtual cohomological dimension $\mathrm{vcd}(G) \leq d$. Let $l \geq 0$ be an integer such that the length $l(H)$ of any finite subgroup $H \subset G$ is bounded by l.*

Then we have $\mathrm{vcd}(G) \leq \dim(\underline{E}G)$ for any model for $\underline{E}G$ and there is a G-CW-model for $\underline{E}G$ such that for any finite subgroup $H \subset G$

$$\dim(\underline{E}G^H) = \max\{3, d\} + l - l(H)$$

holds. In particular there exists a model for $\underline{E}G$ of dimension $\max\{3, d\} + l$.

Theorem 5.24 (Discrete subgroups of Lie groups). *Let L be a Lie group with finitely many path components. Then L contains a maximal compact subgroup K which is unique up to conjugation. Let $G \subseteq L$ be a discrete subgroup of L. Then L/K with the left G-action is a model for $\underline{E}G$.*

Suppose additionally that G contains a torsionfree subgroup $\Delta \subseteq G$ of finite index. Then we have

$$\mathrm{vcd}(G) \leq \dim(L/K)$$

and equality holds if and only if $G\backslash L$ is compact.

Proof. We have already mentioned in Theorem 4.4 that L/K is a model for $\underline{E}G$. The restriction of $\underline{E}G$ to Δ is a Δ-CW-model for $\underline{E}\Delta$ and hence $\Delta\backslash\underline{E}G$ is a CW-model for $B\Delta$. This implies $\mathrm{vcd}(G) := \mathrm{cd}(\Delta) \leq \dim(L/K)$. Obviously $\Delta\backslash L/K$ is a manifold without boundary. Suppose that $\Delta\backslash L/K$ is compact. Then $\Delta\backslash L/K$ is a closed manifold and hence its homology with $\mathbb{Z}/2$-coefficients in the top dimension is non-trivial. This implies $\mathrm{cd}(\Delta) \geq \dim(\Delta\backslash L/K)$ and hence $\mathrm{vcd}(G) = \dim(L/K)$. If $\Delta\backslash L/K$ is not compact, it contains a CW-complex $X \subseteq \Delta\backslash L/K$ of dimension smaller than $\Delta\backslash L/K$ such that the inclusion of X into $\Delta\backslash L/K$ is a homotopy equivalence. Hence X is another model for $B\Delta$. This implies $\mathrm{cd}(\Delta) < \dim(L/K)$ and hence $\mathrm{vcd}(G) < \dim(L/K)$. $\qquad\qquad\qquad\square$

Remark 5.25. An often useful strategy to find smaller models for $E_{\mathcal{F}}(G)$ is to look for a G-CW-subcomplex $X \subseteq E_{\mathcal{F}}(G)$ such that there exists a G-retraction $r\colon E_{\mathcal{F}}(G) \to X$, i.e. a G-map r with $r|_X = \mathrm{id}_X$. Then X is automatically another model for $E_{\mathcal{F}}(G)$. We have seen this already in the case $SL_2(\mathbb{Z})$, where we found a tree inside $\mathbb{H}^2 = SL_2(\mathbb{R})/SO(2)$ as explained in Subsection 4.13. This method can be used to construct a model for $\underline{E}SL_n(\mathbb{Z})$ of dimension $\frac{n(n-1)}{2}$ and to show that the virtual cohomological dimension of $SL_n(\mathbb{Z})$ is $\frac{n(n-1)}{2}$. Notice that $SL_n(\mathbb{R})/SO(n)$ is also a model for $\underline{E}SL_n(\mathbb{Z})$ by Theorem 4.4 but has dimension $\frac{n(n+1)}{2} - 1$.

Example 5.26 (Virtually poly-cyclic groups). Let the group Δ be *virtually poly-cyclic*, i.e. Δ contains a subgroup Δ' of finite index for which there is a finite sequence $\{1\} = \Delta'_0 \subseteq \Delta'_1 \subseteq \ldots \subseteq \Delta'_n = \Delta'$ of subgroups such that Δ'_{i-1} is normal in Δ'_i with cyclic quotient Δ'_i/Δ'_{i-1} for $i = 1, 2, \ldots, n$. Denote by r the number of elements $i \in \{1, 2, \ldots, n\}$ with $\Delta'_i/\Delta'_{i-1} \cong \mathbb{Z}$. The number r is called the *Hirsch rank*. The group Δ contains a torsionfree subgroup of finite index. We call Δ' *poly-$\mathbb{Z}$* if $r = n$, i.e. all quotients Δ'_i/Δ'_{i-1} are infinite cyclic. We want to show:

(i) $r = \mathrm{vcd}(\Delta)$.
(ii) $r = \max\{i \mid H_i(\Delta'; \mathbb{Z}/2) \neq 0\}$ for one (and hence all) poly-$\mathbb{Z}$ subgroup $\Delta' \subset \Delta$ of finite index.
(iii) There exists a finite r-dimensional model for $\underline{E}\Delta$ and for any model $\underline{E}\Delta$ we have $\dim(\underline{E}\Delta) \geq r$.

We use induction over the number r. If $r = 0$, then Δ is finite and all the claims are obviously true. Next we explain the induction step from $(r-1)$ to $r \geq 1$. We can choose an extension $1 \to \Delta_0 \to \Delta \xrightarrow{p} V \to 1$ for some virtually poly-cyclic group Δ_0 with $r(\Delta_0) = r(\Delta) - 1$ and some group V which contains $\mathbb{Z}$ as subgroup of finite index. The induction hypothesis applies to any group Γ which contains Δ_0 as subgroup of finite index. Since V maps surjectively to $\mathbb{Z}$ or the infinite dihedral group D_∞ with finite kernel and both $\mathbb{Z}$ and D_∞ have 1-dimensional models for their classifying space for proper group actions, there is a 1-dimensional model for $\underline{E}V$. We conclude from Theorem 5.16 that there is a r-dimensional model for $\underline{E}\Delta$.

The existence of a r-dimensional model for $\underline{E}\Delta$ implies $\mathrm{vcd}(\Delta) \leq r$. For any torsionfree subgroup $\Delta' \subset \Delta$ of finite index we have $\max\{i \mid H_i(\Delta'; \mathbb{Z}/2) \neq 0\} \leq \mathrm{vcd}(\Delta)$.

It is not hard to check by induction over r that we can find a sequence of torsionfree subgroups $\{1\} \subseteq \Delta_0 \subseteq \Delta_1 \subseteq \ldots \subseteq \Delta_r \subseteq \Delta$ such that Δ_{i-1} is normal in Δ_i with $\Delta_i/\Delta_{i-1} \cong \mathbb{Z}$ for $i \in \{1, 2, \ldots, r\}$ and Δ_r has finite index in Δ. We show by induction over i that $H_i(\Delta_i; \mathbb{Z}/2) = \mathbb{Z}/2$ for $i = 0, 1, \ldots, r$. The induction beginning $i = 0$ is trivial. The induction step from $(i-1)$ to i follows from the part of the long exact Wang sequence

$$H_i(\Delta_{i-1}; \mathbb{Z}/2) = 0 \to H_i(\Delta_i; \mathbb{Z}/2) \to H_{i-1}(\Delta_{i-1}; \mathbb{Z}/2) = \mathbb{Z}/2$$
$$\xrightarrow{\mathrm{id} - H_{i-1}(f; \mathbb{Z}/2) = 0} H_{i-1}(\Delta_{i-1}; \mathbb{Z}/2)$$

which comes from the Hochschild-Serre spectral sequence associated to the extension $1 \to \Delta_{i-1} \to \Delta_i \to \mathbb{Z} \to 1$ for $f \colon \Delta_{i-1} \to \Delta_{i-1}$ the automorphism induced by conjugation with some preimage in Δ_i of the generator of $\mathbb{Z}$. This implies

$$r = \max\{i \mid H_i(\Delta_r; \mathbb{Z}/2) \neq 0\} = \mathrm{cd}(\Delta_r) = \mathrm{vcd}(\Delta).$$

Now the claim follows.

The existence of a r-dimensional model for $\underline{E}G$ is proved for finitely generated nilpotent groups with $\mathrm{vcd}(G) \leq r$ for $r \neq 2$ in [66], where also not necessarily finitely generated nilpotent groups are studied.

The work of Dekimpe-Igodt [24] or Wilking [85, Theorem 3] implies that there is a model for $E_{\mathcal{FIN}}(\Delta)$ whose underlying space is $\mathbb{R}^r$.

5.8. Counterexamples

The following problem is stated by Brown [15, page 32]. It created a lot of activities and many of the results stated above were motivated by it.

Problem 5.27. *For which discrete groups G, which contain a torsionfree subgroup of finite index and has virtual cohomological dimension $\leq d$, does there exist a d-dimensional G-CW-model for $\underline{E}G$?*

The following four problems for discrete groups G are stated in the problem lists appearing in [49] and [82].

Problem 5.28. *Let $H \subseteq G$ be a subgroup of finite index. Suppose that $\underline{E}H$ has a H-CW-model of finite type or a finite H-CW-model respectively. Does then $\underline{E}G$ have a G-CW-model of finite type or a finite G-CW-model respectively?*

Problem 5.29. *If the group G contains a subgroup of finite index H which has a H-CW-model of finite type for $\underline{E}H$, does then G contain only finitely many conjugacy classes of finite subgroups?*

Problem 5.30. *Let G be a group such that BG has a model of finite type. Is then $BW_G H$ of finite type for any finite subgroup $H \subset G$?*

Problem 5.31. *Let* $1 \to \Delta \xrightarrow{i} G \xrightarrow{p} \pi \to 1$ *be an exact sequence of groups. Suppose that there is a* Δ-*CW-model of finite type for* $\underline{E}\Delta$ *and a* G-*CW-model of finite type for* $\underline{E}G$. *Is then there a* π-*CW-model of finite type for* $\underline{E}\pi$?

Leary and Nucinkis [47] have constructed many very interesting examples of discrete groups some of which are listed below. Their main technical input is an equivariant version of the constructions due to Bestvina and Brady [9]. These examples show that the answer to the Problems 5.27, 5.28, 5.29, 5.30 and 5.31 above is *not* positive in general. A group G is *of type VF* if it contains a subgroup $H \subseteq G$ of finite index for which there is a finite model for BH.

(i)   For any positive integer d there exist a group G of type VF which has virtually cohomological dimension $\leq 3d$, but for which any model for $\underline{E}G$ has dimension $\geq 4d$.

(ii) There exists a group G with a finite cyclic subgroup $H \subseteq G$ such that G is of type VF but the centralizer $C_G H$ of H in G is not of type FP_∞.

(iii) There exists a group G of type VF which contains infinitely many conjugacy classes of finite subgroups.

(iv) There exists an extension $1 \to \Delta \to G \to \pi \to 1$ such that $\underline{E}\Delta$ and $\underline{E}G$ have finite G-CW-models but there is no G-CW-model for $\underline{E}\pi$ of finite type.

6. The Orbit Space of $\underline{E}G$

We will see that in many computations of the group (co-)homology, of the algebraic K- and L-theory of the group ring or the topological K-theory of the reduced C^*-algebra of a discrete group G a key problem is to determine the homotopy type of the quotient space $G\backslash\underline{E}G$ of $\underline{E}G$. The following result shows that this is a difficult problem in general and can only be solved in special cases. It was proved by Leary and Nucinkis [46] based on ideas due to Baumslag-Dyer-Heller [8] and Kan and Thurston [40].

Theorem 6.1 (The homotopy type of $G\backslash\underline{E}G$). *Let X be a connected CW-complex. Then there exists a group G such that $G\backslash\underline{E}G$ is homotopy equivalent to X.*

There are some cases, where the quotient $G\backslash\underline{E}G$ has been determined explicitly using geometric input. We mention a few examples.

(i) Let G be a planar group (sometimes also called NEC) group, i.e. a discontinuous group of isometries of the two-sphere S^2, the Euclidean plane $\mathbb{R}^2$, or the hyperbolic plane $\mathbb{H}^2$. Examples are Fuchsian groups and two-dimensional crystallographic groups. If G acts on $\mathbb{R}^2$ or $\mathbb{H}^2$ and the action is cocompact, then $\mathbb{R}^2$ or $\mathbb{H}^2$ is a model for $\underline{E}G$ and the quotient space $G\backslash\underline{E}G$ is a compact 2-dimensional surface. The number of boundary components, its genus and the answer to the question, whether $G\backslash\underline{E}G$ is orientable, can be read off from an explicit presentation of G. A summary of these details can be found in [58, Section 4], where further references to papers containing proofs of the stated facts are given.

(ii) Let $G = \langle (q_i)_{i \in I} \mid r \rangle$ be a one-relator group. Let F be the free group on the letters $\{q_i \mid i \in I\}$. Then r is an element in F. There exists an element $s \in F$ and an integer $m \geq 1$ such that $r = s^m$, the cyclic subgroup C generated by the class $\bar{s} \in G$ represented by s has order m, any finite subgroup of G is subconjugated to C and for any $g \in G$ the implication $g^{-1}Cg \cap C \neq \{1\} \Rightarrow g \in C$ holds (see [59, Propositions 5.17, 5.18 and 5.19 in II.5 on pages 107 and 108]).

In the sequel we use the two-dimensional model for $\underline{E}G$ described in Subsection 4.12. Let us compute the integral homology of BG and $G\backslash\underline{E}G$. Since $G\backslash\underline{E}G$ has precisely one 2-cell and is two-dimensional, $H_2(G\backslash\underline{E}G)$ is either trivial or infinite cyclic and $H_k(G\backslash\underline{E}G) = 0$ for $k \geq 3$. We obtain the short exact sequence

$$0 \to H_2(BG) \xrightarrow{H_2(q)} H_2(G\backslash\underline{E}G) \xrightarrow{\partial_2} H_1(BC) \xrightarrow{H_1(Bi)} H_1(BG)$$
$$\xrightarrow{H_1(q)} H_1(G\backslash\underline{E}G) \to 0$$

and for $k \geq 3$ isomorphisms

$$H_k(Bi)\colon H_k(BC) \xrightarrow{\cong} H_k(BG)$$

from the pushout coming from (4.17)

$$\begin{array}{ccc} BC & \xrightarrow{i} & BG \\ \downarrow & & \downarrow \\ \{pt.\} & \longrightarrow & G\backslash\underline{E}G \end{array}$$

Hence $H_2(G\backslash\underline{E}G) = 0$ and the sequence

$$0 \to H_1(BC) \xrightarrow{H_1(Bi)} H_1(BG) \xrightarrow{H_1(q)} H_1(G\backslash\underline{E}G) \to 0$$

is exact, provided that $H_2(BG) = 0$. Suppose that $H_2(BG) \neq 0$. Hopf's Theorem says that $H_2(BG) \cong R \cap [F,F]/[F,R]$ if R is the subgroup of G normally generated by $r \in F$ (see [16, Theorem 5.3 in II.5 on page 42]). For every element in $R \cap [F,F]/[F,R]$ there exists $n \in \mathbb{Z}$ such that r^n belongs to $[F,F]$ and the element is represented by r^n. Hence there is $n \geq 1$ such that r^n does belong to $[F,F]$. Since $F/[F,F]$ is torsionfree, also s and r belong to $[F,F]$. We conclude that both $H_2(BG)$ and $H_2(G\backslash\underline{E}G)$ are infinite cyclic groups, $H_1(BC) \to H_1(BG)$ is trivial and $H_1(q)\colon H_1(BG) \xrightarrow{\cong} H_1(G\backslash\underline{E}G)$ is bijective. We also see that $H_2(BG) = 0$ if and only if r does not belong to $[F,F]$.

(iii) Let Hei be the three-dimensional discrete Heisenberg group which is the subgroup of $GL_3(\mathbb{Z})$ consisting of upper triangular matrices with 1 on the diag-
onals. Consider the $\mathbb{Z}/4$-action given by

$$
\begin{pmatrix} 1 & x & y \\ 0 & 1 & z \\ 0 & 0 & 1 \end{pmatrix} \mapsto \begin{pmatrix} 1 & -z & y - xz \\ 0 & 1 & x \\ 0 & 0 & 1 \end{pmatrix}.
$$

Then a key result in [53] is that $G\backslash\underline{E}G$ is homeomorphic to S^3 for $G = $
Hei $\rtimes \mathbb{Z}/4$.

(iv) A key result in [70, Corollary on page 8] implies that for $G = SL_3(\mathbb{Z})$ the
quotient space $G\backslash\underline{E}G$ is contractible.

7. Relevance and Applications of Classifying Spaces for Families

In this section we discuss some theoretical aspects which involve and rely on the
notion of a classifying space for a family of subgroups.

7.1. Baum-Connes Conjecture

Let G be a locally compact second countable Hausdorff group. Using the equivariant KK-theory due to Kasparov one can assign to a $\mathcal{COM}$-numerable G-space X its
equivariant K-theory $K_n^G(X)$. Let $C_r^*(G)$ be the reduced group C^*-algebra associated to G. The goal of the Baum-Connes Conjecture is to compute the topological
K-theory $K_p(C_r^*(G))$. The following formulation is taken from [7, Conjecture 3.15].

Conjecture 7.1 (Baum-Connes Conjecture). *The assembly map defined by taking
the equivariant index*

$$
\mathrm{asmb}\colon K_n^G(\underline{J}G) \xrightarrow{\cong} K_n(C_r^*(G))
$$

is bijective for all $n \in \mathbb{Z}$.

More information about this conjecture and its relation and application to
other conjectures and problems can be found for instance in [7], [34], [57], [63],
[80].

7.2. Farrell-Jones Conjecture

Let G be a discrete group. Let R be a associative ring with unit. One can construct
a G-homology theory $\mathcal{H}_*^G(X;\mathbf{K})$ graded over the integers and defined for G-CW-
complexes X such that for any subgroup $H \subseteq G$ the abelian group $\mathcal{H}_n^G(G/H;\mathbf{K})$
is isomorphic to the algebraic K-groups $K_n(RH)$ for $n \in \mathbb{Z}$. If R comes with an
involution of rings, one can also construct a G-homology theory $\mathcal{H}_*^G(X;\mathbf{L}^{\langle-\infty\rangle})$
graded over the integers and defined for G-CW-complexes X such that for any
subgroup $H \subseteq G$ the abelian group $\mathcal{H}_n^G(G/H;\mathbf{L}^{\langle-\infty\rangle})$ is isomorphic to the algebraic L-groups $L_n^{-\infty}(RH)$ for $n \in \mathbb{Z}$. Let $\mathcal{VCYC}$ be the family of virtually cyclic
subgroups of G. The goal of the Farrell-Jones Conjecture is to compute the al-
gebraic K-groups $K_n(RH)$ and the algebraic L-groups $L_n^{-\infty}(RG)$. The following
formulation is equivalent to the original one appearing in [28, 1.6 on page 257].

Conjecture 7.2 (Farrell-Jones Conjecture). *The assembly maps induced by the projection* $E_{\mathcal{VC}yc}(G) \to G/G$

$$\text{asmb}\colon \mathcal{H}_n^G(E_{\mathcal{VC}yc}(G), \mathbf{K}) \quad\to\quad \mathcal{H}_n^G(G/G, \mathbf{K}) = K_n(RG); \qquad (7.3)$$
$$\text{asmb}\colon \mathcal{H}_n^G(E_{\mathcal{VC}yc}(G), \mathbf{L}^{-\infty}) \quad\to\quad \mathcal{H}_n^G(G/G, \mathbf{L}^{-\infty}) = L_n^{-\infty}(RG), \qquad (7.4)$$

are bijective for all $n \in \mathbb{Z}$.

More information about this conjecture and its relation and application to other conjectures and problems can be found for instance in [28] and [57].

We mention that for a discrete group G one can formulate the Baum-Connes Conjecture in a similar fashion. Namely, one can also construct a G-homology theory $\mathcal{H}_*^G(X; \mathbf{K}^{\text{top}})$ graded over the integers and defined for G-CW-complexes X such that for any subgroup $H \subseteq G$ the abelian group $\mathcal{H}_n^G(G/H; \mathbf{K}^{\text{top}})$ is isomorphic to the topological K-groups $K_n(C_r^*(H))$ for $n \in \mathbb{Z}$ and the assembly map appearing in the Baum-Connes Conjecture can be identified with the map induced by the projection $\underline{J}G = \underline{E}G \to G/G$ (see [22], [31]). If the ring R is regular and contains $\mathbb{Q}$ as subring, then one can replace in the Farrell-Jones Conjecture 7.2 $E_{\mathcal{VC}yc}(G)$ by $\underline{E}G$ but this is not possible for arbitrary rings such as $R = \mathbb{Z}$. This comes from the appearance of Nil-terms in the Bass-Heller-Swan decomposition which do not occur in the context of the topological K-theory of reduced C^*-algebras.

Both the Baum-Connes Conjecture 7.1 and the Farrell-Jones Conjecture 7.2 allow to reduce the computation of certain K-and L-groups of the group ring or the reduced C^*-algebra of a group G to the computation of certain G-homology theories applied to $\underline{J}G$, $\underline{E}G$ or $E_{\mathcal{VC}yc}(G)$. Hence it is important to find good models for these spaces or to make predictions about their dimension or whether they are finite or of finite type.

7.3. Completion Theorem

Let G be a discrete group. For a proper finite G-CW-complex let $K_G^*(X)$ be its equivariant K-theory defined in terms of equivariant finite dimensional complex vector bundles over X (see [56, Theorem 3.2]). It is a G-cohomology theory with a multiplicative structure. Assume that $\underline{E}G$ has a finite G-CW-model. Let $I \subseteq K_G^0(\underline{E}G)$ be the augmentation ideal, i.e. the kernel of the map $K^0(\underline{E}G) \to \mathbb{Z}$ sending the class of an equivariant complex vector bundle to its complex dimension. Let $K_G^*(\underline{E}G)\widehat{_I}$ be the I-adic completion of $K_G^*(\underline{E}G)$ and let $K^*(BG)$ be the topological K-theory of BG.

Theorem 7.5 (Completion Theorem for discrete groups). *Let G be a discrete group such that there exists a finite model for $\underline{E}G$. Then there is a canonical isomorphism*

$$K^*(BG) \xrightarrow{\cong} K_G^*(\underline{E}G)\widehat{_I}.$$

This result is proved in [56, Theorem 4.4], where a more general statement is given provided that there is a finite dimensional model for $\underline{E}G$ and an upper bound on the orders of finite subgroups of G. In the case where G is finite, Theorem 7.5 reduces to the Completion Theorem due to Atiyah and Segal [3], [4]. A Cocompletion Theorem for equivariant K-homology will appear in [38].

7.4. Classifying Spaces for Equivariant Bundles

In [55] the equivariant K-theory for finite proper G-CW-complexes appearing in Subsection 7.3 above is extended to arbitrary proper G-CW-complexes (including the multiplicative structure) using Γ-spaces in the sense of Segal and involving classifying spaces for equivariant vector bundles. These classifying spaces for equivariant vector bundles are again classifying spaces of certain Lie groups and certain families (see [78, Section 8 and 9 in Chapter I], [56, Lemma 2.4]).

7.5. Equivariant Homology and Cohomology

Classifying spaces for families play a role in computations of equivariant homology and cohomology for compact Lie groups such as equivariant bordism as explained in [77, Chapter 7], [78, Chapter III]. Rational computations of equivariant (co-)-homology groups are possible in general using Chern characters for discrete groups and proper G-CW-complexes (see [50], [51], [52]).

8. Computations using Classifying Spaces for Families

In this section we discuss some computations which involve and rely on the notion of a classifying space for a family of subgroups. These computations are possible since one understands in the cases of interest the geometry of $\underline{E}G$ and $G\backslash\underline{E}G$. We focus on the case described in Subsection 4.11, namely of a discrete group G satisfying the conditions (M) and (NM). Let $s\colon EG \to \underline{E}G$ be the up to G-homotopy unique G-map. Denote by $j_i\colon M_i \to G$ the inclusion.

8.1. Group Homology

We begin with the group homology $H_n(BG)$ (with integer coefficients). Let $\widetilde{H}_p(X)$ be the reduced homology, i.e. the kernel of the map $H_n(X) \to H_n(\{\text{pt.}\})$ induced by the projection $X \to \{\text{pt.}\}$. The Mayer-Vietoris sequence applied to the pushout, which is obtained from the G-pushout (4.17) by dividing out the G-action, yields the long Mayer-Vietoris sequence

$$\ldots \to H_{p+1}(G\backslash\underline{E}G)) \xrightarrow{\partial_{p+1}} \bigoplus_{i\in I} \widetilde{H}_p(BM_i) \xrightarrow{\oplus_{i\in I} H_p(Bj_i)} H_p(BG)$$

$$\xrightarrow{H_p(G\backslash s)} H_p(G\backslash\underline{E}G) \xrightarrow{\partial_p} \ldots \quad (8.1)$$

In particular we obtain an isomorphism for $p \geq \dim(\underline{E}G) + 2$

$$\bigoplus_{i\in I} H_p(Bj_i)\colon \bigoplus_{i\in I} \widetilde{H}_p(BM_i) \xrightarrow{\cong} H_p(BG). \quad (8.2)$$

This example and the forthcoming ones show why it is important to get upper bounds on the dimension of $\underline{E}G$ and to understand the quotient space $G\backslash\underline{E}G$. For

Fuchsian groups and for one-relator groups we have $\dim(G\backslash \underline{E}G) \leq 2$ and it is easy to compute the homology of $G\backslash \underline{E}G$ in this case as explained in Section 6.

8.2. Topological K-Theory of Group C^*-Algebras

Analogously one can compute the source of the assembly map appearing in the Baum-Connes Conjecture 7.1. Namely, the Mayer-Vietoris sequence associated to the G-pushout (4.17) and the one associated to its quotient under the G-action look like

$$\ldots \to K_{p+1}^G(\underline{E}G) \to \bigoplus_{i\in I} K_p^G(G \times_{M_i} EM_i)$$

$$\to \left(\bigoplus_{i\in I} K_p^G(G/M_i) \right) \bigoplus K_p^G(EG) \to K_p^G(\underline{E}G) \to \ldots \quad (8.3)$$

and

$$\ldots \to K_{p+1}(G\backslash \underline{E}G) \to \bigoplus_{i\in I} K_p(BM_i)$$

$$\to \left(\bigoplus_{i\in I} K_p(\{\text{pt.}\}) \right) \bigoplus K_p(BG) \to \bigoplus_{i\in I} K_p(G\backslash \underline{E}G) \to \ldots \quad (8.4)$$

Notice that for a free G-CW-complex X there is a canonical isomorphism $K_p^G(X) \cong K_p(G\backslash X)$. We can splice these sequences together and obtain the long exact sequence

$$\ldots \to K_{p+1}(G\backslash \underline{E}G) \to \bigoplus_{i\in I} K_p^G(G/M_i) \to \bigoplus_{i\in I} K_p(\{\text{pt.}\}) \bigoplus K_p^G(\underline{E}G)$$

$$\to K_p(G\backslash \underline{E}G) \to \ldots \quad (8.5)$$

There are identifications of $K_0^G(G/M_i)$ with the complex representation ring $R_{\mathbb{C}}(M_i)$ of the finite group M_i and of $K_0(\{\text{pt.}\})$ with $\mathbb{Z}$. Under these identification the map $K_0^G(G/M_i) \to K_0(\{\text{pt.}\})$ becomes the split surjective map $\epsilon \colon R_{\mathbb{C}}(M_i) \to \mathbb{Z}$ which sends the class of a complex M_i-representation V to the complex dimension of $\mathbb{C} \otimes_{\mathbb{C}[M_i]} V$. The kernel of this map is denoted by $\widetilde{R}_{\mathbb{C}}(M_i)$. The groups $K_1^G(G/M_i)$ and $K_1(\{\text{pt.}\})$ vanish. The abelian group $R_{\mathbb{C}}(M_i)$ and hence also $\widetilde{R}_{\mathbb{C}}(M_i)$ are finitely generated free abelian groups. If $\mathbb{Z} \subseteq \Lambda \subseteq \mathbb{Q}$ is ring such that the order of any finite subgroup of G is invertible in Λ, then the map

$$\Lambda \otimes_{\mathbb{Z}} K_p^G(s) \colon \Lambda \otimes_{\mathbb{Z}} K_p^G(EG) \to \Lambda \otimes_{\mathbb{Z}} K_p(G\backslash \underline{E}G)$$

is an isomorphism for all $p \in \mathbb{Z}$ [58, Lemma 2.8 (a)]. Hence the long exact sequence (8.5) splits after applying $\Lambda \otimes_{\mathbb{Z}} -$. We conclude from the long exact sequence (8.5) since the representation ring of a finite group is torsionfree.

Theorem 8.6. *Let G be a discrete group which satisfies the conditions (M) and (NM) appearing in Subsection 4.11. Suppose that the Baum-Connes Conjecture 7.1 is true for G. Let $\{(M_i) \mid i \in I\}$ be the set of conjugacy classes of maximal finite subgroups of G. Then there is an isomorphism*

$$K_1(C_r^*(G)) \xrightarrow{\cong} K_1(G \backslash \underline{E}G)$$

and a short exact sequence

$$0 \to \bigoplus_{i \in I} \widetilde{R}_{\mathbb{C}}(M_i) \to K_0(C_r^*(G)) \to K_0(G \backslash \underline{E}G) \to 0,$$

which splits if we invert the orders of all finite subgroups of G.

8.3. Algebraic K-and L-Theory of Group Rings

Suppose that G satisfies the Farrell-Jones Conjecture 7.2. Then the computation of the relevant groups $K_n(RG)$ or $L_n^{\langle -\infty \rangle}(RG)$ respectively is equivalent to the computation of $\mathcal{H}_n^G(E_{\mathcal{VCYC}}(G), \mathbf{K})$ or $\mathcal{H}_n^G(E_{\mathcal{VCYC}}(G), \mathbf{L}^{-\infty})$ respectively. The following result is due to Bartels [5]. Recall that $\underline{E}G$ is the same as $E_{\mathcal{FIN}}(G)$.

Theorem 8.7. (i) *For every group G, every ring R and every $n \in \mathbb{Z}$ the up to G-homotopy unique G-map $f \colon E_{\mathcal{FIN}}(G) \to E_{\mathcal{VCYC}}(G)$ induces a split injection*

$$H_n^G(f; \mathbf{K}_R) \colon H_n^G(E_{\mathcal{FIN}}(G); \mathbf{K}_R) \to H_n^G(E_{\mathcal{VCYC}}(G); \mathbf{K}_R).$$

 (ii) *Suppose R is such that $K_{-i}(RV) = 0$ for all virtually cyclic subgroups V of G and for sufficiently large i (for example $R = \mathbb{Z}$ will do). Then we get a split injection*

$$H_n^G(f; \mathbf{L}_R^{\langle -\infty \rangle}) \colon H_n^G(E_{\mathcal{FIN}}(G); \mathbf{L}_R^{\langle -\infty \rangle}) \to H_n^G(E_{\mathcal{VCYC}}(G); \mathbf{L}_R^{\langle -\infty \rangle}).$$

It remains to compute $H_n^G(E_{\mathcal{FIN}}(G); \mathbf{K})$ and $H_n^G(E_{\mathcal{VCYC}}(G), E_{\mathcal{FIN}}(G); \mathbf{K})$, if we arrange f to be a G-cofibration and think of $E_{\mathcal{FIN}}(G)$ as a G-CW-subcomplex of $E_{\mathcal{VCYC}}(G)$. Namely, we get from the Farrell-Jones Conjecture 7.2 and Theorem 8.7 an isomorphism

$$H_n^G(E_{\mathcal{FIN}}(G); \mathbf{K}) \bigoplus H_n^G(E_{\mathcal{VCYC}}(G), E_{\mathcal{FIN}}(G); \mathbf{K}) \xrightarrow{\cong} K_n(RG).$$

The analogous statement holds for $\mathbf{L}_R^{\langle -\infty \rangle}$), provided R satisfies the conditions appearing in Theorem 8.7 (ii).

Analogously to Theorem 8.6 one obtains

Theorem 8.8. *Let G be a discrete group which satisfies the conditions (M) and (NM) appearing in Subsection 4.11. Let $\{(M_i) \mid i \in I\}$ be the set of conjugacy classes of maximal finite subgroups of G. Then*

 (i) *There is a long exact sequence*

$$\ldots \to H_{p+1}(G \backslash E_{\mathcal{FIN}}(G); \mathbf{K}(R)) \to \bigoplus_{i \in I} K_p(R[M_i])$$

$$\to \bigoplus_{i \in I} K_p(R) \bigoplus H_p^G(E_{\mathcal{FIN}}(G); \mathbf{K}_\mathbf{R}) \to \bigoplus_{i \in I} H_p(G \backslash E_{\mathcal{FIN}}(G); \mathbf{K}(R)) \to \ldots$$

and analogously for $\mathbf{L}_R^{\langle -\infty \rangle}$.

(ii) For $R = \mathbb{Z}$ there are isomorphisms

$$\bigoplus_{i \in I} \mathrm{Wh}_n(M_i) \bigoplus H_n^G(E_{\mathcal{FIN}}(G), E_{\mathcal{VCYC}}(G); \mathbf{K}_{\mathbb{Z}}) \xrightarrow{\cong} \mathrm{Wh}_n(G).$$

Remark 8.9. These results about groups satisfying conditions (M) and (NM) are extended in [53] to groups which map surjectively to groups satisfying conditions (M) and (NM) with special focus on the semi-direct product of the discrete three-dimensional Heisenberg group with $\mathbb{Z}/4$.

Remark 8.10. In [70] a special model for $\underline{E}SL_3(\mathbb{Z})$ is presented which allows to compute the integral group homology. Information about the algebraic K-theory of $SL_3(\mathbb{Z})$ can be found in [72, Chapter 7], [79].

The analysis of the other term $H_n^G(E_{\mathcal{VCYC}}(G), E_{\mathcal{FIN}}(G); \mathbf{K})$ simplifies considerably under certain assumptions on G.

Theorem 8.11 (On the structure of $E_{\mathcal{VCYC}}(G)$). *Suppose that G satisfies the following conditions:*

- *Every infinite cyclic subgroup $C \subseteq G$ has finite index in its centralizer $C_G C$.*
- *There is an upper bound on the orders of finite subgroups.*

(Each word-hyperbolic group satisfies these two conditions.) Then

(i) *For an infinite virtually cyclic subgroup $V \subseteq G$ define*

$$V_{\max} = \bigcup \{ N_G C \mid C \subset V \text{ infinite cyclic normal} \}.$$

Then

  (a) *$V_{\max}$ is an infinite virtually cyclic subgroup of G and contains V.*

  (b) *If $V \subseteq W \subseteq G$ are infinite virtually cyclic subgroups of G, then $V_{\max} = W_{\max}$.*

  (c) *Each infinite virtually cyclic subgroup V is contained in a unique maximal infinite virtually cyclic subgroup, namely $V_{\max}$, and $N_G V_{\max} = V_{\max}$.*

(ii) *Let $\{V_i \mid i \in I\}$ be a complete system of representatives of conjugacy classes of maximal infinite virtually cyclic subgroups. Then there exists a G-pushout*

$$\coprod_{i \in I} G \times_{V_i} E_{\mathcal{FIN}}(V_i) \longrightarrow E_{\mathcal{FIN}}(G)$$

$$\mathrm{pr} \downarrow \qquad\qquad\qquad\qquad \downarrow$$

$$\coprod_{i \in I} G/V_i \longrightarrow E_{\mathcal{VCYC}}(G)$$

whose upper horizontal arrow is an inclusion of G-CW-complexes.

(iii) *There are natural isomorphisms*

$$\bigoplus_{i \in I} H_n^{V_i}(E_{\mathcal{VCYC}}(V_i), E_{\mathcal{FIN}}(V_i); \mathbf{K}_R) \xrightarrow{\cong} H_n^G(E_{\mathcal{VCYC}}(G), E_{\mathcal{FIN}}(G); \mathbf{K}_R)$$

$$\bigoplus_{i \in I} H_n^{V_i}(E_{\mathcal{VCYC}}(V_i), E_{\mathcal{FIN}}(V_i); \mathbf{L}_R^{\langle -\infty \rangle}) \xrightarrow{\cong} H_n^G(E_{\mathcal{VCYC}}(G), E_{\mathcal{FIN}}(G); \mathbf{L}_R^{\langle -\infty \rangle}).$$

Proof. Each word-hyperbolic group G satisfies these two conditions by [13, Theorem 3.2 in III.Γ.3 on page 459 and Corollary 3.10 in III.Γ.3 on page 462].

(i) Let V be an infinite virtually cyclic subgroup $V \subseteq G$. Fix a normal infinite cyclic subgroup $C \subseteq V$. Let b be a common multiple of the orders of finite subgroups of G. Put $d := b \cdot b!$. Let e be the index of the infinite cyclic group $dC = \{d \cdot x \mid x \in C\}$ in its centralizer $C_G dC$. Let $D \subset dC$ be any non-trivial subgroup. Obviously $dC \subseteq C_G D$. We want to show

$$[C_G D : dC] \leq b \cdot e^2. \tag{8.12}$$

Since D is central in $C_G D$ and $C_G D$ is virtually cyclic and hence $|C_G D/D| < \infty$, the spectral sequence associated to the extension $1 \to D \to C_G D \to C_G D/D \to 1$ implies that the map $D = H_1(D) \to H_1(C_G D)$ is injective and has finite cokernel. In particular the quotient of $H_1(C_G D)$ by its torsion subgroup $H_1(C_G D)/\text{tors}$ is an infinite cyclic group. Let $p_{C_G D}: C_G D \to H_1(C_G D)/\text{tors}$ be the canonical epimorphism. Its kernel is a finite normal subgroup. The following diagram commutes and has exact rows

$$
\begin{array}{ccccccccc}
1 & \longrightarrow & \ker(p_C) & \longrightarrow & C_G C & \xrightarrow{p_C} & H_1(C_G C)/\text{tors} & \longrightarrow & 1 \\
& & \downarrow & & \downarrow & & \downarrow & & \\
1 & \longrightarrow & \ker(p_D) & \longrightarrow & C_G D & \xrightarrow{p_D} & H_1(C_G D)/\text{tors} & \longrightarrow & 1
\end{array}
$$

where the vertical maps are induced by the inclusions $C_G C \subseteq C_G D$. All vertical maps are injections with finite cokernel. Fix elements $z_C \in C_G C$ and $z_D \in C_G D$ such that $p_C(z_C)$ and $p_D(z_D)$ are generators. Choose $l \in \mathbb{Z}$ such that $p_C(z_C)$ is send to $l \cdot p_D(z_D)$. Then there is $k \in \ker(p_D)$ with $z_C = k \cdot z_D^l$. The order of $\ker(p_D)$ divides b by assumption. If $\phi: \ker(p_D) \to \ker(p_D)$ is any automorphism, then $\phi^{b!} = \text{id}$. This implies for any element $k \in \ker(p_D)$ that

$$\prod_{i=0}^{d-1} \phi^i(k) = \left(\prod_{i=0}^{b!-1} \phi^i(k)\right)^b = 1.$$

Hence we get in $C_G D$ if ϕ is conjugation with z_D^l

$$z_C^d = (k \cdot z_D^l)^d = \prod_{i=0}^{d-1} \phi^i(k) \cdot z_D^{dl} = z_D^{dl}.$$

Obviously $z_D \in C_G dC$ since $z_C^d = z_D^{dl}$ generates dC. Hence z_D^e lies in dC and we get $z_D^e = z_C^{df}$ for some integer f. This implies $z_D^e = z_D^{ldf}$ and hence that l divides e. We conclude that the cokernel of the map $H_1(C_G C)/\text{tors} \to H_1(C_G D)/\text{tors}$ is bounded by e. Hence the index $[C_G D : C_G C]$ is bounded by $b \cdot e$ since the order of $\ker(p_D)$ divides b. Since $dC \subseteq C_G C \subseteq C_G dC \subseteq C_G D$ holds, equation 8.12 follows.

Next we show that there is a normal infinite cyclic subgroup $C_0 \subseteq V$ such that $V_{\max} = N_G C_0$ holds. If C' and C'' are infinite cyclic normal subgroups of V, then both $C_G C'$ and $C_G C''$ are contained in $C_G(C' \cap C'')$ and $C' \cap C''$ is again an infinite cyclic normal subgroup. Hence there is a sequence of normal infinite cyclic subgroups of V

$$dC \supseteq C_1 \supseteq C_2 \supseteq C_3 \supseteq \ldots$$

which yields a sequence $C_G dC \subseteq C_G C_1 \subseteq C_G C_2 \subset \ldots$ satisfying

$$\bigcup \{C_G C_n \mid n \geq 1\} = \bigcup \{C_G C \mid C \subset V \text{ infinite cyclic normal}\}.$$

Because of 8.12 there is an upper bound on $[C_G C_n : C_G dC]$ which is independent of n. Hence there is an index n_0 with

$$C_G C_{n_0} = \bigcup \{C_G C \mid C \subset V \text{ infinite cyclic normal}\}.$$

For any infinite cyclic subgroup $C \subseteq G$ the index of $C_G C$ in $N_G C$ is 1 or 2. Hence there is an index n_1 with

$$N_G C_{n_1} = \bigcup \{N_G C \mid C \subset V \text{ infinite cyclic normal}\}.$$

Thus we have shown the existence of a normal infinite cyclic subgroup $C \subseteq V$ with $V_{\max} = C$. Now assertion (i)a follows.

We conclude assertion (i)b from the fact that for an inclusion of infinite virtually cyclic group $V \subseteq W$ there exists a normal infinite cyclic subgroup $C \subseteq W$ such that $C \subseteq V$ holds. Assertion (i)c is now obviously true. This finishes the proof of assertion (i).

(ii) Construct a G-pushout

$$\begin{array}{ccc}
\coprod_{i \in I} G \times_{V_i} E_{\mathcal{FIN}}(V_i) & \xrightarrow{\;j\;} & E_{\mathcal{FIN}}(G) \\
{\scriptstyle \mathrm{pr}} \downarrow & & \downarrow \\
\coprod_{i \in I} G/V_i & \longrightarrow & X
\end{array}$$

with j an inclusion of G-CW-complexes. Obviously X is a G-CW-complex whose isotropy groups are virtually cyclic. It remains to prove for virtually cyclic $H \subseteq G$ that X^H is contractible.

Given a V_i-space Y and a subgroup $H \subseteq G$, there is after a choice of a map of sets $s \colon G/V_i \to G$, whose composition with the projection $G \to G/V_i$ is the identity, a G-homeomorphism

$$\coprod_{\substack{w \in G/V_i \\ s(w)^{-1} H s(w) \subseteq V_i}} Y^{s(w)^{-1} H s(w)} \xrightarrow{\;\cong\;} (G \times_{V_i} Y)^H, \qquad (8.13)$$

which sends $y \in Y^{s(w)^{-1} H s(w)}$ to $(s(w), y)$.

If H is infinite, the H-fixed point set of the upper right and upper left corner is empty and of the lower left corner is the one-point space because of assertion (i)c and equation 8.13. Hence X^H is a point for an infinite virtually cyclic subgroup $H \subseteq G$.

If H is finite, one checks using equation 8.13 that the left vertical map induces a homotopy equivalence on the H-fixed point set. Since the upper horizontal arrow induces a cofibration on the H-fixed point set, the right vertical arrow induces a homotopy equivalence on the H-fixed point sets. Hence X^H is contractible for finite $H \subseteq G$. This shows that X is a model for $E_{\mathcal{VCYC}}(G)$.

(iii) follows from excision and the induction structure. This finishes the proof of Theorem 8.11. $\qquad\square$

Theorem 8.11 has also been proved by Daniel Juan-Pineda and Ian Leary [39] under the stronger condition that every infinite subgroup of G, which is not virtually cyclic, contains a non-abelian free subgroup. The case, where G is the fundamental group of a closed Riemannian manifold with negative sectional curvature is treated in [6].

Remark 8.14. In Theorem 8.11 the terms $H_n^{V_i}(E_{\mathcal{VCYC}}(V_i), E_{\mathcal{FIN}}(V_i); \mathbf{K}_R)$ and $H_n^{V_i}(E_{\mathcal{VCYC}}(V_i), E_{\mathcal{FIN}}(V_i); \mathbf{L}_R^{\langle -\infty \rangle})$ occur. They also appear in the direct sum decomposition

$$K_n(RV_i) \cong H_n^{V_i}(E_{\mathcal{FIN}}(V_i); \mathbf{K}_R) \bigoplus H_n^{V_i}(E_{\mathcal{VCYC}}(V_i), E_{\mathcal{FIN}}(V_i); \mathbf{K}_R);$$

$$L_n(RV_i) \cong H_n^{V_i}(E_{\mathcal{FIN}}(V_i); \mathbf{L}_R^{\langle -\infty \rangle}) \bigoplus H_n^{V_i}(E_{\mathcal{VCYC}}(V_i), E_{\mathcal{FIN}}(V_i); \mathbf{L}_R^{\langle -\infty \rangle}).$$

They can be analysed further and contain information about and are build from the Nil and UNIL-terms in algebraic K-theory and L-theory of the infinite virtually cyclic group V_i. They vanish for L-theory after inverting 2 by results of [19]. For $R = \mathbb{Z}$ they vanishes rationally for algebraic K-theory by results of [45].

References

[1] H. Abels. A universal proper G-space. *Math. Z.*, 159(2):143–158, 1978.

[2] A. Adem and Y. Ruan. Twisted orbifold K-theory. preprint, to appear in Comm. Math. Phys., 2001.

[3] M. F. Atiyah. Characters and cohomology of finite groups. *Inst. Hautes Études Sci. Publ. Math.*, (9):23–64, 1961.

[4] M. F. Atiyah and G. B. Segal. Equivariant K-theory and completion. *J. Differential Geometry*, 3:1–18, 1969.

[5] A. Bartels. On the left hand side of the assembly map in algebraic K-theory. Preprint-reihe SFB 478 — Geometrische Strukturen in der Mathematik, Heft 257, Münster, 2003.

[6] A. Bartels and H. Reich. On the Farrell-Jones conjecture for higher algebraic K-theory. Preprintreihe SFB 478 — Geometrische Strukturen in der Mathematik, Heft 282, Münster, 2003.

[7] P. Baum, A. Connes, and N. Higson. Classifying space for proper actions and K-theory of group C^*-algebras. In C^*-*algebras: 1943–1993 (San Antonio, TX, 1993)*, pages 240–291. Amer. Math. Soc., Providence, RI, 1994.

[8] G. Baumslag, E. Dyer, and A. Heller. The topology of discrete groups. *J. Pure Appl. Algebra*, 16(1):1–47, 1980.

[9] M. Bestvina and N. Brady. Morse theory and finiteness properties of groups. *Invent. Math.*, 129(3):445–470, 1997.

[10] A. Borel and J.-P. Serre. Corners and arithmetic groups. *Comment. Math. Helv.*, 48:436–491, 1973. Avec un appendice: Arrondissement des variétés à coins, par A. Douady et L. Hérault.

[11] N. Brady, I. J. Leary, and B. E. A. Nucinkis. On algebraic and geometric dimensions for groups with torsion. *J. London Math. Soc. (2)*, 64(2):489–500, 2001.

[12] G. E. Bredon. *Introduction to compact transformation groups*. Academic Press, New York, 1972. Pure and Applied Mathematics, Vol. 46.

[13] M. R. Bridson and A. Haefliger. *Metric spaces of non-positive curvature*. Springer-Verlag, Berlin, 1999. Die Grundlehren der mathematischen Wissenschaften, Band 319.

[14] M. R. Bridson and K. Vogtmann. The symmetries of outer space. *Duke Math. J.*, 106(2):391–409, 2001.

[15] K. Brown. Groups of virtually finite dimension. In *Proceedings* "Homological group theory", *editor: Wall, C.T.C., LMS Lecture Notes Series 36*, pages 27–70. Cambridge University Press, 1979.

[16] K. S. Brown. *Cohomology of groups*, volume 87 of *Graduate Texts in Mathematics*. Springer-Verlag, New York, 1982.

[17] K. S. Brown. *Buildings*. Springer-Verlag, New York, 1998. Reprint of the 1989 original.

[18] K. S. Brown and R. Geoghegan. An infinite-dimensional torsion-free fp_∞ group. *Invent. Math.*, 77(2):367–381, 1984.

[19] S. E. Cappell. Manifolds with fundamental group a generalized free product. I. *Bull. Amer. Math. Soc.*, 80:1193–1198, 1974.

[20] F. X. Connolly and T. Koźniewski. Finiteness properties of classifying spaces of proper γ-actions. *J. Pure Appl. Algebra*, 41(1):17–36, 1986.

[21] M. Culler and K. Vogtmann. Moduli of graphs and automorphisms of free groups. *Invent. Math.*, 84(1):91–119, 1986.

[22] J. F. Davis and W. Lück. Spaces over a category and assembly maps in isomorphism conjectures in K- and L-theory. *K-Theory*, 15(3):201–252, 1998.

[23] M. Davis and I. Leary. Some examples of discrete group actions on aspherical manifolds. In T. Farrell, L. Göttsche, and W. Lück, editors, *Topology of high-dimensional manifolds*, pages 139–150. Abdus Salam International Centre for Theoretical Physics, Trieste, World Scientific, 2003. Proceedings of a conference in Trieste in June 2001.

[24] K. Dekimpe and P. Igodt. Polycyclic-by-finite groups admit a bounded-degree polynomial structure. *Invent. Math.*, 129(1):121–140, 1997.

[25] W. Dicks and M. J. Dunwoody. *Groups acting on graphs*. Cambridge University Press, Cambridge, 1989.

[26] A. Dold. Partitions of unity in the theory of fibrations. *Ann. of Math. (2)*, 78:223–255, 1963.

[27] M. J. Dunwoody. Accessibility and groups of cohomological dimension one. *Proc. London Math. Soc. (3)*, 38(2):193–215, 1979.

[28] F. T. Farrell and L. E. Jones. Isomorphism conjectures in algebraic K-theory. *J. Amer. Math. Soc.*, 6(2):249–297, 1993.

[29] É. Ghys and P. de la Harpe, editors. *Sur les groupes hyperboliques d'après Mikhael Gromov.* Birkhäuser Boston Inc., Boston, MA, 1990. Papers from the Swiss Seminar on Hyperbolic Groups held in Bern, 1988.

[30] M. Gromov. Hyperbolic groups. In *Essays in group theory*, pages 75–263. Springer-Verlag, New York, 1987.

[31] I. Hambleton and E. K. Pedersen. Identifying assembly maps in K- and L-theory. Preprint, to appear in Math. Ann., 2003.

[32] J. L. Harer. The virtual cohomological dimension of the mapping class group of an orientable surface. *Invent. Math.*, 84(1):157–176, 1986.

[33] E. Hewitt and K. A. Ross. *Abstract harmonic analysis. Vol. I.* Springer-Verlag, Berlin, second edition, 1979. Structure of topological groups, integration theory, group representations.

[34] N. Higson. The Baum-Connes conjecture. In *Proceedings of the International Congress of Mathematicians, Vol. II (Berlin, 1998)*, pages 637–646 (electronic), 1998.

[35] D. Husemoller. *Fibre bundles.* McGraw-Hill Book Co., New York, 1966.

[36] S. Illman. Existence and uniqueness of equivariant triangulations of smooth proper G-manifolds with some applications to equivariant Whitehead torsion. *J. Reine Angew. Math.*, 524:129–183, 2000.

[37] N. V. Ivanov. Mapping class groups. In *Handbook of geometric topology*, pages 523–633. North-Holland, Amsterdam, 2002.

[38] M. Joachim and W. Lück. The topological (co)homology of classifying spaces of discrete groups. in preparation, 2004.

[39] D. Juna-Pineda and I. Leary. On classifying spaces for the family of virtually cyclic subgroups. in preparation, 2003.

[40] D. M. Kan and W. P. Thurston. Every connected space has the homology of a $K(\pi, 1)$. *Topology*, 15(3):253–258, 1976.

[41] A. Karrass, A. Pietrowski, and D. Solitar. Finite and infinite cyclic extensions of free groups. *J. Austral. Math. Soc.*, 16:458–466, 1973. Collection of articles dedicated to the memory of Hanna Neumann, IV.

[42] S. P. Kerckhoff. The Nielsen realization problem. *Ann. of Math. (2)*, 117(2):235–265, 1983.

[43] P. H. Kropholler and G. Mislin. Groups acting on finite-dimensional spaces with finite stabilizers. *Comment. Math. Helv.*, 73(1):122–136, 1998.

[44] S. Krstić and K. Vogtmann. Equivariant outer space and automorphisms of free-by-finite groups. *Comment. Math. Helv.*, 68(2):216–262, 1993.

[45] A. O. Kuku and G. Tang. Higher K-theory of group-rings of virtually infinite cyclic groups. *Math. Ann.*, 325(4):711–726, 2003.

[46] I. J. Leary and B. E. A. Nucinkis. Every CW-complex is a classifying space for proper bundles. *Topology*, 40(3):539–550, 2001.

[47] I. J. Leary and B. E. A. Nucinkis. Some groups of type VF. *Invent. Math.*, 151(1):135–165, 2003.

320   Wolfgang Lück

[48] W. Lück. *Transformation groups and algebraic K-theory.* Springer-Verlag, Berlin, 1989. Mathematica Gottingensis.

[49] W. Lück. The type of the classifying space for a family of subgroups. *J. Pure Appl. Algebra*, 149(2):177–203, 2000.

[50] W. Lück. Chern characters for proper equivariant homology theories and applications to K- and L-theory. *J. Reine Angew. Math.*, 543:193–234, 2002.

[51] W. Lück. The relation between the Baum-Connes conjecture and the trace conjecture. *Invent. Math.*, 149(1):123–152, 2002.

[52] W. Lück. Equivariant cohomological Chern characters. Preprintreihe SFB 478 — Geometrische Strukturen in der Mathematik, Heft 309, Münster, arXiv:math.GT/0401047, 2004.

[53] W. Lück. K-and L-theory of the semi-direct product discrete three-dimensional Heisenberg group by $\mathbb{Z}/4$. in preparation, 2004.

[54] W. Lück and D. Meintrup. On the universal space for group actions with compact isotropy. In *Geometry and topology: Aarhus (1998)*, pages 293–305. Amer. Math. Soc., Providence, RI, 2000.

[55] W. Lück and B. Oliver. Chern characters for the equivariant K-theory of proper G-CW-complexes. In *Cohomological methods in homotopy theory (Bellaterra, 1998)*, pages 217–247. Birkhäuser, Basel, 2001.

[56] W. Lück and B. Oliver. The completion theorem in K-theory for proper actions of a discrete group. *Topology*, 40(3):585–616, 2001.

[57] W. Lück and H. Reich. The Baum-Connes and the Farrell-Jones conjectures in K- and L-theory. Preprintreihe SFB 478 — Geometrische Strukturen in der Mathematik, Heft 324, Münster, arXiv:math.GT/0402405, to appear in the K-theory-handbook, 2004.

[58] W. Lück and R. Stamm. Computations of K- and L-theory of cocompact planar groups. *K-Theory*, 21(3):249–292, 2000.

[59] R. C. Lyndon and P. E. Schupp. *Combinatorial group theory.* Springer-Verlag, Berlin, 1977. Ergebnisse der Mathematik und ihrer Grenzgebiete, Band 89.

[60] D. Meintrup. *On the Type of the Universal Space for a Family of Subgroups.* PhD thesis, Westfälische Wilhelms-Universität Münster, 2000.

[61] D. Meintrup and T. Schick. A model for the universal space for proper actions of a hyperbolic group. *New York J. Math.*, 8:1–7 (electronic), 2002.

[62] J. Milnor. On spaces having the homotopy type of cw-complex. *Trans. Amer. Math. Soc.*, 90:272–280, 1959.

[63] G. Mislin and A. Valette. *Proper group actions and the Baum-Connes Conjecture.* Advanced Courses in Mathematics CRM Barcelona. Birkhäuser, 2003.

[64] H. Miyazaki. The paracompactness of CW-complexes. *Tôhoku Math. J*, 4:309–313, 1952.

[65] P. S. Mostert. Local cross sections in locally compact groups. *Proc. Amer. Math. Soc.*, 4:645–649, 1953.

[66] B. A. Nucinkis. On dimensions in Bredon homology. preprint, 2003.

[67] B. E. A. Nucinkis. Is there an easy algebraic characterisation of universal proper G-spaces? *Manuscripta Math.*, 102(3):335–345, 2000.

[68] J.-P. Serre. Arithmetic groups. In *Homological group theory (Proc. Sympos., Durham, 1977)*, volume 36 of *London Math. Soc. Lecture Note Ser.*, pages 105–136. Cambridge Univ. Press, Cambridge, 1979.

[69] J.-P. Serre. *Trees*. Springer-Verlag, Berlin, 1980. Translated from the French by J. Stillwell.

[70] C. Soulé. The cohomology of $SL_3(\mathbf{Z})$. *Topology*, 17(1):1–22, 1978.

[71] J. R. Stallings. On torsion-free groups with infinitely many ends. *Ann. of Math. (2)*, 88:312–334, 1968.

[72] R. Stamm. *The K- and L-theory of certain discrete groups*. Ph. D. thesis, Universität Münster, 1999.

[73] N. E. Steenrod. A convenient category of topological spaces. *Michigan Math. J.*, 14:133–152, 1967.

[74] R. G. Swan. Groups of cohomological dimension one. *J. Algebra*, 12:585–610, 1969.

[75] J. Tits. On buildings and their applications. In *Proceedings of the International Congress of Mathematicians (Vancouver, B. C., 1974), Vol. 1*, pages 209–220. Canad. Math. Congress, Montreal, Que., 1975.

[76] J. Tits. Reductive groups over local fields. In *Automorphic forms, representations and L-functions (Proc. Sympos. Pure Math., Oregon State Univ., Corvallis, Ore., 1977), Part 1*, Proc. Sympos. Pure Math., XXXIII, pages 29–69. Amer. Math. Soc., Providence, R.I., 1979.

[77] T. tom Dieck. *Transformation groups and representation theory*. Springer-Verlag, Berlin, 1979.

[78] T. tom Dieck. *Transformation groups*. Walter de Gruyter & Co., Berlin, 1987.

[79] S. Upadhyay. Controlled algebraic K-theory of integral group ring of $SL(3, \mathbf{Z})$. *K-Theory*, 10(4):413–418, 1996.

[80] A. Valette. *Introduction to the Baum-Connes conjecture*. Birkhäuser Verlag, Basel, 2002. From notes taken by Indira Chatterji, With an appendix by Guido Mislin.

[81] K. Vogtmann. Automorphisms of free groups and outer space. To appear in the special issue of Geometriae Dedicata for the June, 2000 Haifa conference, 2003.

[82] C. T. C. Wall, editor. *Homological group theory*, volume 36 of *London Mathematical Society Lecture Note Series*, Cambridge, 1979. Cambridge University Press.

[83] T. White. Fixed points of finite groups of free group automorphisms. *Proc. Amer. Math. Soc.*, 118(3):681–688, 1993.

[84] G. W. Whitehead. *Elements of homotopy theory*. Springer-Verlag, New York, 1978.

[85] B. Wilking. On fundamental groups of manifolds of nonnegative curvature. *Differential Geom. Appl.*, 13(2):129–165, 2000.

Notation

Wolfgang Lück
Fachbereich Mathematik
Universität Münster
Einsteinstr. 62
48149 Münster
Germany
e-mail: lueck@math.uni-muenster.de
URL: www: http://www.math.uni-muenster.de/u/lueck/

Progress in Mathematics, Vol. 248, 323–362
© 2005 Birkhäuser Verlag Basel/Switzerland

Are Unitarizable Groups Amenable?

Gilles Pisier

Abstract. We give a new formulation of some of our recent results on the following problem: if all uniformly bounded representations on a discrete group G are similar to unitary ones, is the group amenable? In §5, we give a new proof of Haagerup's theorem that, on non-commutative free groups, there are Herz-Schur multipliers that are not coefficients of uniformly bounded representations. We actually prove a refinement of this result involving a generalization of the class of Herz-Schur multipliers, namely the class $M_d(G)$ which is formed of all the functions $f \colon G \to \mathbb{C}$ such that there are bounded functions $\xi_i \colon G \to B(H_i, H_{i-1})$ (H_i Hilbert) with $H_0 = \mathbb{C}$, $H_d = \mathbb{C}$ such that

$$f(t_1 t_2 \ldots t_d) = \xi_1(t_1)\xi_2(t_2)\ldots\xi_d(t_d). \qquad \forall\, t_i \in G$$

We prove that if G is a non-commutative free group, for any $d \geq 1$, we have

$$M_d(G) \neq M_{d+1}(G),$$

and hence there are elements of $M_d(G)$ which are not coefficients of uniformly bounded representations. In the case $d = 2$, Haagerup's theorem implies that $M_2(G) \neq M_4(G)$.

Mathematics Subject Classification (2000). 43A07, 43A22, 43A65, 47D03, 47L25, 47L55.

Keywords. Uniformly bounded representation, unitarizable, amenable, Herz–Schur multiplier, similarity problem.

Contents

Supported in part by NSF and by the Texas Advanced Research Program 010366-163.

0. Introduction

The starting point for this presentation is the following result proved in the particular case $G = \mathbb{Z}$ by Sz.-Nagy (1947).

Theorem 0.1 (Day, Dixmier 1950). *Let G be a locally compact group. If G is amenable, then every uniformly bounded (u.b. in short) representation $\pi\colon G \to B(H)$ (H Hilbert) is unitarizable. More precisely, if we define*

$$|\pi| = \sup\{\|\pi(t)\|_{B(H)} \mid t \in G\},$$

then, if $|\pi| < \infty$, there exists $S\colon H \to H$ invertible with $\|S\|\,\|S^{-1}\| \leq |\pi|^2$ such that $t \to S^{-1}\pi(t)S$ is a unitary representation.

Note. We say that $\pi\colon G \to B(H)$ (H Hilbert) is unitarizable if there exists $S\colon H \to H$ invertible such that $t \to S^{-1}\pi(t)S$ is a unitary representation. We will mostly restrict to discrete groups, but otherwise all representations $\pi\colon G \to B(H)$ are implicitly assumed continuous on G with respect to the strong operator topology on $B(H)$.

Definition 0.2. *We will say that a locally compact group G is unitarizable if every uniformly bounded (u.b. in short) representation $\pi\colon G \to B(H)$ is unitarizable.*

In his 1950 paper, Dixmier [19] asked two questions which can be rephrased as follows:

Q1: Is every G unitarizable?
Q2: If not, is it true that conversely unitarizable $\Rightarrow$ amenable?

In 1955, Ehrenpreis and Mautner answered Q1; they showed that $G = \mathrm{SL}_2(\mathbb{R})$ is not unitarizable. Their work was clarified and amplified in 1960 by Kunze-Stein [35]. See Remark 0.7 below for more recent work in this direction. Here of course $G = \mathrm{SL}_2(\mathbb{R})$ is viewed as a Lie group, but a fortiori the discrete group G_d underlying $\mathrm{SL}_2(\mathbb{R})$ fails to be unitarizable, and since every group is a quotient of a free group and "unitarizable" obviously passes to quotients, it follows (implicitly) that there is a non-unitarizable free group, from which it is easy to deduce (since unitarizable passes to subgroups, see Proposition 0.5 below) that $\mathbb{F}_2$ the free group with 2 generators is not unitarizable. In the 80's, many authors, notably Mantero-Zappa [46]–[47], Pytlik-Szwarc [70], Bożejko–Fendler [9], Bożejko [8], ..., and also Młotkowski [44], Szwarc [75]–[76], Wysoczański [81] (for free products of groups), produced explicit constructions of u.b. non-unitarizable representations on $\mathbb{F}_2$ or on $\mathbb{F}_\infty$ (free group with countably infinitely many generators), see [45] for a synthesis between the Italian approach and the Polish one. See also Valette's papers [78]–[79] for the viewpoint of groups acting on trees, (combining Pimsner [68] and [70]) and [32] for recent work on Coxeter groups.

This was partly motivated by the potential applications in Harmonic Analysis of the resulting explicit formulae (see e.g. [24] and [10]). For instance, if we denote by $|t|$ the length of an element in $\mathbb{F}_\infty$ (or in $\mathbb{F}_2$), they constructed an analytic family $(\pi_z)_{z \in D}$ indexed by the unit disc in $\mathbb{C}$ of u.b. representations such that, denoting by δ_e the unit basis vector of $\ell_2(G)$ at e (unit element), we have

$$\forall\, t \in G \qquad z^{|t|} = \langle \pi_z(t)\delta_e, \delta_e\rangle.$$

Thus the function $\varphi_z\colon t \to z^{|t|}$ is a coefficient of a u.b. representation on $G = \mathbb{F}_\infty$. However, it can be shown that for $z \notin \mathbb{R}$, the function φ_z is *not* the coefficient of any unitary representation, whence $\pi_z(\cdot)$ cannot be unitarizable. A similar analysis can be made for the so-called spherical functions.

Since unitarizable passes to subgroups (by "induction of representations", see Proposition 0.5 below) this implies

Corollary 0.3. *Any discrete group G containing $\mathbb{F}_2$ as a subgroup is not unitarizable.*

Remark 0.4. Thus if there is a discrete group G which is unitarizable but not amenable, this is a non-amenable group not containing $\mathbb{F}_2$. The existence of such groups remained a fundamental open problem for many years until Olshanskii [51]–[52] established it in 1980, using the solution by Adian–Novikov (see [1]) of the famous Burnside problem, and also Grigorchuk's cogrowth criterion ([25]). Later, Adian (see [2]) showed that the Burnside group $B(m,n)$ (defined as the universal group with m generators such that every group element x satisfies the relation $x^n = e$) are all non-amenable when $m \geq 2$ and odd $n \geq 665$. Obviously, since every element is periodic, such groups cannot contain any free infinite subgroup. We should also mention Gromov's examples ([26, §5.5]) of infinite discrete groups with Kazhdan's property T (hence "very much" non-amenable) and still without any free subgroup. In any case, it is natural to wonder whether the infinite Burnside groups are counterexamples to the above Q2, whence the following.

Question. Are the Burnside groups $B(m,n)$ unitarizable?

In the next statement, we list the main stability properties of unitarizable groups.

Proposition 0.5. *Let G be a discrete group and let Γ be a subgroup.*

(i) *If G is unitarizable, then Γ also is.*

(ii) *If Γ is normal, then G is unitarizable only if both Γ and G/Γ are unitarizable.*

Proof. (i) Consider a u.b. representation $\pi\colon \Gamma \to B(H)$. By Mackey's induction, we have an "induced" representation $\hat{\pi}\colon G \to B(\hat{H})$ with $\hat{H} \supset H$ that is still u.b. (with the same bound) and hence is unitarizable. Moreover, for any t in Γ, $\hat{\pi}(t)$ leaves H invariant, and $\hat{\pi}(t)_{|H} = \pi(t)$. Hence, the original representation π must also be unitarizable. (See [60, p. 43] for full details).

(ii) Let $q\colon G \to G/\Gamma$ be the quotient map and let $\pi\colon G/\Gamma \to B(H)$ be any representation. Then, trivially, π is u.b. (resp. unitarizable) iff the same is true for $\pi \circ q$. Hence G unitarizable implies G/Γ unitarizable. $\qquad\square$

Remark 0.6. In (ii) above, we could not prove that conversely if Γ and G/Γ are unitarizable then G is (although the analogous fact for ideals in an operator algebra is true, see [66, Exercise 27.1]). In particular, we could not verify that the product of two unitarizable groups is unitarizable, however, it is known, and even for semi-direct products, if one of the groups is amenable, see [50], see also [66] for related questions. Of course this should be true if unitarizable is the same as amenable. Similarly, it is not clear that a directed increasing union of a family $(G_i)_{i\in I}$ of unitarizable groups is unitarizable. Actually, we doubt that this is true in general. However, it is true (and easy to check) if the family $(G_i)_{i\in I}$ is "uniformly" unitarizable, in the following sense: there is a function $F : \mathbb{R}_+ \to \mathbb{R}_+$ such that, for any $i \in I$ and any u.b. representation $\pi\colon G_i \to B(H)$, there is an invertible operator $S\colon H \to H$ with $\|S\|\,\|S^{-1}\| \le F(|\pi|)$ such that $t \to S^{-1}\pi(t)S$ is a unitary representation.

Remark 0.7. There is an extensive literature continuing Kunze and Stein's work first on $\mathrm{SL}(2,\mathbb{R})$ [35] and later on $\mathrm{SL}(n,\mathbb{C})$ [36]–[38], and devoted (among other things) to the construction of non-unitarizable uniformly bounded (continuous) representations on more general Lie groups. We should mention P. Sally [72]–[73] for SL_2 over local fields (see also [46]) and the universal covering group of $\mathrm{SL}(2,\mathbb{R})$, Lipsman [39]–[40] for the Lorentz groups $\mathrm{SO}_e(n,1)$ and for $\mathrm{SL}(2,\mathbb{C})$. See the next remark for a synthesis of the current state of knowledge. We refer the reader to Cowling's papers ([13, 14]) for more recent work and a much more comprehensive treatment of uniformly bounded representations on continuous groups. See also Lohoué's paper [43]. All in all, it seems there is a consensus among specialists that discrete groups should be where to look primarily for a counterexample (i.e. unitarizable but not amenable), if it exists. The next remark hopefully should explain why.

Remark 0.8. (Communicated by Michael Cowling). For an almost connected locally compact group G (that is, G/G_e is compact, where G_e is the connected component of the identity e), unitarizability implies amenability. The first step of the argument for this is based on structure theory. The group G has a compact normal subgroup N such that G/N is a finite extension of a connected Lie group (see [48, p. 175]). Suppose that G is unitarizable. Then a fortiori G/N is unitarizable. If we can show that G/N is amenable, then G will be amenable, and we are done. So we may suppose that G is a finite extension of a connected Lie group. A similar argument reduces to the case where G is a connected Lie group, and a third reduction (factoring out the maximal connected normal amenable subgroup) leads to the case where G is semisimple and non-compact. It now suffices to show that a non-compact connected semisimple Lie group G (which is certainly non-amenable) is not unitarizable.

So let G be a non-compact connected semisimple Lie group. We consider the representations π_λ of G unitarily induced from the characters $man \mapsto \exp(i\lambda \log a)$ of a minimal parabolic subgroup MAN. When λ is real, π_λ is unitary, and, according to B. Kostant [34], π_λ is unitarizable only if there is an element w of the

Weyl group $(\mathfrak{g}, \mathfrak{a})$ such that $w\lambda = \bar{\lambda}$. Take a simple root α. If z is a complex number and there exists w in the Weyl group such that $w(z\alpha) = (z\alpha)^- = \bar{z}\alpha$, then z is either purely real and $w\alpha = \alpha$ or z is purely imaginary and $w\alpha = -\alpha$. Thus if z is neither real nor imaginary, then $\pi_{z\alpha}$ is not unitarizable. However, if the imaginary part of z is small enough, then $\pi_{z\alpha}$ is uniformly bounded. Indeed, using the induction in stages construction (see [14], and also [3]), we can make the representation uniformly bounded at the first stage, which involves a real rank one group only (see [13] for the construction of the relevant Hilbert space) and then induce unitarily thereafter to obtain a uniformly bounded representation.

The contents of this paper are as follows. In §1, we describe our contribution on the above problem Q2, namely Theorem 1.1 which says that if we assume unitarizability with a specific quantitative bound then amenability follows. We explain the main ideas of the proof in §2. There we introduce our main objects of study in this paper namely the spaces $M_d(G)$. The latter are closely related on one hand to the space of "multipliers of the Fourier algebra," (which in our notation corresponds to $d = 2$) and on the other hand to the space $UB(G)$ of coefficients of *uniformly bounded* representations on G, that we compare with the space $B(G)$ of coefficients of *unitary* representations on G. We have, for all $d \geq 2$

$$B(G) \subset UB(G) \subset M_d(G) \subset M_2(G).$$

Our methods lead naturally to a new invariant of G, namely the smallest d such that $M_d(G) = B(G)$, that we denote by $d_1(G)$ (we set $d_1(G) = \infty$ if there is no such d). We have $d_1(G) = 1$ iff G is finite and $d_1(G) = 2$ iff G is infinite and amenable (see Theorem 2.3). Moreover, we have $d_1(G) = \infty$ when G is any non Abelian free group. Unfortunately, we cannot produce any group with $2 < d_1(G) < \infty$, and indeed such an example would provide a negative answer to the above Q2. While the main part of the paper is partially expository, §5 contains a new result. We prove there that if $G = \mathbb{F}_\infty$ (free group with countably infinitely many generators) then $M_d(G) \neq M_{d+1}(G)$ for all $d \geq 2$. As a corollary we obtain a completely different proof of Haagerup's unpublished result that $M_2(G) \neq UB(G)$.

Let H be a Hilbert space. Actually, although it is less elementary, it is more natural to work with the $B(H)$-valued (or say $B(\ell_2)$-valued) analogue of the spaces $M_d(G)$. The space $M_d(G)$ corresponds to $\dim(H) = 1$ using $\mathbb{C} \simeq B(\mathbb{C})$. In the $B(\ell_2)$-valued case, the analogue of $d_1(G)$ is denoted simply by $d(G)$. We have obviously $d_1(G) \leq d(G)$ (but we do not have examples where this inequality is strict). Our results for $d(G)$ run parallel to those for $d_1(G)$.

Although our methods (especially in the $B(\ell_2)$-valued case) are inspired by the techniques of "operator space theory" and "completely bounded maps" (see e.g. [22], [54] or [67]), we have strived to make our presentation accessible to a reader unfamiliar with those techniques. This explains in particular why we present the scalar valued (i.e. $\dim(H) = 1$) case first.

We recall merely that a linear map $u \colon A \to B(H)$ defined on a C^*-algebra A is called completely bounded (c.b. in short) if the maps $u_n : M_n(A) \to M_n(B(H))$

defined by

$$u_n([a_{ij}]) = [u(a_{ij})] \quad \forall [a_{ij}] \in M_n(A)$$

are bounded uniformly over n when $M_n(A)$ and $M_n(B(H))$ are each equipped with their unique C^*-norm, i.e. the norm in the space of bounded operators acting on $H \oplus \cdots \oplus H$ (n-times).

We also recall that, for any locally compact group G, the C^*-algebra of G (sometimes called "full" or "maximal" to distinguish it from the "reduced" case) is defined as the completion of the space $L_1(G)$ for the norm defined by

$$\|f\| = \sup \left\| \int_G f(t)\pi(t)dt \right\| \quad \forall f \in L_1(G),$$

where the supremum runs over all (continuous) unitary representations π on G.

In particular, the following result essentially due to Haagerup ([28]) provides a useful (although somewhat abstract) characterization of unitarizable group representations.

Theorem 0.9. *Let G be a locally compact group and let $C^*(G)$ denote the (full) C^*-algebra of G. Let $\pi\colon G \to B(H)$ be a uniformly bounded (continuous) representation. The following are equivalent:*

(i) *π is unitarizable.*

(ii) *The mapping $\tilde{\pi}\colon f \to \int f(t)\pi(t)dt$ from $L_1(G)$ to $B(H)$ extends to a completely bounded map from $C^*(G)$ to $B(H)$.*

More generally, for an arbitrary bounded continuous function $\varphi\colon G \to B(H)$, the following are equivalent:

(i)' *There is a unitary representation $\sigma\colon G \to B(H_\sigma)$ and operators $\xi, \eta\colon H \to H_\sigma$ such that*

$$\varphi(t) = \xi^*\sigma(t)\eta \quad \forall t \in G.$$

(ii)' *The mapping $\tilde{\varphi}\colon f \to \int f(t)\varphi(t)dt$ extends to a completely bounded map from $C^*(G)$ to $B(H)$.*

Proof. If π is unitarizable, say we have $\pi(\cdot) = \xi\sigma(\cdot)\xi^{-1}$ with σ unitary, then, by definition of $C^*(G), \sigma$ extends to a C^*-algebra representation $\hat{\sigma}$ from $C^*(G)$ to $B(H)$. Then, if we set $\hat{\pi}(\cdot) = \xi\hat{\sigma}(\cdot)\xi^{-1}$, $\hat{\pi}$ extends π and satisfies (ii). Thus (i) $\Rightarrow$ (ii). Conversely, if we have a completely bounded extension $\hat{\pi}\colon C^*(G) \to B(H)$, then by [28, Th. 1.10], there is ξ invertible on H such that $\xi^{-1}\hat{\pi}(\cdot)\xi$ is a $*$-homomorphism (in other words a C^*-algebra morphism) and in particular $t \to \xi^{-1}\hat{\pi}(t)\xi$ is a unitary representation, hence π is unitarizable. The proof of the equivalence of (i)' and (ii)' is analogous. That (ii)' $\Rightarrow$ (i)' follows from the fundamental factorization of c.b. maps (see e.g. [60, Chapt. 3], [63, p. 23], or [22]). The converse is obvious. $\square$

In particular, this tells us that unitarizability is a countably determined property:

Corollary 0.10. *Let $\pi\colon G \to B(H)$ be a uniformly bounded representation on a discrete group G. Then π is unitarizable iff its restriction to any countable subgroup $\Gamma \subset G$ is unitarizable.*

Proof. If π is not unitarizable, then, by Theorem 0.9, there is a sequence a^n with $a^n \in M_n(C^*(G))$ and $\|a^n\| \leq 1$, $a^n_{ij} \in \ell_1(G)$ and such that $\|[\tilde{\pi}(a^n_{ij})]\|_{M_n(B(H))} \to \infty$ when $n \to \infty$. Since each entry a^n_{ij} is countably supported, there is a countable subgroup $\Gamma \subset G$ such that all the entries $\{a^n_{ij} \mid n \geq 1, 1 \leq i, j \leq n\}$ are supported on Γ. This implies (by Theorem 0.9 again) that $\pi_{|\Gamma}$ is not unitarizable. This proves the "if" part. The converse is trivial. $\qquad\square$

Corollary 0.11. *If all the countable subgroups of a discrete group G are unitarizable, then G is unitarizable.*

Proof. This is an immediate consequence of the preceding corollary. $\qquad\square$

Remark. Let G be a locally compact group and let $\Gamma \subset G$ be a closed subgroup, we will say that Γ is σ-compactly generated if there is a countable union of compact subsets of G that generates Γ as a closed subgroup. Then the preceding argument suitably modified shows that, in the setting of Theorem 0.9, if $\pi_{|\Gamma}$ is unitarizable for any σ-compactly generated closed subgroup $\Gamma \subset G$, then π is unitarizable.

1. Coefficients of uniformly bounded representations

It will be useful to introduce the space $B(G)$ of "coefficients of unitary representations" on a (discrete) group G defined classically as follows.

We denote by $B(G)$ the space of all functions $f\colon G \to \mathbb{C}$ for which there are a unitary representation $\pi\colon G \to B(H)$ and vectors $\xi, \eta \in H$ such that

$$\forall\, t \in G \qquad f(t) = \langle \pi(t)\xi, \eta \rangle, \tag{1.1}$$

This space can be equipped with the norm

$$\|f\|_{B(G)} = \inf\{\|\xi\|\, \|\eta\|\}$$

where the infimum runs over all possible π, ξ, η as above. As is well known, $B(G)$ is a Banach algebra for the pointwise product. Moreover, $B(G)$ can be identified with the dual of the "full" C^*-algebra of G, denoted by $C^*(G)$.

More generally, let $c \geq 1$ and let G be a semi-group with unit. In that case, we may replace "representations' by unital semi-group homomorphisms. Indeed, note that (since a unitary operator is nothing but an invertible contraction with contractive inverse) a unitary representation on a group G is nothing but a unital semi-group homomorphism $\pi\colon G \to B(H)$ such that $\sup\{\|\pi(t)\| \mid t \in G\} = 1$. For any semi-group homomorphism $\pi\colon G \to B(H)$, we again denote

$$|\pi| = \sup\{\|\pi(t)\| \mid t \in G\}.$$

In the sequel, unless specified otherwise, G will denote a semi-group with unit.

We denote by $B_c(G)$ the space of all functions $f\colon G \to \mathbb{C}$ for which there is a unital semi-group homomorphism $\pi\colon G \to B(H)$ with $|\pi| \leq c$ together with vectors ξ, η in H such that (1.1) holds. Moreover, we denote

$$\|f\|_{B_c(G)} = \inf\{\|\xi\|\, \|\eta\| \mid f(\cdot) = \langle \pi(\cdot)\xi, \eta \rangle \text{ with } |\pi| \leq c\}.$$

Note that when $c = 1$ and G is a group, $B_1(G) = B(G)$ with the same norm, since $|\pi| = 1$ iff π is a unitary representation. For convenience of notation, we set $B(G) = B_1(G)$ also for semi-groups. In the group case, $B_c(G)$ appears as the space of coefficients of u.b. representations with bound $\leq c$.

The space $B_c(G)$ is a Banach space (for the above norm). Moreover, for any $c' \geq 1$ we have (since $\langle \pi(t)\xi, \eta \rangle \langle \pi'(t)\xi', \eta' \rangle = \langle \pi \otimes \pi'(t)\xi \otimes \xi', \eta \otimes \eta' \rangle$)

$$f \in B_c(G), g \in B_{c'}(G) \Rightarrow f \cdot g \in B_{cc'}(G).$$

Note moreover that if $c \geq c'$ we have a norm one inclusion

$$B_{c'}(G) \subset B_c(G)$$

and in particular if $c' = 1$ we find

$$\forall f \in B(G) \qquad \|f\|_{B_c(G)} \leq \|f\|_{B(G)}.$$

We will denote by $UB(G)$ the space of coefficients of uniformly bounded representations on G; in other words, we set:

$$UB(G) = \bigcup_{c>1} B_c(G).$$

The following result partially answers Dixmier's question Q2.

Theorem 1.1 ([61]). *The following properties of a discrete group G are equivalent.*

(i) G *is amenable.*

(ii) $\exists K \, \exists \alpha < 3$ *such that for every u.b. representation* $\pi\colon G \to B(H) \, \exists S\colon H \to H$ *invertible with* $\|S\| \, \|S^{-1}\| \leq K|\pi|^\alpha$ *such that* $S^{-1}\pi(\cdot)S$ *is a unitary representation.*

(ii)' *Same as* (ii) *with* $\alpha = 2$ *and* $K = 1$.

(iii) $\exists K \, \exists \alpha < 3$ *such that for any* $c > 1$ $B_c(G) \subset B(G)$ *and we have*

$$\forall f \in B_c(G) \qquad \|f\|_{B(G)} \leq Kc^\alpha \|f\|_{B_c(G)}.$$

(iii)' *Same as* (iii) *with* $\alpha = 2$ *and* $K = 1$.

Remark. Actually, the preceding result remains valid for a general locally compact group. Indeed, as Z.J. Ruan told us, the Bożejko criterion [7] which we use to prove Theorem 1.1 for discrete groups (which says, with the notation explained below, that $M_2(G) = B(G)$ iff G is amenable) remains valid in the general case. Z.J. Ruan checked that Losert's proof of a similar but a priori weaker statement specifically for the non-discrete case (see [43]) can be modified to yield this, and apparently (cf. [71]) this fact was already known to Losert (unpublished). Our proof that (iii) above implies $M_2(G) = B(G)$ does not really use the discreteness of the group, whence the result in full generality.

This observation, concerning the extension to general locally compact groups, was also made independently by Nico Spronk [74].

We would like to emphasize that there are two separate, very different arguments: one for the discrete case and one for the non-discrete one. This is slightly surprising. A unified approach would be interesting.

First part of the proof of Theorem 1.1. That (i) $\Rightarrow$ (ii)$'$ is the Dixmier-Day result mentioned at the beginning. The implications (ii)$'$ $\Rightarrow$ (ii) and (iii)$'$ $\Rightarrow$ (iii) are trivial. Moreover, (ii) $\Rightarrow$ (iii) and (ii)$'$ $\Rightarrow$ (iii)$'$ are easy to check: indeed consider $f(\cdot) = \langle \pi(\cdot)\xi, \eta \rangle$ with $|\pi| \leq c$. Then, if S is such that $\hat{\pi} = S\pi(\cdot)S^{-1}$ is a unitary representation, we have $f(\cdot) = \langle \pi(\cdot)\xi, \eta \rangle = \langle \hat{\pi}(\cdot)S\xi, (S^{-1})^*\eta \rangle$, hence the coefficients of π are coefficients of $\hat{\pi}$ and

$$\|f\|_{B(G)} \leq \|S\xi\| \, \|(S^{-1})^*\eta\| \leq \|S\| \, \|S^{-1}\| \, \|\xi\| \, \|\eta\|$$

whence

$$\|f\|_{B(G)} \leq \|S\| \, \|S^{-1}\| \, \|f\|_{B_c(G)}.$$

Now if (ii) holds we can find S as above with $\|S\| \, \|S^{-1}\| \leq Kc^\alpha$, thus (ii) $\Rightarrow$ (iii), and similarly (ii)$'$ $\Rightarrow$ (iii)$'$. Thus it only remains to prove that (iii) $\Rightarrow$ (i). $\qquad\square$

2. The spaces of multipliers $M_d(G)$

To explain the proof of Theorem 1.1, we will need some additional notation.

Notation. Let $d \geq 1$ be an integer. Let G be a a semigroup with unit. We are mainly interested in the group case, but we could also take $G = \mathbb{N}$.

Let $M_d(G)$ be the space of all functions $f\colon G \to \mathbb{C}$ such that there are bounded functions $\xi_i\colon G \to B(H_i, H_{i-1})$ (H_i Hilbert) with $H_0 = \mathbb{C}$, $H_d = \mathbb{C}$ such that

$$\forall\, t_i \in G \qquad f(t_1 t_2 \ldots t_d) = \xi_1(t_1)\xi_2(t_2)\ldots\xi_d(t_d). \tag{2.1}$$

Here of course we use the identification $B(H_0, H_d) = B(\mathbb{C}, \mathbb{C}) \simeq \mathbb{C}$. We define

$$\|f\|_{M_d(G)} = \inf\{\sup_{t_1 \in G} \|\xi_1(t_1)\| \ldots \sup_{t_d \in G} \|\xi_d(t_d)\|\}$$

where the infimum runs over all possible ways to write f as in (2.1).

The definition of the spaces $M_d(G)$ and of the more general spaces $M_d(G; H)$ appearing below is motivated by the work of Christensen-Sinclair on "completely bounded multilinear maps" and the so-called Haagerup tensor product (see [12]). The connection is explained in detail in [61], and is important for the proofs of all the results below, but we prefer to skip this in the present exposition (see however §4 below).

When $d = 2$, and G is a group, the space $M_2(G)$ is the classical space of "Herz-Schur multipliers" on G. This space also coincides (see [9] or [60, p. 110]) with the space of all c.b. "Fourier multipliers" on the reduced C^*-algebra $C_\lambda^*(G)$. The question whether the space $M_2(G)$ coincides with the space of coefficients of u.b. representations (namely $\bigcup_{c>1} B_c(G)$) remained open for a while but Haagerup [27] showed that it is not the case. More precisely, he showed that if $G = \mathbb{F}_\infty$, we have

$$\forall\, c > 1 \qquad B_c(G) \underset{\neq}{\subset} M_2(G).$$

We give a different proof of a more precise statement in §5 below.

For $d > 2$, in the group case, the spaces $M_d(G)$ are not so naturally interpreted in terms of "Fourier" multipliers. In particular, in spite of the strong analogy with the multilinear multipliers introduced in [21] (those are complex valued functions on G^d), there does not seem to be any significant connection.

In the case $G = \mathbb{N}$, the space $M_3(G)$ is characterized in [65] as the space of "completely shift bounded" Fourier multipliers on the Hardy space H_1, but this interpretation is restricted to $d = 3$ and uses the commutativity.

Remark 2.1. Note the following easily checked inclusions, valid when G is a group or a semigroup with unit:

$$B(G) = B_1(G) \subset UB(G) = \bigcup_{c>1} B_c(G) \subset M_d(G) \subset M_{d-1}(G) \subset \cdots$$

$$\cdots \subset M_2(G) \subset M_1(G) = \ell_\infty(G),$$

and we have clearly

$$\forall m \le d \qquad \|f\|_{M_m(G)} \le \|f\|_{M_d(G)}. \tag{2.2}$$

Moreover, we have

$$\forall f \in B_c(G) \qquad \|f\|_{M_d(G)} \le c^d \|f\|_{B_c(G)}. \tag{2.3}$$

Indeed, if $f(\cdot) = \langle \pi(\cdot)\xi, \eta \rangle$ with $|\pi| \le c$, then we can write

$$f(t_1 t_2 \ldots t_d) = \langle \pi(t_1) \ldots \pi(t_d)\xi, \eta \rangle$$

$$= \xi_1(t_1)\xi_2(t_2) \ldots \xi_d(t_d)$$

where $\xi_1(t_1) \in B(H_\pi, \mathbb{C})$, $\xi_d(t_d) \in B(\mathbb{C}, H_\pi)$ and $\xi_i(t_i) \in B(H_\pi, H_\pi)$ $(1 < i < d)$ are defined by $\xi_1(t_1)h = \langle \pi_1(t_1)h, \eta \rangle$ $(h \in H_\pi)$ $\xi_d(t_d)\lambda = \lambda\pi(t_d)\xi$ $(\lambda \in \mathbb{C})$ and $\xi_i(t_i) = \pi(t_i)$ $(1 < i < d)$. Therefore, we have

$$\|f\|_{M_d(G)} \le \sup \|\xi_1\| \sup \|\xi_2\| \ldots \sup \|\xi_d\|$$

$$\le |\pi|^d \|\xi\| \, \|\eta\| \le c^d \|\xi\| \, \|\eta\|$$

whence the announced inequality (2.3).

Remark 2.2. It is easy to see (using tensor products) that $M_d(G)$ is a unital Banach algebra for the pointwise product of functions on G: for any f, g in $M_d(G)$ we have

$$\|fg\|_{M_d(G)} \le \|f\|_{M_d(G)}\|g\|_{M_d(G)}.$$

The function identically equal to 1 on G is the unit and has norm 1.

Remark. Let $h\colon \Gamma \to G$ be a unital homomorphism between two groups (or two semi-groups with unit). Then for any f in $M_d(G)$ the composition $f \circ h$ is in $M_d(\Gamma)$ with $\|f \circ h\|_{M_d(\Gamma)} \le \|f\|_{M_d(G)}$. The proof is obvious.

In particular, if $\Gamma \subset G$ is a subgroup we have $\|f_{|\Gamma}\|_{M_d(\Gamma)} \le \|f\|_{M_d(G)}$. Moreover, if Γ is a normal subgroup in a group G and if $q\colon G \to G/\Gamma$ is the quotient map, then $\|f\|_{M_d(G/\Gamma)} = \|f \circ q\|_{M_d(G)}$. (Indeed, the equality can be proved easily using an arbitrary pointwise lifting $\rho\colon G/\Gamma \to G$.) Let $\Gamma \subset G$ be again an arbitrary subgroup of a group G. Given a function $f\colon \Gamma \to \mathbb{C}$, we let $\tilde{f}\colon G \to \mathbb{C}$

be the extension of f vanishing outside Γ. Then, it is rather easy to see that $\|\tilde{f}\|_{M_2(G)} = \|f\|_{M_2(\Gamma)}$ (but the analogue of this for $d > 2$ seems unclear). It is well known that 1_Γ is in the unit ball of $B(G)$ (hence a fortiori of $M_d(G)$) hence by Remark 2.2 we have for any f in $M_d(G)$:

$$\|f \cdot 1_\Gamma\|_{M_d(G)} \leq \|f\|_{M_d(G)}.$$

The proof of (iii) $\Rightarrow$ (i) in Theorem 1.1 uses the following criterion for amenability due to Marek Bożejko [7].

Theorem 2.3. *Let G be a discrete group. Then G is amenable iff $B(G) = M_2(G)$.*

Remark. We do not know whether $B(G) = M_3(G) \Rightarrow G$ amenable.

Sketch of proof of Theorem 2.3. The only if part is quite easy. Let us sketch the proof of the "if" part. Assume $B(G) = M_2(G)$. Then there is a constant K such that, for any f in the space $\mathbb{C}[G]$ of all finitely supported functions $f \colon G \to \mathbb{C}$, we have

$$\|f\|_{B(G)} \leq K\|f\|_{M_2(G)}.$$

Let $\varepsilon \colon G \to \{-1, 1\}$ be a "random choice of signs" indexed by G, and let $\mathbb{E}$ denote the expectation with respect to the corresponding probability. We will estimate the average of the norms of the pointwise product εf. More precisely we claim that there are numerical constants C' and C'' (independent of f) such that

$$\left(\sum_{t \in G} |f(t)|^2 \right)^{1/2} \leq C' \mathbb{E}\|\varepsilon f\|_{B(G)} \tag{2.4}$$

$$\mathbb{E}\|\varepsilon f\|_{M_2(G)} \leq C'' \left\| \sum |f(t)|^2 \lambda(t) \right\|_{C_\lambda^*}^{1/2}. \tag{2.5}$$

Using this it is easy to conclude: indeed we have

$$\sum |f(t)|^2 \leq (C' K C'')^2 \left\| \sum |f(t)|^2 \lambda(t) \right\|_{C_\lambda^*}$$

and by the well known Kesten-Hulanicki criterion (cf. e.g. [60, Th. 2.4]), this implies that G is amenable.

We now return to the above claims. The inequality (2.4) can be seen as a consequence of the fact (due to N. Tomczak-Jaegermann [77]) that $B(G)$ is of cotype 2 (a Banach space B is called of cotype 2 if there is a constant C such that for any finite sequence (x_i) in B, the following inequality holds $(\sum \|x_i\|^2)^{1/2} \leq C \mathrm{Average}_\pm \| \sum \pm x_i \|)$.

As for (2.5), it is proved in [7] using an idea due to Varopoulos [80]. However, more recently the following result was proved in [59]: Consider all possible ways to have the following decomposition

$$f(t_1 t_2) = \alpha(t_1, t_2) + \beta(t_1, t_2) \qquad \forall\, t_1, t_2 \in G \tag{2.6}$$

and let

$$|||f||| = \inf \left\{ \sup_{t_1} \left(\sum_{t_2} |\alpha(t_1,t_2)|^2 \right)^{1/2} + \sup_{t_2} \left(\sum_{t_1} |\beta(t_1,t_2)|^2 \right)^{1/2} \right\}$$

where the infimum runs over all possible decompositions as in (2.6).
Then (cf. [59]) there is a numerical constant $\delta > 0$ such that

$$\delta |||f||| \leq \mathbb{E}\|\varepsilon f\|_{M_2(G)} \leq |||f||| \qquad \forall\, f \in \mathbb{C}[G]. \tag{2.7}$$

Note that the right-hand side of (2.7) is an immediate consequence of the
following inequality

$$\|f\|_{M_2(G)} \leq |||f|||, \tag{2.8}$$

and the latter is easy: we simply write

$$f(t_1 t_2) = \langle \xi_1(t_1), \xi_2(t_2) \rangle + \langle \eta_1(t_1), \eta_2(t_2) \rangle$$

where

$$\xi_1(t_1) = \sum_{t_2} \alpha(t_1,t_2)\delta_{t_2}, \quad \xi_2(t_2) = \delta_{t_2}$$

and

$$\eta_2(t_2) = \sum_{t_1} \overline{\beta(t_1,t_2)}\delta_{t_1}, \quad \eta_1(t_1) = \delta_{t_1},$$

and (2.8) follows.

Moreover, we also have

$$|||f||| \leq \left\| \sum |f(t)|^2 \lambda(t) \right\|_{C_\lambda^*}^{1/2}, \tag{2.9}$$

therefore (2.5) follows from (2.7) and (2.9) with $C'' = 1$. The inequality (2.9)
follows from the following observation: $\| \sum |f(t)|^2 \lambda(t)\|_{C_\lambda^*} \leq 1$ iff there is a decomposition of the form

$$|f(t_1 t_2)| = |a(t_1,t_2)|^{1/2} \cdot |b(t_1,t_2)|^{1/2} \qquad \forall t_1, t_2 \in G \tag{2.10}$$

for kernels a, b on $G \times G$ such that

$$\sup_{t_1} \sum_{t_2} |a(t_1,t_2)|^2 \leq 1 \quad \text{and} \quad \sup_{t_2} \sum_{t_1} |b(t_1,t_2)|^2 \leq 1. \tag{2.11}$$

Then (2.10) and (2.11) imply that $|||f||| \leq 1$, and hence (2.9) follows by homogeneity. Indeed, by a compactness argument, these assertions are immediate
consequences of the following Lemma.   □

This Lemma gives a converse to Schur's classical criterion for boundedness
on ℓ_2 of matrices with positive entries (we include the proof for lack of a suitable
reference). See [64] for more information on this.

Lemma 2.4. *Let $n \geq 1$. Let $\{f_{ij} \mid 1 \leq i, j \leq n\}$ be complex scalars such that the matrix $[|f_{ij}|^2]$ has norm ≤ 1 as an operator on the Euclidean space ℓ_2^n. Then there are (a_{ij}) and (b_{ij}) with*

$$\sup_i \sum_j |a_{ij}|^2 \leq 1 \quad \text{and} \quad \sup_j \sum_i |b_{ij}|^2 \leq 1$$

such that $|f_{ij}| = |a_{ij}|^{1/2}|b_{ij}|^{1/2}$. Therefore, there are (α_{ij}) and (β_{ij}) with

$$\sup_i (\sum_j |\alpha_{ij}|^2)^{1/2} + \sup_j (\sum_i |\beta_{ij}|^2)^{1/2} \leq 1$$

such that $f_{ij} = \alpha_{ij} + \beta_{ij}$.

Proof. By perturbation and compactness arguments, we can assume that $|f_{ij}| > 0$ for all i, j. Let $T = [|f_{ij}|^2]$. We may assume $\|T\| = 1$. Let $\xi = (\xi_i)$ be a Perron–Frobenius vector for T^*T so that $\xi_i > 0$ for all i and $T^*T\xi = \xi$. Let $\eta = T\xi$, so that $T^*\eta = \xi$. If we then set $|a_{ij}|^2 = |f_{ij}|^2 \xi_j \eta_i^{-1}$ and $|b_{ij}|^2 = |f_{ij}|^2 \xi_j^{-1} \eta_i$ we obtain the first assertion. By the arithmetic-geometric mean inequality we have $|f_{ij}| \leq g_{ij}$ with $g_{ij} = 2^{-1}(|a_{ij}| + |b_{ij}|)$. If we then set

$$\alpha_{ij} = 2^{-1}|a_{ij}| f_{ij} g_{ij}^{-1} \quad \text{and} \quad \beta_{ij} = 2^{-1}|b_{ij}| f_{ij} g_{ij}^{-1},$$

we obtain the second assertion. $\qquad\square$

Remark. The proof of Theorem 2.3 sketched above shows that G is amenable if $\exists K \ \forall f \in \mathbb{C}[G]$

$$\|f\|_{B(G)} \leq K\|f\|_{M_2(G)}.$$

Remark. Note that (2.7) and (2.8) show that

$$\sup_\varepsilon \|\varepsilon f\|_{M_2(G)} \leq \delta^{-1} \mathbb{E}\|\varepsilon f\|_{M_2(G)}.$$

The proof of the implication (iii) $\Rightarrow$ (i) in Theorem 1.1 rests on the following.

Key Lemma 2.5 (Implicit in [61]). *Let $f \in B(G)$. Fix $d \geq 1$. Then, for any $c \geq 2$, we have*

$$\|f\|_{B_c(G)} \leq 2\|f\|_{M_d(G)} + 2c^{-(d+1)}\|f\|_{B(G)}.$$

More generally, for any $1 \leq \theta < c$, we have for any f in $B_\theta(G)$ and any $d \geq 1$

$$\|f\|_{B_c(G)} \leq \left(\sum_{m=0}^d (\theta/c)^m\right) \cdot \|f\|_{M_m(G)} + \left(\sum_{m>d} (\theta/c)^m\right) \cdot \|f\|_{B_\theta(G)}.$$

Remark. The proof of the key lemma uses ideas from two remarkable papers due to Peller [56] and Blecher and Paulsen [5].

Proof of (iii) $\Rightarrow$ (i) in Theorem 1.1. Assume (iii). Then using the key lemma with $d = 2$ (and $\theta = 1$), we have for all f in $B(G)$ and all $c \geq 2$

$$\|f\|_{B(G)} \leq Kc^\alpha \|f\|_{B_c(G)}$$
$$\leq 2Kc^\alpha \|f\|_{M_2(G)} + 2Kc^{\alpha-3}\|f\|_{B(G)}.$$

But we can choose $c = c(K, \alpha)$ large enough so that $2Kc^{\alpha-3} = 1/2$ (say) and then we obtain

$$\left(1 - \frac{1}{2}\right) \|f\|_{B(G)} \leq 2Kc^{\alpha}\|f\|_{M_2(G)}$$

so that we conclude

$$\|f\|_{B(G)} \leq 4Kc(K, \alpha)^{\alpha}\|f\|_{M_2(G)},$$

hence, by Theorem 2.3, G is amenable.   □

The proof of the key lemma is based on the following result (of independent interest) which is "almost" a characterization of $B_c(G)$.

Theorem 2.6. *Fix a number $c \geq 1$. Consider $f \in \bigcap_{m \geq 1} M_m(G)$ such that*

$$\sum_m c^{-m}\|f\|_{M_m(G)} < \infty.$$

Then $f \in B_c(G)$ and moreover

$$\|f\|_{B_c(G)} \leq |f(e)| + \sum_{m \geq 1} c^{-m}\|f\|_{M_m(G)}. \qquad (2.12)$$

Conversely, for all f in $B_c(G)$, we have

$$\sup_{m \geq 1} c^{-m}\|f\|_{M_m(G)} \leq \|f\|_{B_c(G)}. \qquad (2.13)$$

Note. (2.13) is easy and has been proved already (see (2.2)). The main point is (2.12).

Proof. This is essentially [61, Theorem 1.12] and the remark following it. For the convenience of the reader, we give some more details.
In [61] the natural predual of $B_c(G)$ is considered and denoted by $\tilde{A}_c$. By [61, Th. 1.7], any x in the open unit ball of $\tilde{A}_c$ can be written as

$$x = \sum_{m=0}^{\infty} c^{-m} x_m$$

where each x_m is an element of $C^*(G)$ which is the image, under the natural product map of an element X_m in the unit ball of $\ell_1(G) \otimes_h \cdots \otimes_h \ell_1(G)$ (m times). This implies by duality, for all $m > 0$

$$|\langle f, x_m \rangle| \leq \|f\|_{M_m(G)}.$$

In the particular case $m = 0$, x_0 is a multiple of the unit δ_e by a scalar of modulus ≤ 1. Whence

$$|\langle f, x \rangle| \leq |f(e)| + \sum_{m=1}^{\infty} c^{-m}\|f\|_{M_m(G)}$$

and a fortiori, by (2.2) and (2.3)

$$\leq |f(e)| + \sum_{m=1}^{d} c^{-m}\|f\|_{M_d(G)} + \sum_{m=d+1}^{\infty} c^{-m}\|f\|_{B(G)}. \qquad \square$$

Corollary 2.7. *Let $UB(G) = \bigcup_{c>1} B_c(G)$ be the space of coefficients of u.b. representations on G. Then $f \in UB(G)$ iff $\sup_{m\geq 1} \|f\|_{M_m(G)}^{1/m} < \infty$. More precisely, let $c(f)$ denote the infimum of the numbers $c \geq 1$ for which $f \in B_c(G)$. Then, we have*

$$c(f) = \limsup_{m\to\infty} \|f\|_{M_m(G)}^{1/m}.$$

Proof of Key Lemma 2.5. This is an easy consequence of (2.12), (2.13) and the obvious inequalities

$$|f(e)| \leq \|f\|_{\ell_\infty(G)} \leq \|f\|_{M_2(G)} \leq \cdots \leq \|f\|_{M_d(G)}$$
$$\leq \cdots \leq \|f\|_{B(G)}. \qquad \square$$

In the case $c = 1$, Theorem 2.6 seems to degenerate but actually the following "limiting case" can be established, as a rather simple dualization of a result in [5].

Proposition 2.8. *Consider a function $f\colon G \to \mathbb{C}$. Then $f \in B(G)$ iff $f \in \bigcap_{m\geq 1} M_m(G)$ with $\sup_m \|f\|_{M_m(G)} < \infty$. Moreover we have*

$$\|f\|_{B(G)} = \sup_{m\geq 1} \|f\|_{M_m(G)}.$$

Remark. The same argument shows the following. Given a real number $\alpha \geq 0$, we say that G satisfies the condition (C_α) if there is $K \geq 0$ such that for any f in $B(G)$ we have

$$\forall\, c > 1 \qquad \|f\|_{B(G)} \leq Kc^\alpha \|f\|_{B_c(G)}.$$

Then the preceding argument shows that if $d \leq \alpha < d+1$, (C_α) implies that $B(G) = M_d(G)$ (with equivalent norms).

We are thus led to define the following quantities:

$$\alpha_1(G) = \inf\{\alpha \geq 0 \mid G \text{ satisfies } (C_\alpha)\}$$
$$d_1(G) = \inf\{d \in \mathbb{N} \mid M_d(G) = B(G)\}.$$

With this notation, the preceding argument shows that $d_1(G) \leq \alpha_1(G)$. A priori $\alpha_1(G)$ is a real number, but (although we have no direct argument for this) it turns out that it is an integer:

Theorem 2.9. *Assume $B(G) = \bigcup_{c>1} B_c(G)$. Then $\alpha_1(G) < \infty$ and moreover*

$$\alpha_1(G) = d_1(G).$$

In particular, we have

$$M_d(G) = M_{d+1}(G) \quad \forall d \geq \alpha_1(G).$$

Actually, for the last assertion to hold, it suffices to have much less:

Theorem 2.10 ([62]). *Let G be a semigroup with unit. Assume that there are $1 \le \theta < c$ such that $B_\theta(G) = B_c(G)$. Then there is an integer D such that $B_\theta(G) = M_D(G)$, and in particular, we have*

$$M_d(G) = M_{d+1}(G) \quad \forall d \ge D.$$

Remark. Let G be a locally compact group. Let G_d be G equipped with the discrete topology. In [29], Haagerup proves that if a function $\phi \colon G \to \mathbb{C}$ belongs to $M_2(G_d)$ and is continuous, then it belongs to $M_2(G)$ (with the same norm). We do not know if the analogous statement is valid for $M_3(G)$ or $M_d(G)$ when $d \ge 3$.

Remark 2.11. Let $I_1, \ldots, I_d$ be arbitrary sets. We will denote by $M_d(I_1, \ldots, I_d)$ the space of all functions $f \colon I_1 \times \cdots \times I_d \to \mathbb{C}$ for which there are bounded functions f_i

$$f_i \colon I_i \to B(H_i, H_{i-1}) \quad \text{(here } H_i \text{ are Hilbert spaces with } H_d = H_0 = \mathbb{C}) \text{ such that}$$

$$\forall \, b_i \in I_i \qquad f(b_1, \ldots, b_d) = f_1(b_1) \ldots f_d(b_d).$$

We equip this space with the norm

$$\|f\| = \inf \left\{ \prod_{i=1}^d \sup_{b \in I_i} \|f_i(b)\| \right\}$$

where the infimum runs over all possible such factorizations.

In particular, if $I_1 = I_2 = \cdots = I_d = G$, then for any function φ in $M_d(G)$, we have

$$\|\varphi\|_{M_d(G)} = \|\Phi\|_{M_d(G, \ldots, G)}$$

where Φ is defined by

$$\Phi(t_1, \ldots, t_d) = \varphi(t_1 t_2 \ldots t_d).$$

By a well known trick, one can check that $f \to \|f\|_{M_d(I_1, \ldots, I_d)}$ is subadditive (and hence is a norm) on $M_d(I_1, \ldots, I_d)$, i.e. that we have for all f, g in $M_d(I_1, \ldots, I_d)$:

$$\|f + g\|_{M_d(I_1, \ldots, I_d)} \le \|f\|_{M_d(I_1, \ldots, I_d)} + \|g\|_{M_d(I_1, \ldots, I_d)} \tag{2.14}$$

Let us quickly sketch this: Let f, g be in the open unit ball of $M_d(I_1, \ldots, I_d)$. Then by homogeneity we can assume

$$f(b_1, \ldots, b_d) = f_1(b_1) \ldots f_d(b_d) \quad \text{and} \quad g(b_1, \ldots, b_d) = g_1(b_1) \ldots g_d(b_d) \tag{2.15}$$

with $\sup \|f_j(b)\| < 1$ and $\sup \|g_j(b)\| < 1$ for all j. Then we can write for any $0 \le \alpha \le 1$

$$(\alpha f + (1-\alpha)g)(b_1, \ldots, b_d) = F_1(b_1) \ldots F_d(b_d)$$

where

$$F_1(b_1) = [\alpha^{1/2} f_1(b_1) \quad (1-\alpha)^{1/2} g_1(b_1)] \quad \text{(row matrix with operator entries)}$$

$$F_j(b_j) = \begin{bmatrix} f_j(b_j) & 0 \\ 0 & g_j(b_j) \end{bmatrix} \qquad 2 \le j \le d-1$$

and

$$F_d(b_d) = \begin{bmatrix} \alpha^{1/2} f_d(b_d) \\ (1-\alpha)^{1/2} g_d(b_d) \end{bmatrix} \quad \text{(column matrix with operator entries)}.$$

Then it is easy to check that $\sup_b \|F_j(b)\| < 1$ for all $1 \le j \le d$ and hence we obtain

$$\|\alpha f + (1-\alpha)g\|_{M_d(I_1,\dots,I_d)} < 1.$$

Moreover, $M_d(I_1,\dots,I_d)$ is a unital Banach algebra for the pointwise product, i.e. for any f,g in $M_d(I_1,\dots,I_d)$ we have

$$\|f.g\|_{M_d(I_1,\dots,I_d)} \le \|f\|_{M_d(I_1,\dots,I_d)} \|g\|_{M_d(I_1,\dots,I_d)}. \tag{2.16}$$

Indeed, if we assume (2.15) then we have

$$(f.g)(b_1,\dots,b_d) = (f_1(b_1) \otimes g_1(b_1)) \dots (f_d(b_d) \otimes g_d(b_d)),$$

and (2.16) follows easily from this. Obviously, the function identically equal to 1 on $I_1 \times \dots \times I_d$ is a unit for this algebra and it has norm 1 in $M_d(I_1,\dots,I_d)$.

Example 2.12. To illustrate the preceding concepts, we recover here the following result from [70]: Let $G = \mathbb{F}_\infty$ and let $W(1) \subset G$ be the subset of all the words of length 1. Then the indicator function of $W(1)$ is in $UB(G)$. Indeed, we claim that for any bounded function φ with support in $W(1)$ we have

$$\forall\, d \ge 1 \qquad \|\varphi\|_{M_d(G)} \le 2^d \|\varphi\|_{\ell_\infty(G)}. \tag{2.17}$$

Thus (by Corollary 2.7) we have $\varphi \in B_c(G)$ for $c > 2$ (actually, this is known for all $c > 1$).

However it can be shown that for any such function we have

$$\left(\sum_{t \in G} |\varphi(t)|^2 \right)^{1/2} \le 2\|\varphi\|_{B(G)}. \tag{2.18}$$

Thus for instance the indicator function of $W(1)$ is in $B_c(G)$ for all $c > 2$ but not in $B(G)$ (in particular this shows that G is not unitarizable). Note however that since $1_{W(1)}$ belongs to $M_d(G)$ for all $d \ge 1$, this does not distinguish the various classes $M_d(G)$ or $B_c(G)$, but this task is completed in §5.

Proof of (2.18). First it suffices to prove this for a finitely supported φ, with support in $W(1)$. Then (2.18) is an immediate consequence of an inequality proved first by Leinert [42], and generalized by Haagerup [30, Lemma 1.4]: Any ψ finitely supported, with support in $W(1)$ satisfies $\|\sum \lambda(t)\psi(t)\| \le 2(\sum |\psi(t)|^2)^{1/2}$. The inequality (2.18) can be deduced from this by duality, using the fact that, if φ is finitely supported, then $\|\varphi\|_{B(G)} = \sup |\langle \varphi, \psi \rangle|$ where the sup runs over all ψ finitely supported on G such that $\|\sum \lambda(t)\psi(t)\| \le 1$. $\qquad \square$

Proof of (2.17). Let φ be a function with support in $W(1)$. Consider the set

$$\Omega = \{(t_1,\dots,t_d) \in G^d \mid t_1 t_2 \dots t_d \in W(1)\}.$$

Clearly, when $t_1 t_2 \ldots t_d$ has length one, it reduces to a single letter (i.e. a generator or its inverse). Clearly this letter must "come" from either $t_1, t_2, \ldots$ or t_d. Thus we have

$$\Omega = \Omega_1 \cup \ldots \cup \Omega_d$$

where Ω_j is the set of $(t_1, \ldots, t_d)$ in Ω such that the single "letter" left after reduction comes from t_j. Hence we have

$$1_\Omega = \sum_j 1_{\Omega_j} \prod_{i<j} [1 - 1_{\Omega_i}]. \tag{2.19}$$

For any θ in G, we introduce the operator $\xi(\theta) \in B(\ell_2(G))$ defined as follows: Assume $\theta = a_1 a_2 \ldots a_k$ (reduced word where $a_q \in W(1)$ for all q), with $k \geq 1$, then we set $a_0 = a_{k+1} = e$ and

$$\xi(\theta) = \sum_{q=1}^{k} \varphi(a_q) e_{a_1 \ldots a_{q-1}, (a_{q+1} \ldots a_k)^{-1}}$$

where, as usual, $e_{s,t}$ denotes the operator defined by $e_{s,t}(\delta_t) = \delta_s$ and $e_{s,t}(\delta_x) = 0$ whenever $x \neq t$. Moreover, if $\theta = e$ (empty word, corresponding to $k = 0$), we set $\xi(\theta) = 0$. Then it is a simple verification that

$$\langle \lambda(t_1) \ldots \lambda(t_{j-1}) \xi(t_j) \lambda(t_{j+1}) \ldots \lambda(t_d) \delta_e, \delta_e \rangle = \varphi(t_1 t_2 \ldots t_d) 1_{\{(t_1, \ldots, t_d) \in \Omega_j\}}.$$

A moment of reflection shows that $\|\xi(\theta)\| = \sup_q |\varphi(a_q)|$ hence $\sup_{\theta \in G} \|\xi(\theta)\| \leq \|\varphi\|_{\ell_\infty(G)}$. This shows with the notation introduced in Remark 2.11, that if we set

$$\Phi_j(t_1, \ldots, t_d) = \varphi(t_1 t_2 \ldots t_d) 1_{\{(t_1, t_2, \ldots, t_d) \in \Omega_j\}}$$

we have

$$\|\Phi_j\|_{M_d(G, \ldots, G)} \leq \sup_{t \in G} \|\xi(t)\| \leq \|\varphi\|_{\ell_\infty(G)},$$

and hence with $\varphi = 1$ identically, we find $\|1_{\Omega_j}\|_{M_d(G, \ldots, G)} \leq 1$, and

$$\|1 - 1_{\Omega_j}\|_{M_d(G, \ldots, G)} \leq 2$$

hence by Remark 2.11, (2.14) and (2.16), we have

$$\|\varphi\|_{M_d(G)} = \|\Phi\|_{M_d(G, \ldots, G)} = \left\| \sum_j \Phi_j \prod_{i<j} [1 - 1_{\Omega_i}] \right\|_{M_d(G, \ldots, G)} \leq 2^d \|\varphi\|_{\ell_\infty(G)}$$

which completes the proof of (2.17). □

Example 2.13. Let G be a free group.

(i) Let $\psi_d \colon G^d \to \{0, 1\}$ be the indicator function of the set formed by all the d-tuples $(t_1, \ldots, t_d)$ of reduced words such that $t_i \neq e$ for all i and the product $t_1 t_2 \ldots t_d$ allows no reduction. Then

$$\|\psi_d\|_{M_d(G, \ldots, G)} \leq 5^{d-1}$$

(ii) A fortiori, for any subsets $I_1 \subset G, \ldots, I_d \subset G$, we have

$$\|\psi_{d|I_1 \times \cdots \times I_d}\|_{M_d(I_1, \ldots, I_d)} \leq 5^{d-1}.$$

Proof. Fix $1 \leq j \leq d-1$. Let A_j be the subset of G^d formed of all $(t_1, \ldots, t_d)$ in G^d such that $t_j \neq e$, $t_{j+1} \neq e$ and such that $t_j t_{j+1}$ does reduce, i.e. $|t_j t_{j+1}| < |t_j| + |t_{j+1}|$. Also let $B_j = \{t \in G^d \mid t_j \neq e, t_{j+1} \neq e\}$. We will use the fact that

$$\psi_d = \prod_{j=1}^{d-1} 1_{B_j} - 1_{A_j}. \tag{2.20}$$

Observe that for all $t = (t_j) \in G^d$

$$1_{B_j}(t) = (1 - 1_{\{t_j = e\}})(1 - 1_{\{t_{j+1} = e\}})$$

and using $1_{\{t_j = e\}} = \langle \delta_{t_j}, \delta_e \rangle$, it is easy to deduce from this with (2.14) and (2.16) that

$$\|1_{B_j}\|_{M_d(G, \ldots, G)} \leq 4.$$

Now, for any x in G with $x \neq e$ let us denote by $F(x)$ and $L(x)$ respectively the first and last letter of x (i.e. $F(x)$ and $L(x)$ are equal to a generator or the inverse of one). Then it is easy to check that for any $t = (t_j)$ in G^d we have

$$1_{A_j}(t) = \langle \alpha(t_j), \beta(t_{j+1}) \rangle$$

where $\alpha(t) = \delta_{L(t)}$ and $\beta(s) = \delta_{F(s)^{-1}}$ if both $|t| > 0$ and $|s| > 0$ and $\alpha(e) = \beta(e) = 0$. This implies immediately that

$$\|1_{A_j}\|_{M_d(G, \ldots, G)} \leq 1,$$

hence $\|1_{B_j} - 1_{A_j}\|_{M_d(G, \ldots, G)} \leq 5$, and since, by (2.16), $M_d(G, \ldots, G)$ is a Banach algebra, by (2.20) we obtain

$$\|\psi_d\|_{M_d(G, \ldots, G)} \leq 5^{d-1}. \qquad \square$$

3. The predual $X_d(G)$ of $M_d(G)$

The definition of the spaces $B_c(G)$ and $M_d(G)$ shows that they are dual spaces. There is a natural duality between these spaces and the group algebra $\mathbb{C}[G]$ which we view as the convolution algebra of finitely supported functions on G. Indeed, for any function $f \colon G \to \mathbb{C}$ and any g in $\mathbb{C}[G]$, we set

$$< g, f >= \sum_{t \in G} g(t) f(t).$$

Then we define the spaces $X_d(G)$ and $\tilde{A}_c$ respectively as the completion of $\mathbb{C}[G]$ for the respective norms

$$\|g\|_{X_d(G)} = \sup\{| < g, f > | \mid f \in M_d(G), \ \|f\|_{M_d(G)} \leq 1\}$$

and

$$\|g\|_{\tilde{A}_c} = \sup\{| < g, f > | \mid f \in M_d(G), \ \|f\|_{B_c(G)} \leq 1\}.$$

Obviously, we can also write

$$\|g\|_{\tilde{A}_c} = \sup\{\|\sum g(t)\pi(t)\| \mid \pi\colon G \to B(H), \ |\pi| \le c\}.$$

This last formula shows that $\tilde{A}_c$ is naturally equipped with a Banach algebra structure under convolution: we have $\|g_1 * g_2\|_{\tilde{A}_c} \le \|g_1\|_{\tilde{A}_c}\|g_2\|_{\tilde{A}_c}$.
However, the analog for the spaces $X_d(G)$ fails in general. This was the basic idea used by Haagerup [27] to prove that $M_2(\mathbb{F}_\infty) \ne UB(\mathbb{F}_\infty)$. Indeed, Haagerup used spherical functions to show that $X_2(\mathbb{F}_\infty)$ is not a Banach algebra under convolution (see Remark 3.2 below), which implies by the preceding remarks that $X_2(\mathbb{F}_\infty) \ne \tilde{A}_c$ for any c, hence $M_2(\mathbb{F}_\infty) \ne B_c(\mathbb{F}_\infty)$ for any c, from which $M_2(\mathbb{F}_\infty) \ne UB(\mathbb{F}_\infty)$ follows easily by Baire's classical theorem.

Note that, in sharp contrast, for $G = \mathbb{N}$, it is known that $X_2(G)$ is a Banach algebra (due to G. Bennett), but not an operator algebra (see [65] for details).
Although $X_d(G)$ is not in general a Banach algebra under convolution, it satifies the following property: if $g_1 \in X_d(G)$ and $g_2 \in X_k(G)$, then $g_1 * g_2 \in X_{d+k}(G)$ and

$$\|g_1 * g_2\|_{X_{d+k}(G)} \le \|g_1\|_{X_d(G)}\|g_2\|_{X_k(G)}. \tag{3.1}$$

Therefore, Haagerup's result in [27] implies that $X_2(\mathbb{F}_\infty) \ne X_4(\mathbb{F}_\infty)$ (equivalently $M_2(\mathbb{F}_\infty) \ne M_4(\mathbb{F}_\infty)$), since otherwise $X_2(\mathbb{F}_\infty)$ would be a Banach algebra under convolution.

To verify (3.1), we will need an alternate description of the space $X_d(G)$, which uses the Haagerup tensor product and the known results on multilinear cb maps (*cf.* [12, 55]). These results show that $X_d(G)$ may be identified with a quotient (modulo the kernel of the natural product map) of the Haagerup tensor product $\ell_1(G) \otimes_h \ldots \otimes_h \ell_1(G)$ of d copies of $\ell_1(G)$ equipped with its "maximal operator space structure". More explicitly, one can prove that the space $X_d(G)$ coincides with the space of all functions $g\colon G \to \mathbb{C}$ for which there is an element $\hat{g} = \sum_{G^d} \hat{g}(t_1, \ldots, t_d)\delta_{t_1} \otimes \ldots \otimes \delta_{t_d}$ in $\ell_1(G) \otimes_h \ldots \otimes_h \ell_1(G)$ such that

$$\forall t \in G \qquad g(t) = \sum_{t_1 \ldots t_d = t} \hat{g}(t_1, \ldots, t_d)$$

and moreover we have

$$\|g\|_{X_d(G)} = \inf\{\|\hat{g}\|_{\ell_1(G) \otimes_h \ldots \otimes_h \ell_1(G)}\}. \tag{3.2}$$

In addition the norm of an element $\hat{g}$ in the space $\ell_1(G) \otimes_h \ldots \otimes_h \ell_1(G)$ can also be explicited as follows:

$$\|\hat{g}\|_{\ell_1(G) \otimes_h \ldots \otimes_h \ell_1(G)} = \sup\left\{\left\|\sum_{G^d} \hat{g}(t_1, \ldots, t_d)x_{t_1}^1 \ldots x_{t_d}^d\right\|\right\}$$

where the supremum runs over all families $(x_t^1)_{t \in G}, \ldots, (x_t^d)_{t \in G}$ in the unit ball of $B(\ell_2)$. Actually (by e.g. [67, prop. 6.6]), the supremum remains the same if we restrict it to the case when the d families actually coincide with a single family $(x_t)_{t \in G}$ in the unit ball of $B(\ell_2)$.

Clearly, if $\hat{g}_1 \in \ell_1(G) \otimes_h \ldots \otimes_h \ell_1(G)$ (d times) and $\hat{g}_2 \in \ell_1(G) \otimes_h \ldots \otimes_h \ell_1(G)$ (k times) we have $\hat{g}_1 \otimes \hat{g}_2 \in \ell_1(G) \otimes_h \ldots \otimes_h \ell_1(G)$ ($d + k$ times) and $\|\hat{g}_1 \otimes \hat{g}_2\| \leq \|\hat{g}_1\| \|\hat{g}_2\|$. From this, (3.1) follows easily using (3.2).

Remark. Assume that $M_d(G) = M_{2d}(G)$. Then passing to the preduals, $X_d(G) = X_{2d}(G)$ with equivalent norms. By (3.1) with $k = d$, this implies that $X_d(G)$ is (up to isomorphism) a Banach algebra under convolution. Moreover, since the product in $X_d(G)$ is "induced" by the Haagerup tensor product, Blecher's characterization of operator algebras (see [4] which extends [6]) shows that $X_d(G)$ must be (unitally) isomorphic to a (unital) operator algebra. Combined with Theorem 2.10, this implies

Theorem 3.1. *In the situation of Theorem 2.10, the following assertions are equivalent:*

(i) *There is a $\theta \geq 1$ such that $B_\theta(G) = B_c(G)$ for all $c > \theta$.*
(ii) *There are $\theta \geq 1$ and an integer d such that $B_\theta(G) = M_d(G)$.*
(iii) *There is an integer d such that $M_d(G) = M_{2d}(G)$.*
(iv) *There is an integer d such that $X_d(G)$ is (up to isomorphism) a unital operator algebra under convolution.*

Proof. By Theorem 2.10, (i) implies (ii). By (2.2) and (2.3), (ii) implies (iii). The preceding remark shows that (iii) implies (iv). Finally, assume (iv). Then there is a unital operator algebra $A \subset B(H)$ and a unital isomorphism $u\colon X_d(G) \to A$. Let $\theta = \|u\|$ and $K = \|u^{-1}\|$. Clearly u restricted to the group elements defines a unital homomorphism π with $|\pi| \leq \theta$. By the very definition of $\|g\|_{\tilde{A}_\theta}$, this implies $\|u(g)\| \leq \|g\|_{\tilde{A}_\theta}$ for all finitely supported g, hence $\|g\|_{X_d(G)} \leq K\|u(g)\| \leq K\|g\|_{\tilde{A}_\theta}$. Conversely, we trivially have (see (2.3)) $\|g\|_{\tilde{A}_\theta} \leq \theta^d \|g\|_{X_d(G)}$. Thus we obtain $X_d(G) = \tilde{A}_\theta$, hence by duality $M_d(G) = B_\theta(G)$, which (recalling the basic inclusions (2.2) and (2.3))) implies (i). $\qquad\square$

Remark 3.2. Haagerup's proof in [27] that $M_2(\mathbb{F}_\infty) \neq UB(\mathbb{F}_\infty)$ can be outlined as follows. Let $G = \mathbb{F}_n$ with $2 \leq n < \infty$. Assume $M_2(G) = UB(G)$.

Step 1: By Baire's theorem, there exists $c > 1$ such that $M_2(G) = B_c(G)$ with equivalent norms.

Step 2: This implies that $X_2(G)$ is a Banach algebra under convolution (because the predual of $B_c(G)$ is clearly an operator algebra, see Theorem 3.1 above). Hence, there is $C > 0$ such that for all f, g finitely supported we have $\|f * g\|_{X_2(G)} \leq C\|f\|_{X_2(G)}\|g\|_{X_2(G)}$.

Step 3: By an averaging argument, the radial projection $f \to f_R$ defined by

$$f_R(t) = \sum_{s:\ |s|=|t|} f(s) \cdot [\text{card}\{s \mid |s| = |t|\}]^{-1}$$

is bounded on $M_2(G)$ so that $\|f_R\|_{M_2(G)} \leq \|f\|_{M_2(G)}$ for any f in $M_2(G)$.

Step 4: Let φ_z be the spherical function on G equal to z on words of length 1 (cf. e.g. [24]). This means that $\varphi_z(t) = \varphi(|t|)$ where φ is determined inductively by: $\varphi(0) = 1$, $\varphi(1) = z$ and

$$\varphi(k+1) = \frac{2n}{2n-1}\varphi(1)\varphi(k) - \frac{1}{2n-1}\varphi(k-1)$$

for all $k \geq 2$. The spherical property of φ_z implies that for any finitely supported radial function f we have $\varphi_z * f = \langle \varphi_z, f \rangle \varphi_z$, and hence if g is another finitely supported radial function, we have

$$\langle \varphi_z, f * g \rangle = \langle \varphi_z, f \rangle \langle \varphi_z, g \rangle.$$

Moreover, if $|z| < 1$ then $\varphi_z \in M_2(G)$ (actually $\varphi_z \in UB(G)$, see [45]). Thus, in short, although Step 5 below says it is unbounded, $\|\varphi_z\|_{M_2(G)}$ is finite whenever $|z| < 1$. Therefore, if $|z| < 1$, $f \to \langle \varphi_z, f \rangle$ defines a continuous multiplicative unital functional on the Banach subalgebra which is the closure of the set of finitely supported radial functions in $X_2(G)$. Clearly, this implies that $\langle \varphi_z, f \rangle$ is in the spectrum of f, hence its modulus is majorized by the spectral radius of f in the latter Banach algebra, and this is $\leq C\|f\|_{X_2(G)}$ by Step 2. Thus we obtain for f radial $|\langle \varphi_z, f \rangle| \leq C\|f\|_{X_2(G)}$. Now, for f finitely supported but not necessarily radial, we have

$$|\langle \varphi_z, f \rangle| = |\langle \varphi_z, f_R \rangle|$$
$$\leq C\|f_R\|_{X_2(G)}$$

hence by Step 3

$$\leq C\|f\|_{X_2(G)}.$$

This implies $\|\varphi_z\|_{M_2(G)} \leq C$. But this contradicts the next and final step proved in [27]:

Step 5: $\sup_{|z|<1} \|\varphi_z\|_{M_2(G)} = \infty$.

It is not clear to us how to extend this argument to M_d in place of M_2. The analogue of Step 3 is not clear to us (but seems likely to be true). Moreover, note that the analogue of Step 2 would require assuming $X_d(G) = X_{2d}(G)$, so it would seem that the argument would lead, at best, to $M_d(G) \neq M_{2d}(G)$.

4. The $B(H)$-valued case

Up to now, we have mainly concentrated on properties of spaces of coefficients or of analogous spaces of complex valued functions on G. We now turn to the more general $B(H)$-valued case which is entirely similar to the preceding treatment (corresponding to $\dim(H) = 1$). More generally, for any u.b. representation $\pi\colon G \to B(H)$ let us define

$$\mathrm{Sim}(\pi) = \inf\{\|S\|\,\|S^{-1}\| \mid S^{-1}\pi(\cdot)S \text{ is a unitary representation}\},$$

and let

$$\alpha(G) = \inf\{\alpha \geq 0 \mid \exists K \ \forall \pi \colon \ G \to B(H) \text{ u.b. } \mathrm{Sim}(\pi) \leq K|\pi|^\alpha\}.$$

Then again the same phenomenon arises:

Theorem 4.1. *Let G be a discrete group. If G is unitarizable then $\alpha(G) < \infty$. Moreover $\alpha(G) \in \mathbb{N}$.*

We will now explain what replaces $d_1(G)$ in this case.

First, we need to generalize the space $B(G)$, from complex values to operator values. Let H be a Hilbert space and let G be a semi-group with unit. We denote by $B_c(G; H)$ the space of all functions $f \colon G \to B(H)$ for which there are a u.b. unital homomorphism $\pi \colon G \to B(H_\pi)$ with $|\pi| \leq c$ and operators $\xi \colon H_\pi \to H$ and $\eta \colon H \to H_\pi$ such that

$$\forall\, t \in G \qquad\qquad f(t) = \xi\pi(t)\eta.$$

We define

$$\|f\|_{B_c(G;H)} = \inf\{\|\xi\| \ \|\eta\|\}$$

where the infimum runs over all possible such representations.
Here again, in the group case we will denote $B_1(G; H)$ simply by $B(G; H)$ to emphasize that $|\pi| \leq 1$ means that π is a unitary representation.

Similarly, we denote by $M_d(G; H)$ the space of functions $f \colon G \to B(H)$ for which there are bounded functions $\xi_i \colon G \to B(H_i, H_{i-1})$, $1 \leq i \leq d$, with $H_0 = H_d = H$, H_i Hilbert such that

$$\forall (t_1, \ldots, t_d) \in G^d \qquad f(t_1 t_2 \ldots t_d) = \xi_1(t_1)\xi_2(t_2) \ldots \xi_d(t_d).$$

We equip this space with the norm

$$\|f\|_{M_d(G;H)} = \inf\{ \sup_{t_1 \in G} \|\xi_1(t_1)\| \ldots \sup_{t_d \in G} \|\xi_d(t_d)\| \}.$$

Note that there is also an obvious $B(H)$-valued generalization of the spaces $M_d(I_1, \ldots, I_d)$ introduced in Remark 2.11 above. Let us denote it by $M_d(I_1, \ldots, I_d; H)$. Then, as before, for any $\varphi \colon G \to B(H)$, let $\Phi \colon G^d \to B(H)$ be defined by $\Phi(t_1, \ldots, t_d) = \phi(t_1 t_2 \ldots t_d)$. Then we have

$$\|\varphi\|_{M_d(G;H)} = \|\Phi\|_{M_d(G,\ldots,G;H)}.$$

Clearly, when $\dim(H) = 1$, we recover the previous spaces $B(G)$, $B_c(G)$ and $M_d(G)$. The following extensions of the previous results can be proved:

Theorem 4.2. *Consider a function $f \colon G \to B(H)$. Then $f \in B_1(G; H)$ iff $f \in \bigcap_{m \geq 1} M_m(G; H)$ and $\sup_m \|f\|_{M_m(G;H)} < \infty$. Moreover we have*

$$\|f\|_{B_1(G;H)} = \sup_{m \geq 1} \|f\|_{M_m(G;H)}.$$

On the other hand, $f \in \bigcup_{c>1} B_c(G;H)$ iff $f \in \bigcap_m M_m(G;H)$ and $\sup_{m\geq 1} \|f\|_{M_m(G;H)}^{1/m} < \infty$. In addition

$$\limsup_{m\to\infty} \|f\|_{M_m(G;H)}^{1/m} = \inf\{c \mid f \in B_c(G;H)\}.$$

Notation. Let us denote by $d(G)$ the smallest d such that

$$B(G;\ell_2) = M_d(G;\ell_2).$$

Then we have:

Theorem 4.3. *For any unitarizable group, we have*

$$\alpha(G) = d(G).$$

Warning. Unfortunately no example is known of G with $3 \leq \alpha(G) < \infty$.

Remark. Theorem 4.3 (with Theorems 2.3 and 0.1) shows that $\alpha(G) < 3$ iff G is amenable.

We now turn to the $B(H)$-valued variant of the space $X_d(G)$. Here we will use explicitly the Haagerup tensor product for operator spaces. We refer the reader to [12, 55] for more on this notion.

Let H be an infinite dimensional Hilbert space. We denote by $K(H)$ the space of compact operators on H.

Let E_1, E_2 be operator spaces. Let $x_1 \in K(H) \otimes E_1$, $x_2 \in K(H) \otimes E_2$. We will denote by $(x_1, x_2) \to x_1 \odot x_2$ the bilinear mapping from $(K(H)\otimes E_1)\times(K(H)\otimes E_2)$ to $K(H) \otimes (E_1 \otimes E_2)$ which is defined on rank one tensors by

$$(k_1 \otimes e_1) \odot (k_2 \otimes e_2) = (k_1 k_2) \otimes (e_1 \otimes e_2).$$

The Haagerup tensor product $E_1 \otimes_h E_2$ can be characterized as the unique operator space which is a completion of the algebraic tensor product and is such that for any $x \in K(H) \otimes [E_1 \otimes_h E_2]$ we have

$$\|x\|_{\min} = \inf\{\|x_1\|_{\min}\|x_2\|_{\min}\}$$

where the infimum runs over all factorization of the form

$$x = x_1 \odot x_2$$

with $x_1 \in K(H) \otimes E_1$ and $x_2 \in K(H) \otimes E_2$.

By definition of the Haagerup tensor product, $(x_1, x_2) \to x_1 \odot x_2$ extends to a contractive bilinear mapping from $(K(H) \otimes_{\min} E_1) \times (K(H) \otimes_{\min} E_2)$ to $K(H) \otimes_{\min} [E_1 \otimes_h E_2]$. We will still denote by $(x_1, x_2) \to x_1 \odot x_2$ this extension, and similarly for d-fold tensor products.

The space $X_d(G;H)$ is defined as the space of all functions $g\colon G \to K(H)$ for which there is an element $\hat{g} = \sum_{G^d} \hat{g}(t_1,\ldots,t_d) \otimes \delta_{t_1} \otimes \ldots \otimes \delta_{t_d}$ in $K(H) \otimes_{\min} [\ell_1(G) \otimes_h \ldots \otimes_h \ell_1(G)]$ such that

$$\forall t \in G \qquad g(t) = \sum_{t_1\ldots t_d=t} \hat{g}(t_1,\ldots,t_d)$$

and moreover we have

$$\|g\|_{X_d(G;H)} = \inf\{\|\hat{g}\|_{K(H)\otimes_{\min}[\ell_1(G)\otimes_h \ldots \otimes_h \ell_1(G)]}\}. \tag{4.1}$$

In addition the norm of an element $\hat{g}$ in the space $K(H)\otimes_{\min}[\ell_1(G)\otimes_h \ldots \otimes_h \ell_1(G)]$ can also be explicited as follows:

$$\|\hat{g}\|_{K(H)\otimes_{\min}[\ell_1(G)\otimes_h \ldots \otimes_h \ell_1(G)]} = \sup\left\{\left\|\sum_{G^d} \hat{g}(t_1,\ldots,t_d)\otimes x^1_{t_1}\ldots x^d_{t_d}\right\|_{B(H\otimes \ell_2)}\right\}$$

where the supremum runs over all families $(x^1_t)_{t\in G},\ldots,(x^d_t)_{t\in G}$ in the unit ball of $B(\ell_2)$. Here again (by e.g. [67, prop. 6.6]) the supremum is the same if we restrict the supremum to the case when the d families are all equal to a single family $(x_t)_{t\in G}$ in the unit ball of $B(\ell_2)$.

By definition of the Haagerup tensor product, we have also

$$\|\hat{g}\|_{K(H)\otimes_{\min}[\ell_1(G)\otimes_h \ldots \otimes_h \ell_1(G)]} = \inf\left\{\|g_1\|_{K(H)\otimes_{\min}\ell_1(G)}\cdots\|g_d\|_{K(H)\otimes_{\min}\ell_1(G)}\right\} \tag{4.2}$$

where the infimum runs over all factorizations of $\hat{g}$ of the form

$$\hat{g} = g_1 \odot g_2 \odot \cdots \odot g_d, \tag{4.3}$$

with $g_1, g_2, \cdots, g_d \in K(H)\otimes_{\min}\ell_1(G)$. Equivalently, (4.3) means that for all (t_i) in G^d we have

$$\hat{g}(t_1,\ldots,t_d) = g_1(t_1)g_2(t_2)\cdots g_d(t_d),$$

where the product is in $K(H)$.

From this, we deduce that

$$\|g\|_{X_d(G;H)} = \inf\left\{\|g_1\|_{K(H)\otimes_{\min}\ell_1(G)}\cdots\|g_d\|_{K(H)\otimes_{\min}\ell_1(G)}\right\}, \tag{4.4}$$

where the infimum runs over all factorizations of g (as a generalized d-fold convolution) of the form

$$g(t) = \sum_{t_1\ldots t_d=t} g_1(t_1)g_2(t_2)\cdots g_d(t_d) \qquad (g_i \in K(H)\otimes_{\min}\ell_1(G)).$$

In particular, (4.4) implies that for any integers d, k, we have

$$\|g\|_{X_{d+k}(G;H)} = \inf\{\|x\|_{\min}\|y\|_{\min}\}, \tag{4.5}$$

where the infimum runs over all pairs $x \in X_d(G;H)$ $y \in X_k(G;H)$ such that
$g(t) = \sum_{t_1 t_2=t} x(t_1)y(t_2)$ ($K(H)$-valued convolution).

The next statement is the $B(H)$-valued analogue of Theorem 3.1.

Theorem 4.4 ([62]). *Let G be a semigroup with unit. Let $H = \ell_2$. The following assertions are equivalent:*

(i) *There is a $\theta \geq 1$ such that $B_\theta(G;H) = B_c(G;H)$ for all $c > 0$.*
(i)' *There is a $\theta \geq 1$ such that $B_\theta(G;H) = B_c(G;H)$ for some $c > 0$.*
(ii) *There are $\theta \geq 1$ and an integer d such that $B_\theta(G;H) = M_d(G;H)$.*
(iii) *There is an integer d such that $M_d(G;H) = M_{d+1}(G;H)$.*

(iv) *There is an integer d such that $X_d(G)$ is (up to complete isomorphism) a unital operator algebra under convolution.*

Proof. In [61] the above Key Lemma 2.5 is actually proved in the $B(H)$-valued case. Therefore all the preceding statements numbered between 2.5 and 2.9 remain valid in the $B(H)$-valued case. Thus exactly the same rasoning as for Theorem 3.1 yields the equivalence of (i), (i)' and (ii). Clearly (ii) implies (iii).

Assume (iii). Then, passing to the preduals (here we mean the preduals of $M_d(G)$ and $M_{d+1}(G)$ in the operator space sense), we find $X_d(G;H) = X_{d+1}(G;H)$. This implies $X_{d+1}(G;H) = X_{d+2}(G;H)$. Indeed, by (4.5) for any g in the open unit ball of $X_{d+1}(G;H)$, we can find x in the unit ball of $X_d(G;H)$ and y in the unit ball of $X_1(G;H)$ such that

$$g(t) = \sum_{t_1 t_2 = t} x(t_1) y(t_2).$$

Now since $X_d(G;H) = X_{d+1}(G;H)$, $x \in X_{d+1}(G;H)$ hence by (4.5) g must be in $X_{d+2}(G;H)$. Now from $X_{d+1}(G;H) = X_{d+2}(G;H)$ we deduce $X_{d+2}(G;H) = X_{d+3}(G;H)$, and so on.., so that we must have $X_d(G;H) = X_{2d}(G;H)$, or equivalently $X_d(G) = X_{2d}(G)$ completely isomorphically. Note that, by (4.4) or (4.5), the convolution product defines a completely contractive linear map p from $X_d(G) \otimes_h X_d(G)$ to $X_{2d}(G)$, hence since $X_d(G) = X_{2d}(G)$ completely isomorphically, p is c.b. from $X_d(G) \otimes_h X_d(G)$ to $X_d(G)$, which implies by Blecher's result in [4] that $X_d(G)$ is completely isomorphic to an operator algebra. This proves that (iii) implies (iv).

Finally, assume (iv). Then, there are a unital subalgebra $A \subset B(\mathcal{H})$ and a unital homorphism $u\colon X_d(G) \to A$ which is also a complete isomorphism. Let $\theta = \|u\|_{cb}$ and $C = \|u^{-1}\|_{cb}$. Let $\pi(t) = u(\delta_t)$ $(t \in G)$. Then π is a u.b. representation of G with $|\pi| \leq \theta$. By the maximality of $\tilde{A}_\theta$, for any $x \in \mathbb{C}[G]$, we must have

$$\|u(x)\| \leq \|x\|_{\tilde{A}_\theta},$$

hence $\|x\|_{X_d(G)} \leq C\|x\|_{\tilde{A}_\theta}$. By duality, this implies that for all φ in $M_d(G)$ we have

$$\|\varphi\|_{B_\theta(G)} \leq C\|\varphi\|_{M_d(G)}.$$

Moreover, the same arguments with coefficients in $B(H)$ yield the c.b. version of this, so that we obtain, for all φ in $M_d(G;H)$

$$\|\varphi\|_{B_\theta(G;H)} \leq C\|\varphi\|_{M_d(G;H)}.$$

Thus we obtain (ii) and hence also (i), establishing (iv) $\Rightarrow$ (i). $\quad\square$

Remark. The preceding argument shows that (iii) and (iv) are equivalent for the same d.

5. A case study: The free groups

We wish to prove here the following.

Theorem 5.1. *For any* $d \geq 2$,

$$M_d(\mathbf{F}_\infty) \neq M_{d-1}(\mathbf{F}_\infty).$$

More precisely let $\{g_1, g_2, \ldots\}$ *be the free generators of* $\mathbf{F}_\infty$, *and for any* n *let* $W_{d,n}$ *be the subset of* $\mathbf{F}_\infty$ *formed of all the words* w *(of length* d*) of the form* $w = g_{i_1} g_{i_2} \ldots g_{i_d}$ *with* $1 \leq i_j \leq n$ *for any* $1 \leq j \leq d$. *Then, for any* n, *there is a function* $f_{d,n} \colon \mathbf{F}_\infty \to \mathbb{C}$ *supported on* $W_{d,n}$ *and unimodular on* $W_{d,n}$ *such that*

$$n^{\frac{d-1}{2}} \leq \|f_{d,n}\|_{M_d(\mathbf{F}_\infty)} \quad \text{and} \quad \|f\|_{M_{d-1}(\mathbf{F}_\infty)} \leq C(d) n^{\frac{d-2}{2}},$$

where $C(d)$ *is a constant depending only on* d.

Let

$$UB(G) = \bigcup_{c>1} B_c(G).$$

Since we have obviously inclusions $UB(G) \subset M_d(G) \subset M_{d-1}(G)$ for any group G, this implies

Corollary 5.2. *For any* $d \geq 1$,

$$M_d(\mathbf{F}_\infty) \neq UB(\mathbf{F}_\infty).$$

For $d = 2$ this is the main result of [27]. Note however that Theorem 5.1 yields a function f supported in the words of length 3 that is in $M_2(G)$ but not in $M_3(G)$ and hence not in $UB(G)$. It is easy to see that 3 is minimal here, i.e. any function supported in the words of length 2 that is in $M_2(G)$ must be in $UB(G)$ (see Proposition 5.8 below).

Let $I_1, \ldots, I_d$ be arbitrary sets. Recall the notation from Remark 2.11: We denote by $M_d(I_1, \ldots, I_d)$ the space of all functions $f \colon I_1 \times \cdots \times I_d \to \mathbb{C}$ for which there are bounded functions f_i

$$f_i \colon I_i \to B(H_i, H_{i-1}) \quad \text{(here } H_i \text{ are Hilbert spaces with } H_d = H_0 = \mathbb{C}\text{) such that}$$

$$f(b_1, \ldots, b_d) = f_1(b_1) \ldots f_d(b_d) \qquad \forall \, b_i \in I_i.$$

We equip this space with the norm

$$\|f\| = \inf \left\{ \prod_{i=1}^{d} \sup_{b \in I_i} \|f_i(b)\| \right\}$$

where the infimum runs over all possible such factorizations.

Let $J_i \subset I_i$ be arbitrary subsets. Note that we obviously have

$$\|f_{|J_1 \times \cdots \times J_d}\|_{M_d(J_1, \ldots, J_d)} \leq \|f\|_{M_d(I_1, \ldots, I_d)}.$$

Moreover, for any function $g \colon J_1 \times \cdots \times J_d \to \mathbb{C}$ let $\tilde{g} \colon I_1 \times \cdots \times I_d \to \mathbb{C}$ be the extension of $\tilde{g}$ equal to zero outside $J_1 \times \cdots \times J_d$. Then it is easy to check that

$$\|\tilde{g}\|_{M_d(I_1, \ldots, I_d)} = \|g\|_{M_d(J_1, \ldots, J_d)}. \tag{5.1}$$

We will relate these spaces to $M_d(\mathbb{F}_\infty)$ via the following observation. Given a function $\varphi\colon \mathbb{F}_\infty \to \mathbb{C}$ supported by $W_{d,n}$, we can define $f\colon [1,\ldots,n]^d \to \mathbb{C}$ by

$$\forall i_j \in [1,\ldots,n] \qquad f(i_1,i_2,\ldots,i_d) = \varphi(g_{i_1} g_{i_2} \ldots g_{i_d}).$$

We have then obviously if $I = [1,\ldots,n]$

$$\|f\|_{M_d(I,\ldots,I)} \le \|\varphi\|_{M_d(\mathbb{F}_\infty)}. \tag{5.2}$$

The main idea for the proof of Theorem 5.1 is to compare $\|\varphi\|_{M_{d-1}(\mathbb{F}_\infty)}$ with certain norms of f of the form $M_{d-1}(I_1,\ldots,I_{d-1})$ when f is viewed as depending on less than d variables, by blocking together certain variables, so that $I_1 = I^{p_1}$, $I_2 = I^{p_2},\ldots$ with $p_1 + p_2 + \cdots + p_{d-1} = d$.

Remark. With the notation used in operator space theory, the space $M_d(I_1,\ldots,I_d)$ can be identified with the dual of the Haagerup tensor product $\ell_1(I_1) \otimes_h \cdots \otimes_h \ell_1(I_d)$, where the spaces $\ell_1(I_j)$ are equipped (as usual) with their maximal operator space structure in the sense of e.g. [22] or [67].

Consider now a partition $\pi = (\alpha_1,\ldots,\alpha_k)$ of $[1,\ldots,d]$ into disjoint intervals (="blocks") with $k < d$, so that at least one α_i has $|\alpha_i| > 1$. Let $I(\alpha_i) = \prod_{q \in \alpha_i} I_q$.

We have a natural mapping from $M_d(I_1,\ldots,I_d)$ to $M_k(I(\alpha_1),\ldots,I(\alpha_k))$ associated to the canonical identification

$$I_1 \times \cdots \times I_d = I(\alpha_1) \times \cdots \times I(\alpha_k).$$

It is easy to check that this mapping is contractive. For simplicity of notation, we denote

$$M(\pi) = M_k(I(\alpha_1),\ldots,I(\alpha_k)).$$

Moreover, it is useful to observe that if π' is another partition of $[1,\ldots,d]$ that is finer than π (i.e. such that every block of π is a union of certain blocks of π'), then we have $M(\pi') \subset M(\pi)$ and for any $f\colon [1,\ldots,n]^d \to \mathbb{C}$

$$\|f\|_{M(\pi)} \le \|f\|_{M(\pi')}. \tag{5.3}$$

Note however that since the set of all partitions is only *partially* ordered (and not totally ordered), the intersection $\bigcap_\pi M(\pi)$ over all partitions with k blocks does not reduce to one of the $M(\pi)$. We equip this intersection $\bigcap_\pi M(\pi)$ with its natural norm, namely :

$$\|f\| = \max_\pi \|f\|_{M(\pi)},$$

where the maximum runs over all π with at most $d - 1$ blocks.

The main point in the proof of Theorem 5.1 is the following.

Lemma 5.3. *Assuming $I_1,\ldots,I_d$ are infinite sets, then for any $d > 1$ the natural mapping*

$$\Phi\colon M_d(I_1,\ldots,I_d) \longrightarrow \bigcap_\pi M(\pi)$$

is not an isomorphism.

To prove this, we will use two more lemmas.

Lemma 5.4. *For any f in $M_d(I_1, \ldots, I_d)$ and any fixed j in $[1, 2, \ldots, d]$ we have*

$$\|f\|_{M_d(I_1,\ldots,I_d)} \leq \sup_{I_1 \times \cdots \times I_d} |f| \cdot \left[\prod_{m \neq j} |I_m| \right]^{1/2}.$$

Proof. Let us write

$$f(i_1, \ldots, i_d) = f(a, i_j, b).$$

Then for each fixed j, the matrix $(f(a, i_j, b))_{a,b}$ defines an operator $\xi(i_j)$ from $H = \ell_2(I_{j+1}) \otimes \cdots \otimes \ell_2(I_d)$ to $K = \ell_2(I_1) \otimes \cdots \otimes \ell_2(I_{j-1})$ and its norm can be majorized as follows (observe that an $n \times m$ matrix (a_{ij}) has norm bounded by $\sup_{ij} |a_{ij}| \sqrt{n} \sqrt{m}$)

$$\|\xi(i_j)\| \leq \sup |f| \cdot \left(\prod_{m \neq j} |I_m| \right)^{1/2}.$$

Moreover we have

$$f(i_1, \ldots, i_d) = a_1(i_1) \ldots a_j(i_{j-1}) \xi(i_j) b_{j+1}(i_{j+1}) \ldots b_d(i_d)$$

with $a_m \colon K \to K$ and $b_m \colon H \to H$ defined by $a_m(i_m) = 1 \otimes \cdots 1 \otimes e_{1i_m} \otimes 1 \cdots \otimes 1$ and $b_m(i_m) = 1 \otimes \cdots 1 \otimes e_{i_m 1} \otimes 1 \cdots \otimes 1$, where the middle term is in the place of index m. $\square$

Lemma 5.5 (Marius Junge). *Assume $|I_1| = \cdots = |I_d| = n$. Then the natural identity map from $\ell_\infty(n^d)$ to $M_d(I_1, \ldots, I_d)$ has norm $n^{\frac{d-1}{2}}$. Equivalently, we have*

$$\sup \left\{ \left\| \sum z_{i_1 i_2 \ldots i_d} \, e_{i_1} \otimes \cdots \otimes e_{i_d} \right\|_{M_d(I_1,\ldots,I_d)} \ \Big| \ |z_{i_1 i_2 \ldots i_d}| \leq 1 \right\} = n^{\frac{d-1}{2}}. \quad (5.4)$$

Proof. I am grateful to Marius Junge for kindly providing this lemma in answer to a question of mine for $d = 3$. Let C be the left side of (5.4). The fact that $C \leq n^{\frac{d-1}{2}}$ follows from Lemma 5.4. The main point is the converse. To prove this, consider the function

$$\psi(i_1, \ldots, i_d) = a_{i_1 i_2} a_{i_2 i_3} \ldots a_{i_{d-1} i_d}$$

where (a_{ij}) is an $n \times n$ unitary matrix with $|a_{ij}| = n^{-1/2}$. Let $\xi_i(x) = \langle x, e_i \rangle$. Then we can write

$$\psi(i_1, \ldots, i_d) = \langle a e_{i_2 i_2} a e_{i_3 i_3} \ldots e_{i_{d-1} i_{d-1}} a e_{i_d}, e_{i_1} \rangle = \xi_{i_1} \left(a e_{i_2 i_2} a e_{i_3 i_3} \ldots e_{i_{d-1} i_{d-1}} a e_{i_d} \right)$$

where a appears $d - 1$ times, from which it follows that

$$\|\psi\|_{M_d(I_1,\ldots,I_d)^*} \leq n. \quad (5.5)$$

Indeed, if $\|f\|_{M_d(I_1,\ldots,I_d)} < 1$. We may assume $f(b_1,\ldots,b_d) = f_1(b_1)\ldots f_d(b_d)$ $\forall\, b_i \in I_i$ with $\|f_i(b_i)\| < 1$ $(b_i \in I_i)$ for all i. Then we have

$$\sum \psi(i_1,\ldots,i_d)f(i_1,\ldots,i_d) = [\sum_{i_1} \xi_{i_1} \otimes f_1(i_1)][a \otimes I][\sum_{i_2} e_{i_2 i_2} \otimes f_2(i_2)]\ldots$$

hence

$$|\sum \psi(i_1,\ldots,i_d)f(i_1,\ldots,i_d)| \leq \|\sum \xi_{i_1} \otimes f_1(i_1)\|\|a\|^{d-1}\|\sum f_d(i_d) \otimes e_{i_d}\|$$
$$\leq \sqrt{n}\|a\|^{d-1}\sqrt{n} \leq n,$$

which establishes (5.5) Therefore we must have

$$\sum |\psi(i_1,\ldots,i_d)| \leq Cn,$$

whence

$$n^d(n^{-1/2})^{d-1} \leq Cn$$

which yields $C \geq n^{\frac{d-1}{2}}$ as announced.

Note: The preceding proof uses implicitly ideas from operator space theory namely the identity $M_d(I_1,\ldots,I_d) = \ell_\infty^n \otimes_h \cdots \otimes_h \ell_\infty^n$ (d times), for which we refer to e.g. [22] or [67]. $\qquad\square$

Remark. Let

$$\varepsilon(p,q) = \exp\{ipq/n\},$$

and let $a_{pq} = \varepsilon(p,q)n^{-1/2}$. Thus, the $n \times n$ unitary matrix $a = (a_{pq})$ represents the Fourier transform on the group $\mathbb{Z}/n\mathbb{Z}$. Let

$$F_{d,n}(i_1,\ldots,i_d) = \varepsilon(i_1,i_2)\varepsilon(i_2,i_3)\ldots\varepsilon(i_{d-1},i_d).$$

Then the preceding proof yields

$$\|F_{d,n}\|_{M_d(I_1,\ldots,I_d)} = n^{\frac{d-1}{2}}.$$

Proof of Lemma 5.3. By (5.1), it suffices to show that if $|I_1| = \cdots = |I_d| = n$ then $\|\Phi^{-1}\| \geq \sqrt{n}$ for all n. Thus we now assume $|I_1| = \cdots = |I_d| = n$ throughout this proof. By Lemma 5.4, for any π we have $\|\Phi^{-1}\colon \ell_\infty(n^d) \to M(\pi)\| \leq n^{\frac{d-2}{2}}$. (Indeed, we can choose α_j with $|\alpha_j| \geq 2$, hence $|I(\alpha_j)| \geq n^2$.) It follows that

$$\left\|\Phi^{-1}\colon \ell_\infty(n^d) \longrightarrow \bigcap_\pi M(\pi)\right\| \leq n^{\frac{d-2}{2}}.$$

Note that for the mapping underlying Φ^{-1} we have

$$\|\ell_\infty(n^d) \to M_d(I_1,\ldots,I_d)\| \leq \|\ell_\infty(n^d) \longrightarrow \bigcap_\pi M(\pi)\| \times \|\bigcap_\pi M(\pi)$$
$$\longrightarrow M_d(I_1,\ldots,I_d)\|$$

Thus the above estimate together with Lemma 5.5 implies

$$\sqrt{n} \leq \|\Phi^{-1}\colon \bigcap_\pi M(\pi) \longrightarrow M_d(I_1,\ldots,I_d)\|. \qquad\square$$

Let $G = \mathbb{F}_n$ with $2 \leq n \leq \infty$ and let I denote the set of generators of G. Let W_d be the set of all elements of G which are a product of exactly d generators. Let $F\colon I^d \to \mathbb{C}$ be a function and let $f\colon G \to \mathbb{C}$ be the function defined on W_d by

$$f(i_1 i_2 \ldots i_d) = F(i_1, i_2, \ldots, i_d) \tag{5.6}$$

and equal to zero outside W_d.

Lemma 5.6. *With the above notation, we have*

$$\|F\|_{M_d(I,\ldots,I)} \leq \|f\|_{M_d(\mathbb{F}_n)} \quad \text{and} \quad \|f\|_{M_{d-1}(\mathbb{F}_n)} \leq C(d) \sup_\pi \|F\|_{M(\pi)}$$

where the supremum runs over all (nontrivial) partitions of $[1, \ldots, d]$ into K disjoint intervals (= blocks), with $K \leq d-1$ and where $C(d)$ is a constant depending only on d.

Proof. The inequality $\|F\|_{M_d(I,\ldots,I)} \leq \|f\|_{M_d(\mathbb{F}_n)}$ is essentially obvious by going back to the definitions, so we will now concentrate on the converse direction. Consider $t_1, \ldots, t_{d-1}$ in $G = \mathbb{F}_n$ such that their product $t_1 t_2 \ldots, t_{d-1}$ belongs to W_d i.e. $t_1 t_2 \ldots t_{d-1}$ can be written as a reduced word of the form $g_{i_1} g_{i_2} \ldots g_{i_d}$ where $\{g_i \mid i \in I\}$ denotes the free generators of $\mathbb{F}_n$. Since the letters $g_{i_1}, \ldots, g_{i_d}$ remain after successive reductions in the product $t_1 t_2 \ldots t_{d-1}$ it is easy to check that each t_i contributes a block of p_i letters in x with $\sum_1^{d-1} p_i = d$ (we allow $p_i = 0$).

This means that when $p_i > 0$, t_i can be written as a reduced word $x_i a_i y_i^{-1}$ with $|a_i| = p_i$ and when $p_i = 0$ we set $a_i = e$, so that

$$t_1 t_2 \ldots t_{d-1} = a_1 a_2 \ldots a_{d-1}. \tag{5.7}$$

Thus to each $t = (t_1, \ldots, t_{d-1})$ as above we can associate $p(t) = (p_1, \ldots, p_{d-1})$. Actually, we have a problem here: this $p(t)$ is unambiguously defined when $d = 3$ (hence we only have to consider products of two elements). But when $d > 3$ (and thus $d - 1 > 2$) there might be several reductions of $t_1 t_2 \ldots t_{d-1}$ leading to the same element of W_d, thus there might be several possibilities for $p(t)$. For instance, when $d - 1 = 3$, denoting the generators by a, b, c, the product $abcd = (ab)(b^{-1}a^{-1}c)(c^{-1}abcd)$ (we mean here $t_1 = ab$, $t_2 = b^{-1}a^{-1}c$, $t_3 = c^{-1}abcd$) allows $p(t) = (0, 0, 4)$ but also $p(t) = (2, 0, 2)$ or $p(t) = (1, 0, 3)$. We prefer to ignore this difficulty for the moment while still treating the general case, so let us assume $d = 3$ so that $p(t)$ is always well defined. Moreover, if we delete the indices for which $p_i = 0$ (and $a_i = e$), we obtain a partition $\pi(t)$ into k blocks $(\alpha_1, \ldots, \alpha_k)$ with $k \leq d - 1$. Then we can rewrite (5.7) as

$$t_1 t_2 \ldots t_{d-1} = b_1 b_2 \ldots b_k \tag{5.8}$$

with $b_m \in W_{|\alpha_m|}$. Here we implicitly mean that the non-reduced product $t_1 t_2 \ldots t_d$ can be viewed (just by adding parenthesis) as a product

$$c_1 b_1 c_2 b_2 c_3 \ldots b_k c_{k+1}$$

where each of the intermediate products $c_1, \ldots, c_{k+1}$ reduces to e. Moreover, the k-tuple $(b_1, \ldots, b_k)$ determines a k-tuple $(\hat{b}_1, \ldots, \hat{b}_k)$ with $\hat{b}_1 \in I^{\alpha_1}, \hat{b}_2 \in I^{\alpha_2}, \ldots, \hat{b}_k \in$

I^{α_k}. Now fix $\varepsilon > 0$. For any partition $\pi = (\alpha_1, \alpha_2, \ldots, \alpha_k)$, we can "factorize" F as follows:

$$F(j_1, \ldots, j_k) = \eta_1^\pi(j_1) \ldots \eta_k^\pi(j_k) \qquad (j_m \in I^{\alpha_m}) \tag{5.9}$$

where η_m^π are $B(H_m, H_{m-1})$-valued functions $(H_k = H_0 = \mathbb{C})$ such that

$$\prod_m \sup \|\eta_m^\pi(j_m)\| \leq \|F\|_{M(\pi)}(1 + \varepsilon). \tag{5.10}$$

Let $\tilde{f}$ be the function defined on G^{d-1} by $\tilde{f}(t_1, \ldots, t_{d-1}) = f(t_1 t_2 \ldots t_{d-1})$. Consider now the disjoint decomposition $\tilde{f} = \sum_p \tilde{f}_p$ where $\tilde{f}_p = \tilde{f} \cdot 1_{\{t: \ p(t) = p\}}$ where the first sum runs over all choices of $p = (p_1, \ldots, p_{d-1})$ with $p_i \geq 0$ and $\sum p_i = d$.

We claim that $\|\tilde{f}_p\|_{M_{d-1}(G, \ldots, G)} \leq \|F\|_{M(\pi)}$ where π is the partition associated to $p = (p_1, \ldots, p_{d-1})$ after removal of the empty blocks.

To prove this claim, we will produce a factorization formula for $\tilde{f}_p(t_1, t_2, \ldots, t_{d-1})$, namely we will show

$$\tilde{f}_p(t_1, t_2, \ldots, t_{d-1}) = \langle \xi_1^p(t_1) \ldots \xi_{d-1}^p(t_{d-1}) \delta_e, \delta_e \rangle.$$

To define $\xi_i^p(t_i)$ we must distinguish whether $p_i = 0$ or not.

If $p_i = 0$ we set $\xi_i^p(\theta) = \lambda(\theta) \otimes 1$ (here $\lambda(\theta)$ denotes as usual left translation by θ on $\ell_2(G)$). On the other hand, if $p_i > 0$ so that i corresponds to a block α_m of π with $|\alpha_m| = p_i$, we write

$$\xi_i^p(\theta) = \sum e_{x,y} \otimes \eta_m^\pi(\hat{a})$$

where the sum runs over all ways to decompose θ as $x \cdot a \cdot y^{-1}$ as a *reduced* product, with a a product of generators such that $|a| = p_i$ and where $\hat{a}$ denotes the element of I^{α_m} corresponding to a (x and y^{-1} being initial and final segments in the reduced word θ; we allow here $x = e$ or $y = e$). In case θ does not admit any such decomposition (i.e. θ does not admit any subword in W_{p_i}), we set $\xi_i^p(\theta) = 0$.
Note that we have $\|\xi_i^p(\theta)\| \leq \sup_a \|\eta_m^\pi(\hat{a})\|$. Indeed, when θ is fixed, in the various ways to write $\theta = x \cdot a \cdot y^{-1}$ as a *reduced* product as above, all the x's appearing will be distinct since they have different length, and similarly all the y's will be distinct, so the various operators $e_{x,y} \otimes \eta_m^\pi(\hat{a})$ have both orthogonal ranges and orthogonal domains, so that the norm of their sum is majorized by the maximum norm of each term.
A (tedious but) straightforward verification shows that if $t_1 t_2 \ldots t_{d-1} \in W_d$ with $p(t) = p$, and if $t_1 t_2 \ldots t_{d-1} = b_1 b_2 \ldots b_k$ as described in (5.8), then we have using (5.9)

$$\langle \xi_1^p(t_1) \ldots \xi_{d-1}^p(t_{d-1}) \delta_e, \delta_e \rangle = \eta_1^\pi(\hat{b}_1) \ldots \eta_k^\pi(\hat{b}_k) \tag{5.11}$$

$$= F(\hat{b}_1, \ldots, \hat{b}_k)$$

whence by (5.6) and (5.8) $\qquad = f(t_1 \ldots t_{d-1}).$

Moreover, if $p(t) \neq p$ the left side of (5.8) vanishes.

Indeed, if that left side is non zero, then we must have

$$t_1 \cdots t_{d-1} = [x_1.a_1.y_1^{-1}][x_2.a_2.y_2^{-1}] \cdots [x_{d-1}.a_{d-1}.y_{d-1}^{-1}] = a_1 a_2 \cdots a_{d-1}$$

with $a_i \in W_{p_i}$ if $p_i > 0$, and $a_i = e$ otherwise, a_i being a subword of t_i, in such a way that the product of all the terms figuring in between two successive a_i's with $p_i > 0$ reduces to e, as well as the product of all the terms preceding the first a_i with $p_i > 0$, and that of all the terms after the last a_i with $p_i > 0$. This implies $p(t) = p$ and deleting the a_i's equal to e we obtain $t_1 \cdots t_{d-1} = b_1 \cdots b_k$ and (5.8) is then easy to check.

Thus we have the announced factorization of $\tilde{f}_p$; the latter implies

$$\|\tilde{f}_p\|_{M_{d-1}(G,..,G)} \le \prod \sup \|\eta_m^\pi\| \le \|F\|_{M(\pi)}(1+\varepsilon).$$

Using $\tilde{f} = \sum \tilde{f}_p$, this yields $\|f\|_{M_{d-1}(G)} = \|\tilde{f}\|_{M_{d-1}(G,..,G)} \le C_d \|F\|_{M(\pi)}(1+\varepsilon)$ (here C_d is the number of possible p's) thus completing the proof of the lemma, at least in the case $d = 3$. Since there are only four possibilities for p (namely $(3.0), (0.3), (1,2)$, and (2.1)) we obtain $C_3 \le 4$.

Now in the general case, the problem is that, for each $t = (t_1, \ldots, t_{d-1}) \in G^{d-1}$ such that $t_1 \ldots t_{d-1} \in W_d$, there is a multiplicity of possible $p(t)'s$ (or of possible associated partitions $\pi(t)$): each such t admits $N(t)$ possible distinct $p(t)'s$. However, we of course have a bound for this: $1 \le N(t) \le N_d$ where the upper bound N_d depends only on d. If $t_1 \ldots t_{d-1} \notin W_d$, we set $N(t) = 0$. Then, we think of $p(t)$ as a multivalued function and we define

$$\tilde{f}_p = \tilde{f} \cdot 1_{\{t:\ p \in p(t)\}}.$$

Then the preceding shows again that

$$\|\sum \tilde{f}_p\|_{M_{d-1}(G,..,G)} \le C_d \|F\|_{M(\pi)}(1+\varepsilon), \tag{5.12}$$

but, since the sum is no longer disjoint, we have

$$\forall t \in G^{d-1} \quad \sum_p \tilde{f}_p(t) = N(t)\tilde{f}(t). \tag{5.13}$$

Consider now the special case when F is identically equal to 1. Note that $\|F\|_{M(\pi)} \le 1$ and $N(t)\tilde{f}(t) = N(t)$ in this case. Thus, the preceding identity and (5.12) shows that the function $N\colon G^{d-1} \longrightarrow \mathbb{R}$ is in $M_{d-1}(G, \ldots, G)$ with norm $\le C_d$. To conclude, we will mupliply (5.13) by a function equal to $1/N$ on the support of N and we will bound its norm in $M_{d-1}(G, \ldots, G)$ by a constant C_d'. Since $M_{d-1}(G, \ldots, G)$ is a Banach algebra for the pointwise product, this will yield the desired result. (Alternately, we could use a disjointification trick, as above for (2.19).)

Let P be a polynomial such that $P(k) = \frac{1}{k}$ for all $k = 1, 2, \ldots, N_d$. To fix ideas, we let P be determined by Lagrange interpolation. Since $M_{d-1}(G, \ldots, G)$ is a Banach algebra, $P(N) \in M_{d-1}(G, \ldots, G)$ and since P depends only on d, we have $\|P(N)\|_{M_{d-1}(G,..,G)} \le C_d'$ for some C_d' depending only on d. Then we can write

$$P(N) \cdot \sum_p \tilde{f}_p = P(N) \cdot N\tilde{f} = \tilde{f}$$

hence we conclude

$$\|f\|_{M_{d-1}(G)} = \|\tilde{f}\|_{M_{d-1}(G,...,G)} \leq \|P(N)\|_{M_{d-1}(G,...,G)} \|\sum_p \tilde{f}_p\|_{M_{d-1}(G,...,G)}$$
$$\leq C_d' C_d \|F\|_{M(\pi)}(1 + \varepsilon). \qquad \square$$

Proof of Theorem 5.1. Assume $M_d(\mathbb{F}_\infty) = M_{d-1}(\mathbb{F}_\infty)$. Then there must exist a constant C' such that for all f in $M_{d-1}(\mathbb{F}_\infty)$ we have

$$\|f\|_{M_d(\mathbb{F}_\infty)} \leq C' \|f\|_{M_{d-1}(\mathbb{F}_\infty)}.$$

Then by Lemma 5.6 we find that $\Phi^{-1} \colon \bigcap_\pi M(\pi) \to M_d(I,\ldots,I)$ is bounded, which contradicts Lemma 5.3 for any $d > 1$. Now, let $F_{d,n}$ be as in the remark following Lemma 5.5, and let $f_{d,n}$ be defined by

$$f_{d,n}(g_{i_1} g_{i_2} \cdots g_{i_d}) = F_{d,n}(i_1, \ldots, i_d)$$

and $f_{d,n}(t) = 0$ if $t \notin W_{d,n}$. Then by the latter remark and by Lemma 5.6 we have

$$n^{\frac{d-1}{2}} \leq \|f_{d,n}\|_{M_d(\mathbb{F}_\infty)},$$

and also by Lemma 5.6 and Lemma 5.4

$$\|f_{d,n}\|_{M_d(\mathbb{F}_\infty)} \leq C(d) \sup_\pi \|F_{d,n}\|_{M(\pi)} \leq C(d) n^{\frac{d-2}{2}}.$$

This completes the proof. $\qquad \square$

Note that in the special case $d = 3$, we obtain a very explicit example: Namely the function $f_{3,n}$ supported on $W_{3,n}$ and defined there for $1 \leq p,q,r \leq n$ by

$$f_{3,n}(g_p g_q g_r) = \exp(i(p+r)q/n),$$

satisfies

$$n \leq \|f_{3,n}\|_{M_3(G)} \quad \text{but} \quad \|f_{3,n}\|_{M_2(G)} \leq 4n^{1/2}.$$

Remark 5.7. The proof of Lemma 5.6 can be modified to show that $\|f\|_{M_d(\mathbb{F}_n)} \leq C'(d)\|F\|_{M_d(I,\ldots,I)}$ for some constant $C'(d)$ depending only on d. In particular, we have $\|1_{W_{d,n}}\|_{M_d(\mathbb{F}_n)} \leq C'(d)$ and hence $\|1_{W_d}\|_{M_d(\mathbb{F}_\infty)} \leq C'(d)$.

Note however:

Proposition 5.8. *Any function $f \colon \mathbb{F}_\infty \to \mathbb{C}$, supported on W_d, that is in $M_d(\mathbb{F}_\infty)$ must necessarily be in $UB(\mathbb{F}_\infty)$.*

Proof. Indeed, $f_{|W_d} \colon W_d \to \mathbb{C}$ admits an extension to a function $\hat{f} \colon \mathbb{F}_\infty \to \mathbb{C}$ that is in $C^*(\mathbb{F}_\infty)^* = B(\mathbb{F}_\infty)$. This follows from [67, Corollary 8.13]. Now, since, by Remark 5.7, 1_{W_d} belongs to $UB(\mathbb{F}_\infty)$, the pointwise product $f = \hat{f} \cdot 1_{W_d}$ also belongs to $UB(\mathbb{F}_\infty)$. $\qquad \square$

This shows in particular that a function in $M_2(\mathbb{F}_\infty)\backslash M_3(\mathbb{F}_\infty)$ cannot be supported on W_2. Thus the above example $f_{3,n}$ supported on W_3 appears somewhat "minimal".

Let G be any free group. Recall that we denote by $\mathcal{W}(d)$ the set of all words of length d in the generators and their inverses. Note that the inclusion $W_d \subset \mathcal{W}(d)$ is strict. We chose to concentrate on W_d (rather than on $\mathcal{W}(d)$) because then the ideas are a bit simpler and Lemma 5.6 is somewhat prettier in that case: Indeed, that lemma identifies the spaces $\{f \in M_d(G) \mid \mathrm{supp}(f) \subset W_d\}$ and $\{f \in M_{d-1}(G) \mid \mathrm{supp}(f) \subset W_d\}$ with two distinct spaces of functions on G^d, thus reducing, in some sense, a problem in harmonic analysis to one in functional analysis. However, most of our results hold with suitable modification for functions with support in $\mathcal{W}(d)$. We will merely describe them with mere indication of proof.

Fix $d \geq 1$ and let $k \leq d$. Let $f\colon G \to \mathbb{C}$ be a function supported on $\mathcal{W}(d)$. Let π be a partition of $[1, \ldots, d]$ in k disjoint consecutive blocks (intervals) $\alpha_1, \ldots, \alpha_k$ so that $|\alpha_1| + \cdots + |\alpha_k| = d$ and $|\alpha_i| \geq 1$ for all $i = 1, \ldots, k$. We will denote by $k(\pi)$ the number of blocks, i.e. we set $k(\pi) = k$. We define $f_\pi\colon \mathcal{W}(|\alpha_1|) \times \cdots \times \mathcal{W}(|\alpha_k|) \to \mathbb{C}$ by

$$f_\pi(x_1, x_2, \ldots, x_k) = f(x_1 x_2 \ldots x_k).$$

Note that $f_\pi(x_1, \ldots, x_k) = 0$ if the product $x_1 x_2 \ldots x_k$ is not a reduced word, since then it has length $< d$. For any function $F\colon \mathcal{W}(|\alpha_1|) \times \cdots \times \mathcal{W}(|\alpha_k|) \to \mathbb{C}$. we denote again

$$\|F\|_{\mathcal{M}(\pi)} = \|F\|_{M(\mathcal{W}(|\alpha_1|), \ldots, \mathcal{W}(|\alpha_k|))}. \tag{5.14}$$

The preceding proofs (mainly Lemma 5.6) then yield

Theorem 5.9. *With the preceding notation, we have for any function f with support in $\mathcal{W}(d)$ and for any integer $K \geq 1$*

$$\sup_{k(\pi) \leq K} \|f_\pi\|_{\mathcal{M}(\pi)} \leq \|f\|_{M_K(G)} \leq C(d, K) \sup_{k(\pi) \leq K} \|f_\pi\|_{\mathcal{M}(\pi)}$$

where $C(d, K)$ is a constant depending only on d and K.

Remark. Let π_0 be the partition of $[1, \ldots, d]$ into singletons, so that $k(\pi_0) = d$. When $K \geq d$, we have for any f as in (5.14)

$$\sup_{k(\pi) \leq K} \|f_\pi\|_{\mathcal{M}(\pi)} = \|f_{\pi_0}\|_{\mathcal{M}(\pi_0)}. \tag{5.15}$$

Indeed, we have $k(\pi_0) \leq K$ and moreover it is easy to see using (5.3) that if a partition π is less fine than another one π' (i.e. every block in π is the union of certain blocks of π') we have

$$\|f_\pi\|_{\mathcal{M}(\pi)} \leq \|f_{\pi'}\|_{\mathcal{M}(\pi')}.$$

Since any π is less fine that π_0, (5.15) follows immediately.

Thus, for all $K \geq d$, the norms $\|f\|_{M_K(G)}$ are equivalent on functions f with support in $\mathcal{W}(d)$. Indeed, by (5.14) they are equivalent to $\|f_{\pi_0}\|_{\mathcal{M}(\pi)}$. In sharp contrast when $K < d$, in particular when $K = d - 1$, they are no longer equivalent.

Corollary 5.10. *Let G be any group containing a (non-Abelian) free subgroup. Then, for any $\theta > 1$ and any $c > \theta$ we have $B_c(G) \neq B_\theta(G)$, consequently there is a representation $\pi\colon G \to B(H)$ with $|\pi| \leq c$ that is not similar to any representation π' with $|\pi'| \leq \theta$.*

Proof. Assume by contradiction that the conclusion fails. Then, a fortiori it must also fail for $G = \mathbb{F}_\infty$, by an "induction" argument as in Proposition 0.5. Hence, by Theorem 2.10, we obtain $M_d(\mathbb{F}_\infty) = M_{d+1}(\mathbb{F}_\infty)$ for some d, contradicting Theorem 5.1.   □

Remark. In the case of $G = \mathrm{SL}_2(\mathbb{R})$, Michael Cowling showed me a very concrete proof (that he attributed to Haagerup) of the conclusion of Corollary 5.10. That proof uses the estimates of Kunze and Stein from [35], the Bruhat decomposition of $\mathrm{SL}_2(\mathbb{R})$ and the amenability of the subgroup of triangular matrices in $\mathrm{SL}_2(\mathbb{R})$.

Remark. In [65], we study the same question as in this section but for $G = \mathbb{N}$. We answer a related question of Peller concerning power bounded operators (=uniformly bounded representations of $G = \mathbb{N}$), by showing

$$M_2(\mathbb{N}) \neq M_3(\mathbb{N}).$$

On the other hand, the main result of [33] implies that if $H = \ell_2$ we have for any d

$$M_d(\mathbb{N}; H) \neq M_{d+1}(\mathbb{N}; H).$$

However, the same question for $H = \mathbb{C}$ remains open when $G = \mathbb{N}$.

Acknowledgement. This is a much expanded version of a manuscript initially based on the notes of my talk at the 1998 Ponza conference for A. Figà-Talamanca's 60-th birthday. I am very grateful to Michael Cowling for useful conversations and for providing Remark 0.8. I am indebted to Marius Junge for Lemma 5.5. I also thank Slava Grigorchuk for useful remarks on the text.

References

[1] S. I. Adian. The Burnside problem and identities in groups. Ergebnisse der Mathematik 95. Springer-Verlag, Berlin-New York, 1979.

[2] _____. Random walks on free periodic groups. *Izv. Akad. Nauk SSSR Ser. Mat.* **46** (1982), 1139–1149.

[3] J.-Ph. Anker. Applications de la p-induction en analyse harmonique, *Comment. Math. Helv.* **58** (1983), no. 4, 622–645.

[4] D. Blecher. A completely bounded characterization of operator algebras. *Math. Ann.* **303** (1995), 227–240.

[5] D. Blecher and V. Paulsen. Explicit constructions of universal operator algebras and applictions to polynomial factorization. *Proc. Amer. Math. Soc.* **112** (1991), 839–850.

[6] D. Blecher, Z. J. Ruan and A. Sinclair. A characterization of operator algebras. *J. Funct. Anal.* **89** (1990), 188–201.

[7] M. Bożejko. Positive definite bounded matrices and a characterization of amenable groups. *Proc. Amer. Math. Soc.* **95** (1985), 357–360.

[8] M. Bożejko. Uniformly bounded representations of free groups. *J. Reine Angew. Math.* **377** (1987), 170–186.

[9] M. Bożejko and G. Fendler. Herz-Schur multipliers and completely bounded multipliers of the Fourier algebra of a locally compact group. *Boll. Unione Mat. Ital.* (7) **3-A** (1984), 297–302.

[10] F. Choucroun. Analyse harmonique des groupes d'automorphismes d'arbres de Bruhat-Tits. (French) *Mem. Soc. Math. France (N.S.)* **58** (1994), 170 pp.

[11] M. Cowling and B. Dorofaeff. Random subgroups of Lie groups. *Rend. Sem. Mat. Fis. Milano* **67** (1997), 95–101.

[12] E. Christensen and A. Sinclair. A survey of completely bounded operators. *Bull. London Math. Soc.* **21** (1989), 417–448.

[13] M. Cowling. Unitary and uniformly bounded representations of some simple Lie groups. CIME Course Lecture Notes. Harmonic analysis and group representations, Liguori, Naples, 1982, pp. 49–128.

[14] _____. The Kunze-Stein phenomenon. *Ann. Math.* **107** (1978), 209–234.

[15] _____. Harmonic analysis on some nilpotent Lie groups (with application to the representation theory of some semisimple Lie groups). Topics in modern harmonic analysis, Vol. I, II (Turin/Milan, 1982), 81–123, Ist. Naz. Alta Mat. Francesco Severi, Rome, 1983.

[16] M. Cowling and U. Haagerup. Completely bounded multipliers of the Fourier algebra of a simple Lie group of real rank one. *Invent. Math.* **96** (1989), 507–549.

[17] M. Day. Means for the bounded functions and ergodicity of the bounded representations of semi-groups. *Trans. Amer. Math. Soc.* **69** (1950), 276–291.

[18] J. de Cannière and U. Haagerup. Multipliers of the Fourier algebras of some simple Lie groups and their discrete subgroups. *Amer. J. Math.* **107** (1985), 455–500.

[19] J. Dixmier. Les moyennes invariantes dans les semi-groupes et leurs applications. *Acta Sci. Math. Szeged* **12** (1950), 213–227.

[20] L. Ehrenpreis and F.I. Mautner. Uniformly bounded representations of groups. *Proc. Nat. Acad. Sc. U.S.A.* **41** (1955), 231–233.

[21] E. Effros and Z. J. Ruan. Multivariable multipliers for groups and their operator algebras. Operator theory: operator algebras and applications, Part 1 (Durham, NH, 1988), 197–218, Proc. Sympos. Pure Math., 51, Part 1, Amer. Math. Soc., Providence, RI, 1990.

[22] _____. E. Effros and Z.J. Ruan, Operator spaces, Oxford Univ. Press, Oxford, 2000.

[23] G. Fendler. A uniformly bounded representation associated to a free set in a discrete group. *Colloq. Math.* **59** (1990), no. 2, 223–229.

[24] A. Figá-Talamanca and M. Picardello. Harmonic Analysis on Free groups. Marcel Dekker, New-York, 1983.

[25] R. I. Grigorchuk. Symmetrical random walks on discrete groups. Multicomponent random systems, pp. 285–325, Adv. Probab. Related Topics **6**, Dekker, New York, 1980.

[26] M. Gromov. Hyperbolic groups, pp. 75-265 in *Essays in Group Theory* (edited by S. Gersten), Springer, New York, 1987.

[27] U. Haagerup. $M_0A(G)$ functions which are not coefficients of uniformly bounded representations. Unpublished manuscript, 1985.

[28] _____. Solution of the similarity problem for cyclic representations of C^*-algebras. *Ann. of Math. (2)* **118** (1983), 215–240.

[29] _____. Unpublished manuscript,

[30] _____. An example of a nonnuclear C^*-algebra, which has the metric approximation property. *Invent. Math.* **50** (1978/79), 279–293.

[31] K.H. Hofmann and P. Mostert. Splitting in topological groups. *Mem. Amer. Math. Soc.* **43** (1963) 75 pp.

[32] T. Januszkiewicz. For Coxeter groups $z^{|g|}$ is a coefficient of a uniformly bounded representation. *Fund. Math.* **174** (2002), no. 1, 79–86.

[33] N. Kalton and C. Le Merdy. Solution of a problem of Peller concerning similarity. *J. Operator Theory* **47** (2002), no. 2, 379–387.

[34] B. Kostant. On the existence and irreducibility of certain series of representations. *Bull. Amer. Math. Soc.* **75** (1969), 627–642.

[35] R.A. Kunze and E. M. Stein. Uniformly bounded representations and Harmonic Analysis of the 2×2 real unimodular group. *Amer. J. Math.* 82 (1960), 1–62.

[36] _____. Uniformly bounded representations. II. Analytic continuation of the principal series of representations of the $n \times n$ complex unimodular group. *Amer. J. Math.* **83** (1961), 723–786.

[37] _____. Uniformly bounded representations. III. Intertwining operators for the principal series on semisimple groups. *Amer. J. Math.* **89** (1967), 385–442.

[38] _____. Uniformly bounded representations. IV. Analytic continuation of the principle series for complex classical groups of types B_n, C_n, D_n. *Advances in Math.* **11** (1973), 1–71.

[39] R.L. Lipsman. Uniformly bounded representations of SL(2, C). *Bull. Amer. Math. Soc.* **73** (1967), 652–655.

[40] _____. Uniformly bounded representations of SL(2, C). *Amer. J. Math.* **91** (1969), 47–66.

[41] N. Lohoué. Sur les représentations uniformément bornées et le théorème de convolution de Kunze-Stein. (French) [Uniformly bounded representations and the Kunze-Stein convolution theorem] *Osaka J. Math.* **18** (1981), no. 2, 465–480.

[42] M. Leinert. Faltungsoperatoren auf gewissen diskreten Gruppen. *Studia Math.* **52** (1974), 149–158.

[43] V. Losert. Properties of the Fourier algebra that are equivalent to amenability. *Proc. Amer. Math. Soc.* **92** (1984), 347–354.

[44] W. Młotkowski. Irreducible representations of free products of infinite groups. *Colloq. Math.* **69** (1995), no. 2, 193–211.

[45] A. M. Mantero, T. Pytlik, R. Szwarc and A. Zappa. Equivalence of two series of spherical representations of a free group. *Ann. di Matematica pura ed applicata (IV)* **CLXV** (1993), 23–28.

[46] A.M. Mantero and A. Zappa. The Poisson transform and representations of a free group. *J. Funct. Anal.* **51** (1983), no. 3, 372–399.

[47] _____. Uniformly bounded representations and L^p-convolution operators on a free group. Harmonic analysis (Cortona, 1982), 333–343, Lecture Notes in Math. **992**, Springer, Berlin, 1983.

[48] D. Montgomery, and L. Zippin. Topological Transformation Groups. Interscience Publishers, New York-London, 1955.

[49] C. Nebbia. Multipliers and asymptotic behaviour of the Fourier algebra of non amenable groups. *Proc. Amer. Math. Soc.* **84** (1982), 549–554.

[50] M. Nagisa and S. Wada. Simultaneous unitarizability and similarity problem. *Scientiae Mathematicae (electr.)* **2** (1999), no. 3, 255–261.

[51] A. Yu. Ol'shanskii. On the problem of the existence of an invariant mean on a group. *Russian Math. Surveys* **35** (1980), 180–181.

[52] _____. Geometry of defining relations in groups. Kluwer Academic Publishers Group, Dordrecht, 1991.

[53] A. Paterson. Amenability. *Amer. Math. Soc. Math. Surveys* **29** (1988).

[54] V. Paulsen. Completely bounded maps and operator algebras. Cambridge Studies in Advanced Mathematics **78**, Cambridge University Press, Cambridge, 2002. xii+300 pp.

[55] V. Paulsen and R. Smith. Multilinear maps and tensor norms on operator systems, *J. Funct. Anal.* **73** (1987), 258–276.

[56] V. Peller. Estimates of functions of power bounded operators on Hilbert space. *J. Oper. Theory* **7** (1982), 341–372.

[57] J.P. Pier. Amenable locally compact groups. Wiley Interscience, New York, 1984.

[58] _____. Amenable Banach algebras. Pitman, Longman, 1988.

[59] G. Pisier. Multipliers and lacunary sets in non amenable groups. *Amer. J. Math.* **117** (1995), 337–376.

[60] _____. Similarity problems and completely bounded maps. Second, Expanded Edition. Springer Lecture Notes **1618** (2001).

[61] _____. The similarity degree of an operator algebra. *St. Petersburg Math. J.* **10** (1999), 103–146.

[62] _____. The similarity degree of an operator algebra. II. *Math. Zeit.* **234** (2000), 53–81.

[63] _____. The volume of Convex Bodies and Banach Space Geometry. Cambridge University Press, 1989.

[64] _____. Complex interpolation and regular operators between Banach lattices. *Arch. Math.* **62** (1994), 261–269.

[65] _____. Multipliers of the Hardy space H^1 and power bounded operators. *Colloq. Math.* **88** (2001), no. 1, 57–73.

[66] _____. Joint similarity problems and the generation of operator algebras with bounded length. *Integr. Equ. Op. Th.* **31** (1998), 353–370.

[67] _____. Introduction to Operator Space Theory. London Math. Soc. Lect. Notes **294**, Cambridge Univ. Press., Cambridge, 2003.

[68] M. Pimsner. Cocycles on trees. *J. Operator Theory* **17** (1987), no. 1, 121–128.

[69] T. Pytlik. Spherical functions and uniformly bounded representations of free groups. *Studia Math.* **100** (1991), no. 3, 237–250.

[70] T. Pytlik and R. Szwarc. An analytic family of uniformly bounded representations of free groups. *Acta Math.* **157** (1986), no. 3–4, 287–309.

[71] Z. J. Ruan. Unpublished manuscript and private communication.

[72] P. J. Sally Jr. Unitary and uniformly bounded representations of the two by two unimodular group over local fields. *Amer. J. Math.* **90** (1968), 406–443.

[73] _____. Uniformly bounded representations of the universal covering group of SL(2, R). *Bull. Amer. Math. Soc.* **72** (1966), 269–273.

[74] N. Spronk. Measurable Schur multipliers and completely bounded multipliers of the Fourier algebras. *Proc. London Math. Soc.* **89** (2004), 161–192.

[75] R. Szwarc. Banach algebras associated with spherical representations of the free group. *Pacific J. Math.* **143** (1990), no. 1, 201–207.

[76] _____. Matrix coefficients of irreducible representations of free products of groups. *Studia Math.* **94** (1989), no. 2, 179–185.

[77] N. Tomczak-Jaegermann. The moduli of convexity and smoothness and the Rademacher averages of trace class S_p. *Studia Math.* **50** (1974), 163–182.

[78] A. Valette. Cocycles d'arbres et représentations uniformément bornées. (French) *C. R. Acad. Sci. Paris Sér. I Math.* **310** (1990), no. 10, 703–708.

[79] _____. Les représentations uniformément bornées associées à un arbre réel. (French) [Uniformly bounded representations associated with a real tree] Algebra, groups and geometry. *Bull. Soc. Math. Belg. Sér. A* **42** (1990), 747–760.

[80] N. Varopoulos. On an inequality of von Neumann and an application of the metric theory of tensor products to Operators Theory. *J. Funct. Anal.* **16** (1974), 83–100.

[81] J. Wysoczański. An analytic family of uniformly bounded representations of a free product of discrete groups. *Pacific J. Math.* **157** (1993), no. 2, 373–387.

Gilles Pisier
Texas A&M University
College Station, TX 77843
USA

and

Université Paris VI
Equipe d'Analyse, Case 186, 75252
Paris Cedex 05
France
e-mail: `pisier@math.tamu.edu`

Progress in Mathematics, Vol. 248, 363–388
© 2005 Birkhäuser Verlag Basel/Switzerland

Probabilistic Group Theory and Fuchsian Groups

Aner Shalev

To Slava Grigorchuk on his 50th birthday

Abstract. In this paper we survey recent advances in probabilistic group theory and related topics, with particular emphasis on Fuchsian groups. Combining a character-theoretic approach with probabilistic ideas we study the spaces of homomorphisms from Fuchsian groups to symmetric groups and to finite simple groups, and obtain a whole range of applications.

In particular we show that a random homomorphism from a Fuchsian group to the alternating group A_n is surjective with probability tending to 1 as $n \to \infty$, and this implies Higman's conjecture that every Fuchsian group surjects to all large enough alternating groups. We also show that a random homomorphism from a Fuchsian group of genus ≥ 2 to any finite simple group G is surjective with probability tending to 1 as $|G| \to \infty$. These results can be viewed as far-reaching extensions of Dixon's conjecture, establishing a similar result for free groups.

Other applications concern counting branched coverings of Riemann surfaces, studying the subgroup growth of Fuchsian groups, and computing the dimensions of representation varieties of Fuchsian groups over fields of arbitrary characteristic.

A main tool in our proofs are results of independent interest which we obtain on the representation growth of symmetric groups and groups of Lie type. These character-theoretic results also enable us to analyze random walks in symmetric groups and in finite simple groups, with certain conjugacy classes as generating sets, and to determine the precise mixing time.

Most of the results outlined here were proved in recent joint works by Liebeck and myself, but we shall also describe related results by other authors.

Mathematics Subject Classification (2000). 20P05, 20H10, 20D06, 20E07, 20C15.

Keywords. Probabilistic methods, Fuchsian groups, finite simple groups, characters, subgroup growth, random walks.

Contents

1. Background

In recent years the probabilistic approach has been instrumental in solving several difficult problems in group theory. These problems involve finite simple groups, finite permutation groups, various infinite groups, as well as related Cayley graphs.

One root of the subject is work by Erdős and Turán (starting with [ET]) on the properties of random permutations. This work enabled Dixon to settle a 19th century conjecture of Netto, proving that two randomly chosen elements of the alternating group A_n generate A_n with probability tending to 1 as $n \to \infty$ [Di]. Dixon then conjectured that a similar result holds for arbitrary (finite nonabelian) simple groups.

Using the Classification of finite simple groups, and information on their maximal subgroups, Dixon's conjecture was eventually proved ([KL], [LiSh1]), and various other results on generation and random generation were established (see [LP], [GKS], [LiSh2], [LiSh3], [Sh1], [Sh2], [LiSh4], [GLSSh], [GK], [Sh4], [LiSh5], [LiSh6], [GSh]). Furthermore, probabilistic and counting methods were used in the solutions of important problems which have nothing to do with probability.

One such longstanding problem, which arose in geometric context, is finding the finite simple quotients of the modular group $\mathrm{PSL}_2(\mathbb{Z})$ (see [LiSh2], [LM] for a solution up to finitely many exceptions). Another example with a geometric flavor is the Guralnick-Thompson conjecture on monodromy groups of genus g covers of the Riemann sphere [GT], which has been settled in [FM] following [LiSh4] and other results by many authors. Other fruits of the probabilistic approach include conjectures of Babai and Cameron concerning bases for primitive permutation groups, partly motivated by computer science applications (see [Py], [GSSh], [LiSh4]).

The study of residual properties of some infinite groups has also benefited from probabilistic ideas. For example, in [DPSSh] we give a new short proof of a conjecture of Magnus regarding residual properties of free groups. The analogue of Magnus conjecture for the modular group has recently been established in [LiSh7], and extensions to more general free products appear in [LiSh8]. For more general background on residually finite groups and some remarkable constructions and growth properties, see Grigorchuk [Gr].

Another root for the research outlined here is the study of random walks on finite groups, where a rich theory has been developed by Diaconis [D1, D2] and others. Here a main challenge is to study the mixing time, namely, how long it takes for the random walk to converge to an approximately uniform distribution. A classical result by Diaconis and Shahshahani determines the mixing time of a random walk on symmetric groups with the set of transpositions as a generating set [DS]. Various extensions have since then appeared, where the generating set is assumed to be a conjugacy class in S_n, or when the symmetric group is replaced by some group of Lie type, though our knowledge of the latter situation is still extremely limited.

In a recent work with Liebeck [LiSh5] we consider a finite simple group G and a conjugacy class C of G, and determine the diameter of the respective Cayley graph up to a multiplicative constant. We also derive some information on the mixing time, though much remains to be done. Results of this type are partly motivated by questions on profinite groups. For example, Serre asked whether every finite index subgroup in a finitely generated profinite group is open. This longstanding problem has just been answered affirmatively by Nikolov and Segal in [NS1, NS2], using counting methods and [LiSh5] among other tools. This is yet another recent example of the power of probabilistic and counting methods in settling a whole range of longstanding problems.

In the next sections we focus on some of the latest advances in this area, which are related to character theory, Fuchsian groups and random walks. Sections 2 and 3 are based on [LiSh9], and discuss homomorphisms from Fuchsian groups to symmetric groups, with applications to random quotients, subgroup growth, and Riemann surfaces. Sections 4 and 5 are based on [LiSh10], and describe homomorphisms of Fuchsian groups to finite groups of Lie type, with applications to random quotients and representation varieties. Section 6 and 7 deal with random walks and characters, and rely, among other things, on [LiSh11]. Section 8, which concludes this paper, focuses on directions for further research, and describes various open problems and conjectures.

For more detailed background on previous phases of probabilistic group theory and its applications, see also the survey paper [Sh3]. The present paper can be considered as an extended and updated version of my lecture at the Gaeta conference in 2003. I would like to take this opportunity and thank the organizers Laurent Bartholdi, Tullio Ceccherini-Silberstein, Tatiana Smirnova-Nagnibeda and Andrej Zuk, as well as Slava Grigorchuk, for this excellent and inspiring conference.

2. Fuchsian groups, I

Many results on generation and random generation of finite simple groups can be viewed in the much more general context of Fuchsian groups.

A *Fuchsian group* is a finitely generated non-elementary discrete group of isometries of the hyperbolic plane. An orientation-preserving Fuchsian group Γ has a presentation of the form

(A) generators: $a_1, b_1, \ldots, a_g, b_g$
$\qquad\qquad\qquad\qquad x_1, \ldots, x_d$
$\qquad\qquad\qquad\qquad y_1, \ldots, y_s$
$\qquad\qquad\qquad\qquad z_1, \ldots, z_t$

$\qquad$ relations: $x_1^{m_1} = \cdots = x_d^{m_d} = 1,$
$\qquad\qquad\qquad\qquad x_1 \cdots x_d\, y_1 \cdots y_s\, z_1 \cdots z_t\, [a_1, b_1] \cdots [a_g, b_g] = 1,$

where $g, d, s, t \geq 0$ and $m_i \geq 2$ for all i. The number g is defined to be the *genus* of Γ. The *measure* $\mu(\Gamma)$ of an orientation-preserving Fuchsian group Γ is defined by

$$\mu(\Gamma) = 2g - 2 + \sum_{i=1}^{d}\left(1 - \frac{1}{m_i}\right) + s + t.$$

It is well known that $\mu(\Gamma) > 0$.

We also study non-orientation-preserving Fuchsian groups, having the following presentations, with $g > 0$:

(B) generators: $a_1, \ldots, a_g$
$\qquad\qquad\qquad\qquad x_1, \ldots, x_d$
$\qquad\qquad\qquad\qquad y_1, \ldots, y_s$
$\qquad\qquad\qquad\qquad z_1, \ldots, z_t$

$\qquad$ relations: $x_1^{m_1} = \cdots = x_d^{m_d} = 1,$
$\qquad\qquad\qquad\qquad x_1 \cdots x_d\, y_1 \cdots y_s\, z_1 \cdots z_t\, a_1^2 \cdots a_g^2 = 1.$

In this case we define the measure $\mu(\Gamma)$ by

$$\mu(\Gamma) = g - 2 + \sum_{i=1}^{d}\left(1 - \frac{1}{m_i}\right) + s + t,$$

and again, $\mu(\Gamma) > 0$.

Fuchsian groups as in (A) are called *oriented*, and those as in (B) *non-oriented*. Note that $\mu(\Gamma)$ coincides with $-\chi(\Gamma)$, where $\chi(\Gamma)$ is the Euler characteristic of Γ.

If $s + t > 0$ then Γ is a free product of cyclic groups. In particular, non-abelian free groups are Fuchsian, as well as free products of finite cyclic groups (excluding $C_2 * C_2$) such as the modular group. Other examples are surface groups (where $d = s = t = 0$), and triangle groups

$$\Delta(m_1, m_2, m_3) = \langle x_1, x_2, x_3 \mid x_1^{m_1} = x_2^{m_2} = x_3^{m_3} = x_1 x_2 x_3 = 1 \rangle,$$

(where $g = s = t = 0$, $d = 3$ and $\sum \frac{1}{m_i} < 1$). Among triangle groups, the Hurwitz group $\Delta(2, 3, 7)$ has received particular attention, partly since its finite images are the finite groups which occur as automorphism groups of Riemann surfaces of genus $h \geq 2$ and have order achieving the Hurwitz upper bound $84(h-1)$ (see [C3, Hu1]).

Dixon's conjecture mentioned in Section 1 amounts to saying that most homomorphisms from the free group F_d of rank $d \geq 2$ to a finite simple group G are onto, and random (2,3)-generation results from [LiSh2] amount to a similar assertion about the modular group (with some assumptions on G). Similarly, results on (r,s)-generation of simple groups [LiSh6] and on Hurwitz generation (see for instance Conder's survey paper [C3]) can be interpreted as saying that various simple groups are quotients of the Fuchsian groups $C_r * C_s$ and $\Delta(2,3,7)$. Can we study simple quotients of general Fuchsian groups?

In the late sixties Higman conjectured that every Fuchsian group has all large enough alternating groups as quotients. At the time this ambitious conjecture had only limited supporting evidence. Important work by Higman and Conder established this for the Hurwitz group [C1], and other triangle groups were subsequently studied in the past 20 years or so. In 2000 Everitt made a breakthrough, and managed to prove Higman's conjecture for all (oriented) Fuchsian groups [Ev].

Can we prove Higman's conjecture using probabilistic arguments? And can we also establish it for non-oriented Fuchsian groups (also known as NEC groups)? The following result provides affirmative answers.

Theorem 2.1. *Let Γ be a Fuchsian group (oriented or non-oriented). Then a random homomorphism from Γ to the alternating group A_n is onto with probability tending to 1 as $n \to \infty$.*

Note that $\mathrm{Hom}(\Gamma, G)$ is finite for any finite group G, so we can regard it as a probability space with respect to the uniform distribution. By taking n large enough in Theorem 2.1 we see that epimorphisms from Γ to A_n do exist.

Corollary 2.2. *Every Fuchsian group surjects to all sufficiently large alternating groups. In other words, Higman's conjecture holds for all Fuchsian groups (including the non-oriented ones).*

A slightly weaker version of Theorem 2.1, which still implies Corollary 2.2, is proved in [LiSh9]; it deals with transitive homomorphisms (namely those whose image is transitive). However, interesting work by Müller and Puchta [MP2] can be used to show that most homomorphisms from Γ to A_n are transitive, and this implies Theorem 2.1.

In fact, in the course of the proof of this theorem we establish a number of results of independent interest, related to subgroup growth, coverings of Riemann surfaces, random walks and other areas. We list some of these below.

Theorem 2.3. *Let Γ be a Fuchsian group. Then*

$$|\mathrm{Hom}(\Gamma, S_n)| = (n!)^{\mu(\Gamma)+1+o(1)}.$$

Here $o(1)$ denotes a quantity which tends to zero as $n \to \infty$. We also obtain similar results for $|\mathrm{Hom}(\Gamma, A_n)|$, which are essential for some applications.

In fact more accurate estimates for $|\text{Hom}(\Gamma, S_n)|$ are obtained in [LiSh9], and an asymptotic expansion is given by Müller and Puchta in [MP2].

One motivation behind the study of homomorphisms from Fuchsian groups to symmetric groups stems from the important theory of *subgroup growth*. For a finitely generated group Γ and a positive integer n, denote by $a_n(\Gamma)$ the number of index n subgroups of Γ. The relation between the function $a_n(\Gamma)$ and the structure of Γ has been the subject of intensive study over the past two decades (see the monograph [LS]). The subgroup growth of the free group F_d of rank d was determined by M. Hall and M. Newman [Ha, Ne]; extensions to the modular group as well as arbitrary free products of cyclic groups were given by Dey and by Newman [Ne]. The subgroup growth of surface groups was determined in [MP1]. It follows from these results that $a_n(F_d) = (n!)^{d-1+o(1)}$, $a_n(PSL_2(\mathbb{Z})) = (n!)^{\frac{1}{6}+o(1)}$, and for an oriented surface group Γ_g of genus $g \geq 2$, $a_n(\Gamma_g) = (n!)^{2g-2+o(1)}$. We show that these results are particular instances of a very general phenomenon:

Theorem 2.4. *For any Fuchsian group* Γ,

$$a_n(\Gamma) = (n!)^{\mu(\Gamma)+o(1)}.$$

Now, $\frac{1}{42}$ is the smallest possible value of $\mu(\Gamma)$ for a Fuchsian group Γ, obtained by Hurwitz group $\Delta(2,3,7)$ (and by no other group). This implies the following.

Corollary 2.5. *Let* Γ *be a Fuchsian group. Then*

$$a_n(\Gamma) \geq (n!)^{\frac{1}{42}+o(1)},$$

with equality if and only if Γ *is the Hurwitz group.*

Theorem 2.4 amounts to saying that

$$\frac{\log a_n(\Gamma)}{\log n!} \to \mu(\Gamma) = -\chi(\Gamma) \text{ as } n \to \infty.$$

These subgroup growth results were also obtained independently in [MP2], where a precise asymptotic expansion for $a_n(\Gamma)$ is given.

To explain the relation between $a_n(\Gamma)$ and homomorphisms of Γ, define

$$\text{Hom}_{trans}(\Gamma, S_n) = \{\phi \in \text{Hom}(\Gamma, S_n) : \phi(\Gamma) \text{ is transitive}\}.$$

It is well known that $a_n(\Gamma) = |\text{Hom}_{trans}(\Gamma, S_n)|/(n-1)!$. Thus Theorem 2.3 immediately yields an upper bound for $a_n(\Gamma)$. To obtain a lower bound, one can show that most homomorphisms in $\text{Hom}(\Gamma, S_n)$ lie in $\text{Hom}_{trans}(\Gamma, S_n)$, which completes the argument.

For further applications it is important for us to estimate how many of the homomorphisms in $\text{Hom}_{trans}(\Gamma, S_n)$ have primitive images in S_n. This is strongly related to studying the *maximal subgroup growth* of Fuchsian groups. Denote by $m_n(\Gamma)$ the number of maximal subgroups of index n in Γ. It turns out that most finite index subgroups of Fuchsian groups are maximal:

Theorem 2.6. *For any Fuchsian group Γ, we have*

$$\frac{m_n(\Gamma)}{a_n(\Gamma)} \to 1 \ as \ n \to \infty.$$

Moreover, $\frac{m_n(\Gamma)}{a_n(\Gamma)} = 1 - O(c^{-n})$, where $c > 1$ is a constant depending on Γ.

This extends previously known results for free groups (see [Di, Lemma 2]) and surface groups [MP1].

Theorem 2.6 shows that almost all transitive homomorphisms from a Fuchsian group to S_n have primitive images. Our next result is the strongest along these lines, and shows that almost all of these homomorphisms have image S_n or A_n. To show this we need to invoke the Classification of finite simple groups, among other tools.

Theorem 2.7. *Let Γ be a Fuchsian group. Then a random homomorphism $\phi \in \mathrm{Hom}_{trans}(\Gamma, S_n)$ satisfies $\phi(\Gamma) = A_n$ or S_n with probability $1 - O(c^{-n})$, where $c > 1$ is some constant depending on Γ.*

The methods of proof of Theorem 2.7 also works for alternating groups, and so we have

Theorem 2.8. *Let Γ be a Fuchsian group. Then the probability that a random homomorphism in $\mathrm{Hom}_{trans}(\Gamma, A_n)$ is an epimorphism tends to 1 as $n \to \infty$. Moreover, this probability is $1 - O(c^{-n})$ for some constant $c > 1$ depending on Γ.*

Our proofs show that in Theorems 2.6, 2.7 and 2.8, any constant c satisfying $1 < c < 2^{\mu(\Gamma)}$ will do.

As remarked earlier, results from [MP2] can be used to show that most homomorphisms from Γ to A_n have transitive image. Combining this with Theorem 2.8 implies the main result of this section, namely Theorem 2.1.

Consider now a natural variation on Higman's conjecture. Which Fuchsian groups surject to all large enough symmetric groups? Obviously such groups would have a quotient of order 2, and not all Fuchsian group have such a quotient. However, using a probabilistic approach we can show that this is in fact the only obstruction, thus establishing the analogue of Higman's conjecture for symmetric quotients.

To state the precise result let Γ be as in (A) or (B), and define

$$d^* = |\{i \ : \ m_i \ \text{even}\}|.$$

Note that if $s + t = g = 0$ and $d^* \leq 1$, or if $s + t = 1$ and $g = d^* = 0$, then Γ is generated by elements of odd order, so it cannot have S_2 as quotient. In all other cases Γ has a quotient of order 2. But what about larger symmetric quotients?

Theorem 2.9. *If a Fuchsian group has S_2 as a quotient, then it has S_n as a quotient for all sufficiently large n. Thus a Fuchsian group Γ surjects to all but finitely many symmetric groups if and only if $s + t > 1$, or $s + t = 1$ and $(g, d^*) \neq (0,0)$, or $s + t = 0$ and $(g, d^*) \neq (0,0), (0,1)$.*

Theorem 2.9, whose proof is probabilistic, is new even for triangle groups.

Corollary 2.10. *Let $m_1, m_2, m_3 \geq 2$ be integers such that $\sum \frac{1}{m_i} < 1$, and suppose at least two of the m_i are even. Then the triangle group $\Delta(m_1, m_2, m_3)$ surjects to all but finitely many symmetric groups.*

The special case of this where $(m_1, m_2, m_3) = (2, 3, k)$ with k even was established by Conder in [C2].

The above results provide yet another demonstration of the power of probabilistic methods in group theory; see also [Sh3] for various additional examples.

3. Coverings of Riemann surfaces

A classical motivation behind the study of $|\mathrm{Hom}(\Gamma, S_n)|$ stems from the theory of branched coverings of Riemann surfaces. Let Y be a compact connected Riemann surface of genus g, and let $y_1, \ldots, y_d \in Y$ be fixed distinct points. Consider index n coverings $\pi : X \to Y$, unramified outside $\{y_1, \ldots, y_d\}$, and with monodromy elements $g_1, \ldots, g_d \in S_n$ around $y_1, \ldots, y_d$ respectively. We identify geometrically equivalent coverings.

For conjugacy classes $C_1, \ldots, C_d$ of S_n, and integers $m_1, \ldots, m_d \geq 2$, set $\mathbf{C} = (C_1, \ldots, C_d)$, $\mathbf{m} = (m_1, \ldots, m_d)$ and define

$$P(\mathbf{C}, n) = \{\pi : X \to Y : g_i \in C_i \text{ for all } i\},$$
$$P(\mathbf{m}, n) = \{\pi : X \to Y : g_i^{m_i} = 1 \text{ for all } i\}.$$

Attempts to count such coverings go back to Hurwitz [Hu2]. Define $\mathrm{Aut}\,\pi$ to be the centralizer in S_n of the monodromy group of π. Following [KK] we call the sums $\sum 1/|\mathrm{Aut}\,\pi|$ over such sets of coverings $P(\mathbf{C}, n)$ the *Eisenstein numbers* of coverings, a term which is attributed to Serre.

When all but one of the classes C_i consist of transpositions these numbers are called *Hurwitz numbers*, which have been studied extensively in view of connections with geometry and physics - see for example [V] and the references therein. See also [EO], where asymptotic results are proved where the C_i consist of cycles of bounded length, and used to study volumes of moduli spaces.

Eisenstein numbers are related to homomorphisms from Fuchsian groups to symmetric groups.

Let $C_i = g_i^{S_n}$ ($1 \leq i \leq d$) be classes in S_n, and let m_i be the order of g_i. Define $\mathrm{sgn}(C_i) = \mathrm{sgn}(g_i)$, and write $\mathbf{C} = (C_1, \ldots, C_d)$. For a group Γ having presentation as in (A) or (B), define

$$\mathrm{Hom}_{\mathbf{C}}(\Gamma, S_n) = \{\phi \in \mathrm{Hom}(\Gamma, S_n) : \phi(x_i) \in C_i \text{ for } 1 \leq i \leq d\}.$$

Note that if $s + t = 0$, and $\mathrm{Hom}_{\mathbf{C}}(\Gamma, S_n) \neq \emptyset$, then $\prod_{i=1}^{d} \mathrm{sgn}(C_i) = 1$. When Γ is a co-compact oriented Fuchsian group (as in (A) with $s = t = 0$), a covering in $P(\mathbf{C}, n)$ corresponds to an S_n-class of homomorphisms in $\mathrm{Hom}_{\mathbf{C}}(\Gamma, S_n)$. If $\pi \in P(\mathbf{C}, n)$ corresponds to the class ϕ^{S_n} (where $\phi \in \mathrm{Hom}_{\mathbf{C}}(\Gamma, S_n)$), then $|\mathrm{Aut}\,\pi| = |C_{S_n}(\phi(\Gamma))|$.

The following formula, which is essentially due to Hurwitz, connects Eisenstein numbers and $|\mathrm{Hom}_{\mathbf{C}}(\Gamma, S_n)|$ with characters of symmetric groups.

$$(C) \quad \sum_{\pi \in P(\mathbf{C}, n)} \frac{1}{|\mathrm{Aut}\,\pi|} = \frac{|\mathrm{Hom}_{\mathbf{C}}(\Gamma, S_n)|}{n!} = \frac{|C_1| \ldots |C_d|}{(n!)^{2-2g}} \sum_{\chi \in Irr(S_n)} \frac{\chi(g_1) \cdots \chi(g_d)}{\chi(1)^{d-2+2g}}.$$

This formula includes the case where $d = 0$; here Γ is a surface group, the coverings are unramified, $\mathrm{Hom}_{\mathbf{C}} = \mathrm{Hom}$, and empty products are taken to be 1. While the formula (C) has been around for a century or so, it has not been used extensively, partly due to the difficulty in dealing with the character-theoretic sum on the right hand side; geometric and combinatorial methods have often been applied instead.

There is special interest in the case where the classes C_i are all homogeneous, namely, of cycle shape (m^a) (where $n = ma$). Indeed, the Eisenstein numbers corresponding to such classes are studied in [KK], mainly in the case where Γ is Euclidean or spherical, and formulae are obtained in [KK, Section 3] using geometric methods. As stated in [KK, p.414], the most interesting case is the hyperbolic one, in which the group Γ is Fuchsian. For this case, even allowing the permutations in C_i to have boundedly many fixed points, we prove the following result.

Theorem 3.1. *Fix integers $g \geq 0$ and $m_1, \ldots, m_d \geq 2$. Let $\mu = 2g - 2 + \sum_{i=1}^{d}(1 - \frac{1}{m_i})$ and suppose $\mu > 0$.*

  (i) *For $1 \leq i \leq d$ let C_i be a conjugacy class in S_n having cycle-shape $(m_i^{a_i}, 1^{f_i})$, and assume $\prod_{i=1}^{d} \mathrm{sgn}(C_i) = 1$. Then for f_i bounded and $n \to \infty$, we have*

$$\sum_{\pi \in P(\mathbf{C}, \mu)} \frac{1}{|\mathrm{Aut}\,\pi|} = (2 + O(n^{-\mu}))|C_1| \cdots |C_d| \cdot (n!)^{2g-2}$$
$$\sim (n!)^{\mu} \cdot n^{\sum \frac{f_i}{m_i} - \frac{1}{2}(1 - \frac{1}{m_i})}.$$

  (ii) *$\sum_{\pi \in P(\mathbf{m}, n)} \frac{1}{|\mathrm{Aut}\,\pi|} = (n!)^{\mu + o(1)}$.*

Here, for functions f_1, f_2, we write $f_1 \sim f_2$ if there are positive constants c_1, c_2 such that $c_1 f_2 \leq f_1 \leq c_2 f_2$. We call classes C_i as in (i) with f_i bounded *almost homogeneous* classes of S_n. We can show that most coverings in $P(\mathbf{C}, n)$ are connected, and so the estimates in Theorem 3.1 also hold for the numbers of connected coverings.

Next, our results on random homomorphisms from Fuchsian groups to symmetric groups have applications for the monodromy groups of branched coverings of Riemann surfaces. Any finite set of index n coverings can be naturally viewed as a probability space, where the probability assigned to a covering π is proportional to $1/|\mathrm{Aut}\,\pi|$.

Theorem 3.2. *Fix integers $g \geq 0$ and $m_1, \ldots, m_d \geq 2$. Let $\mu = 2g - 2 + \sum_{i=1}^{d}(1 - \frac{1}{m_i})$ and suppose $\mu > 0$.*

  (i) *The probability that a randomly chosen connected covering $\pi \in P(\mathbf{m}, n)$ has monodromy group A_n or S_n tends to 1 as $n \to \infty$; moreover, this probability is $1 - O(c^{-n})$ for some constant $c > 1$.*

(ii) *For $1 \leq i \leq d$ let C_i be a conjugacy class in S_n having cycle-shape $(m_i^{a_i}, 1^{f_i})$ with f_i bounded, and assume $\prod_{i=1}^{d} \operatorname{sgn}(C_i) = 1$. Write $\mathbf{C} = (C_1, \ldots, C_d)$. If $\pi \in P(\mathbf{C}, n)$ is randomly chosen, then the probability that the monodromy group of π is A_n or S_n is $1 - O(n^{-\mu})$.*

We note that while our Proof of Theorem 3.1, counting coverings, is elementary, the proof of Theorem 3.2 on monodromy groups invokes the Classification of finite simple groups.

4. Fuchsian groups, II

In Section 2 we studied homomorphisms from a Fuchsian group to symmetric and alternating groups, and derived various applications. The purpose of this section is to analyze homomorphisms from a Fuchsian group to finite groups of Lie type, deriving applications to probabilistic problems and to representation varieties. Our main probabilistic result is the following.

Theorem 4.1. *Let Γ be a Fuchsian group of genus $g \geq 2$ ($g \geq 3$ if Γ is non-oriented). Then a random homomorphism from Γ to a finite simple group G is onto with probability tending to 1 as $|G| \to \infty$.*

Theorem 4.1 is new even for surface groups, where it takes the following form.

Corollary 4.2. *Let Γ be a surface group which is not virtually abelian, and let G be a finite simple group. Then the probability that a randomly chosen homomorphism in $\operatorname{Hom}(\Gamma, G)$ is an epimorphism tends to 1 as $|G| \to \infty$.*

Theorem 4.1 extends Theorem 2.1, which yields the conclusion in the case where $G = A_n$, without genus assumptions. We note that in Theorem 4.1 some assumption on the genus is essential, since there are Fuchsian groups of genus 0 or 1 which do not have all large enough finite simple groups as quotients. Examples include triangle groups of genus 0 such as the Hurwitz group, and genus 1 groups of the form (A) with $d = 1$ and m_1 an odd prime (since there are infinitely many finite simple groups containing no element of order m_1).

Still, it would be interesting to find partial extensions of Theorem 4.1 to Fuchsian groups of genus 0 or 1. See Conjecture 3 in Section 8.

Again, a crucial step in proving Theorem 4.1 is counting homomorphisms from Fuchsian groups to finite simple groups.

For a finite group G and a positive integer m, let $j_m(G)$ be the number of solutions in G of the equation $x^m = 1$. If Γ is a non co-compact, say oriented, Fuchsian group, then Γ decomposes as a free product of $2g + s + t - 1$ copies of $\mathbb{Z}$ and cyclic groups of orders $m_1, \ldots, m_d$, and hence for any finite group G, we have $|\operatorname{Hom}(\Gamma, G)| = |G|^{2g+s+t-1} \cdot \prod_{i=1}^{d} j_{m_i}(G)$.

We show below that, if G is a finite simple group, a rather similar estimate holds also for co-compact groups (i.e. when $s + t = 0$), provided the genus is not too small. For simplicity, we only state the case of oriented groups Γ. However, we allow the finite group G to be quasisimple.

Theorem 4.3. *Let Γ be a co-compact Fuchsian group of genus $g \geq 2$ as in (A), and let G be a finite quasisimple group. Then*

$$|\mathrm{Hom}(\Gamma, G)| = (1 + o(1))|G|^{2g-1} \cdot \prod_{i=1}^{d} j_{m_i}(G).$$

The case of non-oriented group is more complicated, see [LiSh10] for details.

Theorem 4.3 takes a particularly simple form when $d = 0$, that is, Γ is a surface group:

Corollary 4.4. *Let Γ be an oriented surface group of genus $g \geq 2$, and let G be a finite quasisimple group. Then*

$$|\mathrm{Hom}(\Gamma, G)| = (1 + o(1))|G|^{2g-1}.$$

To apply Theorem 4.3 we need information on the values of $j_m(G)$ for G quasisimple. Such information can be found in [LiSh9, W] for $G = A_n$, in [LiSh2, LiSh6] for G classical and m prime, and in [GLSSh] for G exceptional and $m \leq 5$. Lawther [La] obtained tight estimates for the dimension of the variety $J_m(X) = \{x \in X : x^m = 1\}$, where X is any connected simple algebraic group. Using this we prove the following.

Theorem 4.5. *Let $G = G(q)$ be a finite quasisimple group of Lie type over $\mathbb{F}_q$ of rank r, and let $m \geq 2$ be an integer. Then*

$$j_m(G) = |G|^{1 - \frac{1}{m} + \epsilon(r)},$$

where $|\epsilon(r)| = O(r^{-1})$.

Combining Theorems 4.3 and 4.5 we obtain:

Theorem 4.6. *Let Γ be an oriented Fuchsian group of genus $g \geq 2$, and let G be a finite classical quasisimple group of rank r. Then*

$$|\mathrm{Hom}(\Gamma, G)| = |G|^{\mu(\Gamma)+1+\delta(r)},$$

where $|\delta(r)| = O(r^{-1})$.

Hence, if G_n is a sequence of finite quasisimple classical groups whose ranks tend to infinity, then

$$\lim_{n \to \infty} \frac{\log |\mathrm{Hom}(\Gamma, G_n)|}{\log |G_n|} = \mu(\Gamma) + 1. \tag{1}$$

This also holds for alternating groups $G_n = A_n$, and for non-oriented Fuchsian groups Γ. It would be interesting to find some sort of "explanation" for this general phenomenon.

5. Representation varieties

In this section we apply our results on $\mathrm{Hom}(\Gamma, G)$ for G finite of Lie type to the study of representation varieties of Γ in reductive algebraic groups over algebraically closed fields. See [LuMa] for general background on representation varieties. For a Fuchsian group Γ, an algebraically closed field K, and a positive integer n, define

$$R_{n,K}(\Gamma) = \mathrm{Hom}(\Gamma, GL_n(K)).$$

Then $R_{n,K}(\Gamma)$ is an algebraic variety defined over the prime subfield of K, and has been extensively studied in the case where K has characteristic zero and Γ is a surface group (see for instance [RBC]). However, less is known in the case of positive characteristic. Our first result addresses this case for surface groups.

Theorem 5.1. *Let Γ be an oriented surface group of genus $g \geq 2$, and let K be an algebraically closed field of characteristic $p > 0$.*
      Then $\dim R_{n,K}(\Gamma) = (2g-1)n^2 + 1$ and $R_{n,K}(\Gamma)$ has a unique irreducible component of highest dimension.

This dimension agrees with the one given for the characteristic zero case in [RBC]. It is known that the dimension of a variety in characteristic zero coincides with the dimension of its reduction modulo p for all large primes p, and so Theorem 5.1 provides an alternative proof of the characteristic zero dimension results in [RBC].
      Our methods also provide the values of $\dim R_{n,K}(\Gamma)$ for arbitrary Fuchsian groups. We need some notation. For positive integers n and $m_1, \ldots, m_d$, all at least 2, write $n = k_i m_i + l_i$ with $0 \leq l_i < m_i$, and $\mathbf{m} = (m_1, \ldots, m_d)$, and define

$$c(n, \mathbf{m}) = \sum_{i=1}^{d} l_i \left(1 - \frac{l_i}{m_i}\right).$$

Note that $c(n, \mathbf{m})$ is bounded in terms of $\mathbf{m}$ only.

Theorem 5.2. *Let Γ be a co-compact Fuchsian group as in (A), of genus $g \geq 2$. Set $E = \{i : m_i \text{ even }\}$, $\mu = \mu(\Gamma)$, let $n \geq 2$, and let K be an algebraically closed field of arbitrary characteristic. Then*

$$\dim R_{n,K}(\Gamma) = (\mu + 1)n^2 - c(n, \mathbf{m}) + \delta,$$

where $\delta = 1$ unless $\mathrm{char}(K) \neq 2$, $m_i | n$ for all $i \in E$, and $\sum_{i \in E} \frac{n}{m_i}(m_i + 1)$ is odd, in which case $\delta = -1$.

As a result we see that $\dim R_{n,K}(\Gamma) = (\mu(\Gamma) + 1)n^2 + O(1)$; in particular,

$$\frac{\dim R_{n,K}(\Gamma)}{n^2} \to \mu(\Gamma) + 1 \text{ as } n \to \infty. \tag{2}$$

This can be regarded as a representation-theoretic analogue of (1), providing yet another indication of the central role played by the measure $\mu(\Gamma)$ of Γ.
      As suggested in [RBC], it is interesting to extend these results to representation varieties $\mathrm{Hom}(\Gamma, \bar{G})$ for other algebraic groups $\bar{G}$. We focus naturally on

the case where $\bar{G}$ is a connected simple algebraic group. For a positive integer m, define

$$J_m(\bar{G}) = \{x \in \bar{G} : x^m = 1\}.$$

Then $J_m(\bar{G})$ is an algebraic variety. Information about its dimension can be found in [La].

Theorem 5.3. *Let Γ be an oriented Fuchsian group of genus $g \geq 2$, and let $\bar{G}$ be a connected simple algebraic group over an algebraically closed field K of arbitrary characteristic. Then*

(i) $\dim \mathrm{Hom}(\Gamma, \bar{G}) = (2g - 1) \dim \bar{G} + \sum_{i=1}^d \dim J_{m_i}(\bar{G});$

(ii) $\frac{\dim \mathrm{Hom}(\Gamma, \bar{G})}{\dim \bar{G}} \to \mu(\Gamma) + 1$ *as* $\dim \bar{G} \to \infty$.

For surface groups we obtain more detailed information. For a simple algebraic group $\bar{G}$ denote by $\pi_1(\bar{G})$ the fundamental group of $\bar{G}$, that is, the kernel of the canonical map from the simply connected cover of $\bar{G}$ onto $\bar{G}$.

Corollary 5.4. *Let Γ be an oriented surface group of genus $g \geq 2$, and let $\bar{G}$ be a connected simple algebraic group over an algebraically closed field K of arbitrary characteristic.*

(i) *We have* $\dim \mathrm{Hom}(\Gamma, \bar{G}) = (2g - 1) \dim \bar{G}$.

(ii) *The number of irreducible components of highest dimension in $\mathrm{Hom}(\Gamma, \bar{G})$ is equal to $|\pi_1(\bar{G})|$; in particular if $\bar{G}$ is simply connected, this number is 1.*

It is interesting to note that while results for finite groups are often deduced from corresponding results for algebraic groups, the deductions here are in the reverse direction. More specifically, the philosophy is to pass from the algebraic group $\bar{G}$ to the corresponding finite groups of Lie type $G(q)$ over $\mathbb{F}_q$. The space $\mathrm{Hom}(\Gamma, G(q))$ can essentially be regarded as the set of q-rational points of the representation variety $\mathrm{Hom}(\Gamma, \bar{G})$. Combining our results on $|\mathrm{Hom}(\Gamma, G(q))|$ from the previous section with Lang-Weil estimates for the number of q-rational points in algebraic varieties [LW], we obtain the dimensions of these representation varieties.

6. Random walks

The ideas discussed here also have applications to certain random walks on symmetric groups and on finite groups of Lie type.

Given a finite group G and a generating set X one defines a random walk on G as follows. Start with 1; when you are at an element $g \in G$, choose a random generator $x \in X$ with uniform distribution, and move from g to gx, and so on.

Suppose that the probability of arriving at the element g after t steps is $P^t(g)$. Then P^t is a probability distribution on G, and one can measure its distance from the uniform distribution U on G. Here one often uses the distance function $||P^t - U||$, where the norm is the ℓ_1 norm (or some scalar multiple of it). The *mixing time* of the random walk is the minimal t for which P^t is 'sufficiently uniform', in the sense that $||P^t - U||$ is small enough (say, less than some fixed ϵ). For general background on this fascinating topic, see Diaconis [D1, D2].

There is a special interest in the case where G is the symmetric group or a finite simple group, and the generating set X is a conjugacy class C, or $C \cup C^{-1}$ to make it symmetric, if needed.

For example, Diaconis and Shahshahani [DS] analyzed the case where $G = S_n$ and C is the set of transpositions. Other conjugacy classes (namely cycle-shapes) C in S_n were considered, see Lulov [Lul] for cycle-shapes (m^a) for fixed m, and Roichman [R] for arbitrary classes of permutations with cn fixed points (c a positive constant). New additional results were proved by Müller and Puchta [MP2], who confirmed a conjecture of Roichman in that area. In particular they show that, if the class C consists of fixed-point-free permutations then the mixing time of the corresponding random walk is bounded (say, by 20). Our aim here is to determine the mixing time precisely in some situations, as well as to consider more general settings where the generating set is allowed to change with time.

In the next result, proved in [LiSh9], we fix certain conjugacy classes $C_1, \ldots, C_t$ in S_n. At time i we choose a random element $x_i \in C_i$ and we analyze the distribution of the resulting product $x_1 x_2 \ldots x_t$. We show that this probability distribution is close to the uniform distribution not only in the l_1-norm (as in the definition of mixing time), but also in the l_∞-norm; this is a stronger statement, which also implies that the random walk hits all elements of the correct signature.

Theorem 6.1. *Let* $m_1, \ldots, m_t \geq 2$ *be integers satisfying* $\sum_{i=1}^{t} \frac{1}{m_i} < t - 2$, *and set* $\mu = t - 2 - \sum_{i=1}^{t} \frac{1}{m_i}$. *For* $1 \leq i \leq t$ *let* C_i *be a conjugacy class in* S_n *having cycle-shape* $(m_i^{a_i}, 1^{f_i})$ *with* f_i *bounded. Set* $\alpha = \prod_{i=1}^{t} \mathrm{sgn}(C_i)$. *Then for any* $h \in S_n$ *satisfying* $\mathrm{sgn}(h) = \alpha$, *and for randomly chosen* $x_i \in C_i$, *we have*

$$\mathrm{Prob}(x_1 \cdots x_t = h) = \frac{2}{n!}(1 + O(n^{-\mu})).$$

In particular, if n *is sufficiently large then, taking* σ *to be any permutation with* $\mathrm{sgn}(\sigma) = \alpha$, *we have* $C_1 C_2 \cdots C_t = A_n \sigma$ *almost uniformly pointwise.*

Moreover, the condition $\sum_{i=1}^{t} \frac{1}{m_i} < t - 2$ *is also necessary: if it is violated then the distribution of* $x_1 \ldots x_t$ *is far from uniform in the* l_∞-*norm.*

Taking $m_1 = \ldots = m_d = m$ and $C_1 = \ldots = C_t = C$, of cycle-shape $(m^a, 1^f)$ with f bounded, we see that a random walk on S_n with generating set C achieves an almost uniform distribution pointwise after t steps, where $t = 3$ if $m \geq 4$, $t = 4$ if $m = 3$ and $t = 5$ if $m = 2$. For the case where C is fixed-point-free (i.e. $f = 0$), Lulov [Lul] shows that the mixing time is 3 if $m = 2$ and 2 otherwise; Our number t of steps is slightly larger, but our distribution is arbitrarily close to uniform in the l_∞-norm, which is stronger than the mixing time condition.

Theorem 6.1 also shows that, if $C_1, C_2, C_3 \subset A_n$ are homogeneous conjugacy classes of elements decomposing into cycles of length a, b, c respectively, then $C_1 C_2 C_3 = A_n$ almost uniformly pointwise if and only if $a^{-1} + b^{-1} + c^{-1} < 1$.

Thus, for example, $(a, b, c) = (2, 3, 7)$ produce an almost uniform distribution, but $(a, b, c) = (2, 3, 6)$ do not.

Let us now turn to random walks on groups of Lie type. Here much less is known, and the challenges are harder, though some results have been obtained, see [G2, Hi, LiSh5]. We focus on the case $G = L$, a simple group of Lie type, and $X = x^L \cup (x^{-1})^L$, a union of one or two conjugacy classes of L. We denote the mixing time of the random walk by $T(L, x)$. The results below are proved in [LiSh11].

Recall that an element x of a group of Lie type is called *regular* if its centralizer in the corresponding simple algebraic group G is of minimal dimension, namely $\mathrm{rank}(G)$. For example, the regular elements of SL_n are those which have a single Jordan block for each eigenvalue.

Theorem 6.2. *Let $L \neq L_2(q)$ be a simple group of Lie type and let x be a regular element of L. Then for $|L|$ sufficiently large, the mixing time $T(L, x)$ is 2.*

For $L = L_2(q)$, our proof shows that $T(L, x) = 2$ unless q is odd and $x \in L$ is unipotent, in which case $T(L, x)$ is at most 3.

Note that for $x \in L$ regular, the order of the centralizer $C_L(x)$ is roughly q^r, where r is the Lie rank of L. Our methods also determine the mixing times for more general classes, consisting of elements with larger centralizers.

Theorem 6.3. *Let $L = L(q)$ be a simple group of Lie type over $\mathbb{F}_q$ and fix $\epsilon > 0$ and an integer $k \geq 2$. Then there is a function $r = r(\epsilon)$ such that if L has rank at least r, and $x \in L$ satisfies $|C_L(x)| \leq cq^{4r(1-\frac{1}{k}-\epsilon)}$, then the mixing time $T(L, x)$ is at most k.*

In particular, if $|C_L(x)| \leq cq^{(2-\epsilon)r}$ and r is sufficiently large, then the mixing time $T(L, x)$ is 2.

Recall that Theorem 1.13 in [LiSh5] shows that $|C_L(x)| \leq q^{(2-\epsilon)r}$ implies $T(L, x) \leq f(\epsilon)$, for some function f. Theorem 6.3 greatly improves this, showing we can take $f(\epsilon) = 2$ for large r.

The above results seem to be the first theorems giving exact bounded mixing times for random walks on finite groups of Lie type.

7. The character connection

The proofs of the results outlined here require many tools. For example, to obtain results on random quotients of Fuchsian groups, the Classification of finite simple groups is invoked, as well as the study of their subgroup structure [KLi]. However, the key tool in proving *all* the results described here is undoubtedly character theory.

The purpose of this section is to indicate connections between information on character degrees and character values, and the topics discussed in previous sections.

It is well known that characters are highly relevant in understanding random walks. For example, when $C = x^G$ is a conjugacy class, and the generating set X is $C \cup C^{-1}$, the upper bound lemma of Diaconis and Shahshahani [DS] yields

$$\|P^t - U\| \le \frac{1}{4} \sum_{1 \ne \chi \in Irr(G)} \chi(1)^2 |\chi(x)/\chi(1)|^{2t}, \tag{3}$$

where $Irr(G)$ denotes the set of irreducible complex characters of G. Hence information on character degrees and character values of finite simple groups could be the key to good bounds on the mixing time.

Connections between characters and random quotients of Fuchsian groups are less obvious, but they can also be established. To illustrate this, let Γ be a Fuchsian group and let G be a finite simple group. Let P be the probability that a random homomorphism $\phi \in \text{Hom}(\Gamma, G)$ is onto. Note that, if a homomorphism in $\text{Hom}(\Gamma, G)$ is not onto, then its image lies in a maximal subgroup M of G, and this happens with probability $\frac{|\text{Hom}(\Gamma, M)|}{|\text{Hom}(\Gamma, G)|}$. Summing over all maximal subgroups M of G we obtain

$$1 - P \le \sum_{M \; max \; G} \frac{|\text{Hom}(\Gamma, M)|}{|\text{Hom}(\Gamma, G)|}.$$

It therefore suffices to show that (possibly with some conditions on Γ and G), the right hand side tends to 0 as the order of G tends to infinity.

We conclude that good estimates on $|\text{Hom}(\Gamma, G)|$, where G is a finite simple group or a maximal subgroup, might help in proving that most homomorphisms as above are surjective.

How can we study $|\text{Hom}(\Gamma, G)|$?

In the case where Γ is a surface group, say oriented of genus g, there is a classical formula expressing $|\text{Hom}(\Gamma, G)|$ in terms of the irreducible complex characters of G, namely

$$|\text{Hom}(\Gamma, G)| = |G|^{2g-1} \sum_{\chi \in Irr(G)} \chi(1)^{-(2g-2)}. \tag{4}$$

In view of this, and many other applications, it seems useful to study a 'zeta function' encoding the character degrees of G, defined by

$$\zeta_G(s) = \sum_{\chi \in Irr(G)} \chi(1)^{-s}.$$

In the more general case where Γ contains torsion generators x_i, the size of the subspace $\text{Hom}_{\mathbf{C}}(\Gamma, G) \subseteq \text{Hom}(\Gamma, G)$, consisting of homomorphisms sending x_i to elements in given conjugacy classes $C_i = g_i^G$ of G, can be expressed in terms of character degrees and values. Indeed formula (C) mentioned in Section 3 holds for any finite groups G (instead of S_n), and shows that

$$|\text{Hom}_{\mathbf{C}}(\Gamma, G)| = |G|^{2g-1} |C_1| \dots |C_d| \sum_{\chi \in Irr(G)} \frac{\chi(g_1) \cdots \chi(g_d)}{\chi(1)^{d-2+2g}}. \tag{5}$$

Hence, knowledge of character degrees and values may be instrumental in studying $|\mathrm{Hom}_{\mathbf{C}}(\Gamma, G)|$. Indeed in many cases we are able to show that the contribution of the non-linear characters to the character sums in (4) and (5) are negligible, thus obtaining good estimates on $|\mathrm{Hom}_{\mathbf{C}}(\Gamma, G)|$. Summing up over all the relevant d-tuples of conjugacy classes we can then estimate $|\mathrm{Hom}(\Gamma, G)|$.

Let us now mention some results on character degrees and the related function $\zeta_G(s)$ defined above. Note that we have

$$\zeta_G(s) = \sum_{n \geq 1} r_n(G) n^{-s},$$

where $r_n(G)$ denotes the number of n-dimensional irreducible representations of G. In recent years there has been active interest in the numbers $r_n(G)$ and the associated zeta function also for some infinite groups G. In fact, the function $\zeta_G(s)$ was originally defined and studied by Witten for Lie groups [Wit] in the context of mathematical physics, and then by Jaikin for pro-p groups, and by Lubotzky and Martin for arithmetic groups [LM2].

Going back to finite groups, we obtain the following.

Theorem 7.1. *Fix a real number $s > 0$.*

(i) *If $G = S_n$ then $\zeta_G(s) = 2 + O(n^{-s})$.*
(ii) *If $G = A_n$ then $\zeta_G(s) = 1 + O(n^{-s})$.*

Thus, for $G = S_n$ or A_n, the degree 1 characters dominate $\zeta_G(s)$ for any real number $s > 0$, and the contributions of the other characters is negligible. In particular it follows that for any fixed $s > 0$ and for all sufficiently large n we have

$$r_n(G) \leq n^s.$$

When $G = S_n$ and $s \geq 1$ is an integer this was originally proved by Lulov in his unpublished thesis [Lul]; the result in its full generality is proved in [LiSh9]. The proof is quite combinatorial, based on analysis of Young tableaux. Theorem 7.1 is crucial for many of our results, and we sometimes need it for rather small values of s, for example $s = \frac{1}{42}$.

Are there analogues of this result for simple groups of Lie type? We discuss this in [LiSh11]. It turns out that, when the Lie rank tends to infinity, a similar phenomenon occurs.

Theorem 7.2. *Fix a real number $s > 0$. Then there exists a number $c = c(s)$ such that, if G is a finite simple group of Lie type of rank $r \geq c(s)$, then*

$$\zeta_G(s) \to 1 \ as \ |G| \to \infty.$$

We see that the conclusion holds not only when the rank of G tends to infinity, but also when the rank is fixed and large enough, and the field size tends to infinity.

For simple groups of given Lie type we can prove a more refined result, as follows.

Theorem 7.3. *Let G be a finite simple group of Lie type of rank r, and let u denote the number of positive roots in the root system of the algebraic group corresponding to G. Then for any $s > r/u$ we have*

$$\zeta_G(s) \to 1 \ as \ |G| \to \infty.$$

Moreover, the above conclusion does not hold for $s \leq r/u$.

This result, combined with (3), is a main tool in our proof of the random walks theorems 6.2 and 6.3.

We can think of Theorem 7.3 as determining the precise "abscissa of convergence" of $\zeta_G(s)$. Of course this notion only makes sense for an infinite group G, where it is defined as the minimal s_0 such that $\zeta_G(s)$ converges for all $s > s_0$. However, when G ranges over an infinite family of finite simple groups we may look for the minimal s_0 such that for all $s > s_0$ we have $\zeta_G(s) \to 1$, and regard it as the abscissa of convergence. For example, if $G = PSL_d(q)$ for fixed d and $q \to \infty$, then, in this sense, the abscissa of convergence of $\zeta_G(s)$ is $2/d$, as follows from Theorem 7.3 above.

Theorems 7.2 and 7.3 are established using the Deligne-Lusztig theory (see [Lus]), and their proof is rather technical. The above results imply

Corollary 7.4. *Fix a real number $s > 1$. Then for any finite simple group G we have $\zeta_G(s) \to 1$ as $|G| \to \infty$.*

This is best possible, as the case $G = PSL_2(q)$ shows. Excluding this case we obtain the same conclusion under the weaker assumption $s > 2/3$.

These results on character degrees already give us some useful information, and in the case of surface groups they suffice to obtain some of the theorems stated in previous sections. For example, let Γ_g be an oriented surface group of genus $g \geq 2$, and let G be a finite simple group. As mentioned in (4) above, we have

$$|\text{Hom}(\Gamma_g, G)| = |G|^{2g-1}\zeta_G(2g-2).$$

Now, we have $2g - 2 \geq 2$ and so by Corollary 7.4 we have $\zeta_G(2g-2) = 1 + o(1)$, giving

$$|\text{Hom}(\Gamma_g, G)| = (1 + o(1))|G|^{2g-1}.$$

This proves Corollary 4.4 for G simple.

However, in more general cases, results on character values are also needed. In case of symmetric groups, a theorem of Fomin and Lulov [FL] plays a major role here; it implies that if $x \in S_n$ has cycle-shape (m^a) (where m is fixed and $n = ma$) then

$$|\chi(x)| < cn^{1/2}\chi(1)^{1/m}.$$

For groups $G = G(q)$ of Lie type one can sometimes apply estimates on character ratios proved by Gluck [G1], showing that for every nontrivial $x \in G$ and $\chi \in Irr(G)$ we have

$$|\chi(x)/\chi(1)| \leq cq^{-1/2},$$

where q is the field size. However, for many applications one needs stronger estimates on character ratios, both for groups of Lie type and for symmetric groups. See the next section for more details.

8. Open problems

In this section we provide directions for further research, and discuss open problems and conjectures. We begin with some challenges related to Fuchsian groups.

Problem 1: Let Γ be a Fuchsian group. Find the finite simple quotients of Γ.

For the modular group this is a classical and much studied problem, and a full asymptotic solution has been given, see [LiSh2], [LM].

For a general Fuchsian group, Higman's conjecture (now a theorem) provides an (asymptotic) answer for alternating groups. It is easy to see that oriented Fuchsian groups of genus $g \geq 2$ and non-oriented Fuchsian groups of genus $g \geq 3$ have all finite simple groups as quotients. Therefore, in tackling Problem 1, it remains to focus on oriented groups of genus $0, 1$ and non-oriented groups of genus $1, 2$. In these cases not all finite simple groups need to be obtained as quotients, and the answer seems to depend on various factors, such as the orders of the torsion elements in Γ and G.

Problem 2: Let Γ be a Fuchsian group, and let S be an infinite set of finite simple groups. When can we deduce that a random homomorphism from Γ to $G \in S$ is onto with probability tending to 1 as $|G| \to \infty$?

Positive answers to Problem 2 would lead to positive answers to Problem 1 (showing that almost all simple groups in S are quotients of Γ). By Theorem 4.1 above the answer to Problem 2 is always positive for oriented Fuchsian groups of genus $g \geq 2$ and non-oriented Fuchsian groups of genus $g \geq 3$. Hence again the challenge is to prove probabilistic results for certain families S and for oriented groups of genus $0, 1$ and non-oriented groups of genus $1, 2$. Let me make the following conjecture, which can be regarded as a Lie type analogue of Higman's conjecture for alternating groups.

Conjecture 3: For every Fuchsian group Γ there is a positive integer $r = r(\Gamma)$ such that if S is an infinite set of finite simple groups of Lie type of rank at least r, then a random homomorphism from Γ to $G \in S$ is onto with probability tending to 1 as $|G| \to \infty$.

It would be important to settle this even for specific Fuchsian groups, such as the Hurwitz group (as this would imply new results on Hurwitz generation).

It can be shown that the conjecture holds for non co-compact Fuchsian groups. Since these groups are free products of cyclic groups, some cases are already covered by existing results; the proof is completed by establishing that the simple groups in question are also randomly (r, s)-generated when r, s are not both prime.

The proof of the conjecture for co-compact Fuchsian groups (of small genus) seems to require strong bounds on character ratios $|\chi(x)/\chi(1)|$ for $\chi \in Irr(G)$ and for elements $x \in G$ of given order. Some bounds on these ratios do exist (see for instance [G1]), with many interesting applications, but these bounds are not sufficient to settle our conjecture, and substantial refinements will be required.

Let us now discuss residual properties. As shown in [DPSSh] probabilistic ideas can be used to study residual properties of free groups, thus providing a new short proof of a conjecture of Magnus. The main idea was to show that (for $d \geq 2$) most d-tuples in a large simple group G do not satisfy a given relation. Using the solution to Dixon's conjecture it then follows that the free group F_d is residually S for any infinite set S of finite simple groups, as conjectured by Magnus.

It is now clear that probabilistic methods can establish residual properties of various other infinite groups, and lead to new results which were not obtained by other methods. Examples include the modular group [LiSh7], and other free products of finite groups [LiSh8] (see also [PSh]).

We propose the analogue of Magnus conjecture for general Fuchsian groups.

Problem 4: Let Γ be a Fuchsian group, and let S be an infinite set of finite simple groups. When can we deduce that Γ is residually S?

Of course a first stage would be to realize groups in S as quotients of Γ (see Problem 1), but we also need to verify that the intersection of the kernels is trivial.

The probabilistic approach here seems highly relevant. Indeed, assuming a positive answer to Problem 2 for Γ and S, the question is reduced to the following

Problem 5: Let Γ and S be as above, and let $1 \neq \gamma \in \Gamma$. Show that a random homomorphism ϕ from Γ to $G \in S$ satisfies $\phi(\gamma) \neq 1$ with probability tending to 1 as $|G| \to \infty$.

Thus, affirmative answers to Problems 2 and 5 (for given Γ and S) would yield an affirmative answer for Problem 4.

It would be interesting to study the finite quotients and the residual properties of other classes of infinite groups. These might include groups with one relator, free products with amalgamation, Kleinian groups, and possibly hyperbolic groups.

There are related famous open problems concerning such groups. For example, Baumslag conjectured that one-relator groups with torsion are residually finite, and in spite of significant progress this is still open. Another challenge, proposed by Gromov, is to find out whether all word hyperbolic groups are residually finite.

This part of the project is more speculative: it is still unclear whether our probabilistic approach could help settle such problems, but this seems worth consideration.

The next problem is related to subgroup growth. Results from Section 2 show that the subgroup growth of Fuchsian groups is super-exponential, and has the form $(n!)^{\alpha \mid o(1)}$ for some rational numbers α.

Problem 6: Study finitely generated residually finite groups of subgroup growth $(n!)^{\alpha+o(1)}$. Can α be irrational?

Let me now turn to problems on random walks.

Problem 7: Let G be a finite simple group and let $C \neq 1$ be a conjugacy class of G. Find the mixing time of the random walk on G with $C \cup C^{-1}$ as a generating set.

Various results have been obtained (see Section 6 above), but the general problem is still very much open. For G alternating we propose the following.

Conjecture 8: Let $C \subset A_n$ be a conjugacy class of fixed-point-free permutations. Then the mixing time of the corresponding random walk is 2 or 3.

Consider now the case where the generating set may change with time and we aim at an almost uniform distribution pointwise (namely in the l_∞ norm). We propose the following generalization of Theorem 6.1.

Conjecture 9: Let $C_1, \ldots, C_t \subset A_n$ be conjugacy classes, and let m_i be the average size of a cycle in an element of C_i ($i = 1, \ldots, t$). Suppose m_i are fixed. Then $C_1 \cdots C_t = A_n$ almost uniformly pointwise if and only if $\sum_{i=1}^{t} m_i^{-1} < t - 2$. In particular, for a conjugacy class C of permutations whose average cycle length is some fixed real number $m > 1$, we have $C^t = A_n$ almost uniformly pointwise if and only if $t > 2m/(m-1)$.

We also make a general asymptotic conjecture for groups of Lie type.

Conjecture 10: There exists an absolute constant c such that, if G is a simple group of Lie type, and $C \subset G$ is a non-identity conjugacy class, then the mixing time of the random walk on G with a generating set $C \cup C^{-1}$ is at most $c \log |G| / \log |C|$.

Note that this result does not hold for $G = A_n$. For example, if C is a class of a cycle of length approximately $n/2$, then $\log |G| / \log |C|$ is roughly 2, whereas the mixing time is roughly $\log n$, which is unbounded. Hence, if Conjecture 10 is true, then groups of Lie type behave "better" than alternating groups in this sense.

Note also that in [LiSh5] we prove a similar statement about the diameter of the respective Cayley graph.

There is some evidence supporting Conjecture 10; for example, it is true for classes of regular elements (where the mixing time is 2 or 3, see Theorem 6.2) and for classes of transvections (where the mixing time is roughly n, see [Hi]).

It follows from (3) and Gluck's bounds on character ratios that, if $G = G_r(q)$ is a simple group of Lie type of rank r and field size q, and C is any non-identity conjugacy class of G, then the mixing time of the corresponding random walk on G is at most cr^2. We conjecture that a linear bound holds here.

Conjecture 11: There exists an absolute constant c such that, if G is a simple group of Lie type of rank r, and $C \subset G$ is any non-identity conjugacy class, then the mixing time of the random walk on G with a generating set $C \cup C^{-1}$ is at most cr.

Note that this follows from Conjecture 10.

Proving these conjectures requires deeper understanding of character degrees and values, using the Deligne-Lusztig theory; progress in this direction might also help in studying finite quotients of Fuchsian groups, as well as their residual properties. We propose the following tasks.

Problem 12: For a finite simple group G, study the degrees of characters of G, the growth rate of $r_n(G)$, and the respective function $\zeta_G(s)$.

We have already seen that, for G alternating or symmetric, we have $r_n(G) \leq n^\epsilon$ for every $\epsilon > 0$ and $n \geq n(\epsilon)$. However, it would be interesting to address the following.

Problem 13: Is there an absolute constant c such that $r_n(G) \leq c$ for all symmetric groups G and for all n?

For simple groups G of Lie type we observe in [LiSh11] that

$$r_n(G) \leq c \text{ for all } n \text{ implies } |G| \leq f(c), \qquad (6)$$

where f is a suitable function. Moretó [Mo] and Jaikin-Zapirain [J] showed that (6) also holds for finite solvable groups G. Moreover, it is shown in [Mo] that if (6) holds for symmetric groups then it holds for all finite groups. Hence it would be nice to show that Problem 13 has a negative answer, by finding unboundedly many characters of the same degree. The formula expressing the degree of a character of S_n as a function of the hook lengths of its respective Young tableau reduces this to a very concrete question of combinatorial and number-theoretic flavor.

We now turn to questions concerning character values.

Problem 14: For a finite simple group G, and for $\chi \in Irr(G)$ and $x \in G$, find good upper bounds on character values $|\chi(x)|$ and on character ratios $|\chi(x)/\chi(1)|$, in terms of relevant invariants of x. For example, for suitable $\epsilon > 0$, show that $|\chi(x)| \leq \chi(1)^{1-\epsilon}$ with some prescribed exceptions.

Important work by Gluck (see for instance [G1]) provides general bounds which are essentially best possible. However, under some extra assumptions on x and/or on χ better bounds should exist, which might be analogous to bounds on fixed-point ratios obtained in [LiSh4]. For example, for symmetric and alternating groups, it might be that the most important invariant in estimating $|\chi(x)|$ is the average cycle-length in the decomposition of x into cycles (where cycles of length 1 are also counted). We propose the following

Conjecture 15: Let $G = S_n$ or A_n and let $x \in G$ be a permutation with average cycle-length m. Fix m and $\epsilon > 0$. Then for all large n and $\chi \in Irr(G)$ we have

$$|\chi(x)| \leq \chi(1)^{1/m+\epsilon}.$$

This conjecture could be viewed as a useful generalization of the Fomin-Lulov theorem [FL], and it might serve as the main tool in proving Conjectures 8 and 9 above.

Finally, we note that results on $\zeta_G(s)$ for finite groups G of Lie type may be relevant in studying the representation growth of arithmetic groups Γ. By a result of Lubotzky and Martin [LM2] $r_n(\Gamma)$ grows polynomially if and only if Γ has the congruence subgroup property. Suppose this is the case. The following is still very much open.

Problem 16: Let Γ be an arithmetic group with the congruence subgroup property, e.g. $\Gamma = SL_d(\mathbb{Z})$ $(d \geq 3)$. Find the abscissa of convergence of $\zeta_\Gamma(s)$.

Using our results on finite groups of Lie type we can show that this abscissa of convergence is greater than r/u, where r and u are the rank and the number of positive roots of the corresponding algebraic group, respectively. While Theorem 7.3 above may serve as one of the tools in solving Problem 16, a highly delicate analysis is still required.

References

[C1]   M.D.E. Conder, Generators for alternating and symmetric groups, *J. London Math. Soc.* **22** (1980), 75–86.

[C2]   M.D.E. Conder, More on generators for alternating and symmetric groups, *Quart. J. Math. Oxford Ser.* **32** (1981), 137–163.

[C3]   M.D.E. Conder, Hurwitz groups: a brief survey, *Bull. Amer. Math. Soc.* **23** (1990), 359–370.

[D1]   P. Diaconis, *Group Representations in Probability and Statistics*, Institute of Mathematical Statistics Lecture Notes – Monograph Series, Vol. 11, 1988.

[D2]   P. Diaconis, Random walks on groups: characters and geometry, in *Groups St Andrews 2001 in Oxford*, London Math. Soc. Lecture Note Series **304**, Cambridge Univ. Press, Cambridge, 2003, pp. 120–142.

[DS]   P. Diaconis and M. Shahshahani, Generating a random permutation with random transpositions, *Z. Wahrscheinlichkeitstheorie Verw. Gebiete* **57** (1981), 159–179.

[Di]   J.D. Dixon, The probability of generating the symmetric group, *Math. Z.* **110** (1969), 199–205.

[DPSSh] J.D. Dixon, L. Pyber, Á. Seress and A. Shalev, Residual properties of free groups and probabilistic methods, *J. reine angew. Math. (Crelle's Journal)* **556** (2003), 159–172.

[ET]   P. Erdős and P. Turán, On some problems of statistical group theory. I, *Z. Wahrscheinlichkeitstheorie verw. Geb.* **4** (1965), 175–186.

[EO]   A. Eskin and A. Okounkov, Asymptotics of numbers of branched coverings of a torus and volumes of moduli spaces of holomorphic differentials, *Invent. Math.* **145** (2001), 59–103.

[Ev]   B. Everitt, Alternating quotients of Fuchsian groups, *J. Algebra* **223** (2000), 457–476.

[FL]   S.V. Fomin and N. Lulov, On the number of rim hook tableaux, *J. Math. Sci. (New York)* **87** (1997), 4118–4123.

386 A. Shalev

[FM] D. Frohardt and K. Magaard, Composition factors of monodromy groups, *Annals of Math.* **154** (2001), 327–345.

[G1] D. Gluck, Sharper character value estimates for groups of Lie type *J. Algebra* **174** (1995), 229–266.

[G2] D. Gluck, Characters and random walks on finite classical groups, *Adv. Math.* **129** (1997), 46–72.

[GSSh] D. Gluck, Á. Seress and A. Shalev, Bases for primitive permutation groups and a conjecture of Babai, *J. Algebra* **199** (1998), 367–378.

[Gr] R.I. Grigorchuk, Just infinite branch groups, in *New horizons in pro-p groups*, eds: M.P.F. du Sautoy, D. Segal and A. Shalev, Progress in Mathematics, Birkhäuser, 2000, pp. 121–179.

[GK] R.M. Guralnick and W.M. Kantor, Probabilistic generation of finite simple groups, *J. Algebra* **234** (2000) (Wielandt's volume), 743–792.

[GKS] R.M. Guralnick, W.M. Kantor and J. Saxl, The probability of generating a classical group, *Comm. in Algebra* **22** (1994), 1395–1402.

[GLSSh] R.M. Guralnick, M.W. Liebeck, J. Saxl and A. Shalev, Random generation of finite simple groups, *J. Algebra* **219** (1999), 345–355.

[GSh] R.M. Guralnick and A. Shalev, On the spread of finite simple groups, *Combinatorica* **23** (2003) (Érdos volume), 73–87.

[GT] R.M. Guralnick and J.G. Thompson, Finite groups of genus zero, *J. Algebra* **131** (1990), 303–341.

[Ha] M. Hall, Subgroups of finite index in free groups, *Canad. J. Math.* **1** (1949), 187–190.

[Hi] M. Hildebrand, Generating random elements in $SL_n(F_q)$ by random transvections, *J. Alg. Combin.* **1** (1992), 133–150.

[Hu1] A. Hurwitz, Über algebraische Gebilde mit eindeutigen Transformationen in sich, *Math. Ann.* **41** (1893), 408–442.

[Hu2] A. Hurwitz, Über die Anzahl der Riemannschen Flächen mit gegebener Verzweigungspunkten, *Math. Ann.* **55** (1902), 53–66.

[J] A. Jaikin-Zapirain, On two conditions on characters and conjugacy classes in finite soluble groups, *J. Group Theory* **8** (2005), 267–272.

[KL] W.M. Kantor and A. Lubotzky, The probability of generating a finite classical group, *Geom. Ded.* **36** (1990), 67–87.

[KLi] P.B. Kleidman and M.W. Liebeck, *The Subgroup Structure of the Finite Classical Groups*, London Math. Soc. Lecture Note Series **129**, Cambridge University Press, 1990.

[KK] A. Klyachko and E. Kurtaran, Some identities and asymptotics for characters of the symmetric group, *J. Algebra* **206** (1998), 413–437.

[LW] S. Lang and A. Weil, Number of points of varieties over finite fields, *Amer. J. Math.* **76** (1954), 819–827.

[La] R. Lawther, Elements of specified order in simple algebraic groups, *Trans. Amer. Math. Soc.* **357** (2005), 221–245.

[LiSh1] M.W. Liebeck and A. Shalev, The probability of generating a finite simple group, *Geom. Ded.* **56** (1995), 103–113.

[LiSh2]   M.W. Liebeck and A. Shalev, Classical groups, probabilistic methods, and the (2,3)-generation problem, *Annals of Math.* **144** (1996), 77–125.

[LiSh3]   M.W. Liebeck and A. Shalev, Simple groups, probabilistic methods, and a conjecture of Kantor and Lubotzky, *J. Algebra* **184** (1996), 31–57.

[LiSh4]   M.W. Liebeck and A. Shalev, Simple groups, permutation groups, and probability, *J. Amer. Math. Soc.* **12** (1999), 497–520.

[LiSh5]   M.W. Liebeck and A. Shalev, Diameters of finite simple groups: sharp bounds and applications, *Annals of Math.* **154** (2001), 383–406.

[LiSh6]   M.W. Liebeck and A. Shalev, Random (r, s)-generation of finite classical groups, *Bull. London Math. Soc.* **34** (2002), 185–188.

[LiSh7]   M.W. Liebeck and A. Shalev, Residual properties of the modular group and other free products, *J. Algebra* **268** (2003), 264–285.

[LiSh8]   M.W. Liebeck and A. Shalev, Residual properties of free products of finite groups, *J. Algebra* **268** (2003), 286–289.

[LiSh9]   M.W. Liebeck and A. Shalev, Fuchsian groups, coverings of Riemann surfaces, subgroup growth, random quotients and random walks, *J. Algebra* **276** (2004), 552–601.

[LiSh10] M.W. Liebeck and A. Shalev, Fuchsian groups, finite simple groups, and representation varieties, *Invent. Math.* **159** (2005), 317–367.

[LiSh11] M.W. Liebeck and A. Shalev, Character degrees and random walks in finite groups of Lie type, *Proc. London Math. Soc.* **90** (2005), 61–86.

[LuMa]   A. Lubotzky and A.R. Magid, Varieties of representations of finitely generated groups, *Mem. Amer. Math. Soc.* **58** (1985), no. 336.

[LM2]   A. Lubotzky, B. Martin, Polynomial representation growth and the congruence subgroup problem, to appear in *Israel J. of Math.*

[LS]   A. Lubotzky and D. Segal, *Subgroup growth*, Birkhäuser, 2003.

[LM]   F. Lübeck and G. Malle, $(2,3)$-generation of exceptional groups, *J. London Math. Soc.* **59** (1999), 109–122.

[Lul]   N. Lulov, Random walks on symmetric groups generated by conjugacy classes, Ph.D. Thesis, Harvard University, 1996.

[Lus]   G. Lusztig, *Characters of reductive groups over a finite field*, Annals of Mathematics Studies **107**, Princeton University Press, 1984.

[LP]   T. Łuczak and L. Pyber, On random generation of the symmetric group, *Combinatorics, Probability and Computing* **2** (1993), 505–512.

[Mo]   A. Moretó, On the structure of the complex group algebra of finite groups, Preprint, 2003.

[MP1]   T.W. Müller and J-C. Puchta, Character theory of symmetric groups and subgroup growth of surface groups, *J. London Math. Soc.* **66** (2002), 623–640.

[MP2]   T.W. Müller and J-C. Puchta, Character theory of symmetric groups, subgroup growth of Fuchsian groups, and random walks, to appear.

[Ne]   M. Newman, Asymptotic formulas related to free products of cyclic groups, *Math. Comp.* **30** (1976), 838-846.

[NS1]   N. Nikolov and D. Segal, On finitely generated profinite groups, I: strong completeness and uniform bounds, to appear.

[NS2]   N. Nikolov and D. Segal, On finitely generated profinite groups, II: product decompositions of quasisimple groups, to appear.

[Py]   L. Pyber, Asymptotic results for permutation groups, in *Groups and Computation* (eds: L. Finkelstein and W.M. Kantor), DIMACS Series on Discrete Math. and Theor. Computer Science **11** (1993), 197–219.

[PSh]   L. Pyber and A. Shalev, Residual properties of groups and probabilistic methods, *C.R. Acad. Sci. Paris Sr. I Math.* **333** (2001), 275–278.

[RBC]   A.S. Rapinchuk, V.V. Benyash-Krivetz and V.I. Chernousov, Representation varieties of the fundamental groups of compact orientable surfaces, *Israel J. Math.* **93** (1996), 29–71.

[R]   Y. Roichman, Upper bound on the characters of the symmetric groups, *Invent. Math.* **125** (1996), 451–485.

[Sh1]   A. Shalev, Random generation of simple groups by two conjugate elements, *Bull. London Math. Soc.* **29** (1997), 571-576.

[Sh2]   A. Shalev, A theorem on random matrices and some applications, *J. Algebra* **199** (1998), 124–141.

[Sh3]   A. Shalev, Probabilistic group theory, in *Groups St Andrews 1997 in Bath, II*, London Math. Soc. Lecture Note Series **261**, Cambridge University Press, Cambridge, 1999, pp. 648–678.

[Sh4]   A. Shalev, Random generation of finite simple groups by p-regular or p-singular elements, *Israel J. Math.* **125** (2001), 53–60.

[V]   R. Vakil, Genus 0 and 1 Hurwitz numbers: recursions, formulas, and graph-theoretic interpretations, *Trans. Amer. Math. Soc.* **353** (2001), 4025–4038.

[W]   H.S. Wilf, The asymptotics of $e^{P(z)}$ and the number of elements of each order in S_n, *Bull. Amer. Math. Soc.* **15** (1986), 228–232.

[Wit]   E. Witten, On quantum gauge theories in two dimensions, *Comm. Math. Phys.* **141** (1991), 153–209.

Aner Shalev
Institute of Mathematics
Hebrew University
Jerusalem 91904
Israel
e-mail: shalev@math.huji.ac.il

Progress in Mathematics, Vol. 248, 389–402
© 2005 Birkhäuser Verlag Basel/Switzerland

Just Non-(abelian by P-type) Groups

Said Sidki

Abstract. A group K is of *P-type* provided it is isomorphic to a subgroup of a finitely iterated wreath product of copies of P. For any group P, define a rooted regular tree $T(P)$ on which P acts permuting regularly the vertices of the first level. Given a generating set A of P, we define for each element of A an automorphism of $T(P)$ in a manner similar to the procedure in the Magnus embedding of groups. These automorphisms generate $G(P, A)$, an extension of P. For fairly general groups P, we prove that $G(P, A)$ is a weakly branch group whose proper quotients are *abelian by P-type*.

Mathematics Subject Classification (2000). Primary 20F05, 20F10; Secondary 20H05, 68Q68.

Keywords. Automorphisms of trees, Magnus embedding, just-infinite groups, branch groups, automata, bounded growth.

1. Introduction

Groups of automorphisms of rooted trees, called weakly branch, have revealed novel phenomena in group theory, such as intermediate growth [3], just infiniteness [5], [4], [8] and just non-solvability [2]. Recall that a group is just non-$\mathcal{R}$ provided all its proper quotients satisfy $\mathcal{R}$ while the group itself does not satisfy this property.

For which group properties $\mathcal{R}$ there exist weakly branch groups satisfying the just non-$\mathcal{R}$ condition, is an open problem. The present paper treats this problem for the group property "abelian by P-type". Here a group K is said to be of P-*type* for some group P, provided K is isomorphic to a subgroup of $wr^k(P) = ((\ldots)wrP)wrP$, a wreath product of P iterated a finite k number of times.

In case P is finite, we obtain weakly branch groups which are just non-(abelian by finite) but are not branch. This specific result answers a question posed to us by J. Wilson. In this respect, we note the recent work of P. Hardy and J. Wilson

The author thanks Laurent Bartholdi for hospitality and support at École Polytechnique Fédérale- Lausanne during June 2004. The author acknowledges further support from the Brazilian Conselho Nacional de Pesquisa.

extending the structure theory of just infinite groups to the wider class of just non-(abelian by finite) groups (see, [9]). For the definition of branch and weakly branch groups, see Section 4, or [1] for more extensive material.

Our construction is analogous to the Magnus embedding for groups (see, [6]). Let P be a group, $A = \{a_i \mid 1 \leq i \leq d\}$ be one of its generating sets and let P be defined by relations $r_j = r_j(a_i) = e$. Also, let F be the free group freely generated by x_i, where $1 \leq i \leq d$. Then P is a homomorphic image of the free group F with respect to the extension of the map $\phi : x_i \to a_i$. Denote by R the kernel of ϕ and R' its commutator subgroup. One form of the Magnus embedding is the representation of the group $\widehat{P} = \frac{F}{R'}$ into the wreath product $\Lambda wr P$ where Λ is a free abelian group freely generated by $\{\lambda_i | i = 1, \ldots, d\}$ and where the homomorphism is the extension of the map $a_i \to \lambda_i a_i$.

The construction of our group $G(P, A)$ proceeds as follows. First, let $\underline{P} = \{\underline{p} | p \in P\}$ be a set of symbols in one to one correspondence with P by $\underline{p} \to p$ and let $T = T(P)$ be the tree indexed by the free monoid freely generated by the $\underline{P}$. We make the group P act on the set of symbols $\underline{P}$ by $(\underline{p})^a = \underline{pa}$ and extend this action rigidly to the tree; that is, $(\underline{p_1 p_2} \ldots \underline{p_k})^a = (\underline{p_1 a}) .\underline{p_2} \ldots \underline{p_k}$. Next, for every generator a_i of P we define the automorphism $\alpha_i = (\underline{e}\alpha_i) \, a_i$ of the tree T and let $G = G(P, A)$ be the group generated by the set $\{\alpha_i | i = 1, \ldots, d\}$. The generator α_i induces the permutation a_i on $\underline{P}$ and $\alpha_i a_i^{-1}$ induces the automorphism α_i on the subtree $T(\underline{e})$ with root $\underline{e}$; that is, after $T(\underline{e})$ is identified with T. Note that for $P = <a>$ of order 2, the tree is the binary tree and $\alpha = (\underline{e}\alpha) \, a$ is a form of the binary adding machine.

Let $H = stab_G(1)$ denote the stabilizer in G of the first level of the tree. The definition of the generators $\alpha_i = (\underline{e}\alpha_i) \, a_i$ implies that H is the set of words $r(\alpha_i)$ such that $r(a_i) = e$ in P and therefore H is a subgroup of the set of functions $\times_{\underline{P}} G$ with finite support from $\underline{P}$ into G. It follows that

$$G \leq \left(\times_{\underline{P}} G \right) G = \left(\times_{\underline{P}} G \right) P,$$

which establishes the close connection of our construction to the Magnus embedding.

Our first result is

Theorem 1. *If P does not contain free subgroups of rank 2 then neither does $G = G(P, A)$.*

This is simply an application of [7]. For, each generator α_i of the group $G = G(P, A)$ has only one nontrivial state and the corresponding automaton has bounded growth. Therefore G is a subgroup of the group $Pol(0)$ of automorphisms of bounded growth of the tree $T(P)$.

Analysis of the structure of words in the generators α_i produces specific structural information for a wide range of groups P.

Theorem 2. *Let P be a group minimally generated by $A = \{a_i \mid 1 \le i \le d\}$ with $d \ge 3$ such that P/P' is minimally generated by $\{P'a_i \mid 1 \le i \le d\}$. Suppose also the generators a_i and the commutators $a_i^{-1}a_j^{-1}a_ia_j$ have finite orders for all i,j. Then, $G = G(P, A)$ is a weakly branch group with the following properties:*

(i) *$H = stab_G(1)$ is a subdirect product of $\times_{\underline{P}}G$,*
(ii) *$H' = \times_{\underline{P}}G'$,*
(iii) *$\frac{\gamma_2(G)}{\gamma_3(G)}$ is a torsion group,*
(iv) *any proper quotient of G is abelian by P-type.*

In order to obtain just non-$\mathcal{R}$ groups for $\mathcal{R}$ "*abelian by P-type*" it is sufficient to assume that finitely iterated wreath products of P do not contain weakly branch subgroups.

Corollary 1. *Let P be a group which satisfies the conditions of the theorem and such that $wr^k(P)$ is free of weakly branch subgroups for all $k \ge 0$. Then G is just non-(abelian by P-type).*

The situation for abelian groups P is as follows.

Theorem 3. *Let P be a non-cyclic abelian group generated by $A = \{a_i \mid 1 \le i \le d\}$. Suppose $a_i \ne a_j^2, a_i^2 \ne a_j^2$ for all $i \ne j$. Then, H is a subdirect product of $\times_{\underline{P}}G$ and $G'' = \times_{\underline{P}}G' = H'$. Furthermore, G is just non-(abelian by P-type).*

In case $P \cong C_m \times C_m$, we are able to extract further information, especially about $G(P, A)$ affording a quotient isomorphic to the extension $\widehat{P}$.

Theorem 4. *Let P be a finite abelian group presented by $\{a, b \mid a^m = b^m = [a, b] = e\}$. Then, for $G = G(P, A)$, we have $H = sta_G(1)$ is a subdirect product of $\times_{\underline{P}}G$, $H' = \times_{\underline{P}}G'$, $\frac{G}{G'}$ free abelian of rank 2, $\frac{G}{H'} \cong \widehat{P}$ and G torsion-free. Furthermore, G is just non-(abelian by finite), yet is not a branch group.*

We pose the problem of finding conditions on (P, A) such that the quotient group $G = G(P, A)$ modulo $\times_{\underline{P}}G'$ is isomorphic to $\widehat{P}$. The main difficulty in answering this question lies in the lack of general methods for describing abelian quotients of groups acting on trees, even those generated by easily defined automorphisms.

2. Word forms

Let $u(a_j) = a_{i_1}^{\varepsilon_{i_1}} a_{i_2}^{\varepsilon_{i_2}} \dots a_{i_k}^{\varepsilon_{i_k}}$ be a word in P. We say $u(a_j)$ is reduced provided no proper sub-word is the trivial element in P. To establish the form of words in G we use

$$\alpha_i^{\varepsilon_i} = \left(a_{\underline{i}}^{\delta_i}\alpha_i^{\varepsilon_i}\right)a_i^{\varepsilon_i} \text{ where } \varepsilon_i = -1, 1 \text{ and } \delta_i = \frac{|\varepsilon_i - 1|}{2}.$$

Then,

$$w = u(\alpha_j) = \alpha_{i_1}^{\varepsilon_{i_1}} \alpha_{i_2}^{\varepsilon_{i_2}} \ldots \alpha_{i_k}^{\varepsilon_{i_k}}$$

$$= \left(\frac{a_{i_1}^{\delta_{i_1}} \alpha_{i_1}^{\varepsilon_{i_1}} + \ldots + a_{i_s}^{\delta_{i_s}} a_{i_{(s-1)}}^{-\varepsilon_{i_{(s-1)}}} \ldots a_{i_1}^{-\varepsilon_{i_1}} \alpha_{i_s}^{\varepsilon_{i_s}} + \ldots +}{a_{i_k}^{\delta_{i_k}} a_{i_{(k-1)}}^{-\varepsilon_{i_{(k-1)}}} \ldots a_{i_1}^{-\varepsilon_{i_1}} \alpha_{i_k}^{\varepsilon_{i_k}}} \right) u(a_j).$$

The multiplication is performed from left to right and the indices

$$a_{i_1}^{\delta_{i_1}}, \ldots, a_{i_s}^{\delta_{i_s}} a_{i_{(s-1)}}^{-\varepsilon_{i_{(s-1)}}} \ldots a_{i_1}^{-\varepsilon_{i_1}}, \ldots, a_{i_k}^{\delta_{i_k}} a_{i_{(k-1)}}^{-\varepsilon_{i_{(k-1)}}} \ldots a_{i_1}^{-\varepsilon_{i_1}}$$

are recorded in their order of occurrence. This gives us $w = \sum_{1 \le j \le m} p_j w_{p_j}$. The set $\{\alpha_{i_1}^{\varepsilon_{i_1}}, \alpha_{i_2}^{\varepsilon_{i_2}}, \ldots, \alpha_{i_k}^{\varepsilon_{i_k}}\}$ is partitioned among the coefficients w_{p_j}. If the indices equal to p in w occur in order $p_{j_1}, p_{j_2}, \ldots, p_{j_s}$ then we collect the coefficients of these indices and thus produce the reduced form of w where the coefficient of $\underline{p}$ is $w_{p_1} w_{p_2} \ldots w_{p_s}$.

If $u(a_j)$ is a reduced word in P, in the sense that none of its proper sub-words is the trivial element, then we are able to determine the possible coincidences among the indices $\underline{p}$.

Proposition 1. *Suppose $u(a_j)$ is reduced and suppose for some $s < t$ we have the coincidence*

$$a_{i_s}^{\delta_{i_s}} a_{i_{(s-1)}}^{-\varepsilon_{i_{(s-1)}}} \ldots a_{i_1}^{-\varepsilon_{i_1}} = a_{i_t}^{\delta_{i_t}} a_{i_{(t-1)}}^{-\varepsilon_{i_{(t-1)}}} \ldots a_{i_1}^{-\varepsilon_{i_1}}.$$

Then,

$$t = s+1 \text{ and } (\varepsilon_{i_s}, \varepsilon_{i_{s+1}}) = (-1, 1),$$

or

$$t = k, s = 1 \text{ and } (\varepsilon_{i_1}, \varepsilon_{i_k}) = (1, -1).$$

Proof. We conclude from the given coincidence that $a_{i_s}^{\delta_{i_s}} = a_{i_t}^{\delta_{i_t}} a_{i_{(t-1)}}^{-\varepsilon_{i_{(t-1)}}} \ldots a_{i_s}^{-\varepsilon_{i_s}}$ and we have the following cases to consider:

(i) if $(\varepsilon_{i_s}, \varepsilon_{i_t}) = (-1, -1)$ then $a_{i_t}^{-\varepsilon_{i_t}} a_{i_{(t-1)}}^{-\varepsilon_{i_{(t-1)}}} \ldots a_{i_{s+1}}^{-\varepsilon_{i_{s+1}}} = e$ and reduction in length of u follows;

(ii) if $(\varepsilon_{i_s}, \varepsilon_{i_t}) = (-1, 1)$ then $a_{i_{(t-1)}}^{-\varepsilon_{i_{(t-1)}}} \ldots a_{i_s}^{-\varepsilon_{i_s}} = a_{i_s}^{-\varepsilon_{i_s}}$ and reduction in length of u follows, unless $t - 1 = s$;

(iii) if $(\varepsilon_{i_s}, \varepsilon_{i_t}) = (1, -1)$ then $a_{i_t}^{-\varepsilon_{i_t}} a_{i_{(t-1)}}^{-\varepsilon_{i_{(t-1)}}} \ldots a_{i_s}^{-\varepsilon_{i_s}} = e$ and reduction in length of u follows, unless $t = k, s = 1$;

(iv) if $(\varepsilon_{i_s}, \varepsilon_{i_t}) = (1, 1)$ then $a_{i_{(t-1)}}^{-\varepsilon_{i_{(t-1)}}} \ldots a_{i_s}^{-\varepsilon_{i_s}} = e$ and reduction in length of u follows. $\square$

Suppose each generator a_l appears in some reduced relator $r_i(a_j)$. Then we may assume that r_i starts with a_l. The coefficient of $\underline{e}$ in the reduced form of $r_i(\alpha_j)$ is α_l unless $r_i(a_j) = a_l \ldots a_{i_k}^{-1}$, in which case the coefficient of $\underline{e}$ is $\alpha_l \alpha_{i_k}^{-1}$.

Thus, we have

Proposition 2. *Suppose that the presentation of P satisfies the following property: for each generator a_l there exists a generator $a_{l'}$ and a minimal relator $r_l(a_j) = a_{i_1}^{\varepsilon_{i_1}} a_{i_2}^{\varepsilon_{i_2}} \dots a_{i_k}^{\varepsilon_{i_k}}$ with $a_{i_1}^{\varepsilon_{i_1}} = a_l$ and $a_{i_k}^{\varepsilon_{i_k}} = a_{l'}$. Then the coefficient of $\underline{e}$ in $r_i(\alpha_j)$ is α_l. Hence, H is a subdirect product of $\times_P G$ and, in particular, G is transitive on all levels.*

Generators of finite order

Let $m \geq 1$. Then,

$$(\alpha_i)^m = \left(a_i^m \alpha_i^{-1} + \underline{a_i^{m-1}}\alpha_i^{-1} + \dots + \underline{a_i}\alpha_i^{-1} \right) a_i^m$$

for all i. Suppose a_i has finite order m_i for all i. Then,

$$(\alpha_i)^{m_i} = \underline{e}\alpha_i + a_i^{-1}\alpha_i + \dots + \underline{a_i^{-m_i+1}}\alpha_i$$

in reduced form. Let $j \neq i$, then

$$\alpha_j^{-1}(\alpha_i)^{m_i}\alpha_j = \underline{a_j\alpha_j^{-1}}\alpha_i\alpha_j + a_j^{-1}a_j\alpha_i + \dots + \underline{a_i^{-m_i+1}a_j}\alpha_i$$

in reduced form.

If the cyclic groups $< a_i >, < a_j >$ are disjoint having respective finite orders m_i, m_j then, clearly $[\alpha_j^{m_j}, \alpha_i^{m_i}] = \underline{e}[\alpha_i, \alpha_j]$. More generally,

Lemma 1. *Suppose for the generators a_i, a_j there exist $k, l > 0$ with $k+l$ minimum such that $a_i^k = a_j^l$. Then,*

$$\alpha_i^{-k}\alpha_j^l = \underline{a_i^k}\left(\alpha_i^{-1}\alpha_j \right) + \underline{a_i^{k-1}}\alpha_i^{-1} + \dots + \underline{a_i}\alpha_i^{-1}$$
$$+\underline{a_j^{l-1}}\alpha_j + \dots + \underline{a_j}\alpha_j$$

in reduced form. Furthermore, if α_i has finite order $m_i \geq k$ then, $[\alpha_j^l, \alpha_i^{m_i}] = \underline{a_i^k}[\alpha_j, \alpha_i]$.

Proof. From

$$\alpha_i^k = \left(\underline{e}\alpha_i + a_i^{-1}\alpha_i + \dots + \underline{a_i^{-k+1}}\alpha_i \right) a_i^k,$$
$$\alpha_j^l = \left(\underline{e}\alpha_j + a_j^{-1}\alpha_j + \dots + \underline{a_j^{-k+1}}\alpha_j \right) a_j^l,$$

we calculate

$$w = \alpha_i^{-k}\alpha_j^l$$
$$= \underline{a_i^k}\left(\alpha_i^{-1}\alpha_j \right) + \underline{a_i^{k-1}}\alpha_i^{-1} + \dots + \underline{a_i}\alpha_i^{-1} + a_j^{-1}\underline{a_i^k}\alpha_j + \dots + a_j^{-l+1}\underline{a_i^k}\alpha_j$$
$$= \underline{a_i^k}\left(\alpha_i^{-1}\alpha_j \right) + \underline{a_i^{k-1}}\alpha_i^{-1} + \dots + \underline{a_i}\alpha_i^{-1} + \underline{a_j^{l-1}}\alpha_j + \dots + \underline{a_j}\alpha_j,$$

in reduced from. Since

$$\{a_i^k, a_i^{k-1}, \dots, a_i, a_j^{l-1}, \dots, a_j\} \cap \{e, a_i, a_i^2, \dots, a_i^{m_i-1}\} = \{a_i^k, a_i^{k-1}, \dots, a_i\},$$

we conclude that

$$[w, \alpha_i^{m_i}] = \underline{a}_i^k \left[\alpha_i^{-1}\alpha_j, \alpha_i\right] = \underline{a}_i^k [\alpha_j, \alpha_i]. \qquad \square$$

3. Commutators

We formulate certain elementary conditions on P which imply the following 'strong' branchings:

$$\times_{\underline{P}}\gamma_3(G) \leq G'', \times_{\underline{P}}\gamma_2(G) \leq \gamma_3(G).$$

Proposition 3. *Let $a_j \neq a_i$, $s_{ij} = [a_i, a_j]$ and $\sigma_{ij} = [\alpha_i, \alpha_j]$.*

(i) *Then,*

$$\sigma_{ij} = \left(\underline{a}_i\alpha_i^{-1} + \underline{a}_j\underline{a}_i\alpha_j^{-1}\alpha_i + \underline{a}_i^{-1}\underline{a}_j\underline{a}_i\alpha_j\right)s_{ij}$$

in reduced form.

(ii) *Suppose $a_i, a_j, a_i^{-1}a_j \notin < s_{ij} >$ and let $1 \leq n \leq o\,(s_{ij})$. Then,*

$$\sigma_{ij}^n = \left(\sum_{0 \leq k \leq n-1} (\underline{a}_i s_{ji}^k \alpha_i^{-1} + \underline{a}_j\underline{a}_i s_{ji}^k \alpha_j^{-1}\alpha_i + \underline{a}_i^{-1}\underline{a}_j\underline{a}_i s_{ji}^k \alpha_j)\right) s_{ij}^n,$$

in reduced form.

(iii) *Suppose in addition, $o\,(s_{ij}) = n$, $o(\alpha_u) = m_u$, $o(\alpha_v) = m_v$ are finite and $\alpha_u \notin < \alpha_v >$, $\alpha_v \notin < \alpha_u >$. Then,*

$$\left[(\sigma_{ij}^n)^{\alpha_i^{-1}}, \left[[\alpha_u^{-r}\alpha_v^s, \alpha_u^{m_u}]^{\alpha_u^{-r}}, \alpha_v^{m_v}\right]\right] = \underline{e}\left[\alpha_i^{-1}, [\alpha_u^{-1}, \alpha_v]\right],$$

where $\alpha_u^r = \alpha_v^s$, $1 \leq r \leq m_u$, $1 \leq s \leq m_v$ and $r + s$ is minimum such.

(iv) *Let $n = o\,(s_{ij})$ be finite and $a_i, a_j, a_i^{-1}a_j \notin < s_{ij} >$. Suppose there exists α_l such that $o(\alpha_l) = m_l$ is finite and*

$$< a_l > \cap\{e, a_j, a_i^{-1}a_j\} < s_{ji} >^{a_i^{-1}} = \{e\}.$$

Then,

$$\left[(\sigma_{ij}^n)^{\alpha_i^{-1}}, \alpha_l^{m_l}\right] = \underline{e}\left[\alpha_i^{-1}, \alpha_l\right].$$

Proof. The formula in (i) follows readily from the more general one for $w = u(\alpha_j)$ in Section 1.

(ii) We compute

$$\begin{aligned}
\sigma_{ij}^n = \ & (\underline{a}_i\alpha_i^{-1} + \underline{a}_j\underline{a}_i\alpha_j^{-1}\alpha_i + \underline{a}_i^{-1}\underline{a}_j\underline{a}_i\alpha_j \\
& + \underline{a}_i s_{ji}\alpha_i^{-1} + \underline{a}_j\underline{a}_i s_{ji}\alpha_j^{-1}\alpha_i + \underline{a}_i^{-1}\underline{a}_j\underline{a}_i s_{ji}\alpha_j + \ldots)s_{ij}^n,
\end{aligned}$$

with set of indices

$$\{a_i, a_j a_i, a_i^{-1}a_j a_i\}\{s_{ji}^k \mid 0 \leq k \leq n-1\}$$

and analyze the possible coincidences between these indices. Let $k < l \leq n-1$, then:

$$a_i s_{ji}^k = a_i s_{ji}^l \Rightarrow o(s_{ij})|l-k,$$

$$a_i s_{ji}^k = a_j a_i s_{ji}^l \Rightarrow a_i^{-1} a_j^{-1} a_i = s_{ji}^{l-k} \Rightarrow s_{ij} a_j^{-1} = s_{ji}^{l-k} \Rightarrow a_j^{-1} = s_{ji}^{l-k+1},$$

$$a_i s_{ji}^k = a_i^{-1} a_j a_i s_{ji}^l \Rightarrow a_i^{-1} a_j^{-1} a_i^2 = s_{ji}^{l-k} \Rightarrow s_{ji} a_i^{-1} a_j^{-1} a_i^2 = s_{ji}^{l-k+1}$$
$$\Rightarrow a_j^{-1} a_i = s_{ji}^{l-k+1},$$

$$a_j a_i s_{ji}^k = a_j a_i s_{ji}^l \Rightarrow o(s_{ij})|l-k,$$

$$a_j a_i s_{ji}^k = a_i^{-1} a_j a_i s_{ji}^l \Rightarrow a_i^{-1} a_j^{-1} a_i a_j a_i = s_{ji}^{l-k} \Rightarrow s_{ij} a_i = s_{ji}^{l-k} \Rightarrow a_i = s_{ji}^{l-k+1}.$$

Thus, if $n \le o(s_{ij})$ then the only possibilities for coincidences are as follows:

$$a_i s_{ji}^k = a_j a_i s_{ji}^l \text{ when } a_j^{-1} = s_{ji}^{l-k+1},$$

$$a_j a_i s_{ji}^k = a_i^{-1} a_j a_i s_{ji}^l \text{ when } a_i = s_{ji}^{l-k+1}.$$

Only one of $a_i, a_j, a_j^{-1} a_i \in\, <s_{ij}>$ can hold; for otherwise, $s_{ij} = e = a_i = a_j$.

(iii) As $n = o(s_{ij})$ it follows that

$$\sigma_{ij}^n = \sum_{0 \le k \le n-1} (a_i s_{ji}^k \alpha_i^{-1} + a_j a_i s_{ji}^k \alpha_j^{-1} \alpha_i + a_i^{-1} a_j a_i s_{ji}^k \alpha_j).$$

On conjugating the above by α_i^{-1} we obtain

$$\left(\sigma_{ij}^n\right)^{\alpha_i^{-1}} = \underline{e}\alpha_i^{-1} + \underline{a_j \alpha_j^{-1} \alpha_i} + \underline{a_i^{-1} a_j \alpha_j}$$
$$+ \sum_{1 \le k \le n-1} (\underline{a_i s_{ji}^k a_i^{-1} \alpha_i^{-1}} + \underline{a_j a_i s_{ji}^k a_i^{-1} \alpha_j^{-1} \alpha_i} + \underline{a_i^{-1} a_j a_i s_{ji}^k a_i^{-1} \alpha_j}),$$

in reduced form. By Lemma 1, if $u \ne v$ and $\alpha_u^r = \alpha_v^s$ where $1 \le r \le m_u, 1 \le s \le m_v$ and $r+s$ minimum, then $[\alpha_u^{-r} \alpha_v^s, \alpha_u^{m_u}]^{\alpha_u^{-r}} = \left(\alpha_u^k [\alpha_v, \alpha_u]\right)^{\alpha_u^{-r}} = \underline{e}\,[\alpha_u^{-1}, \alpha_v]$ and thus,

$$\left[\left(\sigma_{ij}^n\right)^{\alpha_i^{-1}}, \left[\alpha_u^{-r} \alpha_v^s, \alpha_u^{m_u}\right]^{\alpha_u^{-k}}, \alpha_v^{m_v}\right] = \underline{e}\,[\alpha_i^{-1}, [\alpha_u^{-1}, \alpha_v]].$$

(iv) Since

$$< a_l > \cap \{e, a_j, a_i^{-1} a_j\} < s_{ji} >^{a_i^{-1}} = \{e\},$$
$$\alpha_l^{m_l} = \underline{e}\alpha_l + \ldots + \underline{a_l^{m_l-1} \alpha_l}$$

we obtain

$$\left[\left(\sigma_{ij}^n\right)^{\alpha_i^{-1}}, \alpha_l^{m_l}\right] = \underline{e}\,[\alpha_i^{-1}, \alpha_l]. \qquad \square$$

4. Weakly branch groups

Let M be a group of automorphisms of a regular rooted tree and u be a vertex of the tree. Then $u * M$ denotes the group of automorphisms of the tree which fixes the vertex u as well as all vertices outside the subtree headed by u and furthermore, induces M on this subtree. The kth rigid stabilizer $rist_M(k)$ is the subgroup generated by $(u * M) \cap M$ for all indices u of length k. The group M is said to be *weakly branch* provided M acts transitively on the levels of the tree and $rist_M(k)$ is nontrivial for all k. The weakly branch group M is said to be *branch*, if the index $[M : rist_M(k)]$ is finite for all k. By a theorem of R. Grigorchuk, a branch group is just infinite provided the index $[rist_M(k) : rist_M(k)']$ is finite for all k (see, [1]).

If the generators a_i of P have finite order we obtain G' branching of G as in the following lemma.

Lemma 2. *Let P be generated by $A = \{a_i \mid 1 \leq i \leq d\}$ and $G = G(P, A)$. Suppose $d \geq 2$, $o(a_i)$ is finite for all i and no proper subset of A generates P. Then G is a weakly branch group such that the subgroup $H = stab_G(1)$ is a subdirect product of $\times_P G$ and $\times_P G' = H'$.*

Proof. Denote $o(a_i)$ by m_i for all i. We have

$$(\alpha_i)^{m_i} = \underline{e}\alpha_i + a_i^{-1}\alpha_i + \ldots + a_i^{-m_i+1}\alpha_i \text{ for all } i$$

and therefore, $H = stab_G(1) \leq \times_P G$. From Lemma 1, we have $[\alpha_i^{-k}\alpha_j^l, \alpha_i^{m_i}] = a_i^k[\alpha_j, \alpha_i] \in H'$ for all i, j and therefore, $H' \leq \times_P G'$. Since $\alpha_i^{-k}\alpha_j^l, \alpha_i^{m_i} \in H$ we conclude $\times_P G' \leq H'$ and therefore, $H' = \times_P G'$. □

The following result and its method of proof are standard.

Proposition 4. *Let P be a nontrivial group and $G = G(P, A)$. Suppose $H = stab_G(1)$ is a subdirect product of $\times_P G$ and $\times_P G' \leq G'$. Then, G is a weakly branch group and $C_G(G')$ is trivial. Also, if N is a nontrivial normal subgroup of G then there exists a level k such that $\times_{P^k} G'' \leq N$ and therefore, $\frac{G}{N}$ is metabelian by P-type.*

Proof. Suppose $\gamma = fs \in C_G(G')$ where $f \in \times_P G$, $s \in P$ and $s \neq e$. Let $\beta \in G', \beta \neq e$. Then, $(\underline{e}\beta)^\gamma = \underline{s}\beta'$ implies $s = e$; a contradiction. As we can repeat the same argument at lower levels, this shows that $C_G(G')$ is trivial.

Let $\gamma = fs \in G$ where $f \in \times_P G, s \in P$ and N be the normal closure of $< \gamma >$ in G. If $s \neq e$. Then let $e \neq \beta, \delta \in G'$ and compute

$$[\underline{e}\beta, \gamma] = (\underline{e}\beta^{-1})(\underline{e}\beta)^{fs} = \underline{e}\beta^{-1} + \underline{s}\beta^{f_e},$$

$$[\underline{e}\beta, \gamma, \underline{e}\delta] = \underline{e}[\beta^{-1}, \delta].$$

Therefore, we produce $\underline{e}G'' \leq N$. If $\gamma \in stab_G(k)$ and γ_u is active for an index u of length k then we use $\underline{u}\beta, \underline{u}\delta$ to produce $\underline{u}G'' \leq N$. Now,

$$\times_{P^k} G'' \leq N \leq (\times_{P^k} G) wr^{k-1}(P).$$

implies that $\frac{G}{N}$ is metabelian by P-type. □

Corollary 2. *Let P be as in the previous proposition with $d \geq 2$. If $\times_{P^s} G' \leq G''$ for some $s \geq 1$ then every proper quotient of G is abelian by P-type. If in addition, the group $wr^k(P)$ is free of weakly branch subgroups for all finite $k \geq 0$ then G is just non-(abelian by P-type).*

Proof. Let N the normal subgroup as in the proof of the previous proposition. From the assumption $\times_{\underline{P}^s} G' \leq G''$, we obtain

$$\times_{\underline{P}^{k+s}} G' \leq \times_{\underline{P}^k} G'' \leq N,$$

and as

$$G \leq \left(\times_{\underline{P}^{k+s}} G \right) \left(wr^{k+s-1} P \right),$$

we conclude, $\frac{G}{N}$ is abelian by P-type.

Suppose N is abelian and non-trivial. Then, from $\times_{\underline{P}^{k+1}} G' \leq N$ we conclude that G' is abelian and therefore G'' is trivial. We use again $\times_{\underline{P}^s} G' \leq G''$ to obtain the stronger conclusion, G' is trivial; thus, both G, P are abelian. But given $a_i \neq a_j \in A$, we note that the commutator $\sigma_{ij} = \underline{a_i} \alpha_i^{-1} + \underline{a_i a_j} \alpha_j^{-1} \alpha_i + \underline{a_j} \alpha_j$ cannot be the trivial element; a contradiction.

Therefore, if the group G is abelian by P-type, then G is of P-type; that is, G is a subgroup of $wr^k(P)$ for some finite k. As G is weakly branch, this is ruled out by the second assumption. $\qquad\square$

Proof of Theorem 2. Items (i) and (ii) of the theorem follow from Lemma 2.

(iii) We deduce from Proposition 3 (parts (iii) and (iv)) that

$$\times_{\underline{P}} \gamma_3 \leq G'', \quad \times_{\underline{P}} G' \leq \gamma_3.$$

Therefore,

$$\times_{\underline{P}^2} G' \leq \times_{\underline{P}} \gamma_3 \leq G''.$$

Since $[\alpha_u^{-r} \alpha_v^s, \alpha_u^{m_u}]^{\alpha_u^{-r}} = [\alpha_v, \alpha_u]^{s m_u} = \underline{e} [\alpha_u^{-1}, \alpha_v]$ modulo γ_3 and since $\underline{e} [\alpha_u^{-1}, \alpha_v] \in \gamma_3$, we conclude $[\alpha_v, \alpha_u]^{s m_u} \in \gamma_3$ and thus $\frac{\gamma_2}{\gamma_3}$ is a torsion group.

(iv) This item follows simply from Corollary 2. $\qquad\square$

5. Commutator Quotient

We produce convenient transversals of $< [H, P] >$ in G' and of $< [H, P] >$ in $G' \cap H$, modulo $\times_{\underline{P}} G'$. These are useful in determining $\frac{G}{G'}$ when $\times_{\underline{P}} G' \leq G'$.

As P acts regularly on the first level of the tree, there exists for every $p \in P$ a $v(p) \in \times_{\underline{P}} G$ such that $v(p)p \in G$. Let $S = \{v(p)p | p \in P\}$ be a transversal of H in G. Then, $G = HS$ and G' is generated by $\{[h_1 s_1, h_2 s_2] | h_1, h_2 \in H, s_1, s_2 \in S\}$.

Since

$$[h_1 s_1, h_2 s_2] = [h_1, s_2]^{s_1} [h_1, h_2]^{s_2 s_1} [s_1, s_2] [s_1, h_2]^{s_2},$$

we conclude that

$$G' = H' < [H, < S >] > [S, S].$$

As $H' \leq \times_{\underline{P}} G'$ we have

$$G' =< [H, < S >][S, S] \text{ modulo } \times_{\underline{P}} G'.$$

Let

$$h \in H, s_1 = v(p_1)p_1, s_2 = v(p_2)p_2,$$

then
$$[h, s_1] = [h, v(p_1)p_1] = [h, p_1][h, v(p_1)]^{p_1}$$
which is an element of $(\times_P G') < [H, P] >$. Also, $[s_1, s_2]$ is an element of
$$(\times_P G') [v(p_1), p_2]^{p_1} [p_1, v(p_2)]^{p_2[p_2, p_1]} [p_1, p_2].$$
Therefore, we have modulo $\times_P G'$,
$$\{[v(p_1), p_2]^{p_1} [p_1, v(p_2)]^{p_2[p_2, p_1]} [p_1, p_2] \mid p_1, p_2 \in P\}$$
is a transversal of $< [H, P] >$ in G' and
$$\{[v(p_1), p_2]^{p_1} [p_1, v(p_2)]^{p_2} \mid p_1, p_2 \in P, [p_1, p_2] = e\}$$
is a transversal of $< [H, P] >$ in $G' \cap H$.

The relationship between G and the Magnus embedding is explained as follows. The quotient group $\overline{G} = \frac{(\times_P G)G}{\times_P G'} \leq (\times_P \frac{G}{G'}) P$ and is generated by the elements $(eG'\alpha_i) a_i, i = 1, \ldots, d$. The map $\lambda_i a_i \to (eG'\alpha_i) a_i$ extends to a homomorphism from $\widehat{P}$ onto $\overline{G}$, and thus, $\widehat{P} \cong \overline{G}$ if and only if the commutator quotient G/G' is free abelian, freely generated by $\{G'\alpha_i \mid 1 \leq i \leq k\}$.

Proposition 5. *Suppose* $G = G(P, A)$ *is such that* $\times_P G' \leq G'$ *and* $\frac{G}{G'}$ *is a free abelian group, freely generated by* $\{G'\alpha_i \mid 1 \leq i \leq k\}$. *Then,* $\widehat{P} \cong \frac{G}{\times_P G'}$ *and* G *is torsion-free.*

Proof. The first assertion $\widehat{P} \cong \frac{G}{\times_P G'}$ was shown in the previous discussion. It is well-known that $\widehat{P}$ is torsion-free and therefore $\frac{G'}{\times_P G'}$ is also torsion-free. Hence, $\frac{G}{\times_P G'}$ is torsion-free for all $k \geq 0$. Now, since $\cap_{k \geq 0} (\times_{P^k} G') = \{e\}$ we conclude that G itself is torsion-free. $\square$

The group $G(P, A)$ may fail to have its abelianization $\frac{G}{G'}$ free abelian of rank $|A|$. A simple example is P infinite cyclic with the presentation $< a, b | a^{-1} b = e >$. Here, G is generated by $\alpha = (e\alpha) a, \beta = (e\beta) b$ and we verify that $\alpha^{-1}\beta = \underline{a} (\alpha^{-1}\beta), \alpha^{-1}\beta = e$ and G is infinite cyclic.

6. P abelian

Suppose P is an abelian group. Then,
$$G' \leq stab_G(1) = H, G'' \leq H' \leq \times G'.$$
We recall the commutator
$$\sigma_{ij} = \underline{a_i \alpha_i^{-1}} + \underline{a_i a_j \alpha_j^{-1}} \alpha_i + \underline{a_j \alpha_j}$$
which is reduced provided, $a_i \neq a_j$. Then,
$$\sigma_{ij}^{\alpha_i^{-1}} = \underline{e\alpha_i^{-1}} + \underline{a_j \alpha_j^{-1}} \alpha_i + \underline{a_i^{-1} a_j \alpha_j},$$

which is reduced, provided $a_j \neq a_i^2$. Also,

$$\sigma_{ij}^{\alpha_j^{-1}} = \underline{e}\alpha_j + \underline{a_i}\alpha_j^{-1}\alpha_1 + \underline{a_i a_j^{-1}}\alpha_i^{-1},$$

which is reduced provided $a_i \neq a_j^2$.

Therefore, if $a_j \neq a_i^2$ for all $i \neq j$ then H is a subdirect product of $\times \underline{p}G$. Furthermore,

$$\left[\sigma_{ij}^{\alpha_i^{-1}}, \sigma_{ij}^{\alpha_j^{-1}}\right] = \underline{e}\left[\alpha_i^{-1}, \alpha_j\right] = \underline{e}\sigma_{ij}^{-\alpha_i^{-1}},$$

provided $a_j \neq a_i^2, a_i \neq a_j^2, a_i^2 \neq a_j^2$. With these exceptions, we conclude

$$\times \underline{p}G' \leq G'' \leq H' \leq \times \underline{p}G',$$
$$\times \underline{p}G' = G'' = H'.$$

Proof of Theorem 3. The first two assertions in the theorem follow from the previous paragraph. Since $wr^k(P)$ is solvable for all k, it does not contain weakly branch groups. Corollary 2 then is applied to finish the proof. □

P homogeneous 2-generated

Let $P = \langle a, b \rangle$ be a finite abelian group with the presentation $\langle a, b | a^m = b^m = [a, b] = e \rangle$ and let G be the group generated by $\alpha = (\underline{e}\alpha)\, a, \beta = (\underline{e}\beta)\, b$. We will prove that G/G' is free abelian of rank 2.

First, recall

$$\alpha^m = \underline{e}\alpha + \underline{a}\alpha + \ldots + \underline{a^{m-1}}\alpha,$$
$$\beta^m = \underline{e}\beta + \underline{b}\beta + \ldots + \underline{b^{m-1}}\beta,$$
$$c = [\alpha, \beta] = \underline{a}\alpha^{-1} + \underline{b}\beta + \underline{ab}\beta^{-1}\alpha.$$

Since $\times \underline{p}G' \leq G' \leq H$, we will work modulo $\times \underline{p}G'$ and write the elements of H in additive notation. Then, for all $h \in H, p = a^i b^j, 0 \leq i, j \leq m - 1$, we have $h^{\alpha^i \beta^j} = h^p$. Thus, the conjugates of c can be written as

$$c_{ij} = c^{\alpha^i \beta^j} = \underline{a^{i+1}b^j}(-\alpha) + \underline{a^i b^{j+1}}\beta + \underline{a^{i+1}b^{j+1}}(\alpha - \beta);$$

note the dependence relation

$$\sum_{0 \leq i,j \leq m-1} c_{ij} = 0.$$

Let k, l be integers such that $\alpha^k \beta^l \in G'$. Then $a^k b^l = e$ and therefore $m|k, l$. We will show by induction on q that $m^q|k, l$ for all integers $q > 1$ and thus conclude, $k = l = 0$. Factor $k = mk', l = ml'$.

There exist integers u_{ij} such that

$$\alpha^k \beta^l = \Pi_{0 \leq i,j \leq m-1} c_{ij}^{u_{ij}};$$

by the dependence relation, we may assume $u_{m-1,n-1} = 0$. The following hold modulo $\times \underline{p}G'$,

$$\alpha^k \beta^l = \underline{e}\,(k'\alpha + l'\beta) + \underline{a}\,(k'\alpha) + \ldots + \underline{a^{m-1}}\,(k'\alpha) + \underline{b}\,(l'\beta) + \ldots + \underline{b^{m-1}}\,(l'\beta),$$

$$\Pi_{0\leq i,j\leq m-1}c_{ij}^{u_{ij}} = \sum_{0\leq i,j\leq m-1}\overline{a^i b^j} f_{ij} \text{ where}$$

$$f_{ij} = (-u_{i-1,j} + u_{i-1,j-1})\alpha + (-u_{i,j-1} + u_{i-1,j-1})\beta;$$

the indices i, j of u_{ij} are computed modulo m.

We obtain the equations:

$$
\begin{aligned}
k'\alpha + l'\beta &= (-u_{-1,0} + u_{-1,-1})\alpha + (-u_{0,-1} + u_{-1,-1})\beta; \\
&\quad \text{for } 0 < i \leq m-1, \\
k'\alpha &= (-u_{-i+1,0} + u_{i-1,-1})\alpha + (-u_{i,-1} + u_{i-1,-1})\beta; \\
&\quad \text{for } 0 < j \leq m-1, \\
l'\beta &= (-u_{-1,j} + u_{-1,j-1})\alpha + (-u_{0,j-1} + u_{-1,j-1})\beta; \\
&\quad \text{for } 1 \leq i,j \leq m-1, \\
0 &= (-u_{i-1,j} + u_{i-1,j-1})\alpha + (-u_{i,j-1} + u_{i-1,j-1})\beta.
\end{aligned}
$$

If $u\alpha + v\beta = 0$ for some integers u, v then by induction on q, we get $u, v \equiv 0$ modulo m^q. We will write in the sequel $u = 0$ for $u \equiv 0$ modulo m^q.

The coefficient $u_{m-1,m-1}(= 0)$ appears in the following equations:

$$k'\alpha + l'\beta = (-u_{-1,0} + u_{-1,-1})\alpha + (-u_{0,-1} + u_{-1,-1})\beta$$

and thus,

$$k'\alpha + l'\beta = -u_{-1,0}\alpha + -u_{0,-1}\beta;$$

$$k'\alpha = (-u_{-i+1,0} + u_{i-1,-1})\alpha + (-u_{i,-1} + u_{i-1,-1})\beta,$$

for $i = -1$ and thus,

$$k'\alpha = (-u_{-2,0} + u_{-2,-1})\alpha + u_{-2,-1}\beta,$$

$$u_{-2,-1} = 0, k' = u_{-2,0};$$

$$l'\beta = (-u_{-1,j} + u_{-1,j-1})\alpha + (-u_{0,j-1} + u_{-1,j-1})\beta$$

for $j = -1$ and thus,

$$l'\beta = (u_{-1,-2})\alpha + (-u_{0,-2} + u_{-1,-2})\beta,$$

$$u_{-1,-2} = 0, l' = u_{0,-2}.$$

Suppose $m = 2$. Then,

$$
\begin{aligned}
k'\alpha + l'\beta &= -u_{1,0}\alpha + -u_{0,1}\beta, \\
u_{0,1} &= 0, k' = u_{0,0}, \\
u_{1,0} &= 0, l' = u_{0,0}, \\
0 &= (-u_{01} + u_{00})\alpha + (-u_{01} + u_{00})\beta.
\end{aligned}
$$

Therefore, $0 = u_{00}\alpha + u_{00}\beta$ and $u_{00} = 0$. Hence, 2^q divides l', k' and the proof is finished for this case.

Now suppose $m > 2$. We note that if $i \neq 0$, $j \neq 0$ then

$$0 = (-u_{i-1,j} + u_{i-1,j-1})\alpha + (-u_{i,j-1} + u_{i-1,j-1})\beta$$

implies either all $u_{i-1,j}, u_{i-1,j-1}, u_{i,j-1} = 0$ or all $\neq 0$.

We want to show $u_{-2,0} = u_{0,-2} = 0$. We start with $u_{-2,-1} = 0$. Then, on substituting $i = -1, j = -1$ in the above equation, we obtain $u_{-2,-1}, u_{-2,-2}, u_{-1,-2} = 0$. Similarly, from $u_{-2,-2}$ we obtain $u_{-2,-2}, u_{-2,-3}, u_{-1,-3} = 0$ and from $u_{-2,-3} = 0$, $u_{-2,-3}, u_{-2,-4}, u_{-1,-4} = 0$. Thus we produce the sequence

$$u_{-2,-r}, u_{-2,-(r+1)}, u_{-1,-(r+1)} = 0$$

and at some point we reach $r + 1 = m$, when, $u_{-2,0} = 0 = l'$ and by symmetry, $n_{0,-2} = 0 = k'$. Hence, m^q divides l', k' and the proof is finished.

Since we have shown that $\frac{G}{G'}$ is torsion-free, freely generated by $G'\alpha, G'\beta$, we apply Proposition 5 to obtain G torsion-free and $\frac{G}{H'}$ isomorphic to $\widehat{P}$.

To show that G is not branch, we will prove that $ris_G(1) = \times_P G'$ or, equivalently, if $\underline{e}g \in G$ then $g \in G'$. Suppose $w = w(\alpha, \beta) = \underline{e}g$. Then, $w(\alpha, \beta) = \alpha^k \beta^l w'$ for some $k, l \geq 0, w' \in G'$. As before, we will work modulo $\times_P G'$ and use additive notation. Let then

$$w' = \sum_{0 \leq i,j \leq m-1} \underline{a^i b^j} f_{ij} \text{ where}$$

$$f_{ij} = (-u_{i-1,j} + u_{i-1,j-1})\alpha + (-u_{i,j-1} + u_{i-1,j-1})\beta.$$

We obtain from $\alpha^k \beta^l w' = \underline{e}g$ the following equations:

$$
\begin{aligned}
k'\alpha + l'\beta - g &= (-u_{-1,0} + u_{-1,-1})\alpha + (-u_{0,-1} + u_{-1,-1})\beta; \\
&\quad \text{for } 0 < i \leq m-1, \\
k'\alpha &= -(-u_{-i+1,0} + u_{i-1,-1})\alpha - (-u_{i,-1} + u_{i-1,-1})\beta; \\
&\quad \text{for } 0 < j \leq m-1, \\
l'\beta &= -(-u_{-1,j} + u_{-1,j-1})\alpha - (-u_{0,j-1} + u_{-1,j-1})\beta; \\
&\quad \text{for } 1 \leq i,j \leq m-1, \\
0 &= (-u_{i-1,j} + u_{i-1,j-1})\alpha + (-u_{i,j-1} + u_{i-1,j-1})\beta.
\end{aligned}
$$

Since $\frac{G}{G'}$ is freely generated by $G'\alpha, G'\beta$ we obtain

$$
\begin{aligned}
k' &= -u_{-i+1,0} + u_{i-1,-1}, u_{i,-1} = u_{i-1,-1} \\
&\quad \text{for } 0 < i \leq m-1; \\
l' &= -u_{0,j-1} + u_{-1,j-1}, u_{-1,j} = u_{-1,j-1} \\
&\quad \text{for } 0 < j \leq m-1; \\
u_{i-1,j} &= u_{i-1,j-1} = u_{i,j-1} \text{ for } 1 \leq i,j \leq m-1.
\end{aligned}
$$

The last equation establishes that the $u_{i,j}$'s are all equal except possibly for $u_{-1,-1}(= 0)$. But on substituting $i = -1$ in $u_{i,-1} = u_{i-1,-1}$ (which holds for $0 < i \leq m-1$) we obtain $u_{-1,-1} = u_{-2,-1} = 0$. Therefore, $u_{i,j} = 0$ for all i, j and it follows that $l' = k' = 0$. Therefore, $g \in G'$, as required.

Theorem 3 is now totally established.

References

[1] L. Bartholdi, R. Grigorchuk, Z. Sunik, Branch groups. In *Handbook of algebra,* vol. 31, pp. 989–1112, North-Holland, Amsterdam, 2003, pp. 181–203.

[2] A. M. Brunner, S. Sidki, A. C. Vieira, A just non-solvable torsion-free group defined on the binary tree. *J. Algebra* **211** (1998), 99–114.

[3] R. I. Grigorchuk, The growth degrees of finitely generated groups and the theory of invariant means. *Izv. Akad. Nauk. SSSR Ser. Mat.* **48** (1984), 939–985. (English Transl. *Math. USSR Izv.* **25** (1985), 259–300.)

[4] R. I. Grigorchuk, Just infinite branch groups. In *New horizons in pro-p groups,* Birkhäuser, Boston, 2000, pp. 121–179.

[5] N. Gupta, S. Sidki, Some infinite p-groups, *Algebra i Logica* (1983), 584–598.

[6] N. Gupta, Free Group Rings, Contemporary Mathematics, vol. 66, American Mathematical Society, 1987.

[7] S. Sidki, Finite automata of polynomial growth do not generate a free group. *Geom. Dedicata* **108** (2004), 193–204.

[8] J. S. Wilson, On just infinite abstract and profinite groups. In *New horizons in pro-p groups,* Birkhäuser, Boston, 2000, pp. 181–203.

[9] J. S. Wilson, Structure theory for branch groups. To appear.

Said Sidki
Departamento de Matemática
Universidade de Brasília
70910-900, Brasília-DF
Brazil
e-mail: sidki@mat.unb.br

Progress in Mathematics, Vol. 248, 403–413
© 2005 Birkhäuser Verlag Basel/Switzerland

Infinite Algebras and Pro-p Groups

Efim Zelmanov

To Slava Grigorchuk on his 50th birthday

Abstract. We discuss three areas where infinite dimensional algebras meet pro-p groups: (1) the linearity problem for free pro-p groups, (2) asymptotic properties of Golod-Shafarevich groups, (3) algebras with property (τ).

Mathematics Subject Classification (2000). Primary 20F69; Secondary 16R10, 16R30, 17B01.

Keywords. Pro-p group, polynomial identity, property (τ).

This survey is based on the talk at the international conference on Group Theory in Gaeta, Italy, in June 2003.

We will discuss some interactions between infinite dimensional algebras and pro-p groups. Let's start with the definitions.

Fix a prime number p. For a group G let $\{H_i\}_{i \in I}$ be the family of normal subgroups of G of p-power index. If $\bigcap_{i \in I} H_i = (1)$ then we say that G is a *residually-p* group. The system of subgroups $\{H_i\}_{i \in I}$ can be viewed as a basis of neighborhoods of 1. This makes G a topological group. If this topology is complete, then G is called a *pro-p group*. In any case a residually-p group G is embeddable in its pro-p completion $G_{\hat{p}}$ in the above topology. The pro-p group $G_{\hat{p}}$ is called the pro-p completion of G.

Examples:

(1) Let R be a local commutative complete ring with the maximal ideal M such that R/M is a finite field of characteristic $p > 0$. Consider the exact sequence

$$(1) \to GL^1(n, R) \to GL(n, R) \to GL(n, R/M) \to (1).$$

The (congruence) subgroup $GL^1(n, R)$ is a pro-p group.

(2) Let F_m denote the free group on m free generators $g_1, \ldots, g_m$. For an arbitrary prime p the group F_m is residually-p. The pro-p completion $(F_m)_{\hat{p}}$ is called the free pro-p group on the free generators $g_1, \ldots, g_m$. An arbitrary

mapping of $g_1, \ldots, g_m$ into a pro-p group G uniquely extends to a continuous homomorphism $(F_m)_{\hat{p}} \to G$.

1. Nonlinearity of free pro-p groups

The following question is still open:

> *Is a free pro-p group $(F_m)_{\hat{p}}$, $m \geq 2$, continually embeddable into a pro-p group of the type $GL^1(n, R)$ for some n, R?*

Let $\mathbb{Z}, \mathbb{Z}_p$ denote the ring of integers and p-adic integers respectively. Since $(F_m)_{\hat{p}}$ is not a p-adic analytic group (see [DMSS]) it is not embeddable in $GL^1(n, \mathbb{Z}_p)$.

In 1989 A. Zubkov [Zu] proved that $(F_m)_{\hat{p}}$ is not embeddable in $GL^1(2, R)$ provided that $p > 2$. Using highly nontrivial results of Pink [Pi] Barnea and Larson [BL] proved that $(F_m)_{\hat{p}}$ is not embeddable into $GL^1(n, (\mathbb{Z}/p\mathbb{Z})[[t]])$.

In this paper we sketch the proof of the following theorem.

Theorem 1.1. *There exists a function $\gamma : N \to N$ such that $(F_m)_{\hat{p}}$ is not embeddable into $GL^1(n, R)$ for $p \geq \gamma(n)$.*

Among all m-tuples of $n \times n$ matrices over commutative rings there is a universal one. Consider the algebras

$$\Lambda = \mathbb{Z}[[x_{ij}^{(k)}, 1 \leq i, j \leq n; k = 1, 2, \ldots, m]]$$
$$\Lambda_p = \mathbb{Z}_p[[x_{ij}^{(k)}, 1 \leq i, j \leq n; k = 1, 2, \ldots, m]]$$

of infinite series in mn^2 commuting variables $x_{ij}^{(k)}$ over $\mathbb{Z}$ and $\mathbb{Z}_p$ respectively, $\Lambda \subset \Lambda_p$. The matrices

$$X_k = \left(x_{ij}^{(k)} \right)_{1 \leq i, j \leq n} \in M_n(\Lambda), \qquad k = 1, 2, \ldots, m$$

are referred to as generic matrices.

Proposition 1.2 (Zubkov, [Zu]). *If the free pro-p group $(F_m)_{\hat{p}}$ is embeddable in $GL^1(n, R)$ over some commutative local ring R then $g_i \to 1 + X_i \in GL^1(n, \Lambda_p), 1 \leq i \leq m$, is an embedding.*

Thus to prove that $(F_m)_{\hat{p}}$ is not representable by $n \times n$ matrices we need to establish the existence of a nonidentical element $1 \neq w(g_1, \ldots, g_m) \in (F_m)_{\hat{p}}$ such that $w(1 + X_1, \ldots, 1 + X_m) = 1$, where $w(1 + X_1, \ldots, 1 + X_m)$ is the image of $w(g_1, \ldots, g_m)$ under the homomorphism $(F_m)_{\hat{p}} \to GL^1(n, \Lambda_p), g_i \to 1 + X_i, 1 \leq i \leq m$.

Definition 1.3. Let G be a pro-p group and let $1 \neq w(g_1, \ldots, g_m) \in (F_m)_{\hat{p}}$. We say that G satisfies the identity $w = 1$ (or, that $w = 1$ holds identically on G) if $w(a_1, \ldots, a_m) = 1$ for arbitrary elements $a_1, \ldots, a_m \in G$.

Proposition 1.4. *The following assertions are equivalent:*

(1) *the group $(F_m)_{\hat{p}}$ is not representable by $n \times n$ matrices*
(2) *there exists an element $1 \neq w \in (F_m)_{\hat{p}}$ which identically holds on all pro-p groups $GL^1(n, R)$*
(3) *the group $GL^1(n, \mathbb{Z}_p)$ satisfies same pro-p identity.*

Now we are ready to explain the unfortunate appearance of the function $\gamma(n)$ in Theorem 1.1. For an abelian group A let $\pi(A)$ denote the set of all primes l such that A has a l-torsion, $\pi(A) = \{\text{prime } \ell | \text{ there exists } 0 \neq a \in A \text{ such that } \ell a = 0\}$.

Let $A\langle X \rangle$ be the subring of the associative ring $M_n(\Lambda)$, generated by the generic matrices $X_1, \ldots X_m$. Let $L\langle X \rangle$ be the Lie ring generated by $X_1, \ldots, X_m$. Consider the quotient of the abelian groups $A\langle X \rangle / L\langle X \rangle$.

Theorem 1.5. *The set $\pi(A\langle X \rangle / L\langle X \rangle)$ is finite.*

Example. Let $n = 2$. Then $2 \in \pi(A\langle X \rangle / L\langle X \rangle)$. Indeed, the element $[X_1, X_2]^2 X_3 - [X_1, X_2]X_3[X_1, X_2]$ does not lie in $L\langle X \rangle$, but

$$2([X_1, X_2]^2 X_3 - [X_1, X_2]X_3[X_1, X_2]) = [[X_3, [X_1, X_2]], [X_1, XX_2]] \in L\langle X \rangle.$$

Question. Let R be an associative commutative ring. Choose arbitrary $n \times n$ matrices $a_1, \ldots a_m$ over R. Let $A\langle a_1, \ldots, a_m \rangle$ and $L\langle a_1, \ldots, a_m \rangle$ be the associative ring and the Lie ring, respectively, generated by the elements $a_1, \ldots, a_m$. Is the set $\pi(A\langle a_1, \ldots, a_m \rangle / L\langle a_1, \ldots, a_m \rangle)$ always finite?

Let $\pi(A\langle X \rangle / L\langle X \rangle) = \{p_1, \ldots, p_r\}$. Denote $\mathbb{Z}_\pi = \mathbb{Z}\left[\frac{1}{p_1}, \ldots, \frac{1}{p_r}\right]$. We will recall Malcev's construction of the $\mathbb{Z}_\pi$-completion of the free groups F_m [Mal]. Consider the free associative algebra $\mathbb{Q}\langle Z \rangle, Z = \{z_1, \ldots, z_m\}$ on m free generators $z_1, \ldots, z_m$ over the rational field $\mathbb{Q}$. Let $\mathbb{Q}\langle\langle Z \rangle\rangle$ be the completion of $\mathbb{Q}\langle Z \rangle$ in the degree topology, that is, the algebra of infinite series. Denote by $U\mathbb{Q}\langle\langle Z \rangle\rangle$ the multiplication group of all series having the constant term 1. It is well known that the map $g_i \to 1 + z_i, 1 \leq i \leq m$, defines an embedding of the free group F_m into $U\mathbb{Q}\langle\langle Z \rangle\rangle$. The group $U\mathbb{Q}\langle\langle Z \rangle\rangle$ has a natural topology induced by the total degree in Z.

If $\lambda \in \mathbb{Z}_\pi$ then the binomial coefficients $\begin{pmatrix} \lambda \\ k \end{pmatrix} = \frac{\lambda(\lambda-1)\ldots(\lambda-k+1)}{k!}, k \geq 0$, also lie in $\mathbb{Z}_\pi$. For a series $1 + a \in U\mathbb{Q}\langle\langle Z \rangle\rangle$ define $(1 + a)^\lambda = 1 + \sum_{k=1}^{\infty} \begin{pmatrix} \lambda \\ k \end{pmatrix} a^k$. The group $F_m^{\mathbb{Z}_\pi}$ is defined as the smallest closed subgroup of $U\mathbb{Q}\langle\langle Z \rangle\rangle$ which contains F_m and is closed with respect to all maps $g \to g^\lambda, g \in U\mathbb{Q}\langle\langle Z \rangle\rangle, \lambda \in \mathbb{Z}_\pi$.

Let $\sigma_1, \sigma_2, \ldots$ be a base of the free Lie algebra over Q, which consists of commutators in Z. We assume the degrees $\deg(\sigma_1), \deg(\sigma_2), \ldots$ to increase. Let $\rho_1, \rho_2, \ldots$ be the corresponding group commutators. An arbitrary element $g \in F_m^{\mathbb{Z}_\pi}$ can be uniquely represented as an (infinite) product $g = \rho_1^{\lambda_1} \rho_2^{\lambda_2} \ldots$, where $\lambda_i \in \mathbb{Z}_\pi$.

If a prime integer p is greater than $p_1, \ldots, p_r$ then the Malcev completion $F_m^{\mathbb{Z}_\pi}$ can be embedded in the free pro-p group $(F_m)_{\hat{p}}$.

We will show the existence of a nonidentical element $w(g_1, \ldots, g_m) \in F_m^{\mathbb{Z}_\pi} < (F_m)_{\hat{p}}$ such that $w(1 + X_1, \ldots, 1 + X_m) = 1$.

Let $A\langle X \rangle = \sum\limits_{i=0}^{\infty} A\langle X \rangle_i$ be the decomposition of $A\langle X \rangle$ into a sum of homogeneous components with respect to the total degree in X. Let $\langle 1 + X_1, \ldots, 1 + X_m \rangle$ be the subgroup of $GL(n, \Lambda)$ generated by the elements $1 + X_1, \ldots, 1 + X_m$. An arbitrary element $g \in \langle 1 + X_1, \ldots, 1 + X_m \rangle$ can be represented as an (infinite) sum $g = 1 + \sum\limits_{i=1}^{\infty} a_i, a_i \in M_n(A\langle X \rangle_i)$. Denote $\min(g) = a_k$ if $a_1 = \cdots = a_{k-1} = 0, a_k \neq 0; \min(1) = 0$.

The following proposition has been essentially proved in [Zu].

Proposition 1.6. *For an arbitrary element $g \in \langle 1 + X_1, \ldots, 1 + X_m \rangle$ there exists an integer $s \geq 1$ such that $s \min(g) \in L\langle X \rangle$. Clearly, the integer s can be assumed to be a product of the primes $p_1, \ldots, p_r$. Remark also that the $\mathbb{Z}_\pi$-span of $\min(\langle 1 + X_1, \ldots, 1 + X_m \rangle)$ coincides with the set $\min(\langle 1 + X_1, \ldots, 1 + X_m \rangle^{\mathbb{Z}_\pi})$.*

Assume for convenience that $m = n^2 + 1$. Denote $[a, b] = ab - ba, (g, h) = g^{-1} h^{-1} g h$.

Let us start with a non identity element

$$P_1 = \prod_{\sigma \in S_{n^2}} (\ldots (g_{n^2+1}, g_{\sigma(1)}), g_{\sigma(2)}), \ldots, g_{\sigma(n^2)})^{sgn(\sigma)}$$

of the free group F_m with an arbitrary order of factors on the right hand side. The minimal component $\min P_1(1 + X_1, \ldots, 1 + X_m)$ has degree $> n^2 + 2$ because the element $\sum\limits_{\sigma \in S_{n^2}} (-1)^{sgn(\sigma)} [\ldots [X_{n^2+1}, X_{\sigma(1)}], \ldots, X_{\sigma(n^2)}]$ of degree $n^2 + 1$ which might have been the minimal component is equal to 0. Indeed, any multilinear expression which is skew-symmetric in $n^2 + 1$ variables is identically zero in the algebra of $n \times n$ matrices over a commutative ring.

By Theorem 2 and Proposition 3 there exists an integer s_1 which is a product of $p_1, \ldots, p_r$ and a homogeneous element $u_1 \in L\langle X \rangle$ such that $s_1 \cdot \min(P_1) = u_1$.

Let $u_1 = \Sigma k_i [\ldots [X_{i_1}, X_{i_2}], \ldots, X_{i_q}], k_i \in \mathbb{Z}, 1 \leq i_1, \ldots, i_q \leq m$, and let P_{u_1} be a corresponding group element

$$P_{u_1}(1 + X) = \prod_i (\ldots (1 + X_{i_1}, 1 + X_{i_2}), \ldots, 1 + X_{i_q})^{k_i}$$

with an arbitrary order of factors on the right hand side. Clearly, $\min(P_{u_1}) = u_1$ and $\min\left(P_{u_1}^{1/s_1}\right) = \min P_1(1 + X)$.

Let $P_2 = P_1 \left(P_{u_1}^{1/s_1} \right)^{-1}$. Then $\deg \min(P_2) > \deg \min(P_1)$. Arguing as above we can find an integer $s_2 \geq 1$ and a homogeneous element $u_2 \in L\langle X \rangle$ such that $s_2 \cdot \min(P_2) = u_2$. Furthermore, we can construct an element $P_{u_2} \in F_m$, such that $\min \left(P_{u_2}^{1/s_2} \right) = \min(P_2)$. Let $P_3 = P_2 \left(P_{u_2}^{1/s_2} \right)^{-1}$, $\deg \min(P_3) > \deg \min(P_2)$, and so on.

The sequence $P_i, i \geq 1$, converges to a non identity element in $F_m^{\mathbb{Z}_\pi}$, because all the elements P_i do not lie in the $(n^2 + 2)$-th term of the lower central series. At the same time $P_i(1 + X_1, \ldots, 1 + X_m) \to 1$ as $i \to \infty$.

Now we will sketch the proof of Theorem 2 (the full proof is a lot more complicated). To start with we will introduce a stronger version of the Theorem 2 which will be more appropriate for an induction process.

Let F be a field of zero characteristic. Let A be a finite dimensional algebra and let $\mathcal{G} \subseteq A$ be a Lie subalgebra of A.

Choose a base of $e_1, \ldots, e_q$ of $\mathcal{G}$. Consider the polynomial algebra $A[x_{ij}, 1 \leq i \leq q, 1 \leq j \leq m]$ and the generic elements $x_j = \sum_{i=1}^{q} e_i x_{ij}, 1 \leq j \leq m$. As above let $A\langle x_1, \ldots, x_m \rangle, L\langle x_1, \ldots, x_m \rangle$ be the associative and Lie rings respectively generated by the elements $x_1, \ldots, x_m$.

Theorem 1.7. *The set $\pi(A\langle x_1, \ldots, x_m \rangle / L\langle x_1, \ldots, x_m \rangle)$ is finite.*

In the language of [Ba] Theorem 1.7 refers to relatively free pairs. Recall (see [Ba]) that an element $f(z_1, \ldots, z_k)$ of the free associative algebra is called a *weak identity* of the pair $(\mathcal{G}, A)$ if $f(a_1, \ldots, a_k) = 0$ for arbitrary elements $a_1, \ldots, a_k \in \mathcal{G}$. The ideal of the free associative algebra which consists of weak identities of $(\mathcal{G}, A)$ is called the T-ideal of the pair $(\mathcal{G}, A)$. It is easy to see that an ideal of the free associative algebra is a T-ideal of some pair if and only if it is invariant under all substitutions by Lie elements. The proof of Theorem 1.7 proceeds by reverse induction on the T-ideals of the pair $(\mathcal{G}, A)$. We could have based the induction on the important and difficult result of A. Il'tyakov [I] which states that for an arbitrary finite dimensional pair $(\mathcal{G}, A)$ an arbitrary ascending sequence of T-ideals of pairs $T(\mathcal{G}, A) \subset T_1 \subset T_2 \subset \ldots$ stabilizes (which is based on the celebrated Kemer's solution of the Specht problem [Ke]). However, the T-ideals that arise in the proof of the Theorem 1.7 are fairly explicit, which makes the stabilization straightforward.

Let's return to Theorem 1.5 as the starting point of Theorem 1.7 and describe the first stage of enlarging the T-ideal of weak identities. As above, $X_k = \left(x_{ij}^{(k)} \right), 1 \leq i, j \leq n, k = 1, \ldots, m$, are generic matrices. Let F be the algebraic closure of the field of rational functions $\mathbb{Q}(x_{ij}^{(k)}, 1 \leq i, j \leq n; 1 \leq k \leq m)$. Then the F-span of $A\langle X \rangle$ is $M_n(F)$ and the F-span of $L\langle X \rangle$ is $g\ell_n(F)$. Consider the so called generic matrices of zero trace $X_k' = X_k - \frac{1}{n}\left(\sum_{i=1}^{n} x_{ii}^{(k)}\right) I_n$, I_n is the identical $n \times n$ matrix.

If $a(X_1, \ldots, X_m)$ is a homogeneous element of $A\langle X \rangle$ of degree ≥ 2 and there exists an integer $s \geq 1$ such that $s.a \in L\langle X \rangle$ then $a(X_1, \ldots, X_m) = a(X_1', \ldots, X_m')$.

Let $L\langle X' \rangle, A\langle X' \rangle$ be the Lie ring and the associative ring respectively generated by $X_1', \ldots, X_m'$. The F-span of $L\langle X' \rangle$ is $sl_n(F)$, the F-span of $A\langle X' \rangle$ is $M_n(F)$. The pair $(L\langle X \rangle, A\langle X' \rangle)$ is a free pair in the variety generated by $(sl_n(F), M_n(F))$ (see [Ba]).

Let U be the universal associative enveloping ring of the Lie ring $L\langle X' \rangle$. Clearly, $\mathcal{G} = sl_n(F)$ is a module over $L\langle X' \rangle$ and hence, over U. For an element $u \in U$ let $t(u)$ denote the trace of the action of u on g. Let T be the $\mathbb{Z}\left[\frac{1}{(n^2)!}\right]$-algebra generated by $t(U)$.

Proposition 1.8. (1) T is a finitely generated ring.
(2) The T-span $A\langle X' \rangle T$ is a finitely generated T-module.

E. Formanek [F] and Yu. Razmyslov [R] constructed multilinear central polynomials. A polynomial is said to be n-central if it is not identically zero on the algebra of $n \times n$ matrices but every value of it lies in the center. Denote $d = \dim_F g = n^2 - 1$. Let $f(z_1, \ldots, z_s)$ be a multilinear d-central polynomial. For arbitrary elements $u_1, \ldots, u_s \in U$ the operator $f(u_1, \ldots, u_s)$ acts on g as a scalar. Hence $af(u_1, \ldots, u_s) = \frac{1}{d} t(f(u_1, \ldots, u_s))a, a \in g$.

The following lemma is probably due to Yu. Razmyslov.

Lemma 1.9. Let V be a vector space of dimension $d, \varphi : V \to V$ is a linear transformation, $h : \underbrace{V \otimes \cdots \otimes V}_{d} \to V$ is a skew symmetric linear mapping. Then

$$h(v_1\varphi, v_2, \ldots) + h(v_1, v_2\varphi, v_3, \ldots) + \cdots + h(v_1, \ldots, v_d\varphi) = h(v_1, \ldots, v_d) \cdot Tr(\varphi).$$

There exists a multilinear element $h(z_1, z_2, \ldots, z_\mu), \mu \geq d$, of the free Lie ring, which is skew-symmetric in $z_1, \ldots, z_d$ and $h(g) \neq (0)$. The F-linear span of $h(L\langle X' \rangle)$ is the whole of g. From the lemma above it follows that $h(L\langle X' \rangle)T \subseteq h(L\langle X' \rangle)$. Let U_0 be the subring of U generated by $h(L\langle X' \rangle), C$ is the $\mathbb{Z}\left[\frac{1}{(n^2)!}\right]$-span of the set $\{t(f(u_1, \ldots, u_s))|u_1, \ldots, u_s \in U_0\} \subseteq T$. We proved that C is an ideal in T.

If a Lie element $a(X_1', X_2', \ldots) \in L\langle X' \rangle$ is linear in X_1' and $c = t(f(u_1, \ldots, u_s))$ then $ac = a(X_1' f(u_1, \ldots, u_s), X_2', \ldots)$.

From the Proposition 4 (2) it follows that there exists a finite collection of homogeneous elements $a_1, \ldots, a_q \in L\langle X' \rangle$ such that

$$L\langle X' \rangle \subseteq \sum_{i=1}^{q} a_i T.$$

Let $\ell = \max(\deg a_i, 1 \leq i \leq q)$ and let $\tilde{a}_i$ be the full linearization of the element a_i. Then (cf. [ZSSS]) $(\ell!)^\ell a_i$ is a sum of values of the element $\tilde{a}_i$. Hence $(\ell!)^\ell L\langle X'\rangle C \subseteq \sum_i \tilde{a}_i(L\langle X\rangle)C \subseteq \Sigma\tilde{a}_i(L\langle X'\rangle) \subseteq L\langle X'\rangle$. By the Artin-Rees Lemma there exists $r \geq 1$ such that

$$L\langle X'\rangle \cap A\langle X'\rangle C^r \subseteq L\langle X'\rangle C.$$

Consider a finitely generated T-module

$$M = A\langle X'\rangle C^r / (L\langle X'\rangle \cap A\langle X'\rangle C^r)T.$$

There exists an integer $s_1 \geq 1$ such that for an arbitrary torsion element $a \in M$ we have $s_1 a = 0$.

The T-ideal of the pair $(L\langle X'\rangle + A\langle X'\rangle C^r / A\langle X'\rangle C^r, A\langle X'\rangle / A\langle X'\rangle C^r)$ is greater than the T-ideal of the pair $(L\langle X'\rangle, A\langle X'\rangle)$. Hence by the induction assumption the set $\pi' = \pi(A\langle X'\rangle / L\langle X'\rangle + A\langle X'\rangle C^r)$ is finite. Let $a \in A\langle X'\rangle$ be a torsion element modulo $L\langle X'\rangle$. Then there exists an integer s_2 which is a product of prime numbers from π', such that $s_2 a \in L\langle X'\rangle + A\langle X'\rangle C^r$. Choose $b \in L\langle X'\rangle$ such that $s_2 a - b \in A\langle X'\rangle C^r$. The element $s_2 a - b$ is a torsion element modulo $L\langle X'\rangle \cap A\langle X'\rangle C^r$. Hence $s_1(s_2 a - b) \in (L\langle X'\rangle \cap A\langle X'\rangle C^r)T \subseteq L\langle X'\rangle CT \subseteq L\langle X'\rangle C$. Hence $(\ell!)^\ell s_1(s_2 a - b) \in L\langle X'\rangle$ and thus

$$(\ell!)^\ell s_1 s_2 a \in L\langle X'\rangle.$$

2. Algebras with property (τ)

We will start with a brief introduction to the property (τ) and to the concept of amenability. For an excellent treatment of these subjects cf. [Lu2].

In 1967 D. Kazhdan [Ka] showed that for some finitely generated groups, such as $SL(n, \mathbb{Z}), n \geq 3$, the trivial representation in isolated in the space of all unitary representations. He called this property (T). In 1975 G. Margulis [Mar] made the following observation. Let $G = \langle X\rangle, |X| < \infty, X = X^{-1}$, be a group with property (T). Then there exists a constant $\lambda = \lambda(X) > 1$ such that if G acts transitively on a finite set A and B is a subset of $A, |B| \leq \frac{1}{2}|A|$, then $|B \cup XB| \geq \lambda|B|$. Recall that for a subgroup N of G the Cayley graph $Cay(G/N, X)$ has cosets $gN, g \in G$, as vertices; two vertices $gN, g'N$ are connected if there exists $x \in X$ such that $g'N = xgN$ or $x^{-1}gN$. Now the statement above says that for a group $G = \langle X\rangle, |X| < \infty$, with property (T) the family of Cayley graphs

$$\{Cay(G/N, X) | N < G, |G : N| < \infty\}$$

forms a family of expanders.

Definition 2.1. (cf. [Lu3]) A finitely generated residually finite group G has property (τ) if Margulis' theorem holds for G. In other words, if the family of Cayley graphs $\{Cay(G/N, x) | N < G, |G : N| < \infty\}$ is a family of expanders. This is equivalent to the assertion that all $\lambda_1(G/N, X)$, the smallest positive eigenvalues of the Laplacians of $Cay(G/N, X)$ are bounded away from zero, independent of N (cf. [Lu2]).

Thus the property (T) implies (τ). The reverse is not true.

The opposite to (τ) type of behavior of a group is *amenability*. This property appears in many areas of mathematics and has many equivalent definitions. A finitely generated group G is amenable if it has an invariant finitely additive probabilistic measure. Equivalently, G is amenable if for an arbitrary element $g \in G$,
$$\lim_{S \subseteq G, |S| < \infty} \inf \frac{|gS \cup S|}{|S|} = 1 \text{ (Fölner criterion, cf. [Fo]). Equivalently, } G \text{ is amenable if it}$$
is m-generated and has cogrowth $\leq 2m - 1$ (Grigorchuk, [Gr]).

All groups of subexponential growth are amenable. It is easy to see that a residually finite group is amenable and has property (τ) if and only if it is finite.

Now let us discuss Golod-Shafarevich groups. In 1964 Golod and Shafarevich found a sufficient condition for a group presented by generators and relators to be infinite.

Let p be a prime integer, K_p is the field of p elements. Let $F = (F_m)_{\hat{p}}$. Consider the group algebra $K_p F$ and its augmentation ideal J. The subgroups $F^{(i)} = \{g \in F | 1 - g \in J^i\}$ form a filtration $F = F^{(1)} \supset F^{(2)} \supset \dots$, $\bigcap_{i \geq 1} F^{(i)} = (1)$.

For an element $1 \neq g \in F$ let $\deg(g) = \max\{i | g \in F^{(i)}\}$.

Let R be a subset of $F^{(2)}$ containing r_i elements of degree $i (i \geq 2)$. The Hilbert series of R is defined as $H_R(t) = \sum_{i=2}^{\infty} r_i t^i$.

Definition 2.2. (1) *A pro-p group G is called a Golod-Shafarevich group (GS-for short) if it has a pro-p presentation $\langle x_1, \dots, x_m | R = 1 \rangle$ such that $R \subseteq F^{(2)}$ and there exists $0 < t_0 < 1$ for which $H_R(t)$ converges at t_0 and $1 - mt_0 + H_R(t_0) < 0$.*
(2) *A discrete group is a GS-group if it has a presentation $\langle x_1, \dots, x_m | R = 1 \rangle$ which is GS in the category of pro-p groups.*

A famous result of Golod and Shafarevich [GS] asserts that GS-groups are infinite. In [Z1] it is proved that every pro-p GS-group contains a nonabelian free pro-p subgroup.

Examples.

(1) Golod [Go] used the above criterion to construct infinite finitely generated p-groups, the first counter examples to the General Burnside Problem.
(2) A Lubotzky [Lu1] showed that for almost every prime p the fundamental group of a hyperbolic 3-manifold has a finite index subgroup, which is a GS-group.

(3) Let S be a finite set of primes. Denote the maximal p-extension of $\mathbb{Q}$ unramified outside S (allowing ramification at infinity) as $\mathbb{Q}_S, G_S = Gal(\mathbb{Q}_S/\mathbb{Q})$. I. Shafarevich [Sh] proved that G_S has a presentation $\langle x_1, \ldots, x_m | x_i^{p^{k_i}} = [x_i, a_i], 1 \leq i \leq m \rangle, m = |S|$ where $a_1, \ldots, a_m$ are some elements of the free pro-p group, and therefore is a GS-group for $m > 4$.

A long-standing problem concerning GS-groups is

Problem (A. Vershik): Is there an amenable discrete GS-group?

We will formulate an even stronger conjecture.

Conjecture A. *For an arbitrary $\varepsilon > 0$, arbitrary $0 < t_0 < 1$, there exists a subset R of elements of the free group of degree ≥ 2 such that $H_R(t_0) < \varepsilon$ and the group $F = < x_1, \ldots, x_m | R = 1 >$ is amenable. If true, it would mean that an arbitrary discrete GS-group has an amenable homomorphic image which is still a GS-group.*

A related (and weaker) conjecture is

Conjecture B. *A discrete GS-group does not have property (τ).*

Of particular interest is

Conjecture C. (Lubotzky-Sarnak, cf. [Lu3]) *A lattice in $SL(2, \mathbb{C})$ does not have property (τ).*

Since such a lattice has a subgroup of finite index which is GS for some p (cf. [Lu1]) the Conjecture C follows from the Conjecture B.

It is known that $SL(2, \mathbb{C})$ and all its lattices do not have the stronger property (T) of Kazhdan.

Some recent (still unpublished) work of M. Lackenby suggests the use of expanders and property (τ) for the following long standing question on hyperbolic 3-manifolds.

The virtual Haken Conjecture. *Every irreducible compact 3-dimensional hyperbolic manifold has a finite sheeted cover which is Haken, or equivalently; every cocompact lattice Γ in $SL(2, \mathbb{C})$ has a finite index subgroup which is either a free product with amalgam, or an HNN-construction.*

Lackenby showed that Lubotzky-Sarnak Conjecture (plus another plausible conjecture about Heegaard splitting) would imply the virtual Haken conjecture.

The crucial difficulty of tackling the conjectures A, B, C lies in the fact that the Golod-Shafarevich condition is not formulated in terms of the generators X of the groups $G = \langle X | R = 1 \rangle$. It is rather formulated in terms of the generators $x - 1, x \in X$ of the group algebra $K_p G, K_p = \mathbb{Z}/p\mathbb{Z}$. This puts a spotlight on the group algebra $K_p \langle X | R = 1 \rangle$ which is a GS-algebra (the same definition as for groups).

Definition 2.3. A residually finite algebra A has property (τ) if there exists a finite dimensional generating subspace X of A and a constant $\lambda = \lambda(X) > 1$ such that for an arbitrary finite dimensional module V over A and an arbitrary subspace $W \subset V$ if V does not contain submodule of dimension $< 2\dim W$ then

$$\dim(W + XW) \geq \lambda \dim W.$$

Another motivation for the theory of algebras with property (τ) came from computer science. A. Wigderson raised the following problem.

Problem (A. Wigderson). Given a field K, find for some fixed $0 < m \in N$ and $\lambda > 1$ and for every $n \in N$, m linear transformations $T_1, \ldots, T_m$ from $V = K^n$ to V with the following property: for every subspace W of V of dimension at most $n/2$,

$$\dim\left(\sum_{i=1}^m T_i W\right) \geq \lambda \cdot \dim W.$$

One can show that "random matrices" will do it and the issue is to have explicit examples. Clearly, irreducible finite dimensional representations of algebras with property (τ) provide such examples. Lubotzky and Zelmanov [LZ] proved that the group algebras of the groups $SL(n, (\mathbb{Z}/p\mathbb{Z})[[t]]), n \geq 3$, over a field of zero characteristic have property (τ), thus answering Wigderson's question in the case of zero characteristic. An extension of the theory to fields of positive characteristics could help to resolve both: the Lubotzky-Sarnak conjecture as well as the general case of Wigderson's Problem.

In [E1], [Sa] G. Elek and A. Samet initiated the study of algebras with a Fólner-type condition.

References

[BL] Y. Barnea, M. Larsen, A nonabelian free pro-p group is not linear over a local field, *J. Algebra* **214** (1999), no. 1, 338–341.

[Ba] Yu. Bakhturin, Identical relations in Lie algebras, *MNU Science Press*, Utrecht, 1987.

[E1] G. Elek, The amenability of affine algebras, *J. Algebra* **264** (2003), no. 2, 469–478.

[F] E. Formanek, Central polynomials for matrix rings, *J. Algebra* **23** (1972), 129–132.

[Fo] E. Fólner, Note on groups with and without full Banach mean value, *Math. Scand.* **5** (1957), 5–11.

[Go] E. S. Golod, On nil-algebras and finitely approximable p-groups, *Izv. Akad. Nauk SSSR* **28** (1964), 273–276.

[GS] E. S. Golod, I. R. Shafarevich, On class field towers, *Izv. Akad. Nank SSSR* **28** (1964), 261–272.

[Gr] R. I. Grigorchuk, Symmetrical random walks on discrete groups, Multicomponent random systems, *Dekker, NY,* 1980, 285–325.

[I]   A. V. Iltyakov, On finite basis of identities of Lie algebra representations, *Nova J. Algebra Geom.* (1992), no. 3, 207–259.

[Ka]   D. A. Kazhdan, Connection of the dual space of a group with the structure of its closed subgroups, *Func. Anal. Appl.* **1** (1967), 63–65.

[Ke]   A. R. Kemer, Ideals of identities of associative algebras, *AMS, Providence, RI,* 1991.

[Lu1]   A. Lubotzky, Group presentations, p-adic analytic groups and lattices in $SL_2(\mathbb{C})$, *Annals of Math. (2)* **118** (1983), no. 1, 115–130.

[Lu2]   A. Lubotzky, Discrete groups, expanding graphs and invariant measures, Progress in Math. **125**, *Birkhäuser, Basel,* 1994.

[Lu3]   A. Lubotzky, Eigenvalues of the Laplacian, the first Betti numbers and the congruence subgroups problem, *Annals of Math. (2)* **144** (1996), no. 2, 441–452.

[LS]   A. Lubotzky, A. Shalev, One some Λ-analytic pro-p groups, *Isr. J. Math.* **85** (1994), 307–337.

[LZ]   A. Lubotzky, E. Zelmanov, Algebras with property τ, in preparation.

[Mal]   A. I. Malcev, Nilpotent groups without torsion, *Izvest. Akad. Nauk SSSR, ser. Mat.* **13** (1949), 201–212.

[Mar]   G. A. Margulis, Explicit constructions of concentrators, *Probl. Of Inform. Transm.* **10** (1975), 325–332.

[Pi]   R. Pink, Compact subgroups of linear algebraic groups, *J. Algebra* **206** (1998), 438–504.

[R]   Yu. P. Razmyslov, A certain problem of Kaplansky, *Izv. Akad. Nauk SSSR* **3** (1973), 483–501.

[Sa]   A. Samet, Amenable Algebras, preprint.

[Sh]   I. R. Shafarevich, Extensions with prescribed ramification points, *IHES Publ. Math.* **18** (1964), 71–95.

[Z1]   E. Zelmanov, On groups satisfying the Golod-Shafarevich condition, New horizons in pro-p groups, 223–232, Progress in Math. **184**, *Birkhäuser, Boston,* 2000.

[Z2]   E. Zelmanov, On non-linearity of free pro-p groups, in preparation.

[Zu]   A. Zubkov, Non-abelian free pro-p groups cannot be represented by 2-by-2 matrices, *Siberian Math. J.* **2** (1987), 742–747.

[ZSSS] K. A. Zhevlakov, A. M. Slinko, I. P. Shestakov, A. I . Shirshov, Rings that are nearly associative, *Academic Press, NY,* 1982.

Acknowledgment. The author is grateful to the referee for very helpful remarks.

Efim Zelmanov
Department of Mathematics
University of California at San Diego
9500 Gilman Drive
La Jolla, CA 92093-0112
USA
e-mail: ezelmano@math.ucsd.edu

Progress in Mathematics

PM 250: Unterberger, A.
The Fourfold Way in Real Analysis. An
Alternative to the Metaplectic
Representation (2006)
ISBN 3-7643-7544-2

PM 249: Kock, J. / Vainsencher, I.
An Invitation to Quantum Cohomology.
Kontsevich's Formula for Rational Plane
Curves (2006). ISBN 0-8176-4456-3

PM 248: Bartholdi, L. /
Ceccherini-Silberstein, T. /
Smirnova-Nagnibeda, T. / Zuk, A. (Eds.)
Infinite Groups: Geometric, Combinatorial
and Dynamical Aspects (2006)
ISBN 3-7643-7446-2

PM 247: Baues, H.-J.
The Algebra of Secondary Cohomology
Operations (2006). ISBN 3-7643-7448-9

PM 246: Dragomir, S. / Tomassini, G.
Differential Geometry and Analysis on CR
Manifolds (2006). ISBN 3-7643-4388-5

PM 245: Fels, G. / Huckleberry, A.T. /
J.A. Wolf
Cycle Spaces of Flag Domains. A Complex
Geometric Viewpoint (2006)
ISBN 0-8176-4391-5

PM 244: Etingof, P. / Retakh, V. /
Singer, I.M. (Eds.)
The Unity of Mathematics. In Honor of the
Ninetieth Birthday of I.M. Gelfand (2006)
ISBN 0-8176-4076-2

PM 243: Bernstein, J. / Hinich, V. / A.
Melnikov (Eds.)
Studies in Lie Theory. A Joseph Festschrift
(2006). ISBN 0-8176-4342-7

PM 242: Dufour, J.-P. / Zung, N.T.
Poisson Structures and their Normal Forms
(2005). ISBN 3-7643-7334-2

PM 241: Seade, J.
On the Topology of Isolated Singularities in
Analytic Spaces (2005)
ISBN 3-7643-7322-9

PM 240: Ambrosetti, A. / Malchiodi, A.
Perturbation Methods and Semilinear
Elliptic Problems in R^n (2005)
ISBN 3-7643-7321-0

PM 239: van der Geer, G. / Moonen,
B.J.J. / Schoof, R. (Eds.)
Number Fields and Function Fields – Two
Parallel Worlds (2005)
ISBN 0-8176-4397-4

PM 238: Sabadini, I. / Struppa, D.C. /
Walnut, D.F. (Eds.)
Harmonic Analysis, Signal Processing, and
Complexity. Festschrift in Honor of the 60th
Birthday of Carlos A. Berenstein (2005)
ISBN 0-8176-4358-3

PM 237: Kulish, P.P. / Manojlovic, N. /
Samtleben, H. (Eds.)
Infinite Dimensional Algebras and Quantum
Integrable Systems (2005)
ISBN 3-7643-7215-X

PM 236: Hotta, R. / Takeuchi, K. /
Tanisaki, T.
D-Modules, Perverse Sheaves, and
Representation Theory (due 2006)
ISBN 0-8176-4363-X

PM 235: Bogomolov, F. / Tschinkel, Y.
(Eds.)
Geometric Methods in Algebra and Number
Theory (2005). ISBN 0-8176-4349-4

PM 234: Kowalski, O. / Musso, E. /
Perrone, D. (Eds.)
Complex, Contact and Symmetric
Manifolds. In Honor of L. Vanhecke (2005)
ISBN 0-8176-3850-4

PM 233: David, G.
Singular Sets of Minimizers for the
Mumford–Shah Functional (2005)
ISBN 3-7643-7182-X

PM 232: Marsden, J.E. / Ratiu, T.S.
(Eds.)
The Breadth of Symplectic and Poisson
Geometry. Festschrift in Honor of Alan
Weinstein (2005)
ISBN 0-8176-3565-3

PM 231: Brion, M. / Kumar, S.
Frobenius Splitting Methods in Geometry
and Representation Theory (2005)
ISBN 0-8176-4191-2

PM 230: Anker, J.-P. / Orsted, B. (Eds.)
Lie Theory. Harmonic Analysis on Symmetric
Spaces — General Plancherel Theorems
(2005)
ISBN 0-8176-3777-X

PM 229: Anker, J.-P. / Orsted, B. (Eds.)
Lie Theory. Unitary Representations and
Compactifications of Symmetric Spaces
(2005)
ISBN 0-8176-3526-2

PM 228: Anker, J.-P. / Orsted, B. (Eds.)
Lie Theory. Lie Algebras and
Representations (2004)
ISBN 0-8176-3373-1

PM 227: Lepowski, J. / Li, H.
Introduction to Vertex Operator Algebras
and Their Representations (2004)
ISBN 0-8176-3408-8

PM 226: Poonen, B. / Tschinkel, Y. (Eds.)
Arithmetic of Higher-Dimensional Algebraic
Varieties (2004)
ISBN 0-8176-3259-X

PM 225: Andersson, M. / Passare, M. /
Sigurdsson, R.
Complex Convexity and Analytic Functionals
(2004)
ISBN 3-7643-2420-1

PM 224: Cremona, J. / Lario, J.-C. /
Quer, J. / Ribet, K. (Eds.)
Modular Curves and Abelian Varieties
(2004)
ISBN 3-7643-6586-2

PM 223: Andreu-Vallio, F./Caselles, V./
Mazòn, J.M.
Parabolic Quasilinear Equations Minimizing
Linear Growth Functionals (2004)
ISBN 3-7643-6619-2

PM 222: Ortega, J.-P. / Ratiu, T.S.
Momentum Maps and Hamiltonian
Reduction (2004)
ISBN 0-8176-4307-9

PM 221: Gray, A.
Tubes. Second Edition (2003)
ISBN 3-7643-6907-8

PM 220: Delorme, P. / Vergne, M. (Eds.)
Noncommutative Harmonic Analysis. In
Honor of Jacques Carmona (2003)
ISBN 0-8176-3207-7

PM 219: Suris, Y.B.
The Problem of Integrable Discretization:
Hamiltonian Approach (2003)
ISBN 3-7643-6995-7